HABITATS AND ECOSYSTEMS

An Encyclopedia of Endangered America

HABITATS AND ECOSYSTEMS

An Encyclopedia of Endangered America

Mark Crawford

ABC-CLIO

Santa Barbara, California
Denver, Colorado
Oxford, England

Cataloging-in-Publication Data is available from the Library of Congress.

ISBN 0-87436-997-5

03 02 01 00 99 10 9 8 7 6 5 4 3 2 1

ABC-CLIO, Inc.
130 Cremona Drive, P.O. Box 1911
Santa Barbara, California 93116–1911

This book is printed on acid-free paper.
Manufactured in the United States of America.

CONTENTS

PREFACE

When this book was first conceived in 1996, the idea was to explore how the environment in the United States was at risk. It quickly became clear that two volumes would be required: one to deal with the more than 35,000 toxic waste sites in the country, another to investigate vanishing ecosystems and species. The first volume, *Toxic Waste Sites*, was published in 1997 and focused on the Superfund program. This volume, *Habitats and Ecosystems*, reviews the history of conservation in the United States, identifies current issues and challenges, summarizes the larger ecoregions in each state, and describes parks, sanctuaries, and preserves that shelter special habitats and ecosystems. The goal of this book is preservation: It is hoped the reader will become more aware of conservation and ecology, visit some of the places in his or her state, and become inspired to help save what remains.

ACKNOWLEDGMENTS

This book would not have been possible without the assistance of the hundreds of dedicated men and women I contacted for information and advice. Too numerous to individually name, these members of various conservation groups and government agencies often went out of their way to get me the data I needed, sometimes compiling it for the first time. Their willingness to help with this project is much appreciated. Understaffed, overworked, and underpaid, they work tirelessly to protect our natural resources because they love and understand nature. We owe them great thanks, and I dedicate this book to them with gratitude.

INTRODUCTION

How to Use This Book

The following sections of this introduction describe the history of conservation in the United States. As for the main text, state-by-state sections begin with a description of the larger ecoregions and then lists, in alphabetical order, sites possessing special or unique natural resource values. This was a daunting task in itself. Depending on one's viewpoint, it could be argued that every tree, park, woodland, and wetland is endangered because they are being steadily impacted by urbanization and pollution. It could be argued that a wetland in Nevada is more special than a wetland in northern Minnesota (the Land of 10,000 Lakes). Species may be rare in one state but abundant in a neighboring state. After considerable thought and a few false starts, I decided to include special sites that preserved original, pristine remnants of vanishing ecosystems; preserved high-quality examples of natural communities; preserved natural communities of high biodiversity; or preserved natural communities that are habitats for rare, unique, uncommon, special-concern, threatened, or endangered species. The only other criteria was that the lands were available for public use.

Properties that did not make the cut included those established to protect scenic views, rock formations, paleontological or archeological sites, or species that were at the edge of their range in a particular state (thus being rare there but common, perhaps, only 100 miles away). National parks and wildlife areas are also generally not included, because comprehensive books already describe those lands. Thus, this volume is by no means an exhaustive treatment of state ecoregions and preserves, but it is representative in that the largest ecoregions in the state are described and covered.

It is logical to assume that these special sites are tucked away in the far corners of state and national parks. This is not necessarily true, especially in the eastern United States. I was surprised to learn that many rare and delicate communities exist in urbanized areas, sometimes sectioned off by interstates, railroads, county highways, and subdivisions. Although an isolated patch of wood, bog, or field might seem ordinary, it may very well contain embattled species that are globally rare. In fact, even in the wide expanses of the western United States, endangered species are often more common on lands used by the public than they are in wilderness sanctuaries. The fact that national parks, wilderness areas, and preserves have been established does not mean that endangered species are at less risk; species preservation is an ongoing process.

The number of sites listed for each state largely depends on the size of the state, its dedication to establishing high-quality natural areas, and its willingness to share information. For example, Massachusetts does not disclose data regarding such sites, citing a fear of increased vandalism, degradation, and poaching of rare species (it even refused to disclose the locations of stands of old-growth forest). Accordingly, fewer sites are listed for Massachusetts.

Natural heritage programs are better funded in some states than they are in others; sometimes the information is in a computer database but is not readily compiled and therefore is not included. Descriptions of sites vary according to the amount of available information; sometimes it is just three or four sentences. The approximate acreage for each site is mentioned to give the reader a feeling for how large the area is (the actual acreage may be different due to recent property acquisitions). Most state personnel will agree that gaps remain in much of their coverage, and they continue to diligently explore, map, and document natural communities and species types.

Only through the use of a complete inventory of a state can natural resource management, preservation, and development goals be adequately planned. After all, if we don't know it's there, it can't be protected.

Terms frequently used in this book are *biodiversity, ecosystems, natural communities,* and *habitats.* **Biodiversity** is the number of species that occur within an area; the higher the number of species, the greater that area's biodiversity (biodiversity can also mean the genetic variation *within* a species). An **ecosystem** is a natural community or communities and the surrounding physical environment that occupy a particular area. **Natural communities** are plant and animal groupings that naturally occur in an area. A **habitat** is the natural home in which a species lives. Thus, natural communities make up ecosystems; an ecosystem can range in size from one community to many. Habitats can be single or multiple natural communities or ecosystems, depending on the range of the species.

I also describe (sometimes loosely) species as being *endangered, threatened,* or *of special concern.* A federally "listed" species is **endangered** if it is in danger of extinction in the United States; a federally listed species is **threatened** if its population is rapidly declining but not yet in danger of extinction. A state "listing" is applicable only within that state, not the entire country. Species of **special concern** are those that live in very rare or restricted habitats of limited distribution and that would be at risk if habitats were to be damaged. Global rankings also exist, ranging from G1 (critically imperiled globally) to G5 (globally secure).

Appendices in the rear matter of the book provide useful general references on conservation and ecology; references according to state; addresses and phone numbers for national organizations; addresses and phone numbers for state agencies; and an endangered species list. The sites that are listed in this book are publicly owned and managed by many different agencies. Call or write the managing agency for permission to visit, detailed directions, maps of the area, as well as any regulations and restrictions that may apply.

History of Conservation in the United States

The first prerequisite of intelligent tinkering is to save all the pieces. –Aldo Leopold

When Christopher Columbus first came to the New World, it was not a pristine, virgin wilderness that he saw. Native American cultures had already modified and degraded the landscape through forest reduction, overhunting, and farming, which increased erosion. This, of course, was compounded by the arrival of an industrializing Anglo-European culture. The rapid reduction of natural resources was identified as a problem by none other that Thomas Jefferson, one of America's first conservationists. As Americans pushed relentlessly westward, witnesses to rippling bluegrass prairies that resembled an ocean and were tall enough to hide buffalo, eastern writers such as Henry David Thoreau (*Walden*) began to lament our disassociation with nature. Western lands were quickly settled, farmed, and grazed. Unwise farming practices resulted in the terrible erosion problems that led to the Dust Bowl disaster of the 1930s. One of the earliest activists was naturalist John Muir, who founded the Sierra Club in 1892 and worked to fight development projects such as dams.

A growing awareness by the public and federal government of rapidly changing landscapes and loss of wildlife led to the establishment of Yellowstone National Park in 1872, the nation's first. The Forest Reserve Act of 1891 authorized the president to withdraw certain portions of the public domain for forest reserves. In 1905, President Theodore Roo–sevelt (often counseled by Muir) established the U.S. Forest Service. As a result of the Dust Bowl, widespread drought, and the Great Depression, President Franklin D. Roosevelt established various conservation programs to maintain and restore natural resources. These included the Civilian Conservation Corps, the Tennessee Valley Authority, and the Soil Conservation Service.

After World War II and the Korean War, government efforts again focused on conservation. The U.S. Congress passed the Wilderness Act in 1964, setting aside roughly 9 million acres as wilderness (today wilderness comprises some 110 million acres). Other bills passed in the 1960s included the Clean Air Act, the Federal Water Pollution Control Act, and the National Environmental Policy Act—all geared toward fighting pollution and the toxification of the environment. The Endangered Species Act was passed in 1973.

The idealistic hopes of the environmental movement in the 1970s soon stalled as conservation began to be seen as an enemy to development, jobs, and a stable economy (impeding, for example, industry efforts to log old-growth forests in the Pacific Northwest, habitat for the federally threatened spotted owl). President Ronald Reagan's Republican administration was effective in reducing the budget and manpower at the Environmental Protection Agency and actually accelerated the exploitation of natural resources. His Republican successor, George Bush, was also reluctant to address environmental issues. The administration of Democrat Bill Clinton is trying to push preservation to the forefront of the agenda; he signed the International Biodiversity Treaty in 1994 and has lead efforts to revise the antiquated 1872 Mining Law.

National Conservation Agencies and Organizations

National Forest Service (U.S. Department of Agriculture)

Our national forests are intended to be managed according to the doctrine of *multiple use*, defined in the Federal Land Policy and Management Act (1976) as

> the management of the public lands and their various resource values so that they are utilized in the combination that will best meet the present and future needs of the American people ... for renewable and nonrenewable resources, including, but not limited to, recreation, range, timber, minerals, watershed, wildlife and fish, and natural scenic, scientific, and historical values.

To protect the highest-quality and most pristine communities, parts of the national forests are given wilderness as well as research natural area designations. Most of these areas escaped human impact because they were too steep to log, too dry to cultivate, or too remote to access.

Wilderness Areas are designated according to provisions contained in the 1964 Wilderness Act. Wilderness tracts are maintained as such and are not subjected to road building, *permanent* commercial development, and motorized equipment/vehicles. Thus, certain activities, such as livestock grazing, mining, and water usage, can theoretically occur, but they are too onerous and expensive for individuals and companies to undertake.

Research Natural Areas (RNAs) are lands from 40 to 10,000 acres in size that protect biological diversity and natural habitats. RNAs are intended for forest ecosystem research, observation, and study. They preserve segments of pristine habitat, maintain genetic diversity, prevent environmental damage, and serve as a baseline for successional studies and long-term ecological change. For example, by comparing exploited lands to RNAs, we can quantify the impacts of development, logging, mining, grazing, fires, and other human activities. The first RNA was established in 1927. At present there are about 250, with another 500 sites being considered for RNA designation.

Bureau of Land Management (U.S. Department of Interior)

Areas of Critical Environmental Concern (ACECs) are administered by the Bureau of Land Management (BLM). In passing the 1976 Federal Land Policy and Management Act, Congress authorized the BLM to protect areas on public lands with special resource values (including cultural-archeological sites). Not as restrictive as wilderness designations, ACECs need not be roadless, and development and extraction of resources are permitted if environmental values are adequately protected. Like RNAs, ACECs tend to be pristine, biologically diverse areas that are protected for study and research. There are currently more than 700 ACECs in the United States.

National Park Service (U.S. Department of Interior)

One of the first attempts to protect special places was the National Natural Landmark (NNL) program. Its goal was to increase public awareness of unique ecological and geological sites in the United States. Although there are hundreds of NNLs across the country, and they have strengthened our appreciation of our natural resources, the selection process was inconsistent, and protection is not guaranteed. In fact, in some cases, the increased visitation and lack of proper protection and management led to degradation of the sites.

The Nature Conservancy

The premiere preservation organization in the country is The Nature Conservancy, a privately funded nonprofit established in 1951. Its mission is to preserve plants, animals, and natural communities that represent the diversity of life on earth. State chapters identify and save special habitats. A system of preserves in each state consists of properties (usually open to the public) owned, managed, or leased by The Nature Conservancy (some properties are later transferred to state or federal agencies). Its creative acquisition and management strategies have shown that development and preservation can work together. A recent focus is on community-based conservation (bioreserves), areas where the goal is to protect core natural areas of high biodiversity then encourage human use in the rest of the area that is compatible with the conservancy's preservation goals. The Nature Conservancy has worked closely with the U.S. Forest Service in identifying and establishing RNAs; its early state-by-state activities led to the development of statewide natural heritage programs. The Nature Conservancy has more than 1,400 preserves and has saved about 10 million acres of land in the United States. A useful acquisition tool is the conservation easement, a legal agreement with private landowners that allows the owners to protect the natural resources from development without surrendering ownership of the property. The Nature Conservancy is active globally; the Wings of America program protects habitats in South America vital to U.S. songbird populations—as important as preserving habitats here.

Natural Heritage Programs

Most of the state natural heritage programs are outgrowths of the initial work conducted by The Nature Conservancy. These offices are very active in compiling an up-to-date database that can be used by private industry and public agencies to make informed decisions about development. Natural heritage scientists review tens of thousands of environmental impact assessments every year (these are required before building, excavation, and development can commence). They also make recommendations regarding land purchases and conservation easements from private landowners. Fieldwork includes mapping natural communities, conducting inventories of rarer species, removing exotic species, and introducing natural management processes, such as fire. Forest fires are suppressed to protect economic timber, homes, and towns yet are one of the most effective natural resource management tools. The 1988 fires that ravaged Yellowstone National Park, long considered tragic, are quickly resulting in benefits: The vegetation being reborn is ten times more diverse than the previous vegetation.

Our Priorities Today

Science can teach us cause and effect, but only philosophy can teach us the meaning of kinship. ... The extinction of the dusky seaside sparrow [in 1987] *teaches us that outward journeys—even to the moon—can never take us as far as inward ones; that searching is different from seeking; that facts about the universe are no substitute for a philosophy of caring for life on Earth. They all teach us that to preserve animals we must first preserve the multitude of habitats in which they live.* —Mark Jerome Walters

Some quick facts:

- One-fourth of the vast Everglades marshes in Florida has been destroyed, the remaining degraded by water diversion and pollution;
- Once stretching more than 24 million acres, the primeval "Big Woods" along the Mississippi River has been reduced to less than 4 million acres;
- The longleaf pine ecosystem in the southeastern United States has been reduced from 90 million acres to less than 3 million;
- In the past 25 years, songbird populations have dropped by up to 50 percent;
- Ninety-seven percent of all the old-growth forest east of the Cascades in Oregon has been destroyed;
- More than half of all the wetlands in the United States have disappeared.

These examples show how endangered our ecosystems and natural communities have become—and what little is left. What we see in a national or state forest may seem huge and sprawling, but it is not so when compared to what has been lost—and to what species need in order to thrive. Key issues today include preservation of wetlands, habitat fragmentation, habitat connectivity, ecological restoration, and biodiversity.

Wetlands

The incredibly diverse and fertile wetlands communities are one of the most valuable resources in the country. In addition to harboring a number of unique plant and animal species, wetlands improve the quality of the water passing through them. The water is filtered through the dense vegetation, and many of the inorganic ions and toxics are removed through settling and organic processes. Coastal wetlands are critical spawning and nursing grounds for fish and shellfish species, upon which so many other species survive (including humans). As floodwater storage areas, wetlands in drainage basins protect urban areas and reduce erosion. Without wetlands, rivers are much more destructive as waters surge downstream.

Habitat Fragmentation

Urbanization, as well as the growth of industry and agriculture, continue to impact natural areas, reducing their size and cutting them into smaller pieces, or fragments. These smaller and smaller fragments of habitat become increasing isolated from each other. As a fragment becomes smaller, it cannot support individual animals, many of which die as they cannot safely migrate to another fragment (itself likely to be overpopulated). Smaller patches also have a higher edge-to-area ratio, which increases the chances of predation, nest parasitism, and changes in microclimate, all of which reduce species populations.

Habitat Connectivity

Development also has eliminated many of the natural connections between habitats, such as woods along streams. These linkages, or corridors, allow animals to range freely between patches under cover, permit juveniles to leave their place of birth, and enhance the safer movement of animals within their home ranges. Corridors, especially riparian corridors, are good habitats for many mammals, birds, amphibians, reptiles, and insects, all of which contribute to the health and biodiversity of an ecosystem.

Maintaining large wild areas is one of the most important issues in conservation today. Biodiversity is highest in these wild terrains. Patches of forest do not support species as well as large tracts of forest. Large areas support larger and healthier populations of species, and negative human influences have less impact. For a given total forest area, one large patch supports more native interior species than two or more smaller patches. Large mammals who range widely are safer and reproduce better within large tracts of wilderness. For example, the black bear needs a home range as large as 40,000 acres; a prothonotary warbler needs at least 7,000 acres of interior forest.

Loss of Wilderness

Many biogeographers and conservationists stress that successful management of natural resources today must incorporate the viewpoints and knowledge of local residents, especially in ecologically sensitive areas. According to Arturo Gómez-Pompa and Andrea Kaus (*BioScience*, 1992), "the concept of wilderness as the untouched or untamed land is mostly a [Western] urban perception, the view of people who are far removed from the natural environment they depend on for raw resources ....There is virtually no land on Earth that has not been impacted by humankind. Western policies in the past 50 years regarding conservation have unrealistically viewed wilderness as an area without people, and focused its efforts on saving these areas. Cooperative relationships, with the local residents in ecologically fragile areas are of utmost importance to our understanding of the natural environment." Those who live and work in these environments sometimes have sounder land use and resource management practices than the removed scientist does; they know what works and what does not work. Thus, local collaboration is critical in the exchange of information and for making informed decisions about conservation. Outside policies that do not take into account the needs of the local populace will likely fail in preserving that habitat or ecosystem. "The fundamental challenge," continues Gómez-Pompa and Kaus, "is not to conserve the wilderness, but to tame the myth [of wilderness] with an understanding that humans are not apart from nature." Noted ecologist Reed Noss argues that they are both important challenges: "Large, wild areas—whether or not we call them wilderness—remain among the most important components of a conservation network. I am amazed at how few defenders of wilderness there are among the modern conservation crowd" (*International Journal of Wilderness*, 1996). Thus, maximum success will be achieved if both preservation and economic needs are recognized—and ways are discovered to meet both.

Ecological Restoration

Landscapes that have been damaged need to be healed: wetlands restored, exotic species removed, and native species planted and reintroduced. Whenever possible, natural processes should be used, such as prescribed burns to manage savannas and prairies. Areas for restoration should be selected because they represent diverse ecosystems and species assemblages and have the ability to support viable populations in natural patterns, a critical factor in maintaining biodiveristy. Education is critical so that people will embrace those human activities that are compatible with the ecosystems involved and reject and restrict those that are not. Long-term monitoring and research is important in determining what works and what does not. For example, there are lakes stocked with the wrong fish, forests that are unhealthy because of fire suppression, and other areas that are damaged from overgrazing. Some of the federal programs that are designed to induce the private landowner to restore wetlands, better manage forests, and reduce erosion are the Water Bank Program, Conservation Reserve Program, Forest Stewardship Program, and Wetland Reserve Program.

Maintaining Biodiversity

"Biodiversity and ecological integrity," says Reed Noss, "should be foremost objectives for wilderness designation and management" (*International Journal of Wilderness*, 1996). Conservationists and ecologists have learned much in the past few decades, but not nearly enough about the complicated ways ecosystems work. Each species plays a distinct role in its own biotic community, and the interrelationships can be delicate and complex. When one relationship is broken (for example, the extinction of a species), the strength and integrity of the entire ecosystem is weakened. "By protecting ecological diversity, we can hope to 'capture' genetic diversity without even detecting [or understanding] it at first," says Julian Campbell of The Nature Conservancy. "There are many groups of smaller organisms (other than vascular plants and vertebrates) that will take many decades to fully understand. In the meantime, we know, as a general rule, that rare or unusual environments often support rare or unusual organisms." The preservation of natural communities and ecosystems maximizes the natural selection of genetic types, as well as the genetic biodiversity of individual species.

Extinction is a natural result of evolution as species are replaced by others better suited to the changing environment. Yet the rate of extinction is greatly accelerated by outside factors, such as exotic species, which compete for space and nutrients, as well as human intervention. Of the 500 species and subspecies that have become extinct in the United States since the 16th century, all but one were the result of human interference. As species diminish, their gene pools are greatly reduced, as are their abilities to survive the stresses of changing environments. When communities and species vanish, we have fewer genetic resources from which to acquire new medicines, foods, drought- and disease-resistant crops, and other economic products. A gradual decline in biodiversity could, some believe, lead to a world that cannot support human life.

Cooperation

For conservation to be successful in the 21st century, ecologists and other proponents of conservation must recognize the needs of industry and how the land is intertwined in the local economy and land use. Support from the local populace is critical to the success of conservation projects. Compromise is important. For example, when a ski-area operator near the Telfer RNA in Lincoln National Forest, New Mexico, wanted to add three new lifts, a conflict erupted. After heated discussions, the resulting compromise added one new lift, and the size of the RNA was reduced from 1,300 acres to 797 acres—still a sufficient size to protect rare subalpine communities. The Nature Conservancy manages active cattle ranches that are compatible with preserving the special natural communities and species on the property. Certain methods of farming, forestry, and other industries can be compatible with preservation, such as selective logging. Ecotourism is a rapidly growing industry in the United States. By preserving high-quality ecosystems and habitats for ecotourism, nature's integrity is maintained, jobs are created, people learn more about the environment, and millions of tourist dollars are pumped into local economies. Visits to a single park in Tennessee in 1989, for example, generated revenues of nearly $3 million. With growing populations and urban sprawl, landowner participation—hrough partnerships in the private and public sectors or various incentives—is critical to preserving at-risk species and habitats today.

What You Can Do

We console ourselves with the comfortable fallacy that a single museum piece will do, ignoring the clear dictum of history that a species must be saved in many places if it is to be saved at all. –Aldo Leopold

Protection and conservation can coexist with other human activities, but planning must accompany development. Natural heritage staff scientists depend on information from a variety of sources. Make suggestions or report occurrences of unique plants and habitats—even if you are not sure what they are. The more complete the database, the greater the chance of preventing the inadvertent loss of species and natural communities.

Conservation organizations are always in need of money, materials, and people. Materials can range from copier paper to a used car for fieldwork. If a donation helps operations so that money can be spent in the field, it will be gratefully accepted. Just as important as money and equipment is volunteerism. For example, volunteers for The Nature Conservancy remove exotic species, plant native species, build boardwalks, maintain the preserves, and conduct fieldwork, among many other things.

Learn about preservation issues in your state. Petition your senators and representatives and find out what their stand is. Join local chapters of conservation groups, such as The Nature Conservancy and the Audubon Society.

Visit some of the places in this book. Only by being part of it can anyone truly understand why it is important to save: Look up at a 1,000-year-old cypress tree, lean against an ancient bristlecone pine high in the Sierra Nevadas, look across a rippling ocean of bluestem prairie, or listen to the howl of a wolf.

And most importantly, leave no trace of your visit. Make it hard for others to see or hear you while you are there. And if you packed it in, then pack it out.

Many a man, when I tell him that I have been on to a mountain, asks if I took a glass with me. It was not to see a few particular objects, as if they were near at hand, as I had been accustomed to see them, that I ascended the mountain, but to see an infinite variety far and near in their relation to each other, thus reduced to a single picture. –Henry David Thoreau

ALABAMA

The Lower Coastal Plain of Alabama occupies the southernmost fifth of the state and includes the Atlantic coastline. Generally flat-lying, the region is covered by longleaf pine–turkey oak–wiregrass woodlands with abundant herbs and grasses, including prairie species like Indian grass and little bluestem. Numerous, slow-moving blackwater streams with thick bottomland riparian corridors wind through this area. Important maritime communities edge the coastline.

The Lime Hills are located in the southwestern portion of the state and form a linear, northwest-southeast province. This hilly terrain is covered by spruce pine–beech–white oak–southern magnolia forest. Pockets of cucumber tree–black walnut occur in shaded, cool topographic lows.

The Red Hills occupy about one-fifth of the state in an east-west zone north of the Lower Coastal Plain. Its dominant communities are beech–southern magnolia–American holly forest with a few pine trees. The upper slopes also support communities of mixed pine-oak.

North of the Red Hills is an east-west trending belt of woodland and grassland called the Black Belt. Converted largely to farmland, the Black Belt was once covered by extensive little bluestem–Indian grass–dropseed tallgrass prairie.

The Upper Coastal Plain represents a northwest-southeast swath that crosses through the center of Alabama. Its dominant forest type is loblolly pine–southern red oak–mockernut hickory. It also includes occasional stands of Arkansas oak, a species of concern in the state.

The east-central portion of Alabama is the Piedmont Province. It is rugged and hilly, pine forests cover the ridge tops, and mixed hardwood forest grows along the lower slopes and river corridors. Sporadic large outcrops of barren granite rock harbor rare plant species, such as stonecrop and flatsedge.

The mountainous Blue Ridge, the Ridge and Valley, and Appalachian Plateau occupy much of northeastern Alabama. These rugged areas are dominated by longleaf pine–chestnut oak forest with lowbush blueberry, stands of Virginia pine, oak-hickory forest on dry rock ridges, and beech–sweet gum–umbrella magnolia–bottlebrush buckeye forest. The unique limestone glade community can be found interspersed with some of these woodlands.

The northern fifth of Alabama is part of the Tennessee Valley, a fertile flatland that is edged by bluffs of limestone. It is now an important farming area. Remaining forest types include shagbark hickory–chinkapin oak–white oak–American smoketree, with a varied herbaceous understory, including several rare species. Little bluestem–dropseed prairie is also present.

Special Habitat Sites

Blowing Wind Cave
Jackson County
Just west of Scottsboro, this cave system is located in the forested bottomlands along the Tennesssee River. Mature to old-growth forests include cypress, maple, poplar, and beech trees that are hundreds of years old. The federally endangered Indiana bat and gray bat live in the cave. Nearly a quarter-million bats can be seen when they take flight at dusk to feed.

Buck's Pocket State Park
De Kalb, Jackson, and Marshall Counties
Located a few miles northeast of Grove Oak, this state park supports rare plant assemblages in a cool and moist canyon along South Sauty and Little South Sauty Creeks. Several varieties of ferns grow here, including Christmas fern, rattlesnake fern, and walking fern. Large specimens of blue beech, butternut, and buckeye trees grow along the cliffs. Colorful wildflowers, including the rare Wake Robin trillium, bloom between February and June. Several rare birds and amphibians have also been found in the park.

Conecuh National Forest
Covington and Escambia Counties
Located about 20 miles southwest of Andalusia, this 84,000-acre property includes bottomland forest, canebreaks, longleaf pine forest, bald cypress swamp, marshes, bogs, and riparian corridors. Wetlands contain more than 20 different rare species of carnivorous plants, including pitcher plants and sundews. The underlying cave systems have created a karst topography of shallow sinkhole ponds. Other rare species found in this high-biodiversity area include the dusky gopher frog, Indiana bat, red-cockaded woodpecker, orchids, and needle palm. A few stands of mature to old-growth virgin longleaf pine also remain.

Dauphin Island Bird Sanctuary
Mobile County
A barrier island in the Gulf of Mexico, Dauphin Island is home to this 164-acre Audubon Society sanctuary. Natural communities include tidal flat, beach, dune, marsh, maritime pine–magnolia–live oak–tupelo forest, and scrub oak forest. The island is a critical breeding habitat for several rare birds and a migratory resting and feeding stop for nearly 400 others, many coming from the Yucatán Peninsula, Mexico. The dunes are anchored by prickly pear cactus, sassafras, and various grasses. Alligators, turtles, and amphibians inhabit the beaches.

De Soto State Park
De Kalb County
Located about 8 miles northeast of Fort Payne, this highly diverse, mature, mixed hardwood forest forms an important riparian corridor near the Little River. More than 500 different species of wildflowers, including pink lady's slippers, magenta fire pinks, and lavender cranesbill bloom during spring. Peat bogs support sphagnum moss, the pitcher plant, and other rare bog species. Raptors frequent the rock bluffs above the river, and bald eagles occasionally use the park for winter habitat. The damp cliff faces support several unusual plant communities (mosses, lichens, and ferns).

Grand Bay Savanna
Mobile County
Straddling the Alabama-Mississippi border, this 300-square-mile Nature Conservancy project area encloses one of the largest and most pristine wet pine savannas in the country. More than 20 natural communities and 70 rare species can be found, including longleaf and slash pine savannas, pine flatwood, salt marsh, brackish marsh, and cypress savanna. Unusual species include Chapman's butterwort, Mobile crayfish, yellow-blotched map turtle, the Alabama redbellied turtle, and carnivorous pitcher plants. The productive tidal flats and estuaries are an important food source for a variety of waterbirds.

Lake Guntersville
Marshall and Jackson Counties
Located along the Tennessee River and Lake Guntersville, this elongate area includes Taylor Mountain. Riparian habitats along the creeks and streams support the rare yellow lady's slipper and mountain stewartia. The mountain offers oak-hickory forest, sandstone glade, maple–sweet gum forest, and cliff face natural communities. The mature forests are on an important songbird flyway and are thus an important migratory stopover during spring and fall for rare species, such as the cerulean warbler. Bald eagles also nest in the area.

Little River Canyon
Cherokee, De Kalb, and Etowah Counties
One of the most impressive canyons east of the Rocky Mountains, Little River Canyon south of Fort Payne is part of the Cumberland Plateau. Its canyon walls, talus slopes, vernal pools, and various waterfalls support some unique plant communities, including mosses, ferns, and lichens. The river's waters are exceptionally clean. The canyon has been mostly undisturbed by human activity and contains large specimens of pine and hardwood trees and rare plants in the understory.

Mobile-Tensaw River Bottomlands
Mobile and Baldwin Counties
Located on 185,000 acres, this extensive estuarine habitat includes forested marshy bottomlands along the Mississippi Delta. Other communities include freshwater, brackish, and saltwater marshes and mudflats. The wetlands are an important spawning ground and nursery for an abundance of fish and shellfish and is consequently an important food source for numerous waterbirds. Tens of thousands of birds stop during spring and fall migrations. Larger mammals, such as the rare black bear, can also be spotted in the marshlands.

Monte Sano Preserve
Madison County
Located east of the city of Huntsville, this preserve contains excellent examples of second-growth mixed mesophytic forest, eastern red cedar forest, and large specimens of white oak, shagbark hickory, and sweet buckeye. The rare American smoketree also grows here. Other unusual species include yellowwood, Morefield's clematis, and several wildflowers and orchids. The rocky, dry cedar glades support the diminutive and rare lime stonecrop and leavenworthia. The preserve has one of the most diverse populations of forest birds in the state.

Payne Lake Recreation Area
Hale County
Located about 30 miles southeast of Tuscaloosa, this state recreation area is located in the Talladega National Forest. High-quality natural communities found here include mature pine forest and mixed upland pine–hardwood forest, with occasional wetlands and bottomland forest riparian corridors. The pine forest is an important habitat for nesting red-cockaded woodpeckers.

Sipsey Wilderness Area
Lawrence County
Part of Bankhead National Forest about 30 miles southwest of Decatur, this 26,000-acre parcel protects mature to old-growth remnants of the "Big Woods" that once covered Alabama during presettlement times. Patches of hemlock trees in cool, shaded, moist ravines are relict species from a cooler climate long past. Large specimens of beech, oak, and maple are hundreds of years old and nearly 100 feet tall with broad canopies. Several rare or unusual plant and animal species have been found here.

Weeks Bay National Estuarine Research Reserve
Baldwin County
Located a few miles from Fairhope, this reserve and nature trail along Mobile Bay protects excellent examples of intertidal emergent marsh, wetland, bottomland hardwood forest, and upland pine–oak forest. Other interesting trees that occur here are loblolly pine, wax myrtle, popcorn tree, red cedar, live oak, black gum, and bald cypress. Cordgrass, salt-meadow grass, and sea-myrtle dominate the emergent marsh. Rare ferns and sundews can be found in the boggy areas. These communities are important habitats for a variety of birds, reptiles, and amphibians, some of them rare or unusual.

ALASKA

Unlike the other 49 states, less than 5 percent of Alaska has been permanently altered by human activity. Huge areas of the state are filled with pristine ecosystems. Alaska has lost less than 1 percent of its wetlands, compared to nearly 60 percent in the continental United States. As encouraging as this is, more than 75 percent of Alaska's rare plants occur on unprotected lands. In light of pressures from the logging and oil industries, new roads will likely be built into wilderness areas, which will fragment habitats. Consequently, The Nature Conservancy is working to establish community-based partnerships that promote subtle shifts in the way people use the land.

Alaska is one of the most ecologically diverse states in the country. Its series of mountains and valleys are covered by the following major vegetation types: hemlock-spruce forest, spruce-hardwood forest, tall-shrub thicket, muskeg, tundra, coastal wetlands, and barrier island–lagoon systems. Other habitats include freshwater lakes and rivers, beach–cliff–rocky shore, and glaciers and ice fields.

Hemlock-spruce forest edges the southern border of the state, especially along Cook Inlet, the Kenai Peninsula, the Gulf of Alaska coastline, and the Alexander Archipelago. This forest is dominated by western hemlock and Sitka spruce and often termed the "coastal rain forest." It ranges in elevation from tidewater to nearly 9,000 feet. Some of the biggest trees in the world occur in the old-growth forests; some reach 300 feet and are nearly 1,000 years old.

The dominant community type in the vast central portion of the state is spruce-hardwood forest, which is interspersed with patches of alpine tundra and muskeg. Typical trees are balsam poplar, willow, tamarack, aspen, birch, and white and black spruce; the exact tree associations depend on the wetness and chemistry of the soil. Muskeg is a soft, spongy groundcover of sphagnum mosses, sedges, lichens, and shrubs that cover low, wet, poorly drained areas.

The northern third of the state, the Seward Peninsula, and the southwestern portion of the state, which juts into Norton Sound, the Bering Sea, and Bristol Bay, are dominated by tundra with lesser tall-shrub thickets. The tundra grades from wet to moist to alpine (dry) progressively inward from the coast. Alpine tundra grows above the timberline and consists of low mat plants on rock outcrops or rocky soil; subalpine vegetation includes tall shrubs, such as willow and birch. Moist tundra is found most often in the western and arctic foothills and typically contains cottongrass-sedge, mosses, and low shrubs. Wet tundra occurs in lowlands as a saturated groundcover of lichens, sedges, shrubs, herbs, and grasses. Tall-shrub thickets are very dense and consist of coastal alder, Sitka alder, and birch-alder-willow thickets, often in riparian corridors or on floodplains.

Coastal wetlands consist of tidal flats, saltwater marsh, freshwater marsh, brackish marsh, estuaries that are influenced by tidal action, coastal river deltas, and sandy beaches. The various plant communities depend on the amount of salt water and geographic location.

Barrier islands and lagoons rim many of the northern and western coastlines, especially around the Seward Peninsula and along the Arctic Ocean–Beafort Sea coastline. Comprising masses of sand and gravel deposited by longshore currents, the islands are sparsely covered by beach ryegrass, beach sandwort, and salt marsh plants.

Special Habitat Sites

Admiralty Island National Monument
Unorganized Borough
This island is west of Tracy Arm and south of Juneau. Its old-growth forests are an important habitat for the Alaskan brown bear (nearly 2,000 live on the island) and nesting bald eagles (more than 800). The old-growth spruce and hemlock trees reach 200 feet and are hundreds of years old. Sea otters frequent the island's beaches.

Anchor River/Fritz Creek Critical Habitat Area
Kenai Peninsula Borough
This site protects about 20 miles of brushy, swampy riparian habitat along Fritz Creek and the Anchor River, midway along the Kenai Peninsula in southern Alaska. The wetlands also support tall cottonwood trees and dense spruce forest. It was deemed a critical habitat area by the Department of Fish and Game for being the major moose overwintering area. The moose are attracted by the willow browse and cover along the streams, and moose often calve there during spring. Other species that depend on this habitat are bald eagles, black bears, river otters, and wolves. Several bird species visit during spring and fall migrations.

Anchorage Coastal Wildlife Refuge
Unorganized Borough
Located west of Anchorage, this refuge includes Potter Marsh, a very extensive tidal flat. Other communities include alder-bog forest. More than 150 species of birds use the site during spring and fall migrations, including Pacific loons, swans, Canada geese, loons, and some rare and unusual species. The rare lesser Canada goose nests in the marshes, as do grebes, short-eared owls, yellowlegs, and phalaropes. Moose frequently graze in the marsh.

Blind Slough Trumpeter Swan Management Area
Unorganized Borough
Located about 20 miles south of Petersburg, this coastal wetland-hemlock-spruce forest complex is an important wintering habitat for trumpeter swans; some 100 reside here from about October to April. The wetlands also attract tens of thousands of other waterbirds during spring and fall migrations; bald eagles are frequently seen.

Bristol Bay
Bristol Bay, Lake and Peninsula Borough
Five critical habitat areas have been established by the state of Alaska along the Alaska Peninsula: Egegik, Pilot Point, Cinder River, Port Heiden, and Port Moller. These are all riverine estuaries along Bristol Bay that attract hundreds of thousands of migrating waterfowl during the spring and fall. The extensive tidelands and wetlands serve as an important feeding area and resting stop. Pilot Point is used by the world's largest population of rare cackling Canada geese. Various terns, gulls, and eiders nest at Port Moller. Harbor seals inhabit offshore islands. The rivers are important spawning grounds for salmon and Pacific herring.

Chilkat Bald Eagle Preserve
Haines Borough
This preserve along the Chilkat River is world famous for its outstanding population of bald eagles. Thousands of these majestic birds can be seen during the fall, when they feed on a late run of salmon in the river. The surrounding forests contain an abundance of North American birds and animals, including caribous, bears, and moose.

Copper River Delta Critical Habitat Area
Unorganized Borough
Located east of Cordova, along the northern shore of the Gulf of Alaska, is a sprawling complex of sandbars, tidal flats, and deltaic mudflats at the mouth of the Copper River, meandering streams and sloughs, dense wet-

lands, and coastal rain forest with towering conifers. Located on the Pacific Flyway, the Copper Delta is occupied by up to 15 million shorebirds during spring and fall migrations. The waters are covered with birds—up to 250,000 per square mile, including less common species, such as the sandhill crane. Moose and bears also inhabit the wetlands, which are an important calving habitat for moose during spring. Thriving populations of trout and salmon travel up the Copper River to spawn.

Creamer's Field Waterfowl Refuge
Fairbanks North Star Borough
This 1,776-acre pond and wetland complex north of Fairbanks is a former dairy farm that has been converted into a waterfowl habitat. It is a convenient stop for thousands of waterfowl migrating along the Pacific Flyway during spring and fall. Canada geese, pintails, teals, sandhill cranes, Lapland longspurs, plovers, northern harriers, and eagles all stop to rest and feed before moving on to the their summer nesting sites. More than 150 bird species have been identified here.

Dude Creek Citical Habitat Area
Unorganized Borough
This site encloses the drainage basin for Dude Creek, a 3-mile stream that empties into Icy Passage. These wet flatlands enclose a large expanse of undisturbed wet meadow. Plant communities include shrubs, sedges, horsetail, various mosses, willow thickets, and scrub forest. Located near Gustavus, it is a preferred habitat of the lesser sandhill crane. Other important bird species are the snipe, least sandpiper, savanna sparrow, and bald eagle. The forested Dude Creek corridor serves as a safe passage for bears and moose traveling between Glacier Bay and Icy Passage.

Fox Lake Flats
Kenai Peninsula Borough
Established in 1972, this property encloses the wetlands that surround the meandering mouths of the Fox River and Sheep Creek as they enter Kachemak Bay, a few miles east of Homer. Deemed a critical habitat area by the Department of Fish and Game, the extensive wetland–wet shrubland complex is an important habitat for moose, bears, foxes, wolverines, and wolves. The pristine streams are an important spawning ground for salmon, chum, and Dolly Varden char. Hundreds of thousands of waterbirds use the area as a migratory stopover and feeding ground. Some of the more unusual species include the sandhill crane.

Gates of the Arctic National Preserve
North Slope Borough
Located in the Brooks Range about 60 miles north of Bettles Field, this 8 million–acre preserve comprises mountains, rivers, and valleys, most of it designated wilderness. Alpine tundra dominates the mountain slopes and ridges, and birch-willow-alder-spruce borders the lower riparian corridors. Caribou herds of nearly a half-million animals, and plentiful moose, wolverines, black and grizzly bears, martens, wolverines, and Dall sheep live in the preserve. The high alpine pastures are filled with flowers during spring.

Goose Bay State Game Refuge
Matanceska-Susitna Borough
Located along Upper Cook Inlet across from Anchorage, this dense wetland complex contains slow-moving streams that drain into Goose Bay. The dominant natural community is wet shrubland. The wetland is an important nesting site and migratory stopover for a variety of waterbirds, including trumpeter and tundra swans, teals, yellowlegs, and the rare sandhill crane. The shrubland is also an important calving site for moose during spring.

Icy Cape-Kasegaluk Lagoon
North Slope Borough
One of the largest barrier island–lagoon complexes in North America, Kasegaluk Lagoon rims Alaska's northwestern coast near Wainwright. Natural communities include barrier island, mudflat, coastline beach and wetland, and widespread tundra. Largely undisturbed by man, this area is a key habitat for a variety of bird species, including up to 500,000 king eiders and thousands of other seabirds. Whales frequent the lagoons during summer. Other species of interest include the arctic fox, caribou, snowy owl, snow bunting, Lapland longspur, arctic loon, and ptarmigan.

Kachemak Bay Critical Habitat Area
Kenai Peninsula Borough
This unusual preserve protects Kachemak Bay and its estuaries that penetrate the mainland. Located near Homer, nearly 95 percent of the critical habitat is submerged. Because of water

circulation patterns, the bay is highly productive, with abundant fisheries and shelfish larvae. Plentiful populations of crab, shrimp, mussels, snails, and other marine invertebrates thrive on the mudflats and beaches. Because it is such an excellent food source, great concentrations of waterbirds frequent the area during the spring and fall migrations, as do harbor porpoises, sea lions, and several whale species.

Kalgin Island
Kenai Peninsula Borough
Established in 1972, this state critical habitat area 20 miles southwest of Kenai protects a dense wetland complex along Swamp Creek on Kalgin Island in Cook Inlet, including flanking tidal flats. The stream-wetland complex is an important resting and feeding spot for a variety of bird life during the spring and fall migrations, including greater yellowlegs, northern harriers, and bald eagles. Other animals, such as harbor seals and spawning silver salmon, use Swamp Creek.

Kenai River and Watershed
Kenai Peninsula Borough
This 82-mile watershed in southern Alaska is an internationally renowned habitat for fish and waterfowl. The river empties into Cook Inlet at Kenai. Development along the river has increased dramatically in the last 20 years. Nesting bald eagles and other raptors are common among the lakes and vast wetlands edged by willow forests. Nearly 30 different fish species thrive in the Kenai River, an important spawning ground for the king salmon. Rare or endangered birds that breed along the river are the double-breasted cormorant and Wrangel Island snow goose. The surrounding wilderness is prime habitat for large mammals, such as moose, bears, wolves, and caribou.

Kodiak National Wildlife Refuge
Kodiak Island Borough
Covering almost 2 million acres on four islands and accessible only by plane or boat, this refuge protects critical habitat for the Kodiak brown bear. Other animals are the elk, reindeer, spawning salmon, tundra vole, little brown myotis bat, river otter, and red fox. More than 2 million seabirds inhabit its bays, inlets, and freshwater and saltwater marshes, including 200 pairs of nesting bald eagles. Natural communities include rocky shoreline, rock face, coastal wetland, estuary, salt marsh, alpine tundra, tall-shrub thicket, lake, and riparian corridor. The tidal waters are frequented by harbor seals, sea lions, porpoises, and whales.

Lower Talarik Creek
Lake and Peninsula Borough
Located about 200 miles southwest of Anchorage, Talarik Creek flows into Lake Iliamna. Its heavily forested watershed supports a healthy population of North American mammals, including caribou, grizzly bears, wolves, mink, and red foxes. Rare or threatened bird species found here are bald eagles, teals, whistling swans, and the American widgeon. Talarik Creek is home to the sockeye and coho salmon.

McNeil River Game Sanctuary
Kenai Peninsula Borough
Located south of Pedro Bay along the west side of Cook Inlet, this sanctuary includes part of the drainage basin of McNeil River, which flows through the Aleutians northeastward into Kamishak Bay. These forested hills shelter a population of about 100 brown bears that feed on spawning salmon in the river during summer. Other residents include moose, otters, various voles, wolves, and wolverines. It is also an important bird habitat, with more than 130 species having been identified.

Mendenhall Wetlands State Game Refuge
Juneau Burough
This elongated open water–wetland complex protects part of the Gastineau Channel northwest of Juneau. Communities include freshwater stream mouth, salt marsh, sedge meadow, mudflat, and other intertidal wetlands. The site is an important nesting habitat and migratory stopover for a variety of waterfowl, including Canada geese, goldeneyes, great blue herons, and arctic terns. Bald eagles are frequently seen. It is also a critical nursery rearing area for several fish species, including Dolly Varden char, various salmon, and cutthroat trout.

Nancy Lake Recreation Area
Matanuska-Susitna Burough
Located on 22,685 acres west of Willow, this tract is an important habitat for birds and mammals. Natural communities include lake, stream, riparian corridor, swamp, bog, and lowland spruce–hardwood forest. Big-game

animals, such as bears and moose, depend on this habitat for food and shelter. Several varieties of nesting and migrating birds appear throughout the year, including several species of loons, golden and bald eagles, hawks, ospreys, grebes, and Canada geese.

Palmer Hay Flats State Game Refuge
Matanuska-Susitna Borough
Located northeast of Anchorage, Palmer Hay Flats encloses the mouths of Wasilla and Cottonwood Creeks, which discharge into Knik Arm. Natural communities include tidal flat, marsh, dense shrubby swamp, wet meadow, stream, wooded riparian corridor, and mixed conifer forest. This wet, flat-lying expanse is a critical habitat for moose, bears, and hundreds of thousands of waterbirds that nest here or visit during spring and fall migrations. Interesting species include the rare sandhill crane, snow geese, tundra swans, and bald eagles. The patches of spruce forest along the riparian corridors are important birthing habitats for moose calves during spring; silver salmon spawn up Wasilla and Cottonwood Creeks.

Prince of Wales Island
Unorganized Borough
Located west of Ketchikan, this 1.4-million-acre area is one of the largest islands in the United States. Its natural communities include subtidal flat, cave, saltwater marsh, freshwater marsh, wetland, meadow, and old-growth Sitka spruce–hemlock forest. The many estuaries support large populations of shellfish and crustaceans, an important food source for the hundreds of thousands of waterbirds that nest or visit the island during migrations. Significant areas of towering old-growth forest escaped clear-cutting. Bald eagles, black and brown bears, the Sitka black-tailed deer, endemic Alexander Archipelago wolves, and several rare bats inhabit the interior of the island. Several varieties of sea mammals, such as otters, harbor seals, and sea lions, frequent the coastal waters and bays.

Prudhoe Bay Area
North Slope Borough
Located in northern Alaska along the Beaufort Sea coastline about 450 miles north of Fairbanks, natural communities here include marine coastline, wet and moist tundra, lake, river, and riparian corridor. Large numbers of caribou, arctic foxes, arctic loons, various swans and geese, majestic king eiders, plovers, phalaropes, snowy owl, and rock ptarmigan inhabit the wetlands.

Redoubt Bay
Kenai Peninsula Borough
Located on the western edge of Cook Inlet near Tyonek, this state critical habitat area covers 268 square miles. Natural communities include extensive wetland and wandering stream with wooded riparian corridor. Bears and moose can frequently be seen along the streams. The site is a critical migratory stopover for hundreds of thousands of waterbirds, including cackling and Taverner's Canada geese, trumpeter swans, sandhill cranes, and a huge number of ducks. Bald eagles, yellowlegs, godwits, plovers, and phalaropes all nest in the woodlands.

Sheep Mountain
Unorganized Borough
Located about 100 miles east of Anchorage, this 25,600-acre tract enclosed the rugged alpine communities that cover the upper slopes and peak of Sheep Mountain. This site is one of the prime habitats for herds of Dall sheep. Sheep Mountain is also located along a major raptor flyway, and a variety of eagles and hawks frequent the area, nesting and hunting from its rock ledges.

Stan Price Wildlife Sanctuary
Unorganized Borough
Located 30 miles south of Juneau, this property at the mouth of Pack Creek on Admiralty Island protects critical habitat for brown bears. Pack Creek is surrounded by old-growth spruce-hemlock forest that reaches heights of nearly 200 feet. The intertidal wetlands and tidal flats are frequented by bears, which feed on salmon during summer months. Bald eagles also nest and hunt on the island.

Susita Flats Game Refuge
Matanuska-Susita Borough
Located along Cook Inlet west of Spendard, this large game refuge includes mud tidal flats, stream deltas, shallow lakes and ponds, salt marshes, sedge meadows, and extensive shrubby or spruce-forested wetlands. The wetlands are an important concentration point for migrating waterfowl during spring and fall. Hundreds of thousands of waterbirds nest,

feed, and rest on the property, including sandhill cranes, a variety of swans and plovers, and the more rare Tule goose. Shellfish and other crustaceans in the tidal mudflats are a primary source of food. Moose and bears are frequently seen, and moose calves are born in the thick shrubland during spring.

Tern Lake
Peninsula Borough
Located in the mountains of the Kenai Peninsula north of Seward, this 100-acre property encloses Tern Lake. The dominant natural communities include the lake itself, flanking bog and marsh, and high-elevation mixed conifer forest. Several varieties of birds nest around the lake, including arctic terns, mallards, pintails, and greater yellowlegs. It is also a habitat for large mammals, such as Dall sheep, moose, wolves, black bears, and brown bears. Even at this high elevation, specialized plants, such as sundew, cottongrass, sphagnum moss, and bog rosemary, thrive at the lake's edges.

Teshekpuk Lake
North Slope Borough
Located along the northern coast about 80 miles southeast of Barrow, this expanse of wet tundra dotted with lakes is an important summer habitat for Canada and great white-fronted geese. Other species include plovers, sandpipers, phalaropes, dunlins, loons, gulls, and snowy owls. The stark tundra landscape is a critical habitat for large herds of caribou, arctic foxes, and lemmings.

Trading Bay Game Refuge
Kenai Peninsula Borough
This large game refuge is located along the northern shore of Cook Inlet and includes parts of the drainages of McArthur River and Middle River. Predominantly wetlands and tidal flats, the thick swamps and bushy riparian corridors are ideal habitats for calving moose during spring. Other important mammals that use the area are brown bears, black bears, salmon, mink, otters, and wolves. The wetlands are most known for the huge populations of birds that feed during spring, nest during summer, and migrate during fall. These include swans, cranes, Canada geese, trumpeter swans, goldeneyes, and Tule geese.

Tugidak Island
Kodiak Island Borough
Identified as a state critical habitat area in 1988, this island is part of the Trinity Islands, south of Kodiak and Kodiak Island in the Pacific Ocean. Various natural features include shallow lagoons, a barrier spit complex, and low-lying tundra relief. It is an important nesting environment for tens of thousands of birds, including loons, tundra swans, emperor geese, ground-nesting bald eagles, arctic terns, and ptarmigan. Nearly 25,000 harbor seals use the island beaches to reproduce some 5,000 pups annually. Dense eelgrass beds offshore in the subtidal zone are an important haven for young fish and shellfish.

Walrus Islands Game Sanctuary
Unorganized Borough
This sanctuary is composed of seven rocky islands in northern Bristol Bay, about 5 miles southwest of Kulukak. One of the islands, Round Island, is the only land base in the southern Bering Sea, where up to 12,000 walruses haul out every summer. It is also used by sea lions and harbor seals, and gray whales feed offshore. Almost a half-million seabirds nest every year on Round Island, including some rare and unusual species. It is also an important migratory stopover during spring and fall.

Yakataga State Game Refuge
Unorganized Borough
Located near Cape Yakataga on the Gulf of Alaska, this expansive coastal wetland-lowland complex is an important wildlife habitat. Natural communities include lowland forest, grass flat, shrubland, stream, swamp, and riparian corridor. Connected to the short Tsie and Tsivat Rivers, the wetlands support wolves, wolverine, bear, moose, and Coho and sockeye salmon. Located beneath the Pacific Flyway, it is an important feeding place and migratory stopover for thousands of birds, including some rare species, during spring and fall.

ARIZONA

Arizona is divided into the following biotic regions: desert, grassland, chaparral, woodland, forest, and tundra.

Southwestern desert scrub and Great Basin desert scrub are the two major associations in the desert region. Typical southwestern desert scrub plants are mesquite, paloverde, saltbush, blackbrush, Joshua tree, sandpaperbush, whitethorn, tarbush, and creosotebush. Great Basin desert scrub is dominated by the sagebrush-blackbrush-shadscale assemblage and can be found in the northern part of the state.

Grasslands include the mountain, plains short-grass, and desert grasslands. Mountain grasslands are found in small openings in coniferous forest. Plains grassland (grama grass-muhly-needlegrass-dropseed-sprangletop) can generally be found at elevations between 5,000 and 7,000 feet in the eastern half of the state, especially Santa Cruz, Cochise, and Yavapai Counties. Desert grasslands are found mostly in the southeastern quarter of the state and are often grama grasses and cottongrass; cacti may also be interspersed.

Chaparral is mostly found in the central portion of the state, between elevations of 4,000 and 6,000 feet. This dense shrubby growth can be almost impenetrable and consists of scrub oak, manzanita, mountain mahogany, and turpentine bush. Some grasses may also occur.

Evergreen woodlands and juniper-pinyon woodlands compose the woodland biotic region. Evergreen woodlands include the oak-pine and juniper–pinyon pine assemblages. Deciduous woodlands are dominated by cottonwood, willow, alder, and sycamore trees and grow in riparian corridors. These sometimes grow to heights of 100 feet, with large, spreading crowns.

The forest biotic region is found at elevations between 6,000 and 9,000 feet and comprises different conifer associations, including ponderosa pine (6,000–7,000 feet), Douglas fir (7,500–9,500 feet), limber pine and spruce–alpine fir (9,000–11,500 feet), and quaking aspen (various elevations). Forest type depends primarily on elevation, soil chemistry, and moisture content.

The tundra is a cold, windy, high-elevation environment (11,000–12,000 feet) that is usually treeless. Arizona's alpine tundra occurs above timberline and consists of patches or meadows of sedges, grasses, and herbs with occasional small shrubs.

Special Habitat Sites

Aravaipa Canyon Preserve
Pinal County
Located about 70 miles north of Tucson, this Nature Conservancy tract encloses 7,000 acres along Aravaipa Canyon, including tree-lined Aravaipa Creek. The canyon is home to mountain lions, the ringtail cat, bighorn sheep, and a variety of birds that nest along the rock walls. The creek supports a large number of fish species endemic to Arizona and is one of the most biologically diverse riparian zones in the state. More than 200 birds species, including the rare Mexican black hawk, inhabit the preserve.

Arivaca Cienega
Pima County
Located on nearly 1,000 acres, this *cienega* is the best example of the less than six *cienegas* that occur in Arizona. The *cienega*, a marsh that is filled by groundwater seepage, is one of the most endangered habitats in the state. The large cottonwood-willow forest that surrounds the marsh is an important habitat for a variety of subtropical birds, including gray hawks, kingbirds, and many others. Several rare plant species also occur here.

Babocomari Cienega
Santa Cruz and Cochise Counties
This 1,200-acre property west of Sierra Vista is a lush, 3-mile stretch of the Babocomari River. The river's edge is lined with mature cottonwood trees, willow, and walnut. The flanking communities are desert scrub and short-grass prairie. The stream is a critical habitat for several native fish, including the speckled dace. Other rare species are the Mexican garter snake, a primrose flower, and Canelo ladies' tresses orchid.

Bill Williams River
Mohave County
Located southeast of Lake Havasu City, this 30-mile stretch of high-quality river flows between Alamo Lake State Park and Buckskin Mountain State Park. The river is edged by tall cottonwoods, willows, sedges, grasses, and a broad wetland near its confluence with the Colorado

River. Rare or endangered species that inhabit this site include green kingfishers, the southern bald eagle, Yuma clapper rails, black rails, and other raptors.

Bonita Creek
Graham County
Located north of Safford, this 19-mile stretch of Bonita Creek flows through a steep-walled canyon in the Sonora Desert. The stream has a well-developed riparian corridor that is marked by tall cottonwoods, willows, and grassy patches. Numerous raptors nest in the rock walls, including the threatened Mexican black hawk. The rare Rocky Mountain bighorn sheep also lives in the canyon. The threatened spikedace and loach minnow populate the creek.

Burrow Creek
Mojave County
Located southeast of Kingman, this clear-running stream flows through deep, scenic canyons. Mature cottonwoods and willow trees line the canyon and are an important bird habitat. Biodiversity is high and includes a variety of mammals, birds, reptiles, and amphibians. Some of the rare or threatened species are zone-tailed hawks, Mexican black hawks, southern bald eagles, peregrine falcons, gila monsters, and mountain lions. The canyon is prone to flash floods.

Canelo Hills Cienega Preserve
Cochise County
This 205-acre wetland complex 8 miles northwest of Sierra Vista is one of the most important in the southwestern United States. It is centered on spring-fed O'Donnell Creek. Tall cottonwoods, willows, and walnut trees line the creek. The streambed is surrounded by grassland and oak-juniper woodland. In addition to several endangered fish, the preserve has populations of the rare ladies' tresses orchid. Rare species include the Gila chub and Gila sucker, canyon tree frog, red-spotted toad, and black-tailed rattlesnakes. It is owned and maintained by The Nature Conservancy.

Cave Creek Canyon
Cochise County
Just outside the hamlet of Portal in the Chiricahua Mountains, this property contains red-orange cliff faces and wind caves. Its higher elevations contain Arizona pine forest, alligator juniper woodland, pine-oak woods, stands of Chiricahua and Apache pines, and Engelmann spruces on peaks, the tallest of which grow at nearly 10,000-feet elevation. Rare mammals include elusive jaguars, Apache fox squirrels, mountain lions, bears, and several rare bats. The wooded riparian corridor in the canyon attracts hundreds of bird species, such as trogons, the Mexican chickadee, rare hummingbirds, thick-billed parrots, and raptors.

Chevelon Creek
Navajo and Coconino Counties
Located near Winslow, Chevelon Creek flows from the Mogollon Rim to the Little Colorado River. The water is very clean and supports diverse aquatic life. The stream is bordered by mature Arizona walnut-oak-ash-alder woodland and stands of ponderosa pine. Some of the rare or endangered species that depend on this pristine watercourse are the bluehead mountain sucker, Little Colorado spinedace, speckled dace, black-tailed rattlesnake, bald eagle, prairie falcon, mountain lion, and black bear.

Chiricahua National Monument
Cochise County
Located 35 miles southeast of Willcox in the Chiricahua Mountains, this 12,000-acre property encloses rocky ridges and canyons. The high, windswept ridges have low, twisted oaks, manzanitas, and chaparral. The canyons shelter mixed conifer forest of Arizona cypress and madrone and more than 30 species of lichens, some of them rare or unusual. The rare Cochise rock daisy, endemic to this part of the world, grows here. Several varieties of raptors nest in the canyon, which also harbors the rare Chihuahua spotted whiptail lizard.

Cluff Ranch Wildlife Area
Graham County
Located 5 miles south of Pima, this 1,300-acre state wildlife management area encloses good-quality upland shrub and riparian woodland communities. Upland vegetation is dominated by cacti, creosotebush, jimmy weed, and saltbush. The green riparian corridor is shaded by tall cottonwood, willow, and ash trees as well as mesquite and netleaf hackberry. The water is clean and supports trout. Large mammals that inhabit the site include mule deer, the javelina (wild boar), mountain lions, and a variety of raptors, some of them rare.

Empire Cienega Resource Conservation Area
Pima County
Located about 25 miles southwest of Benson, this 45,000-acre grassland is a rare habitat in the state. The diverse grasses grow to heights of six feet and higher and bloom colorfully during spring. Mature cottonwoods, oak, willows, and ash drape Cienega Creek. In fact, the largest Emory oak tree in the country grows here. The clean waters of the creek support the endangered Gila topminnow, Gila chub, and longfin dace, all rare and endemic species.

Escudilla Mountain Research Natural Area
Apache County
Part of the Apache National Forest, this 909-acre Rocky Mountain montane grassland is located at about 10,800 feet on Escudilla Mountain, roughly 9 miles north of Alpine. Because of its inaccessibility, the grassland is in excellent presettlement condition. Grassland associations are Arizona fescue–mountain muhly, Arizona fescue–tufted hairgrass, and Thurber fescue, a very rare type in Arizona.

Grapevine Gulch
La Paz and Mohave Counties
This 960-acre tract near the confluence of the Big Sandy and Santa Maria Rivers encloses a pristine, mature-to-old-growth bosque of dense mesquite forest. The largest mesquite trees are nearly 30 feet tall, with a dense canopy. This is probably the best example of one of the most threatened natural communities in Arizona.

Hart Prairie Preserve
Yavapai County
Located about 100 miles northwest of Phoenix, this high-altitude complex of meadow and forest is an important habitat for large mammals such as elk. Its Bebb's willow riparian community is one of the largest in the world. It is owned and maintained by The Nature Conservancy.

Hartwell Canyon Preserve
Coconino County
Operated by The Nature Conservancy, this property southwest of Flagstaff preserves rocky canyons and tree-lined riparian habitat. Trees include pinyon, juniper, and oak. It is best known as a nesting ground for the peregrine falcon. Mountain lions also roam the property. Ancient Native American pictographs are also protected.

Hassayampa River Preserve
Maricopa and Yavapai Counties
This Nature Conservancy property near Wickenburg protects a 5-mile stretch of the Hassayampa River. It was established to shelter one of the best stands of mature cottonwood-willow riparian habitat remaining in the state. This special woodland is a critical habitat for zone-tailed hawks, Harris hawks, and the only known population of the endemic Arizona skink.

Hot Springs Canyon
Cochise County
Located northeast of Cascabel, this remote, scenic canyon and riparian strip hosts a variety of interesting, special-concern, and rare species. Mature cottonwoods, willows, and alder shade portions of the canyon. It is known for its healthy populations of rare native fish, including the Gila sucker, mountain sucker, and longfin dace. The rare zone-tailed hawk, other raptors, and yellow-billed cuckoos also reside in the canyon.

House Rock Wildlife Area
Coconino County
This 60,000-acre preserve occurs at elevations of roughly 5,000–7,500 feet at the edge of the Kaibab Plateau, some 15 miles west of Marble Canyon. High-quality communities include cliff face, sagebrush shrub, saltbush shrub, and Utah juniper woodland. A herd of buffalo has been reintroduced on the property and is doing well. Other important species include the antelope, mountain lion, raptors that nest in the cliff walls, and several rare songbirds.

Kachina Peaks Wilderness Area
Coconino County
Part of the volcanic San Francisco Peaks, this property 15 miles north of Flagstaff encompasses a range of elevations. From lowest to highest, natural communities include ponderosa pine woodland, spruce-fir forest with grassy meadows, subalpine bristlecone pine, and alpine tundra. Rare and hardy alpine plants include the endemic San Franciso groundsel. This community is very fragile.

Kaibab Plateau
Coconino County
Located north of the Grand Canyon at an elevation of about 9,000 feet, this topographic feature in northern Arizona is some 45 miles southwest of Page. Forests at lower elevations consist of open ponderosa pine; upper elevations have mixed montane spruce–fir forest and meadows. The ponderosa pine forest is the only habitat in the world for the endemic Kaibab squirrel.

Leslie Canyon
Cochise County
Located on 1,240 acres 18 miles north of Douglas, Leslie Spring originates in the Swisshelm Mountains and has a lush riparian corridor of mature cottonwood-willow-ash-walnut forest. Leslie Creek Canyon contains several endemic species, such as the globally endangered Yaqui chumb and topminnow. This area is an excellent example of what these presettlement natural communities resembled in this part of Arizona.

Lower San Pedro River
Pinal County
This 5-mile stretch of the San Pedro River is located near Dudleyville in Pinal County. It is edged by an excellent example of mature Fremont cottonwood and willow riparian habitat; saltcedar, bulrushes, and sedges also grow in marshy areas. The site is a critical nesting habitat for the Mississippi kite, gray hawk, and yellow-billed cuckoo.

Madera Canyon
Pima and Santa Cruz Counties
With its headwaters in the Santa Rita Mountains about 40 miles south of Tucson, Madera Creek tumbles through ponderosa forest, mixed conifer forest, and scrub oak woodland of the Coronado National Forest. Surrounded by flanking desert, the lush canyon is lined with large cottonwood, ash, and willow trees and several grasses and shrubs. Natural communities, from the highest elevation of 9,400 feet to the lowest of 3,000 feet, proceed from Douglas fir–quaking aspen forest, ponderosa pine forest, Mexican blue oak–alligator juniper–pinyon pine woodland, mesquite shrubland, and saguaro-paloverde lowlands. Nearly 200 bird species inhabit the canyon, including the rare flame-backed tanager, trogan, and hummingbird. It also has well-established populations of mule deer, black bears, and mountain lions.

Mittry Lake Wildlife Area
Yuma County
Located in the floodplain of the Colorado River, this 750-acre wildlife management area is about 15 miles northeast of Yuma. Natural communities include the open lake, its riparian rim of cottonwood–willow–saltcedar, and desert shrubland consisting of ironwood, mesquite, and saguaro. The lake attracts a variety of wildlife, especially migrating birds during spring and summer. Endangered birds found here are the Yuma clapper rail and bald eagle.

Mount Baldy Wilderness Area
Apache County
One of the highest peaks in Arizona, Mount Baldy is some 25 miles southwest of Springerville, just within Fort Apache Indian Reservation. It contains deep canyons, meadows, valleys, thick forests of rare Colorado blue spruce, ponderosoa pine, white pine, Engelmann spruce, and aspen, and the headwaters of four rivers. Elk, mule deer, black bears, mountain lions, and various snakes live in this wildland. Several rare plant and animal species have been documented here.

Muleshoe Ranch
Cochise County
The Nature Conservancy manages Muleshoe Ranch, 49,000 acres of private and federal land south of Wilcox in the foothills of the Galiuro Mountains. Seven pristine streams with endemic fish species flow across the property.

Pajarita Wilderness Area
Santa Cruz County
Located in the Coronado National Forest, this 7,400-acre tract is about 17 miles west of Nogales. Lush canyons, such as Sycamore Canyon, are very diverse, with nearly 700 plant species. This is the only place in the world 17 endemic species can be found, including rare ferns, flowers, and animals.

Patagonia-Sonoita Creek Preserve
Santa Cruz County
Located about 60 miles south of Tucson, this 312-acre preserve encloses a stretch of Sonoita Creek and its cottonwood forest riparian habitat. It is well known as a bird habitat that attracts

nearly 300 different species, including the gray hawk and green kingfisher. Other unusual animals include the desert tortoise, Gila topminnow, and beardless flycatcher. The property is operated by The Nature Conservancy.

Pole Bridge Research Natural Area
Cochise County
Located about 25 miles northeast of Douglas in the Coronado National Forest, this 565-acre tract encloses a rare stand of mature to old-growth southern Arizona pine forest. The typical assemblage is Apache pine–southwestern white pine–border pinyon–Arizona pine–Chihuahua pine. It is highly unusual for all these species to occur in one area.

Ramsey Canyon Preserve
Cochise County
Located 6 miles south of Sierra Vista, this unique preserve encloses the area where ecosystems from the Rocky Mountains and the Chihuahua Desert converge. The wooded (sycamore–maple–Douglas fir) gorge in the Huachuca Mountains contains more than 170 bird species, including 14 species of hummingbirds. The desert flanks are covered by yucca, cacti, and sagebrush. Rare species include the Ramsey Canyon leopard frog, blue-throated hummingbird, and lemon lily. It is owned and maintained by The Nature Conservancy.

Saguaro National Monument
Pima County
Located 10 miles west of Tucson, this forest of saguaro cacti covers 21,000 acres. Some of the larger specimens are up to 200 years old. Other cacti, prickly pear, and paloverde also grow here. The site is an important habitat for the Gila woodpecker and gilded flicker, which nest in holes in the cacti. Kestrels, Lucy's warblers, cactus wrens, and purple martins are other important bird species. Saguaro fruit is an important food source for insects, foxes, rodents, and mammals.

San Francisco River
Green Lee County
This 10-mile stretch of the San Francisco River flows southwest from the town of Clifton to the Gila River. The rocky canyons are lined with tall, water-loving trees, such as cottonwood and willow. It is one of the most important raptor sites in the state and supports the threatened Mexican black hawk and the peregrine falcon. Rare Rocky Mountain bighorn sheep can be seen on the rocky heights. Several rare fish, including the threatened loach minnow, inhabit the river.

San Pedro Riparian Conservation Area
Cochise County
Located about 20 miles south of Benson, this 56,431-acre preserve is on the undammed San Pedro River. Its pristine, mature Fremont cottonwood, Gooding willow, Arizona ash, Arizona walnut, netleaf hackberry, and soapberry riparian corridor is one of the best in the state. Upland slopes are desert scrub: tarbush, creosote, acacia, mesquite, and tall grasses. More than 80 mammals, including mountain lions, and more than 350 bird species have been sighted in the area. Rare birds include the gray hawk, crested caracara, rare green kingfisher, and northern beardless tyrannulet.

Virgin River Gorge
Coconino County
Located in the northwestern corner of Arizona along Interstate 15 about 20 miles south of Saint George, Utah, this scenic, high-quality river flows through desert gorges. The riparian cottonwoods and willows shade the water and canyon. This 20-mile stretch of river is critical habitat for three globally rare fish: the Virgin River roundtail chub, Virgin River spinedace, and woundfin minnow. Golden eagles, Mexican black hawks, and peregrine falcons, all rare species, nest in the rock cliffs of the gorge.

West Clear Creek Research Natural Area
Coconino County
Located near Beaver Creek in Coconino National Forest, this 700-acre site protects a stretch of pristine West Clear Creek. Cut through basalt and limestone, the streambed is bordered by a lush riparian zone of Arizona sycamore, alder thickets, velvet ash, Fremont cottonwood, and netleaf hackberry. Utah juniper woodland can be found on the rocky sideslopes. Biodiversity is high and includes the rare roundtail chub, river otter, and razor-

back sucker. Several types of raptors nest on the rock walls and hunt in the canyon.

Wildcat Canyon Research Natural Area
Navaho County
This 513-acre parcel of relatively undisturbed, older-growth pinyon-juniper woodland borders part of Wildcat Canyon, a few miles from Heber. It is flanked by ponderosa pine forest. It is regarded as one of the best examples of pristine, presettlement-type pinyon-juniper woodland in Arizona. It is an important habitat to a variety of mammals, birds, and reptiles, including mountain lions and raptors such as the golden eagle.

Willcox Playa
Cochise County
Managed by the Arizona Fish and Game Department, this 555-acre wildlife management area is about 7 miles south of Willcox. Its natural communities include a closed-basin playa lake, which is rimmed by semidesert grasslands of saltgrass. Trees that shade the shoreline include cottonwoods, saltcedars, and willows. As an oasis in a desert climate, this body of water attracts a variety of birds and animals, including raptors, several rare amphibians, and sandhill cranes.

ARKANSAS

Arkansas contains six different natural divisions: Ozark, Arkansas Valley, Ouachita Mountains, West Gulf Coastal Plain, Mississippi Alluvial Plain, and Crowley's Ridge.

The Ozark natural division comprises the rugged terrain of the Ozark Mountains in the northern quarter of the state. These hills and ravines are dominated by various types of dry oak–hickory forest. Occasional stands of shortleaf pine, sugar maple, black walnut, flowering dogwood, and pawpaw are also found. Steep, rocky slopes are often covered by sugar maples. The rare and beautiful magnolia forest can be found along cooler, damper, north-facing slopes and deep ravines. Lowland wet forests of pin oak, overcup oak, and sweet gum grow in wetland areas. Floodplain forests along the major river systems consist of elm, maple, cottonwood, and sycamore. Barrens and glades with prairie grasses and forbs, such as big bluestem and Indian grass, can be found in the forest communities. Only tiny pockets of undisturbed forest remain.

Rolling lowlands about 500 feet above sea level make up the Arkansas Valley natural division, which also contains the Arkansas River. It forms an east-west band across the central portion of the state. Occasional flat-topped landforms rise up to about 1,500 feet above the surrounding lowland. The countryside is covered by a mix of wetlands, bottomland forest, upland dry forest, and a few cedar glades and oak glades and savannas. Upland forests are typically oak-hickory or shortleaf pine. Bottomlands support wetter species, such as cherrybark oak, shagbark hickory, swamp chestnut oak, overcup oak, and willow oak. The big wetlands are dominantly cypress-tupelo or cypress swamp. A few remaining tracts contain remnants of the tallgrass prairie that used to be much more extensive in the Arkansas Valley.

A series of parallel ridges and valleys form the heavily forested Ouachita Mountain natural division in west-central Arkansas. The higher rocky slopes are vegetated with scrubby oak and hickory, lichens, and mosses. Forests of shortleaf pine–oak–hickory can be found on the lower, wetter slopes. Rarer beech-magnolia-basswood stands grow on north-facing slopes and deep, narrow ravines, where the microclimate is cooler and more damp. Virtually all of these forest communities are second-growth or younger because of heavy logging in the past.

The West Gulf Coastal Plain natural division occupies the southwestern quarter of the state. It is underlain by flat-lying marine sedimentary rocks and deposits of gravel, sand, and coal. Three big rivers—the Red, Saline, and Ouachita—wind across the Coastal Plain. Dense, wet forests contain overcup hickory–water hickory–cypress–sweet gum–willow oak. Drier oak species and shortleaf and loblolly pine appear on higher slopes. The rare scrubby sandhill oak communities—Arkansas oak, bluejack oak, shortleaf pine–black hickory—can be found here, along with some rare herbs. Savannalike places with tall prairie grasses and forbs and scattered oak and loblolly pine trees are another rare community. Remnants of blacksand prairie are dominated by tallgrass species such as little bluestem.

The Mississippi Alluvial Plains natural division, or the Delta, runs north-south and occupies roughly the eastern third of Arkansas. Very low-lying, it consists of forested bottomlands and loess plains. Various communities include oak-hickory-pawpaw-spicebush forest, swamp, tallgrass prairie, slough, bayou, and impressive bald cypress–water tupelo swamp with buttonbush–southern wild rice–swamp loosestrife in the shrub zone. Backwater basins are forested with overcup oak, water hickory, water locust, and swamp privet. The more rare Nuttall oak can be found on low ridges and levees of sand. Riverbanks are lined with cottonwood, black willow, and silver maple. Remnants of tallgrass prairie are covered with big and little bluestem, Indian grass, switchgrass, and various orchids. Canebreaks can still be found in the Delta near the big rivers. There are virtually no virgin remnants of these communities in the heavily used Delta.

Crowley's Ridge natural division is a narrow, north-south trending topographic feature in the northeastern corner of the state. It is formed by beds of clay and gravel that stand above the surrounding Mississippi Alluvial Plain. The ridge has been cleared and heavily logged. Second-growth or younger woodlands of mixed oak–hickory–shortleaf pine now dominate. Mesic forest types in ravines and along north-facing slopes are forested with beech, tulip, and magnolia trees. Wetlands consist of small scattered sphagnum bogs and shrubby swamps.

Special Habitat Sites

Alcoa Bauxite Natural Area
Saline County
This 1,250-acre Nature Conservancy property a few miles southeast of Benton protects specialized communities that have developed on a syenite bedrock-substrate. Natural communities include rock face, xeric upland oak–ash–hickory–shortleaf pine forest, and sunny glades. A range of mosses, algal mats, and lichens can be found anchored to the rock surfaces in the syenite glade, a globally rare community. Several rare or endangered species grow here, including the small-headed pipewort. Ferns thrive in seepage areas.

Arkansas Oak Natural Area
Nevada County
Located about 5 miles northeast of Rosston, this 40-acre tract protects mixed upland forest and wet seeps with well-developed plant communities. Rare plants here include the Arkansas oak tree, various sedges, scarlet beardtongue, large clammy weed, and smooth twistflower.

Baker Prairie Natural Area
Boone County
Located about a mile west of Harrison in the Ozark natural division, this 71-acre Nature Conservancy preserve protects the largest remnant tract of Ozark tallgrass bluestem prairie in the state. Specialized plants include the Indian paintbrush and compass plant. The prairie grasses are also home to the endangered grasshopper sparrow. Other rare or endangered species include the prairie mole cricket, ornate box turtle, royal catchfly, and the endemic Ozark least trillium. The wildflowers are colorful during spring and summer.

Bear Hollow Natural Area
Madison County
This 394-acre nature preserve in the Ozark natural division is located about 4 miles northeast of Clifty. It encloses most of thickly wooded Bear Hollow and a stretch of pristine Rockhouse Creek. The steep slopes and ravines are covered by mature mixed oak-hickory-beech upland hardwood forest. It is also the home of the Ozark Natural Science Center. The hollow is an important nesting site for a variety of birds.

Benson Creek
Monroe County
Located a few miles northwest of Brinkley, this 302-acre cooperative project between the U.S. Army Corp of Engineers and The Nature Conservancy protects a large wetland complex. The dominant feature is a cypress-tupelo swamp, with tall bald cypress trees that are hundreds of years old. The swamp is surrounded by bottomland forests of cottonwood, willow oak, Nuttall oak, and overcup oak.

Big Creek Natural Area
Cleburne County
Located southeast of Heber Springs, this 1,108-acre area is centered on a 3-mile stretch of Big Creek. Its clean waters support a diverse population of fish and other aquatic species. Part of the Ozark natural division, the creek is bordered by steep limestone bluffs that are forested with mature oak-hickory-beech trees; a rare willow oak forest can be found along the creek. Other rare species are varieties of mock

orange. Interesting ferns, lichens, and mosses can be found on the rock faces.

Big Fork Creek Natural Area
Polk County
Located about 10 miles east of Ink, this 14-acre property includes part of Big Fork Creek, springs, lowland forest, and rocky bluffs. A very rare and fragile plant community has developed around the spring, including species such as the southern lady's slipper, log fern, waterleaf, and leafy bulrush. It is also an important habitat for the threatened Caddo Mountain salamander and southern red-backed salamander.

Big Lake National Wildlife Refuge
Mississippi County
Located near Blytheville, this 11,000-acre tract is a remnant of the Great Swamp of Arkansas. The terrain is mostly dense wetland and bottomland forest. It is one of the most diverse bird habitats in the state. Millions of ducks and geese visit during migrations. Several tracts of bald cypress, some of them hundreds of years old, still grow here. Access is by boat only.

Blue Heaven Cave
Marion County
Located on 5 acres a few miles northwest of Yellville, this small cave system is one of the few in the United States in which the federally endangered Ozark big-eared bat reproduces. At least 300 individual bats live in the cave. Because this such a specialized habitat, access is closed to the public except by special permission from The Nature Conservancy.

Brushy Creek Natural Area
Montgomery County
This 80-acre Nature Conservancy property protects part of the riparian corridor along Brushy Creek, about 30 miles northeast of Pine Ridge. High-quality natural communities include wetland, marsh, sedge meadow, and a small upland hardwood forest. Unique species found here include cardinal flower, several rare sedges, prickly pear cactus, and several interesting wildflowers, such as rattlesnake master and blazing star. Contact The Nature Conservancy regarding access permission.

Byrd Lake Natural Area
Jefferson County
Located a few miles north of Pine Bluff, this 144-acre property shelters a rare oxbow lake community along Bayou Bartholomew in the Mississippi Alluvial Plain. The lake is rimmed by a bald cypress wetland and mixed pine-hardwood forest. Some of the cypress trees are quite large. The lake supports two rare fish: the bluehead shiner and taillight shiner. The wetlands are an important nesting habitat and migratory stopover for a number of different bird species, including herons and other waterbirds.

Cache River Natural Area
Woodruff County
This 937-acre nature preserve is located southwest of Patterson. Part of the Hancock Wildlife Management Area, it includes a stretch of the Cache River as well as pristine wetlands, cypress-tupelo swamp, and bottomland willow oak forest along the banks. In addition to the large trees in these rare wetland communities, the preserve is an important habitat for a variety of wetland birds and amphibians.

Cave Springs Cave Natural Area
Benton County
Located southwest of Bentonville in the Ozark, this 15-acre parcel protects a cave system and surrounding oak-hickory upland forest. The cave is a special habitat for the Ozark cavefish, an endemic species that is found only in the Ozark Mountains. The rare Ozark cave amphipod also lives in the cave system. A population of gray bats roosts and breeds in the cave.

Cherokee Prairie Natural Area
Franklin County
Part of the Arkansas Valley natural division, this 565-acre property located southeast of Vesta protects the largest remnant piece of undisturbed tallgrass prairie left in the state. As a result, its biodiversity is high, with more than 100 native plant species. It is also an important habitat to grassland birds, mammals, and insects. Rare or endangered species found here include the prairie mole cricket, ornate box turtle, tuberous grass-pink, and narrow-leaved puccoon. The wildflowers are abundant in the tall grasses during spring and summer.

Coffee Prairie Natural Area
Ashley County
This 56-acre property is located east of Huttig. It is one of the few remaining pristine remnants of lowland sand prairie in the state. Its typical grasses and forbs, such as bluestem, hide rarer plants, including stylisma. It is also an impor-

tant habitat for grassland birds, insects, mammals, and reptiles.

Cossatot River State Park/Natural Area
Polk County
Located about 5 miles north of Wickes, this 4,447-acre preserve is part of the Ouachita Mountains natural division. Its communities include clear, clean, fast-flowing streams, waterfalls, rocky outcrops and cliff faces, grassland openings, glades, and mature mixed hardwood upland forest. Some tree specimens are quite large. Biodiversity is high, with numerous rare or endangered plant and animal species, including the leopard darter, Ouachita leadplant, milk vetch, several sedges, Moore's larkspur, Ouachita mountain shiner, Georgia holly, grass-of-Parnassus, stonecrop, twistflower, and Virginia saxifrage. The property is an important habitat for a number of woodland birds, mammals, amphibians, and reptiles.

Cove Creek Natural Area
Faulkner County
This 228-acre natural area west of Greenbrier is in the Arkansas Valley natural division. Communities include a mile-long stretch of Cove Creek, rocky gorge, bottomland forest, rare red cedar glade, rare juniper-hardwood forest, and dry upland oak-hickory-beech forest. Various mosses, lichens, and ferns can be found in the cool, damp microclimate of the gorge. Rare or unusual species are the Carolina spring-beauty, alumroot, threadfoot, tall cinquefoil, and juniper tree. It is also an important habitat for a number of birds, including songbirds and woodpeckers.

Cow Shoals Riverfront Forest Natural Area
Cleburne County
This 63-acre natural area is located northeast of Heber Springs in the Ozark natural division. The main natural community is the uncommon river birch—sycamore floodplain forest along Little Red River. The shoals in the river are an important spawning ground for brown trout. The forest is an important habitat for a variety of mammals, birds, fish, and amphibians.

Cut-Off Creek Ravines Natural Area
Drew County
Located west of Jerome in the West Gulf Coastal Plain, this 377-acre tract encloses steep forested ravines. Forest communities are upland pine-oak and oak-beech and bottomland willow oak–water oak–overcup oak–cypress associations. The property is part of the Cut-Off Creek Wildlife Management Area, and its high biodiversity is a reflection of the overlap between the Coastal Plain and Mississippi Delta natural divisions.

Devil's Backbone Natural Area
Izard County
Located southwest of Melbourne, this 499-acre tract is part of the Ozark natural division. Communities include rocky slope, sandstone glade, limestone glade, rare Ashe's juniper woodland, and mixed oak-hickory upland forest. Several rare plants occur here, including the Ashe's juniper, purple beardtongue, Ozark chinquapin oak, silky aster, and lobed spleenwort. Specialized plants grow in cracks in the exposed rocks in cliff faces and glades.

Easley Creek Heron Rookery
Hot Spring County
Located northeast of Arkadelphia on 71 acres owned by International Paper Company, this property is managed by The Nature Conservancy. Mature tall pine trees are nesting sites for about 200 great blue herons and great egrets, which return every year. The nesting birds inhabit the site from roughly February to July.

Falcon Bottoms Natural Area
Columbia and Lafayette Counties
Part of the West Gulf Coastal Plain, this 741-acre preserve is located a few miles west of Waldo. It contains mixed hardwood-pine upland forest, pristine bottomland forest along Dorcheat Bayou, cypress-tupelo swamp, and willow oak forest. Some of the tree specimens are quite tall and probably hundreds of years old. The willow oak forest is a rare community in itself. The wetlands are an important habitat for nesting and migratory birds and include a number of great blue heron rookeries.

Flanagan Prairie Natural Area
Franklin County
Located near Charleston, this 258-acre natural area is an unplowed remnant of what was once the vast presettlement Cherokee Prairie. It is an important habitat for a number of grassland birds, insects, and mammals. Dominant plants are big and little bluestem and Indian grass. Rare or endangered species here are the prairie

mole cricket, ornate box turtle, and tuberous grass-pink. The wildflowers add huge splashes of color during spring and summer months.

Gap Creek Natural Area
Montgomery County
Located near Caddo Gap, this property is part of the Arkansas Valley natural division. Gap Creek, located high in the Ouachita Mountains, is one of the cleanest streams in the state. The tributary is in rugged, isolated, forested terrain that has been relatively undisturbed. The forest is dominated by mixed oak-hickory; a few of the specimens are large, having escaped logging in the past 200 years. The wet edges of the creek support the rare southern lady's slipper orchid and several unusual ferns.

Goose Pond Natural Area
Pope and Conway Counties
Located west of Morrilton, this 392-acre property is part of the Arkansas Valley natural division. Natural communities include open water, shrub swamp, water tupelo swamp, American lotus/water lily marsh, and lowland forest. The rare swamp rabbit and mink can be found along the wetland edges. It is a critical nesting site or migratory stopover for numerous bird species, including herons, interior forest birds, warblers, and raptors. The rare bird-voiced tree frog can also be found here.

Hell Creek Natural Area
Stone County
Located near Mountain View, this 214-acre tract protects a very rare cave habitat in the White River watershed. Hell Creek cave is the only place in the world where an endemic white cave crayfish is known to occur. Rare isopods and a population of gray bats also live in the cave.

Holland Bottoms Willow Oak Forest Preserve
Lonoke County
Located northeast of Jacksonville, this 632-acre parcel protects a rare stand of high-quality lowland willow oak forest. This wetland-forest complex is an important habitat for the rare red-shouldered hawk, Swainson's warbler, yellow-crowned night heron, and certain amphibians.

Holloway Memorial Natural Area
Prairie County
This 75-acre natural area is located north of Des Arc. Part of the Mississippi River Alluvial Plain, it contains a rare, high-quality remnant of mixed oak flatwoods. The understory is composed of muscadine and blueberry. Several rare or unusual ground plants, including ferns, grow here.

Idlewild Seep Natural Area
Pulaski County
Located on 12 acres about a mile southwest of Ferndale, this small, wooded, acidic seep harbors a variety of rare species that are specialized to the groundwater chemistry. Among the dense ferns can be found the rare southern lady's slipper orchid, featherbells, New York fern, and devil's bit. The property is managed by The Nature Conservancy; permission is required.

Iron Mountain Natural Area
Polk County
This 161-acre natural area is located about 8 miles west of Mena. Part of the Ouachita Mountains natural division, its communities include a small stream and mature upland pine-oak-hickory forest. The clean waters of the stream support the rare endemic southern red-backed salamander. Some rare or unusual plants also occur here.

Kings River Falls Natural Area
Madison County
This 596-acre tract is located northeast of Dutton in the Ozark. Natural communities include Kings River and waterfalls, riparian shrub, rock face, mixed pine-hardwood forest, and hardwood forest. An abundance of ferns, mosses, and lichens (some of them rare) can be found on the rock outcrops. Rare or unusual species include the Ozark chinquapin oak. The site is also an important nesting site for a number of birds, including forest interior species such as the pileated woodpecker.

Konecy Prairie Natural Area
Prairie County
Located a few miles west of Ulm, this 29-acre tract protects one of the few undisturbed remnants of what was once the expansive Grand Prairie. Dominant plants are big and little bluestem and Indian grass. Plant diversity is high, including the rare field pussytoes, grass-pink, prairie evening primrose, and sand cherry.

The willow flycatcher and ornate box turtle are among the rare bird and reptile species that live here.

Lake Chicot State Park
Chicot County
Located east of Lake Village and beside the state's largest natural lake, the major natural communities are bald cypress swamp and slough. Some of the larger specimens are hundreds of years old. Bird life is diverse, including egrets, herons, night herons, cormorants, rarer ibis, pied-billed grebes, spoonbills, and wood storks, many of which nest in the swamp.

Logly Natural Area
Columbia County
Part of the West Gulf Coastal Plain, this 200-acre property is located about 15 miles north of Magnolia. It protects excellent examples of loblolly pine forest, oak-hickory forest, beech forest in the moist ravines, springs, and streams. Some of the loblolly pine specimens are very tall and probably exceed 200 years old. It is an important habitat for a variety of mammals, birds, plants, and reptiles.

Lorance Creek Natural Area
Pulaski and Saline Counties
Located about 5 miles southwest of Wrightsville, this 289-acre property encloses a wetland complex. Part of the West Gulf Coastal Plain, it includes a stretch of Lorance Creek and its flanking wetlands, cypress–tupelo gum swamp, and blackgum swamp. Some of the trees are quite large and old. The swamp blackgum is a tree of special concern in Arkansas. The wetland is an important habitat for numerous birds, amphibians, and reptiles. Rare or unusual species found here are the Carolina ash, red-shouldered hawk, and bird-voiced tree frog.

Louisiana Purchase Natural Area
Monroe, Phillips, and Lee Counties
This 35-acre tri-county tract is located 25 miles southeast of Brinkley. A headwaters swamp for Little Cypress Creek, this relatively rare community helps limit flooding. The dominant vegetation is mature cypress–water tupelo bottomland forest. Special species that occur here include tree frogs, various warblers, rare wood ducks, pileated woodpeckers, and the rare swamp cottonwood and bird-voiced tree frog.

Magazine Mountain
Yell County
About 20 miles southeast of Paris, this dominant topographic feature protects a number of rare or unusual plant and animal species. Natural communities include mixed hardwood upland forest. Rare species include a tiny amphipod that is found nowhere else in the world, the maple-leaved oak, and the relict hay-scented fern and Appalachian woodsia on the shaded, rocky north face. The rufous-crowned sparrow nests in the forest.

Miller County Sandhills Natural Area
Miller County
Located about 6 miles north of Brightstar, this 36-acre site shelters remnants of rare, relatively undisturbed sandhill natural communities. These include sandhill prairie and scrubby mixed hardwood forest. Rare or endangered plants include milk vetch, various sedges, various prairie clovers, evening primrose, Riddell's spike moss, climbing milkweed, little-leaved prairie clover, tragia, hairy grama, and spiderwort. It is an important habitat for grassland birds, mammals, and insects, some of them rare.

Mills Park Natural Area
Saline County
Located in the town of Bryant southwest of Little Rock, this 10-acre property protects a moist mixed woodland along a series of seeps and springs that flow into Crooked Creek. The lush ground vegetation of ferns, shrubs, and sedges contain several rare plants, including black chokeberry and devil's bit. The property is owned and managed by the town of Bryant.

Moro Creek Bottoms Natural Area
Cleveland County
Part of the West Gulf Coastal Plain, this 78-acre tract is located a few miles southeast of Fordyce. It contains excellent undisturbed remnants of lowland oak–sweet gum forest and bald cypress swamp communities. Many of the swamp trees, including cherrybark oak, sweet gum, and water oak, are nearly 100 feet tall, three feet in diameter, and hundreds of years old. The wetlands are an important habitat for many plant, bird, amphibian, and reptile species, such as the yellow-crowned night heron and certain sedges.

Nacatoch Ravines Natural Area
Hempstead County
Located about a mile north of McNab, this 206-acre preserve is part of the West Gulf Coastal Plain. It contains excellent, relatively undisturbed remnants of the mixed hardwood upland forests that once covered much of Arkansas. Many of the trees are more than 100 feet tall, more than three feet at breast height, and at least 200 years old. Trees include white oak, red oak, sycamore, and basswood. Rare ground plants that grow here are certain sedges, southern lady's slipper, and eared goldenrod. It is an important habitat for a variety of birds, such as pileated woodpeckers.

Pine City Pine/Forest Natural Area
Monroe County
Located just outside Holly Grove, this 160-acre tract of mature pine-hardwood forest is a critical habitat for an isolated population of the federally endangered red-cockaded woodpecker. This population is even more unique in that it has chosen to live in a relatively small tract of woods that is surrounded by farms and open space.

Pipewort Glade
Saline County
Located on private property but managed by The Nature Conservancy, this 100-acre tract is a few miles southeast of Benton. Within the upland forest of shortleaf pine, white oak, and hickory is a series of glades that developed on outcrops of syenite, a relatively rare rock type. An assemblage of rare plants grows in sunny, rocky patches; they are especially suited to the chemistry of the bedrock and include the small-headed pipwort. Prairie grasses dominate the glades and offer attractive wildflowers, such as fame flower and little bluestem, during spring.

Poison Springs Forest Preserve
Ouachita County
Part of the West Gulf Coastal Plain, this 52-acre property is about 10 miles northwest of Camden. Located in Poison Springs State Forest, this tract shelters rare sandhill woodlands and a mature pine-hardwood forest that has been set aside as a future old-growth forest. The sandhills community contains a number of rare plants (and some insects) that thrive in the dry, sandy soil.

Railroad Prairie State Natural Area
Prairie County
Located a few miles east of Carlisle, this 242-acre property encloses a very rare, complete transect across what was once the Grand Prairie. This strip of virgin prairie, 14 miles long and 150 feet wide, is a tremendously important tract for understanding the makeup of the presettlement prairie. Its plant diversity is quite high, including the rare grass-pink, various sedges, down gentian, prairie evening primrose, sand cherry, and slender marsh-pink. Other rare species found here are the western chicken turtle and prairie mole cricket. Wildflowers are abundant during spring and summer.

Rich Mountain
Polk County
Located near Mena, this property encloses high-quality upland forest of white oak, black walnut, beech, maple, magnolia, and Ohio buckeye. Endemic to the area is Rich Mountain salamander, an extremely rare animal. The forest is also an important bird habitat for various warblers and tanagers.

Rock Creek Natural Area
Sharp County
Located southwest of Williford, this 415-acre property is part of the Harold Alexander Wildlife Management Area in the Ozark natural division. Natural communities include the high-quality Rock Creek and its tributaries, seep, spring, fen, calcareous glade, and mixed hardwood lowland and upland forest. This property has one of the highest concentrations of rare or endangered plants in the state, including milk vetch, porcupine sedge, lax hornpod, western wallflower, false loosestrife, square-stem monkeyflower, purple beardtongue, heart-leaved plantain, Virginia Mountain-mint, low nut-rush, and Riddell's goldenrod. It is also an important habitat for a variety of birds, mammals, amphibians, and reptiles.

Roth Prairie Natural Area
Arkansas County
Located on 41 acres just south of Stuttgart, this remnant of tallgrass prairie is in the Mississippi Alluvial Plain. Its diverse plant population is supported by very acidic underlying soils. Pimple mounds dot the prairie. Rare plants that

grow here are tuberous grass-pink, prairie evening primrose, and pineywoods dropseed. The prairie is an important habitat for certain grassland birds, mammals, and insects.

Saratoga Landing Prairie Natural Area
Hempstead and Howard Counties
Located about 1 mile southwest of Saratoga near Millwood Lake, this 208-acre property contains high-quality remnants of rare dry prairie, dry-mesic calcareous mixed hardwood forest, and dry-calcareous mixed hardwood forest. The diverse prairie grasses are dominated by purple prairie clover, blue salvia, pale purple coneflower, and little bluestem. Rare plant species found here are stenosiphon, Carolina larkspur, and beardtongue. The prairie openings are an important habitat for small grassland birds, mammals, insects, and reptiles.

Searles Prairie Natural Area
Benton County
Located northwest of Bentonville in the Ozark, this 10-acre property protects an undeveloped remnant of what was once the expansive Osage Prairie. More than 100 tallgrass prairie plant species grow here, including big and little bluestem and also some rare or unusual species. Specialized grasses and forbs grow in seepage areas. The abundant wildflowers are colorful during spring. The prairie is an important habitat to grassland birds, insects, and reptiles.

Singer Forest Natural Area
Poinsett County
Located southeast of Payneway, this 519-acre natural area protects a rare high-quality remnant of willow oak bottomland forest and surrounding wetlands. Several rare or unusual plant and animal species can be found here. The property is also an important habitat for both nesting and migrating birds, including herons and other waterbirds.

Slippery Hollow Natural Area
Marion County
This 203-acre site about 10 miles northwest of Summit, in the Ozark natural division, protects the entrance to a fragile cave ecosystem. The cave is unique in that it is the home of a colony of the endemic Ozark big-eared bat. The cave is surrounded by mixed oak-hickory hardwood forest.

Smoke Hole Natural Area
Lonoke and Prairie Counties
Located a few miles northwest of Slovak, this 455-acre property is part of the Mississippi River Alluvial Plain. It encloses a wetland-forest complex containing the rare tupelo-blackgum swamp, mixed bottomland hardwood forest, and upland hardwood forest along Two Prairie Bayou. Several rare or unusual plant and animal species can be found here. It is also an important bird and amphibian habitat.

Stateline Sand Ponds Natural Area
Clay County
This 140-acre property is located about 6 miles northwest of Corning. Part of the Mississippi River Alluvial Plain, this woodland-wetland complex protects a rare and fragile community of plants. The major natural communities are sand pond wetland and forest that has developed on ancient sand dune-swale topography. Rare plants include the federally endangered pondberry, corkwood, leatherwood, and several rare orchids. The threatened mole salamander can also be found here.

Stone Road Glade Natural Area
Howard County
Located about 20 miles northwest of Nashville, this property protects the largest and most pristine limestone glade community in the West Gulf Coastal Plain natural division. The grassy openings support a number of prairie plants, some of them rare, that thrive on the calcium-rich substrate. Dominant plants are purple prairie clover, yellow-puff, little bluestem, and black-eyed Susan. The glade is surrounded by mixed hardwood forest.

Striplin Woods Natural Area
Arkansas County
Located just north of Ethel, this 30-acre parcel encloses a rare, relatively undisturbed hardwood forest that goes from rare bottomland overcup oak–water hickory–willow oak–honey locust forest to upland oak-hardwood-beech forest. Many of the bottomland trees are nearly 100 feet tall, three feet at breast height, and hundreds of years old. The property borders the White River. The rare Swainson's warbler has been sighted here.

Sweden Creek Falls Natural Area
Madison County
Located just outside of Kingston in the Ozark natural division, this 40-acre tract includes an 80-foot waterfall. Other natural communities that are well-represented are flowing stream, cliff face, and mature mixed oak–hickory–beech–cucumber magnolia forest. Several rare or endangered plant and animal species occur here, including certain ferns and lichens, lobed spleenwort, Ozark chinquapin oak, small-headed pipewort, and wood frog. The tract is also an important bird-nesting habitat.

Taylor Woodlands Natural Area
Jefferson County
Located about 10 miles northeast of Pine Bluff, this 137-acre property contains an excellent example of the rarer, closed-canopy, mature to old-growth bald cypress–black willow–ash–red maple–water locust bottomland forest. The more upland portions contain pecan, cedar elm, and overcup oak trees. Several rare plant and animal species can be found on the site.

Terre Noire Natural Area
Clark County
Located southwest of Arkadelphia, this 77-acre property contains one of the best remnants of blackland prairie in the state. The rare community includes patches of high-quality prairie mixed with patches of young oak-hickory trees. Its highly diverse plant species include beaked panicum, Indian grass, rough blazing star, and compassplant. The rare Bachman's sparrow has been sighted here. The property, owned and managed by The Nature Conservancy, is part of the West Gulf Coastal Plain.

Warren Prairie Natural Area
Bradley County
Part of the West Gulf Coastal Plain, this 887-acre tract is located a few miles east of Warren. It encloses a rare prairie-type community that has developed on high-sodium soils. The prairie is dominated by grasses and herbs and supports clumps of dwarf palmetto. Several varieties of rare plant and animal species can be found here, including the northern crawfish frog, gulf crayfish snake, southern shield wood-fern, spike-rush, geocarpon, falling beak rush, Texas sunnybell, and pineywoods dropseed. The prairie is an important habitat for certain grassland birds, reptiles, and insects.

White Cliffs Natural Area
Little River County
This 573-acre tract northeast of Wilton protects a unique plant community that has developed atop and surrounding a steep chalk bluff overlooking Little River. The mixed hardwood forest here contains rare laurel oak and Durand's white oak. Other rare or endangered plant species include fringed-leaved petunia, evening primrose, celestial lily, Ashe's juniper, bladder-pod, cressed coral-root, rabbit-tobacco, and milk vetch. Several varieties of mosses and lichens can be found on the rock faces. The property is also an important habitat for interior forest birds.

Wittsburg Natural Area
Cross County
Part of Crowley's Ridge, this 168-acre tract is located near Parkin. It protects a well-developed mixed beech–sugar maple mesophytic forest that supports a rare group of plant species, including the tulip tree and pennywort. Several of the beech and maple trees are quite old. The natural area is also important in that it encloses the entire watershed for a number of small streams that dissect the rugged slopes. The wildflowers are quite colorful during the spring months.

CALIFORNIA

California is a state of mountains, foothills, valleys, wetlands, deserts, and coastline. The locations of larger natural communities are dependent mostly on elevation, temperature, soil chemistry, and precipitation conditions. In general, from lowest to highest elevations, the major community types are beach-dune complex, coastal prairie and scrub, coastal forest, interior (foothill) grassland-chaparral-woodland, montane forest, mixed subalpine forest, and alpine tundra. Other communities that cross a range of elevations are deserts and wetlands.

Beach-dune complexes line the Pacific coastline along the length of the state. Because of the high-energy conditions of surf and wind, this strip tends to be open and sparsely vegetated. The actual coastline ranges from sand beach to rocky cliffs and headlands. The shifting dunes are anchored by a few hardy plants, such as European beachgrass, which was introduced in 1869.

Inward from the beach communities are the coastal prairie and scrub communities. These are more sheltered from the harsh conditions and salt spray of the beach environment and have better soil development. Major plant species include coyote brush, bush lupine, and California coffeeberry. Most of the native prairie has been converted to fields of milk thistle and Klamath weed because of agricultural and grazing practices. Other exotics here are pampas grass and Scotch broom. The coastal scrub communities are dominated by sagebrush, cacti, and yucca.

California is perhaps most famous for its lush, gigantic coastal forests of redwood, Douglas fir, and other conifers. Although heavily logged in the past, stands of old-growth forest remain that are nearly 1,000 years old and taller than 200 feet. Because of logging, other forest types have intruded, such as tan oak and madrone hardwood forests.

Moving farther inward toward the foothills, moisture levels drop and temperatures become hotter and more extreme. These changing conditions result in a belt of interior grasslands, woodlands, and chaparral. Bunchgrasses dominate the sweeping grasslands. The woodlands that cover many of the hillsides contain various oak trees, such as the coast live oak, blue oak, and California walnut. Chaparral species, such as chamise, Christmas berry, yucca, and manzanita, cover extensive areas in a dense, almost impenetrable blanket.

Between the foothills and the highest mountaintops is the montane zone, which ranges in elevation from about 2,000 feet to roughly 8,000 feet. Much of this area is covered by extensive Douglas fir, ponderosa pine, incense cedar, and sequoia trees at lower elevations. The upper half of the montane zone is dominated by lodgepole pine, Jeffrey pine, red fir, Sierra juniper, and aspen forests. Other communities include lakes and meadows swept with wildflowers.

The subalpine zone extends from roughly 8,000 feet to 10,000 feet and is dominated by western white pine, mountain hemlock, limber pine, bristlecone pine, and lodgepole pine. The bristlecone pine trees are some of the oldest living organisms on the planet, often exceeding 1,500 years.

The alpine tundra is located on the highest mountain peaks; it is rocky, windswept, and experiences extreme temperature changes. With elevations above 10,000 feet, vegetation is limited to patches of hardy plants, such as sedges and grasses, buckwheat, cushion cress, and pussypaws. Several of these species are endemic to California.

Much of southeastern California is covered by desert. Although this terrain is arid, rocky, windy, and dry, it supports a number of thriving communities, such as sagebrush and bunchgrass flats, Joshua tree woodland, blackbrush scrub, creosotebush, burrobrush, cacti, and a variety of wildflowers that bloom during spring, such as the brittlebush shrub.

Special Habitat Sites

Adorni Research Natural Area
Humboldt County
Located in the Six Rivers National Forest, this 600-acre preserve about 3 miles north of Weitchpec encloses low-elevation mesic forest and higher-elevation montane forest. The old-growth forests contain majestic specimens of Port-Orford-cedar, chinquapin oak, tan oak, and Douglas fir that are hundreds of years old. Several rare groundcover plants are also found here. Plant diversity is high, and the acreage is an important habitat for a variety of bird and animal species. The area is threatened by clear-cutting.

Amargosa Canyon Natural Area
San Bernardino County
Located south of Tecopa near Death Valley, this 9,500-acre sanctuary protects a section of the riparian corridor along the Amargosa River. The lush green strip between the steep canyon walls consists of tall cottonwood and black willow trees and tall grasses. The corridor is a critical habitat for the rare endemic Amargosa vole as well as a variety of bird life, including the uncommon or endangered northern pygmy owl, Bell's vireo, and western yellow-billed cuckoo. Herons, sandpipers, ducks, and geese can frequently be seen. The clean river water supports the rare Nevada speckled dace and Amargosa pupfish.

Ancient Bristlecone Pine Forest
Inyo County
Located in the steep, forbidding White Mountains, this 28,000-acre preserve encloses one of the best examples of bristlecone pine forest in the world. Clinging to the upper slopes and flatlands of the mountains, some of these gnarled trees, although unimpressive in size, are more than 5,000 years old. This windswept environment is also home to great horned owls, burrowing owls, long-eared owls, and a variety of raptors, reptiles, and small mammals. Songbirds can be seen during summer.

Angelo Coast Range Preserve
Mendocino County
Located about 4 miles northeast of Branscomb, this 7,520-acre property shelters an impressive stand of old-growth forest. Dominated by tall Douglas fir and redwood trees that are centuries old, the closed canopy creates a shady, cool, and moist microclimate on the forest floor that supports a variety of rare or unusual species. The rare spotted owl, Olympic salamander, and red-backed vole can be found in the forest.

Año Nuevo State Reserve
San Mateo County
Located about 9 miles south of Pescadero, this 4,000-acre preserve encloses a stretch of Pacific beach that is recognized as a critical habitat for the northern elephant seal. Natural communities include beach, dune, tidal pool, and coastal bluff. In addition to being one of the most important breeding areas of the northern elephant seal, other animals that live and breed here are Stellar sea lions and harbor seals. Several varieties of birds, some of them rare, can be seen along the bluffs and grasslands. The property is part of the University of California's Natural Reserve System.

Antelope Creek Lakes
Siskiyou County
Located on 544 acres about 10 miles northeast of Weed, this proposed RNA protects small subalpine meadows along riparian corridors and lake margins. Other natural communities include mountain hemlock–Shasta red fir forest, subalpine whitebark pine forest, high-montane scree vegetation, and montane chaparral. In areas sheltered from avalanches, the hemlock trees are hundreds of years old. Several rare or unusual plant and animal species can be found here, including prairie falcons, wolverines, golden eagles, Pacific tree frogs, and long-toed salamanders.

Audubon Canyon Ranch–Bolinas Lagoon Preserve
Marin County
Located about 10 miles northwest of the Golden Gate Bridge along the Pacific coastline, this 1,200-acre preserve encloses a host of natural communities: beach, salt marsh, tidal mudflat, freshwater marsh, coastal scrub, chaparral, grassland, and lush Douglas fir–coast redwood–California bay upland forest. The variety of habitats has resulted in a rich array of flora and fauna. The estuary shelters shellfish and crustaceans are an important food source for seabirds that flock here, especially during spring and fall migrations. Large colonies of great blue herons and egrets nest nearby in the redwood forest. Several rare or unusual plant species have also been identified here.

Babbit Peak
Sierra County
Located on about 1,800 acres in the Tahoe and Toiyabe National Forests 15 miles northwest of Reno, Nevada, this proposed RNA protects habitats on the tallest part of the Bald Mountain Range. Natural communities include a rare Washoe pine forest, Jeffrey pine–white fir forest, western white pine forest, red fir forest, scrubby mountain mahogany woodland, and sagebrush scrub. Old-growth trees are present over much of the mountain, including Washoe pine and western white pine that are 800–1,000 years old. In addition to its unique forest stands, the rare prairie falcon, goshawk, and mountain lion inhabit the mountain.

Bell Meadow
Calaveras County
Part of Stanislaus National Forest about 2 miles south of Pinecrest, this 655-acre proposed RNA shelters an extensive, well-developed stand of aspen forest. Other communities include Jeffrey pine forest, riparian corridors along streams, white fir–red fir forest, wet and dry meadow, and lodgepole pine forest. The aspen forest and meadows are a critical summer habitat for Sierra mule deer and other mammals and birds, including raptors. Biodiversity in this area is high—nearly 250 plant species have been identified.

Big Morongo Preserve
San Bernardino County
Located on 4,500 acres just east of Morongo Valley, this Nature Conservancy preserve protects a rare Fremont cottonwood–willow wetland and perennial springs in a high-elevation desert climate. Nearly 400 different plant species have been recorded here. The oasis is a critical habitat for animals and birds, including some uncommon species, such as the yellow-breasted chat and Bell's vireo. It is an important resting stop for thousands of migrating birds during spring and fall. The surrounding hills are covered by creosote scrub, catclaw, Mormon tea, and cacti.

Blue Sky Ecological Preserve
San Diego County
Located a few miles from Poway, this 470-acre ecological preserve, run by the California Department of Fish and Game, protects a high-quality sycamore-cottonwood riparian corridor, coastal sage scrub, oak woodland, and chaparral. This site has a relatively high proportion of rare, endangered, or special-concern species, such as the San Diego horned lizard and Engelmann oak. Several varieties of bird species nest and hunt in the preserve, including raptors.

Bodega Marine Reserve
Sonoma County
Owned by the University of California, this 326-acre natural reserve is located about 45 miles northwest of San Francisco, along the Pacific Ocean. Natural communities include harbor tidal flat, beach, saltgrass-pickleweed salt marsh, sedge-bulrush freshwater marsh, coastal prairie, dune, and subtidal zones. These well-preserved and pristine communities support a highly diverse population of flora and fauna, with several rare or endangered species. It is a critical habitat for a variety of nesting and migratory birds, as well as California sea lions and harbor seals.

Bridge Creek Research Natural Area
Siskiyou County
Located on 1,806 acres in the Klamath National Forest about 10 miles northeast of Somes Bar, this property protects a multilayered old-growth Douglas fir forest. Many of the larger specimens are hundreds of years old and more than 200 feet tall. Other rare or unusual species include various shrubs, giant chinquapin, tan oak, and Pacific madrone. Interesting flora and fauna can also be found in the shaded, moist, riparian habitats in deep gullies. Bridge Creek RNA is an excellent example of what the pre-settlement wilderness looked like.

Burns Piñon Ridge Reserve
San Bernardino County
This 303-acre tract owned by the University of California is about 2 miles north of Yucca Valley. It encloses a series of rocky hills and canyons that contain excellent remnants of pinyon-juniper woodland, Joshua tree woodland, scrub oak woodland, montane mountain mahogany chaparral, and freshwater seeps. Although the climate is harsh, a number of desert plants thrive here, including the rare desert mallow, Parish's daisy, juniper mistletoe, and several ferns, mosses, and lichens. Among the 153 vertebrate species that have been documented here, two are federally listed. The threatened desert tortoise and rare grasshopper mouse also live on the property.

Calaveras Big Tree State Park
Calaveras County
Located about 25 miles northeast of Angels Camp on the western flank of the Sierra Nevada Mountains, this 6,000-acre park preserves virgin stands of towering sequoia trees. Circumferences reach 100 feet and heights exceed 200 feet.

Carpinteria Salt Marsh Reserve
Santa Barbara County
Located due west of Carpinteria, this 120-acre property is part of the University of California's Natural Reserve System. It protects excellent examples of southern coastal salt marsh, tidal mudflat, salt flat, brackish marsh, sand beach, and remnant coastal dune communities. It is one of the few relatively undisturbed estuaries that remain in Southern California. The dominant community—salt marsh—is composed of pickleweed–saltgrass–Parish's glasswort. Several rare or endangered plant species can be found here, including salt marsh bird's-beak. Other rare species that depend on these habitats are several clams, hawks and ospreys, Belding's savanna sparrow, and the state-endangered light-footed clapper rail.

Carrizo Plain Natural Area
San Luis Obispo County
This Nature Conservancy property about 25 miles southeast of Simmler consists of 259,000

acres flanking Soda Lake and the surrounding remnant prairie. The grasslands exhibit a high diversity of plants. It is an important habitat for pronghorn antelope, tule elks, various raptors such as the prairie falcon, and waterbirds, including sandhill cranes.

Castle Crags State Park
Shasta County
Located about 5 miles west of Dunsmuir, this state park protects a range of natural features and communities: rock faces and cliffs, mature forests of incense cedar, Douglas fir, ponderosa pine, and black oak, flowery meadows, and riparian corridors along clear-running streams. A few old-growth areas of towering trees remain. The park is a critical habitat for a variety of birds, reptiles, and mammals, some of them rare or endangered, such as the mountain lion, black bear, and certain owls and woodpeckers. The wildflowers in the meadows are especial colorful during spring.

Cedar Basin
Siskiyou County
This approved RNA in the Shasta-Trinity National Forest is southeast of Mount Shasta. Located on 972 acres along the South Fork of the Sacramento River, natural communities include glacial pond, xeric mixed coniferous forest, mountain chaparral, red fir–mountain hemlock forest, Port-Orford-cedar forest, and rock-cliff fern-moss-lichen assemblages. In addition to the tall old-growth trees (stands of Port-Orford-cedar are more than 500 years old), rare or unusual species found here include the spotted owl, various wildflowers, pine martin, mountain lion, and wolverine. A new species of manzanita has also been discovered on the site.

Coachella Valley Preserve
Riverside County
This 20,000-acre preserve is about 10 miles east of Palm Springs. It protects mesa, sand dune, and palm oasis–wetland-woodland natural communities. Its California fan palm forest is one of the largest in the United States. These unique environments are well known for their colorful wildflowers and crowded bird populations during spring and fall migrations. Several rare or threatened plant and animal species can be found here, including the fringe-toed lizard. The property is owned and managed by The Nature Conservancy.

Consumnes River Preserve
Sacramento County
Located about 5 miles south of Bruceville, this Nature Conservancy property protects 6,700 acres of valley oak riparian habitat and freshwater marsh. Waterbirds, including the rare sandhill crane, nest on the property, and thousands of others visit during spring and summer during migration. Annual flooding supports an abundance of various plant species, some of them rare.

Crater Creek
Siskiyou County
This proposed RNA covers about 240 acres in Klamath National Forest, about 17 miles northwest of Mount Shasta. Natural communities include high-density subalpine forest, mountain mahogany woodland, shrub-steppe, and gabbro scrub. Subalpine trees contain foxtail pine and mountain hemlock that are more than 600 years old. Rare, specialized plant assemblages grow on the rocky serpentinite (iron- and magnesium-rich rock)–gabbro soils. It is also an important habitat for a variety of birds, reptiles, and large mammals.

Cub Creek
Tehama County
Located about 12 miles south of Mineral, this 3,922-acre proposed RNA is unique because it encloses the entire watershed of Cub Creek, a major mountain drainage (very few complete watersheds are protected in the United States). The large size of the tract makes it an important breeding site for a variety of interior forest birds and mammals as well as a site for ecological studies. Natural communities include mixed coniferous forest (mostly ponderosa pine and Douglas fir), red fir–white fir forest, montane chaparral, and meadows.

Dawson Los Monos Canyon Reserve
San Diego County
As part of the University of California Natural Reserve System, this 163-acre tract preserves high-quality remnants of perennial coastal stream and flanking sycamore-willow riparian corridor, coast live oak woodland, inland sage scrub, mixed and chamise chaparral, and mixed bunchgrass grasslands. Some of the chaparral is considered old-growth, with ages

of 80 to 120 years. More than 50 birds species breed in this area, including the rare Anna's hummingbird, several swallows, California gnatcatcher, various raptors, black-shouldered kites, and the federally endangered Bell's vireo. An abundance of amphibians can be found in the wet, lush stream drainage, some of them rare.

Desert Tortoise Natural Area
Kern County
Located a few miles northeast of California City on the edge of the Mojave Desert, this 3-square-mile preserve encloses an arid rocky landscape of canyons, gullies, and hills partially covered in creosotebush scrub with burrobrush, cheesebush, Anderson thornbush, and saltbush scrub. Several rare or unusual desert species live here, including the desert tortoise, Mojave ground squirrel, golden eagle, and burrowing owl. Desert flowering plants are well represented—over 160 species have been identified, including the desert candle. It is an important nesting habitat and migratory stopover for a variety of birds. The property is owned and managed by The Nature Conservancy.

Devil's Basin
Tehama County
Located in Mendocino National Forest, this 646-acre proposed RNA is near Paskenta. Natural communities include mature to old-growth California black oak forest, mixed Douglas fir–California black oak forest, blue oak savanna, canyon live oak forest, and California buckeye woodlands. The property is one of the best sections of California black oak forest in the state. Black-tailed deer, California quail, gray squirrels, and band-tailed pigeons depend on the black oak acorns for food; consequently, the forest is an important habitat for these vertebrates. The very rare California black walnut tree also grows here. The larger Douglas fir trees are about 135 years old.

Devil's Canyon (South Fork)
Monterey County
This 542-acre RNA is part of the Los Padres National Forest, about 20 miles southeast of Big Sur. It contains an excellent example of Santa Lucia fir forest, sugar pine forest, and stands of canyon live oak. Bristlecone pine trees that grow here are more than 1,000 years old, and sections of sugar-pine forest are hundreds of years old; some trees reach 150 feet, with breast heights of nearly 5 feet. Other aspects are rocky canyons with riparian corridors. A few rare Pacific madrone trees can also be found here. Several rare or endemic herbaceous plants grow in the understory. The canyon is an important habitat for a variety of animals and birds, including raptors and several unusual reptiles.

Devil's Garden
Modoc County
This 800-acre RNA is in Modoc National Forest about 25 miles northwest of Alturas. Natural communities that occur on this site include western juniper woodland, transmontane sagebrush shrub-steppe, and rocky scabland. The RNA protects a portion of one of the best stands of productive western juniper anywhere in the world. Four rare or endangered plants have also been documented in the area.

Doll Basin
Mendocino County
Recommended for RNA management, this 821-acre portion of the Mendocino National Forest is about 18 miles northeast of Covelo. Natural communities include an excellent example of mature, pristine, Sierran mixed conifer forest, white fir forest, Douglas fir canyon bottom forest, stream riparian habitat, wet meadow, and canyon live oak woodland. The forest is an important habitat for several rare or endangered fauna, including the spotted owl, goshawk, peregrine falcon, blue grouse, pileated woodpecker, mountain lion, and black bear. Several rare plants have also been identified. The larger specimens of Douglas fir are nearly 200 feet tall with breast-height diameters of 6 feet.

Elkhorn Slough
Monterey County
This 800-acre Nature Conservancy property protects part of a high-quality coastal estuary inland from Moss Landing on the Pacific coast. These wetlands include tidal channels, mudflats, salt marsh, and lowland forest. It is one of the largest coastal estuaries in California. Nearly 300 bird species and 100 fish species depend on this large estuary for food and shelter. Rare or endangered species that depend on these habitats include the California clapper rail and California brown pelican, both of which nest on the site.

Fairfield Preserve
Sonoma County
Located on the flanks of Sonoma Mountain, this 210-acre preserve near Sonoma includes a variety of natural communities: chaparral, stream, pond, vernal pool, freshwater marsh, mixed evergreen forest, and oak upland forest. Several varieties of rare or unusual plants grow in this mix of communities, including an abundance of colorful wildflowers such as the giant trillium. The property is owned and managed by The Nature Conservancy.

Falls Canyon
Los Angeles County
Located only 10 miles northeast of Pasadena in the Angeles National Forest, this 2,400-acre tract encloses an outstanding example of mature bigcone Douglas fir and canyon live oak forest. These stands have been protected from forest fires because of their location in Falls Canyon. Some of these are 400–500 years old. Large specimens of white alder and Fremont cottonwood grow along the narrow riparian corridors. Several rare plant species have been discovered growing on the site. The canyon is an important habitat for a variety of species, including owls, woodpeckers, warblers, and raptors.

Fish Slough
Inyo County
A few miles north of Bishop, this unique spring-fed wetland harbors the endangered Owens pupfish and Owens tui chub. Being an oasis in a desert landscape, the slough attracts a wide range of wildlife, including several unique or rare bird species. The rare endemic slough milk vetch, alkali mariposa lily, cordgrass, Hot Spring fimbristylis, King's ivesia, and silverleaf milk vetch grow at the water's edge. The rocky hills and ledges around the slough are an excellent nesting habitat for raptors, which hunt the small mammals that come to the oasis.

Frenzel Creek
Colusa County
Located on 830 acres in Mendocino National Forest, this RNA is about 5 miles southwest of Stonyford. Natural communities include serpentinite chaparral, Sargent cypress woodland, MacNab cypress woodland, serpentinite barren, riparian corridor, and pine-oak woodland. Developed mostly on serpentinite (iron- and magnesium-rich rock) bedrock or serpentinite-derived soils and substrate, all these communities are relatively rare in California. At least 40 serpentinite endemic plants can be found, eight of which are considered rare. Several new species discovered at Frenzel Creek are in the process of being classified. Riparian corridors are shaded by the uncommon California bay and Fremont cottonwood trees.

Granite Mountains Reserve
San Bernardino County
Located in the East Mojave Desert, this 8,958-acre property is about 75 miles east of Barstow. It encloses the edge between the Granite Mountains and the alluvial fans and *bajadas* of the valley floor. Natural communities include sand dune, Joshua tree forest, pinyon-juniper woodland, mountain creosote scrub, yucca scrub, sagebrush hill, springs, and bunchgrass prairie. Rare or unusual species found here include bladderpod, trixis, desert willow, golden eagles, Anna's hummingbirds, loggerhead shrikes, Mexican free-tailed bat, desert shrew, dusky-footed wood rat, and ringtailed cats.

Grass Lake
Eldorado County
Part of the El Dorado National Forest south of Lake Tahoe, this proposed RNA encloses 2,400 acres of bog, peatland, and surrounding forest. Natural communities include Jeffrey pine forest, red fir forest, wet meadow, bog, and shrubby riparian willow–lodgepole pine thicket. Nearly 300 different plant species have been identified here. Grass Lake is unique as one of the largest floating sphagnum bogs in California. This unusual bog community harbors a variety of rare or endangered plants, including the pitcher plant, bladderwort, and various orchids. Other plants include monkeyflower, long- and short-beaked sedge, and Nebraska sedge. Wildflowers in the meadows are abundant during spring.

Gray Davis-Dye Creek Preserve
Tehama County
Located about 12 miles southeast of Red Bluff, this 37,540-acre parcel shelters a piece of the Sacramento River valley and the flanking foothills of the Sierra Nevada and Cascade Mountains. Natural communities include rare blue oak upland forest, pond, wetland, vernal pool, open field, grassy hill, and the riparian corridor along Dye Creek. The largest deer

herd in California moves into this area during the winter months. The wetlands are an important migratory stopover for thousands of waterfowl. Cattle are being used here on ranchland to help restore native grassland habitat. The property is owned and managed by The Nature Conservancy.

Green Island Lake
Plumas County
An extensive moss bog is protected on this 1,185-acre proposed RNA in Lassen National Forest. Located about 15 miles south of Canyondam, it encloses wet meadow, montane chaparral, lodgepole pine and red fir forest, and several lakes, bogs, and streams. This RNA is a critical habitat for the rare wolverine, goshawk, and mountain lion. Several varieties of rare or unusual plant species occur in the wetlands, some of them relicts from a colder glacial climate.

Grizzly Island State Wildlife Area
Solano County
Located east of Vallejo, this 14,300-acre property encloses the vast Suison Marsh. Known as one of the largest estuarine marshes in the country, its salt marshes, ponds, grasslands, tidal flats, and forested uplands are important habitats for a variety of birds, mammals, insects, reptiles, and amphibians. Rare or endangered species that live in the marsh are the salt marsh harvest mouse, golden eagle, and the endemic Suison shrew, sparrow, and aster. Other inhabitants include the interesting native tule elk, river otters, white pelicans, and cinnamon teals. Nearly a quarter-million birds visit the area during migration to rest and feed.

Guadalupe-Nipomo Dunes
Santa Barbara and San Luis Obispo Counties
Located about 5 miles northwest of Guadalupe, this 4,000-acre Nature Conservancy property protects an important stretch of the Pacific coast and a flanking dune system. This rare ecosystem supports at least 18 rare or threatened plants, such as the Indian paintbrush, and is also a breeding site for the rare least tern. Peregrine falcons and ospreys also use the site. More than 200 bird species inhabit the preserve or stop during spring and fall migration. With heights up to 500 feet, the beach dunes here are some of the highest along the Pacific coast.

Guatay Mountain
San Diego County
Located about 3 miles northwest of Pine Valley in Cleveland National Forest, this 3,000-acre property encloses a rare Tecate cypress woodland-chaparral. Most of the specimens are more than 100 years old, a remarkable age for chaparral species. Other communities include California mixed chaparral, southern interior oak woodland, and chamise chaparral. Several rare plant species grow here, including several endemics that adapted to the gabbroic soil and substrate.

Hale Ridge
Lake County
Part of the Mendocino National Forest, this 865-acre tract about 10 miles northeast of Upper Lake preserves knobcone pine forest. Other communities include Douglas fir–canyon live oak forest, chaparral, and chamise chaparral. The knobcone pine stands are considered to be some of the best in California. Serpentinite soils and outcrops support several rare plant species that prefer serpentinite, such as leather oak. Some of the older Douglas firs are about 400 years old and have breast-height diameters of 4 feet. The property is also an important habitat for several bird and mammal species, some of them rare or uncommon.

Harvey Monroe Hall
Mono County
This RNA in Inyo National Forest a few miles west of Lee Vining protects excellent examples of alpine meadow and Sierran mixed subalpine forest. The ecosystem comprises rock face, scree, alpine meadow, subalpine forest, wet and dry meadow, pond, seep, stream, and herb-willow riparian corridor. The variety of communities and rapid changes in elevation result in a high biodiversity: About 400 plant species have been identified, several of which are extremely rare. Fifteen pristine lakes, and the headwaters of eight streams, are within the RNA.

Haypress Meadows
Siskiyou County
This 1,200-acre portion of the Marble Mountain Wilderness in Klamath National Forest is located about 15 miles east of Somes Bar. Natural communities include Shasta red fir forest, mesic and xeric meadows, riparian habitats,

and mountain chaparral. Several exceptional stands of majestic old-growth (greater than 500 years old) red fir remain. The meadows have been only lightly grazed and contain their original diverse plant species, including pussypaws and western needlegrass. About a dozen rare or endangered plants can be found along the forest-meadow ecotone.

Heath and Marjorie Angelo Coast Range Reserve
Mendocino County
This 4,055-acre preserve is located about 20 miles west of Laytonville and is part of the University of California's Natural Reserve System. Located in rugged terrain, the preserve encloses excellent examples of old-growth conifer forest, hardwood-conifer forest, chaparral, and grassland communities. It is a very diverse region, with more than 1,000 species of flora and fauna, some of them rare, unusual, or endemic. Its Douglas fir forest is one of the largest uncut old-growth stands remaining in California. The trees are more than 200 feet tall, about 6 feet at breast height, and 300 to 800 years old. It thus is a critical habitat for old-growth-dependent species. Other special-concern species include the northern goshawk, northern spotted owl, Olympic salamander, and three-spine stickleback fish.

Hennessy Ridge
Trinity County
Part of Six Rivers National Forest, this 1,008-acre area is about one-half mile southeast of Salyer. Natural communities include old-growth mixed Douglas fir–hardwood forest and Douglas fir–tan oak–Oregon grape forests. The largest trees are Douglas fir and tan oak, which are hundreds of years old. Several rare or unusual plant species grow in the shrub layer, which is dominated by poison oak.

Horton (L.E.) Research Natural Area
Del Norte County
A 1,305-acre proposed RNA 6 miles northwest of Gasquet in Six Rivers National Forest, this property encloses bog, old-growth Port-Orford-cedar forest, subalpine dwarf serpentinite forest, and Douglas fir forest. The taller Douglas firs are more than 200 years old; several of the Port-Orford-cedars in the bog complex are about 400 years old. The bog and surrounding bog forest contain one of the highest concentrations of rare plants in California, including the California pitcher plant and several gentians. Unique, stunted vegetation is adapted to the rocky serpentinite soil and substrate.

Hosselkus Limestone
Shasta County
This proposed RNA covers about 6,500 acres in the Shasta-Trinity National Forest, about 25 miles northeast of Redding. Natural communities include limestone-based plant and animal populations, California black oak forest, and canyon live oak forest. The alkaline rocky substrate supports a variety of local endemic plant species, land snails, and the Shasta salamander. Patches of native grasslands occur in the black oak forest, and Douglas fir–mixed evergreen forest (including the uncommon Pacific yew) can be found in the narrow ravines and canyons. The site also contains important archeological and paleontological resources.

Indian Creek
Siskiyou County
Located on 500 acres in the Klamath National Forest about 15 miles northwest of Happy Camp, this pristine site contains one of the best stands of old-growth Brewer spruce in northern California. Three extremely rare ground plants have been identified here. Other communities include red fir forest, rock surface vegetation, and successional montane chaparral. The forest is an important nesting habitat for a variety of bird species.

Indian Creek Research Natural Area
Tehama County
Located on about 3,900 acres 10 miles southeast of Paynes Creek, this parcel protects a regenerating blue oak forest, relatively rare in California. Almost the entire 5-mile drainage of Indian Creek is enclosed within the property. The stream supports a variety of unique and interesting flora and fauna, including rainbow trout, western pond turtles, California newts, and several reptiles. Several rare plants are found throughout the blue oak woods. Other natural communities include digger pine stand, scrub slope, grassland, spring, and seep. Large specimens of canyon live oak can be found in the stream bottoms that have breast-height diameters up to 3 feet.

Indiana Summit
Mono County
Located about 9 miles east of June Lake, this 1,156-acre tract encloses a very rare virgin remnant of Jeffrey pine forest. Other natural communities include lodgepole pine forest, white fir forest, and a variety of grasses and herbs that survive on the rocky flats and slopes. Several rare plant species occur here, as does the threatened prairie falcon.

Irvine Open Space Reserve
Orange County
Located on 17,000 acres in and around Irvine, this Nature Conservancy property protects a number of rare or endangered plant, animal, and insect species, such as the California gnatcatcher. Natural communities include rocky canyon, rocky upland forest, and hills covered with coastal sage and scrub. Access is via hiking, biking, or horseback riding; contact The Nature Conservancy.

James San Jacinto Mountains Reserve
Riverside County
Excellent remnants of high-quality mixed conifer forest, pine-oak woodland, montane chaparral, and riparian corridor communities occur on this 29-acre tract about 8 miles north of Idyllwild. Larger ponderosa pine specimens are almost 200 feet tall. Seeps in the rock outcrops and creeks are rimmed by deergrass, monkeyflowers, lupine, and the rare lemon lily. The exposed rock faces harbor 35 different species of mosses and lichens, some of them rare. Among the many vertebrates that inhabit this mountain terrain are the rare granite night lizard, Anna's hummingbird, spotted owls, bobcats, and mountain lions. This property is part of the Natural Reserve System of the University of California.

Jawbone Ridge
Tuolumne County
Located near Groveland, this excellent example of pristine chamise chaparral is part of the Stanislaus National Forest. The rugged 1,442 acres are covered by chamise-shrub chaparral with the occasional Digger pine, as well as blue oak–California buckeye savanna. The area has one of the highest density of rare or threatened flora and fauna in California, including 16 rare plants, two extremely rare land snails endemic to the Tuolumne Canyon, and the peregrine falcon and bald eagle.

Jepson Prairie Preserve
Solano County
Located on 1,566 acres, this preserve about 15 miles south of Dixon is owned and maintained by the University of California. It shelters one of California's few undisturbed remnants of grassland prairie. Other natural communities include wet prairie and vernal pools. Several rare native grasses are found here, such as Colusa grass and the endemic Solano grass, as well as the endemic Delta green ground beetle. Rare plants found around the vernal pools are the Boggs Lake doddier, peppergrass, and California sandwort. The grasslands are an important habitat to a variety of grassland birds, mammals, and reptiles, some of them rare or endangered, such as the California tiger salamander.

Jug Handle State Reserve
Mendocino County
This 769-acre sanctuary is located about 5 miles south of Fort Bragg. It encloses Pacific beach and a series of ancient beach terraces that climb in elevation inward toward Jackson State Forest. The slight differences in elevation and precipitation have created a distinct plant community on each terrace, including north coast prairie, coast redwood, a pygmy pine–cypress forest, and sphagnum bog-marsh. This has created high biodiversity, including a host of rare or unusual plant and animal species. Some of these are the endangered Mendocino coast paintbrush, dwarf Bishop and Bolander's pine, swamp harebell, several mosses and lichens, and the carnivorous sundew.

Kaweah Oaks Preserve
Tulare County
This 324-acre Nature Conservancy property protects one of the last native valley oak forests in California. Located about 5 miles northeast of Farmersville, the trees are hundreds of years old and surrounded by native grasslands. In addition to the large trees, it is well known for its spring wildflowers and migrating bird populations.

Kendall-Frost Mission Bay Marsh Reserve
San Diego County
Owned by the University of California, this 16-acre tract protects excellent examples of south coastal salt marsh (cordgrass-pickleweed), salt

flats, eelgrass beds, and tidal channels. The marsh is distinctive in that nearly 80 percent of its flora is native to Southern California. Most of all it is critical habitat for a number of nesting birds as well as those migrating along the Pacific Flyway during spring and fall. Endangered or special-concern species include the California brown pelican, light-footed clapper rail, and Belding's savanna sparrow.

Kern River Preserve
Kern County
Located on 1,127 acres, this property just northwest of Weldon protects an important stretch of forested riparian habitat along the Kern River. Some of the trees are hundreds of years old. The surrounding hills are covered with saltgrass, native wild ryegrass, and cactus. It is an important habitat for a number of rapidly disappearing songbirds, including the yellow-billed cuckoo. Other rare birds include the willow flycatcher, yellow-headed blackbird, and Lawrence's goldfinch. The plant diversity along the river is high, including some rare or unusual species, such as the alkali mariposa lily. More than 100 different butterflies have been identified, including the rare Eunus skipper. The property is owned and maintained by The Nature Conservancy.

King Creek Research Natural Area
San Diego County
Part of the Cleveland National Forest, this 750-acre RNA is about 21 miles northeast of El Cajon. It was established to protect the extremely rare and endemic Cuyamaca cypress. Natural communities include chamise-manzanita chaparral, chamise–scrub oak chaparral, dry meadow, cypress woodland, and stands of coast live oak. In addition to the cypress trees, several other rare plant species grow on the property.

King Range National Conservation Area
Humboldt County
Just north of Shelter Cove on the Pacific Ocean, this large expanse between the Mattole River and Whale Gulch protects coastal habitats as well as parts of the forested King Range. Natural communities include tidal zone, beach, rocky canyon, cliff face, and lowland and upland forest. The old-growth Douglas fir forests are impressively tall and a critical habitat for the northern spotted owl and American bald eagle. The beach environment supports a diverse population of seabirds, harbor seals, and sea lions.

Kopta Slough Preserve
Tehama County
Located on 706 acres a few miles west of Vina, this Nature Conservancy property shelters an important wetland complex along the Sacramento River. Communities include freshwater slough, swamp, riparian forest, and upland forest. The wetlands are an important habitat for a variety of plants, animals, and birds. It is also a resting and feeding stop for migrating waterfowl and other species during spring and fall.

Landels-Hill Big Creek Reserve
Monterey County
This 3,858-acre property is part of the University of California Natural Reserve System. It is located roughly midway between Big Sur and Cambria on the Pacific coast. It is most known for the rapid ascent from beachfront to rugged mountains over a linear distance of less than 3 miles. The transition from beach to highland comprises more than 20 different plant communities, from beach-dune grasses and coastal sage scrub to forested mountainsides (coast redwood, coast live oak–mixed hardwood). Its biodiversity is one of the highest in the state and includes more than 500 plant species, eight of which are rare or endangered, such as the Santa Lucia fir, Hoover's manzanita, and phantom orchid. Other species are unusual because they are relicts from previous climates. More than 100 bird species depend on this reserve as a nesting habitat or migratory stopover, including the rare spotted owl, peregrine falcon, and golden eagle. Wildflowers are colorful during spring.

Last Chance Meadow Research Natural Area
Inyo County
This RNA in Inyo National Forest covers about 1,900 acres some 10 miles southwest of Lone Pine. It was established to protect a unique population of subalpine old-growth foxtail pine forest. These hardy trees, which grow on top of granitic bedrock, are as old as 1,500 years. Its understory includes several rare or unusual plants. Other communities include wet and dry meadows and lodgepole pine forest.

Limekiln Creek
Monterey County
Located on 5,478 acres in Los Padres National Forest along the Pacific Ocean, 2 miles southeast of Lucia, this RNA preserves one of the most naturally diverse areas in California. Natural communities include mixed evergreen forest, redwood forest, chaparral, scrub, grassland, riparian corridor, and California coast live oak forest. Nearly 400 different plant species are represented here. Rare or unusual plant and animal species that inhabit the property include the Santa Lucia fir, sugar pine, several salamanders, and the sagebrush lizard.

Los Peñasquitos Lagoon–Torrey Pines State Reserve
San Diego County
A wetland complex covers this 600-acre parcel about 2 miles north of Del Mar on the Pacific coastline. Important natural communities include coastal salt marsh (pickleweed–sea lavender–salt grass), tidal mudflat, salt flat, coastal bluff grassland and pine forest, chaparral, and cliff face. Several rare plant and animal species depend on these critical habitats, including the California least tern, light-footed clapper rail, Belding's savanna sparrow, and the very rare Torrey pine tree.

Manila Beach and Dunes
Humboldt County
A relatively undisturbed section of Pacific coastal dune is preserved on this tract a few miles from Arcata. Natural communities include beach, sand dune, and beach pine–Sitka spruce–bearberry forest. Some of the dunes are nearly 100 feet tall, anchored by scattered patches of vegetation, including evening primroses and rare mustard plants found only in California. The thick coastal forest harbors several rare species, including the relict reindeer lichen, several orchids, and a host of fungi species.

Manzanita Creek
Trinity County
This 6,160-acre proposed RNA in the Shasta-Trinity National Forest is about a mile west of Big Bar. It contains the entire watershed of Manzanita Creek, riparian white alder–dogwood forest, digger pine–chaparral assemblages, Douglas fir–mixed oak forest, white fir–Douglas fir forest, and specialized shrubs, herbs, mosses, and lichens that grow on the high, windswept ridge tops. The high number of communities result in diverse flora and fauna, including several rare or threatened species. Some stands of old-growth forest remain.

McCloud River Preserve
Shasta County
Located on 2,300 acres, this Nature Conservancy property about 16 miles south of McCloud protects important habitat along the pristine McCloud River. This 6-mile stretch offers a variety of riparian communities and harbors many plant and animal species, including the black bear. The high quality of the river water supports a healthy population of trout.

Millard Canyon
Riverside County
This future RNA is part of San Bernardino National Forest, a few miles northeast of Banning. Its 793 acres protects a well-developed interior live oak forest endemic to California. Typically more like a shrub, here it is developed into a taller woodland. Other significant communities are bigcone Douglas fir forest, chaparral, chamise chaparral, sage scrub, and rock outcrop. Several rare plants have been found in the groundcover. Black bears and mountain lions, both diminishing species, live in Millard Canyon.

Mott Rimrock Reserve
Riverside County
Owned and managed by the University of California, this 1,224-acre reserve about a mile northwest of Perris protects high-quality interior coastal sage scrub (brittlebush–California buckwheat–sage), grassland, chaparral, and spring-riparian communities. These are critical habitats for a variety of fauna, including the rare spadefoot toad, coast horned lizard, orange-throated whiptail, black-shouldered kite, burrowing owl, black-tailed gnatcatcher, and the federally endangered Stephen's kangaroo rat.

Mount Eddy
Siskiyou County
This 940-acre tract in Shasta-Trinity National Forest about 9 miles west of Mount Shasta City has been nominated for RNA status. Natural communities include pristine foxtail pine forest, western white pine forest, and herbaceous meadow. The old-growth foxtail pine forests are more than 300 years old. Several varieties of rare or endangered plant species are found

on the property, some of them endemic to this part of California.

Mount Pleasant
Plumas County
Recommended for RNA management, this 1,785-acre property in Plumas National Forest protects an excellent example of red fir forest. Located about 2 miles west of Silver Lake, its natural communities include red fir forest, mountain chaparral, mountain riparian, meadow, rock face, and bog-fen. About 250 plant species have been identified, including some rare bog and rock face plants. The larger red fir trees are nearly 200 feet tall, up to 5 feet in breast-height diameter, and hundreds of years old. It is an important habitat for a variety of birds and mammals, including raptors and woodpeckers.

Mud Lake
Plumas County
Part of Plumas National Forest 10 miles east of Greenville, this 400-acre tract is well known for its excellent stand of Baker cypress. Several of these trees are 150–300 years old with 4-foot breast-height diameters. Other community types include mixed conifer forest (white fir–red fir), mountain chaparral, sagebrush scrub, seeps, meadow, montane freshwater marsh, and riparian corridors along streams.

North Coast Dunes Preserves
Humboldt County
Located on 473 acres along the Pacific coast south of McKinleyville, this preserve shelters undisturbed coastal communities, such as open beach, dune, beach-pine forest, and Sitka spruce forest. The dunes reach heights of 80 feet and are anchored by hardy but beautiful wildflowers, such as Menzies wallflower. The flowers are most colorful during summer. Rare or unusual plants have also been found in this specialized coastline environment. The property is owned and maintained by The Nature Conservancy.

North Trinity Mountain
Humboldt County
Part of Six Rivers National Forest, this 640-acre tract 10 miles northeast of Hoopa has been nominated for RNA status. The property shelters an excellent stand of Noble fir montane forest, white fir forest, montane chaparral, and high-elevation meadows. The stands of Noble fir are considered to be some of the best in the state. Older white fir trees are hundreds of years old and more than 100 feet tall. Shrub layers are well developed and highly diverse. Several rare flora species can be found here.

Onion Creek
Placer County
Two miles north of Norden is Onion Creek, a well-preserved example of white fir–red fir forest. Part of the Plumas National Forest, this 1,035-acre site was considered special enough to be nominated as an RNA about 20 years ago. Other tree types infiltrating the forest are sugar pine, mountain hemlock, and Jeffrey pine. Larger specimens are hundreds of years old. The ecosystem supports a variety of large mammal and bird species, including black bears, woodpeckers, and raptors.

Organ Valley
San Diego County
An excellent forest of Engelmann oak, a California endemic tree, will be protected within this future RNA. Located on a tract of nearly 600 acres about 10 miles north of Ramona, healthy populations of oak are found throughout the property. Other communities include chamise chaparral, mixed chaparral, and riparian oak woodland. Nearly 300 plant species have been identified on this relatively small tract, which reflects its high biodiversity. Five other plant species on the property have been identified as rare or threatened.

Pearch Creek
Humboldt County
Once nominated as an RNA, this 2,800-acre tract in the Six Rivers National Forest about 3 miles east of Orleans contains well-developed Douglas fir–tan oak–Pacific madrone forest. Other communities include Noble fir forest, montane chaparral, and delicate flora that live on rocky outcrops. Giant chinquapin trees can also be found here. Some Douglas fir trees exceed 200 years old. Small stands of Brewer spruce on rocky heights are nearly 500 years old. Biodiversity is high, and several rare or endangered plant species occur in the ground and rock cover.

Point Lobos State Reserve
Monterey County
Located about 5 miles south of Monterey on the Pacific coastline, this 2,000-acre property protects a variety of important natural commu-

nities, including beachfront, meadow, cypress-pine forest, and cliff face. Its rocky headlands are a popular congregation site for California sea lions. Several very rare endemic plants grow here, including the endangered Gowen cypress and Monterey pine. Offshore kelp beds are an important habitat for the diminished southern sea otter. Several varieties of birds, including raptors and swallows, nest and hunt along the cliff faces.

Prairie Creek Redwoods State Park
Humboldt County
This large state park, located along the Pacific coast about 30 miles south of Crescent City, contains ocean beach, dune, redwood forest, cliff face, and riparian corridor habitats. Old-growth sections of forest include a 367-foot tall redwood—the tallest tree in the world. The giant redwoods are more than 2,000 years old and shelter an understory of rhododendron and sword ferns. The park is known for its high biodiversity and contains a number of rare or unusual plant species. A herd of Roosevelt elk also inhabits the park.

Preacher Meadows
Trinity County
This 600-acre tract in Shasta-Trinity National Forest, about 6 miles west of Trinity Center, is an excellent example of mature midelevation mixed conifer forest. Dominant tree species are Douglas fir, sugar pine, white fir, and Shasta red fir. Many of the taller specimens are hundreds of years old. Several rare or unique plant species can be found in the mesic and xeric meadows interspersed with the forest. Other species that prefer cool, moist microclimates, such as ferns and lichens, can be found along the steep Swift Creek gorge.

Pygmy Forest Reserve
Mendocino County
Located about 2 miles southeast of Mendocino, this 70-acre University of California property protects a very rare pygmy forest. A result of climatic factors and acidic soil conditions, the short, twisted, stunted Bolander's pine and Mendocino cypress trees are thought to be the best example of pristine pygmy forest in the world. An excellent stand of mature Bishop pine forest is also on the property. The extremely rare Helfer's blind weevil is endemic to the reserve.

Ring Mountain Preserve
Marin County
This 377-acre Nature Conservancy property southwest of Corte Madera protects very rare communities that grow on serpentinite bedrock and serpentinite-derived soil. They are uniquely adapted to the high levels of magnesium, iron, nickel, and chromium in the substrate. Many of the plant species are globally rare because they grow only under these specific conditions.

Rock Creek Butte
Siskiyou County
Nominated as a federal RNA, this 500-acre property is located in the Siskiyou Mountains, about 20 miles northwest of Orleans. Natural communities include mountain chaparral, old-growth Brewer spruce forest, enriched conifer forest, Noble fir forest, and Douglas fir–sugar pine forest. Rocky exposures in this high-elevation environment harbor a variety of mosses and lichens. The site is relatively undisturbed and pristine, and Brewer spruce trees approach 400 years old. Several rare or unusual plants can be found in the understory and groundcover, especially at springs and seeps.

Ruth Research Natural Area
Trinity County
Located in Six Rivers National Forest about 3 miles from Forest Glen, this 630-acre proposed RNA preserves an excellent stand of Douglas fir forest. Other natural communities include riparian corridor along narrow stream, Ponderosa pine forest, mixed coniferous forest, montane chaparral, and oak-conifer forest. Some trees are nearly 150 feet tall and up to 8 feet in diameter at breast height. Occasional large old-growth sugar pines can also be found. Stands of incense cedar grow in ravines and stream bottoms. Several rare or unusual plant and animal species can also be found on the property.

Santa Cruz Island Preserve
Santa Barbara County
This large, 54,500-acre property, which covers most of Santa Cruz Island in the Channel Islands off the Santa Barbara–Ventura coastline, is owned by The Nature Conservancy. Natural communities include undeveloped coastline and beach, coastal sage scrub, chaparral, grassland, coast live oak and endemic island oak forest, an unusual ironwood forest, closed-

cone pine forest, and some freshwater wetland. More than 500 of the 680 plant species are native, and eight are endemic to Santa Cruz Island. These restricted ecosystems are host to a diverse abundance of rare endemic plant and animals species, such as the Santa Cruz Island scrub jay and Santa Cruz Island fox.

Santa Monica Mountains National Recreation Area
Ventura County
Located about 16 miles east of Oxnard, this area encloses high-quality habitat along the front of the Santa Monica Mountains. Natural communities include beachfront, saltwater marsh, freshwater marsh, tidal pool, grassland, chaparral, and mature upland oak forest. Only 30 miles from Los Angeles, its wooded hills, rocky canyons, and lush riparian corridors are critical habitats for large animals, such as deer, bears, and mountain lions. The wildflowers, especially fields of California poppies, are dazzling during spring. Thousands of birds are attracted to the wetlands, where they find food and shelter, especially during spring and fall migrations.

Santa Rosa Plateau Ecological Reserve
Riverside County
Located about 75 miles north of San Diego and just west of Murrieta, this 7,350-acre property preserves an abundance of natural communities, including vernal pool, grassland, and groves of rare Engelmann oak trees. It is an important nesting habitat and migratory stopover for thousands of birds during spring and fall migrations. The rare red-legged frog can also be found here.

Scripps Coastal Reserve
San Diego County
Owned and managed by the University of California, this 943-acre property lies just outside San Diego. Natural communities include coastal sage and maritime succulent scrub, sea bluff succulent scrub, grassland, coastal strand, beach, and submerged sandy plain. The very rare sea dahlia grows on the sea bluff. Forests of giant kelp are protected on the submerged sandy plain. Several varieties of invertebrates, fish, birds, and mammals can be found here, several of them rare, endangered, or of some special concern. The site is an important migratory stopover for waterbirds during the spring and fall migrations.

Sentinel Meadow
Mono County
This RNA covers 3,840 acres in the Inyo National Forest, about 15 miles northeast of Mammoth Lakes. Natural communities include mixed coniferous forest, lodgepole pine forest, Jeffrey pine forest, sagebrush scrub, and mountain mahogany woodland. Large Jeffrey pine specimens are about 200 years old, stunted, and wind-sheared. Several rare or unusual plant species have been discovered in the groundcover layer.

Shasta Red Fir
Siskiyou County
Nominated as an RNA, this 1,800-acre site 8 miles north of McCloud is part of the Shasta-Trinity National Forest. It is known for its exceptional stands of old-growth Shasta red fir forest, some of which are nearly 700 years old. The wet subalpine meadows that are interspersed with the woodlands are glacier-fed. Subordinate stands of western hemlock are nearly 700 years old. Other natural communities include mountain buckwheat–western needlegrass xeric meadow, scree slopes, pinemat manzanita forest, alpine sedge rockland, pennyroyal shrubland, and wet sedge meadow. The differences in climate and elevation contribute to a high biodiversity, including several very rare plant species.

Sinkyone Wilderness State Park
Mendocino County
Located on the Pacific coast about 15 miles southeast of Shelter Cove, this 7,000-acre preserve encloses beach, tidal pool, coastal bluff, grassland, and lush old-growth Douglas fir and coast redwood forests. The largest specimens approach 250 feet and are nearly 3,000 years old. The forest is an important habitat for a variety of birds and mammals, including the Roosevelt elk, violet-green swallows, western sandpipers, and western kingbirds. Several rare or unusual plant species have been identified in the forest understory, including a number of fungi.

Smoky Creek
Trinity County
This nominated RNA covers about 600 acres 5 miles south of Forest Glen. The rugged, forested terrain contains old-growth Pacific ponderosa pine–Douglas fir forest, Jeffrey pine forest, white alder riparian habitat, sedge ponds,

and meadows. The variety of communities results in a high diversity of flora and fauna. In addition to the old mixed conifer forest, the area is known for five species of rare plants that grow only in the soil and rocky substrate developed from serpentinite-rich bedrock. Part of the Jeffrey pine forest also grows in this serpentinite domain.

Soldier Research Natural Area
Trinity County
This 709-acre proposed RNA is located a few miles north of Zenia in Six Rivers National Forest. The dominant natural community is a mature Oregon white oak forest; other communities include digger pine forest, upland Douglas fir forest, nonnative grassland, and bald hills prairie. The white oak forest is largely a product of fire succession. Stands of Douglas fir are hundreds of years old. Rare plants have been documented in the groundcover.

South Fork Mountain
Trinity County
Located on nearly 3,300 acres in Shasta-Trinity National Forest, this proposed RNA is about 16 miles southeast of Forest Glen. It protects an excellent example of old-growth Douglas fir forest and giant chinquapin forest. Douglas firs reach heights of 200 feet and a diameter breast height of 12 feet. Other tree types include impressive specimens of California black oak and Pacific madrone. A narrow riparian corridor along the South Fork of the Trinity River protects old-growth red alder and white alder and the rare Oregon ruffed grouse. The river is an important spawning habitat for steelhead trout and salmon. Other rare species found on the tract include the spotted owl, fisher, and mountain lion.

Starr Ranch Sanctuary
Orange County
Owned and managed by the Audubon Society, this 4,000-acre sanctuary is in the foothills of the Santa Ana Mountains. High-quality communities here are oak woodland, oak-sycamore riparian forest, coastal sage scrub, grassland, and chaparral along Trabuco Canyon. Biodiversity is high, including a number of endemic species. The rare coastal sage scrub community is the only habitat in which the California gnatcatcher will reproduce.

Station Creek
El Dorado County
Proposed as an RNA, this 738-acre site in El Dorado National Forest 3 miles west of Strawberry protects a superb example of transitional sugar pine-white fir forest. These old-growth trees are hundreds of years old and nearly 200 feet tall. Subordinate trees are incense cedar and Douglas fir. Several rare or unusual plant species have been reported here.

Sugar Creek
Siskiyou County
Nominated as an RNA, this roughly 4,000-acre section of the Klamath National Forest is about 6 miles west of Callahan. It protects lower montane-subalpine old-growth stands of one of the most diverse assemblages of conifer forest in the world. Because of rapid changes in elevation and moisture levels, and the presence of some relict species, seventeen different conifers have been identified within a single square mile. The incredible biodiversity is reflected by the documentation of more than 400 plants, several of which are rare or endangered, such as the subalpine fire and Engelmann spruce. In addition to the various coniferous forest types, other communities include mountain mahogany forest, wet montane meadow, and montane freshwater marsh.

Tijuana River National Estuarine Research Reserve
San Diego County
Located on 2,531 acres about 5 miles south of San Diego, this property encloses tidally connected wetland, riparian, and upland natural communities. The productive salt marshes and mudflats attract tens of thousands of birds along the Pacific Flyway. More than 400 species have been sighted, including the rare Belding's savanna sparrow, California least tern, light-footed clapper rail, Bell's vireo, peregrine falcon, western brown pelicans, and the threatened snowy plover. The sand dunes are anchored by ambrosia, beach evening primrose, and sea rocket. Upland areas are cloaked in sagebrush, buckwheat, and alkali heath (coastal sage scrub).

Upper Goose Creek
Del Norte County
An excellent example of Douglas fir–western hemlock forest–Port-Orford-cedar forest covers this 460-acre tract in Six Rivers National Forest,

about 7 miles northeast of Klamath Glen. Much of this kind of ecosystem has already been destroyed by logging in the state. The Douglas firs reach heights of 250 feet with breast-height diameters spanning nearly 7 feet. Port-Orford-cedar trees are nearly as large and approach 1,000 years old. Giant chinquapin trees, tan oak, big-leaf maple, Pacific dogwood, and canyon live oak are other forest members. Narrow riparian corridors contain some unique or unusual plant and animal species.

Upper Newport Bay Ecological Reserve
Orange County
Located on some 500 acres in a heavily populated area near Newport Beach, this coastal wetland system encloses saltwater marsh, tidal mudlflats, and coastal sage scrub hills. The wetland is a critical nesting and feeding habitat for the endangered piping plover, brown pelican, American avocet, black rail, and the rare California least tern. The threatened California gnatcatcher lives in the dense sage scrub above the bay. Several varieties of clams, anchovies, and crustaceans live in the bay and serve as an important food source for the waterbirds.

Vina Plains
Tehama County
A part of the verdant and pristine Sacramento Valley is protected on this 1,950-acre Nature Conservancy property about 15 miles northwest of Chico. Natural communities preserved here are undisturbed grassland, basalt vernal pool, and scattered woods. It is one of the few areas in California where the grassland has not been plowed. As a result, the biodiversity is very high—nearly 300 plant species have been identified here, including some rare wildflowers. Some of these include Ahart's paronychia, Hoover's spurge, hairy orcuttia, and Greene's tuctoria. Several varieties of specially adapted invertebrates and tiny shrimp survive in the vernal pools. It is also an important habitat for grassland birds and mammals, some of them rare or threatened. The wildflower displays during spring and summer are some of the best in the state.

Whipporwill Flat
Modoc County
Located on 3,328 acres in Inyo National Forest, this new RNA is about 18 miles south of Big Pine. Natural communities include single-leaf pinyon pine forest, pinyon-juniper woodland, subalpine sagebrush scrub, and bristlecone pine–limber pine forest. The well-developed mature pinyon forest is an excellent habitat for several birds, including gray flycatchers, western tanagers, and hermit thrushes. The less common pinyon mouse also thrives here. The larger bristlecone trees are estimated to be more than 3,000 years old. Several rare or unusual plant species occur in the scrub and forest groundcover.

White Mountain Natural Area
Mono County
This 2,303-acre RNA in the Inyo National Forest is about 16 miles northeast of Bishop. It was established to protect an ancient grove of bristlecone pine trees; those on the windswept subalpine ridges are more than 3,000 years old. Several rare plants are found atop the rocky soils. Biodiversity is high, with nearly 200 plant species recorded from the site.

White Mountain Summit (McAfee Meadow)
Mono County
This proposed RNA about 30 miles northeast of Bishop is in the Inyo National Forest. This property is unique as it encloses the largest continuous area of alpine steppe habitat in the western United States. The dry, sandy, windswept flats support a variety of small plants and shrubs, sagebrush, and meadow grasses, which are often watered by melting snow, even into the summer months. Ancient bristlecone pine trees can also be found here.

Yurok Research Natural Area
Del Norte County
Part of Six Rivers National Forest, this 150-acre RNA preserves an excellent remnant of old-growth coast redwood forest. These majestic trees are typically taller than 250 feet. Other trees include western hemlock, Douglas fir, red alder, tan oak, and big-leaf maple. The understory is dominated by shrubs and herbs. The forest is an excellent habitat for a variety of birds, mammals, and reptiles, some of them rare or unusual.

COLORADO

Because of its location at a major ecotone between the Rocky Mountains and the Great Plains, Colorado supports diverse assemblages of flora and fauna among several major vegetation types: plains grassland, shrubland, pinyon pine–juniper woodland, various elevation-dependent pine forests, and alpine tundra.

Much of the eastern third of Colorado is covered by plains grasslands (relict patches of native grassland are scattered across this region, including parts of Pawnee and Comanche National Grassland). The topography is generally flat to rolling, covered by a mixture of blue grama and buffalo grass with lesser bunchgrass, sand dropseed, sideoats grama, and big and little bluestem. Depending on moisture content and soil chemistry, sagebrush and prickly pear may also occur.

Shrublands can be found in various parts of the state in the bottomlands, foothills, and montane hillsides; they are marked by sagebrush, greasewood, and Gambel oak–serviceberry populations. Other distinctive plant species are saltbush and shadscale. Shrublands dominate in elevations between the grasslands and the pinyon-juniper woodlands, or in sandier sites.

Low mesas, ridges, and lower mountain flanks (4,000 to 7,000 feet of elevation) are typically covered by pinyon pine–juniper woodlands. Shrubs and grasses commonly make up the understory. The upper limits of the pinyon pine–juniper zone (at roughly 9,000-feet elevation) may also include stands of juniper, ponderosa pine, and Douglas fir.

Conifer forests begin to appear at roughly 6,000 feet. Different pine species dominate according to elevation, soil chemistry, and precipitation. The lowermost forests (6,000–9,000 feet) are typically ponderosa pine, which often form open parkland, with shrubs and grasses forming the understory. Other trees are junipers (red cedars). Douglas fir forests also grow in the 6,000–9,000-foot montane zone and are most frequent on north-facing slopes. They may be mixed with stands of ponderosa pine. Both forest types in Colorado have been heavily logged. Quaking aspen stands are common throughout this elevation range and are usually found in canyons and ravines.

Higher-elevation grasslands and meadows are scattered across the mountain ranges. Plant assemblages are dependent on moisture levels and form dry, moist, or wet meadows. Wet meadows contain sedges, rushes, tufted hairgrass, and occasional willow trees and cinquefoil; dry meadows are dominated by various lush bunchgrasses, blue grama, June-grass, wheatgrass, and bluegrass.

The highest-elevation forests consist of lodgepole pine, limber and bristlecone pine, and Engelmann spruce–subalpine fir. Dense lodgepole pine forests are common at the montane–lower subalpine elevation (9,000–10,000 feet). Scattered stands of limber pine and bristlecone pine are kept short by the harsh conditions and strong winds at elevations from about 8,000 feet to timberline. Engelmann spruce–subalpine fir trees dominate the forests from about 10,000 feet to 11,000 feet and are well watered from snowmelt.

Only the hardiest natural communities survive in the alpine tundra. This vegetation type occurs at the highest altitudes (11,200 feet to the mountaintops) and is marked by high winds and great changes in temperature. Tundra plants include shrubs, forbs, sedges, grasses, lichens, mosses, and cushion plants. Typical species include arctic willow, plainleaf willow, alpine bluegrass, tufted hairgrass, Pyrennian sedge, dwarf clover, and the rare arctic gentian and koenigia.

Special Habitat Sites

Aiken Canyon Preserve

El Paso County

Located about 15 miles south of Colorado Springs, this 1,621-acre sanctuary is managed by The Nature Conservancy. It preserves a highly diverse section of the foothills along the Rocky Mountains. The woodland and foothill shrubland communities are important habitats for a number of large mammals, including black bears and mountain lions. It is an important nesting habitat for more than 100 species of birds. Globally rare plant communities can also be found here, such as pinyon pine–one-seeded juniper and Gambel oak–mountain mahogany woodlands.

Beaver Creek Wilderness Area

Fremont County

Located about 5 miles northeast of Canon City, this 26,000-acre tract encloses parts of both

East and West Beaver Creeks. Pinyon-juniper forest and subalpine conifer forest are the dominant communities. Bighorn sheep and mountain lions range the rocky heights. The narrow streams are lined with narrowleaf cottonwoods, willows, alder, and wildroses, which attract a variety of mammals and birds.

Black Canyon of the Gunnison National Monument
Montrose County
Located along the Gunnison River east of Montrose, this property straddles a 12-mile stretch of the steep-walled Black Canyon. The sheer cliff faces are about 2,300 feet tall and provide important nesting habitats for white throated swifts, canyon wrens, swallows, ravens, and peregrine falcons. The cool microclimate has created unique natural communities of stunted Gambel oak and mountain mahogany; stands of Douglas fir and aspen also occur. Sagebrush and patches of grassland occur away from the canyon.

Bonny Prairie Natural Area
Yuma County
Located along the Bonny Reservoir in Bonny Reservoir State Recreational Area, this 50-acre property shelters a remnant of undisturbed loess tallgrass prairie, one of the few remaining in the state. Dominant grasses include little bluestem and sideoats grama. Among the less common species is the extremely rare prairie moonwort. The prairie is also an important habitat for a variety of insects, mammals, birds, and reptiles, some of them rare or threatened.

Campo Research Natural Area
Baca County
Pristine short-grass prairie occurs on this 35-acre tract about 4 miles south of Campo, near the Oklahoma border. Part of the Comanche National Grassland, this virgin grassland of blue grama, sideoats grama, and buffalo grass has never been plowed or grazed. It is an important habitat for grassland birds, insects, mammals, and reptiles, including the rare plain pockets gopher and lesser prairie chicken.

Carpenter Ranch
Routt County
This Nature Conservancy property 4 miles east of Hayden is well known for its high-quality narrowleaf cottonwood riparian habitat along the Yampa River. This habitat attracts more than 80 bird species, including nesting bald eagles and great sandhill cranes. The original buildings of the ranch still stand and are an important historic landmark.

Castlewood Canyon State Park
Douglas County
A rugged canyon along Cherry Creek is the main feature of this 873-acre park about 5 miles south of Franktown. Tall cottonwoods line the stream. The park also contains excellent remnants of northern Black Forest plant communities, including ponderosa pine and Douglas fir forests, oak shrubland, and open grasslands, home to many large mammals. Three rare endemic fish can be found in Cherry Creek. Raptors nest and breed on cliff faces. Unusual features are the rare tassel-eared squirrel, several rare plants, a rare dryland heron rookery, and turkey vulture nesting sites.

Comanche National Grassland
Baca and Las Animas Counties
Located southwest of Springfield, this 435,000-acre prairie is highly diverse. More than 235 species of birds have been spotted here, some of them rare or endangered. Native prairie grasses, including some sections of unplowed prairie, dominate the landscape. Rocky riparian corridors along canyon streams harbor different plant and animal species and are important raptor habitats. The grassland also has one of the largest breeding populations of the endangered lesser prairie chicken in Colorado (in the dense sandsage community). Tanagers, cliff swallows, golden and bald eagles, prairie falcons, bobcats, the rare barred tiger salamander, and the Swift fox are other important species.

Cross Mountain Canyon Natural Area
Moffat County
This 2,080-acre natural area protects a section of the deep Cross Mountain Canyon along the Yampa River. This riparian corridor protects several rare or unusual plant species, and several endangered fish species can be found in the river. The steep canyon walls attract a variety of raptors, including golden eagles and nesting peregrine falcons.

Escalante Canyon Natural Area
Montrose and Delta Counties
Escalante Creek flows through this 1,895-acre section of a canyon 1,300 feet deep. Natural

communities include excellent undisturbed examples of grassland, saltbush barren, pinyon-juniper woodland, saltgrass meadow, and rock face plant assemblages. Numerous rare plants can be found anchored to the canyon's rock walls. Raptors are attracted to the steep rock faces, and large mammals, including mountain lions, hunt in the canyon.

Escalante Creek Research Natural Area
Montrose and Mesa Counties
This 45-acre tract in the Uncompahgre National Forest is about 25 miles southwest of Delta. It was established to protect a rare undisturbed stand of mature to old-growth blue spruce along a 2-mile stretch of the Dry Fork of Escalante Creek. Other trees in this riparian corridor are Engelmann spruce, Douglas fir, and ponderosa pine. It is an important research site for studying the succession of blue spruce forests in the montane riparian zone.

Gothic Research Natural Area
Gunnison County
One of the best Engelmann spruce–subalpine fir forests in Colorado is protected on this 1,490-acre RNA about 10 miles northwest of Crested Butte. Located at elevations of about 10,000 to 12,000 feet on Mount Baldy, natural communities include pristine virgin Engelmann spruce–subalpine fir forest, wet meadow, and rock face assemblages. It is an important habitat for large mammals, such as black bears, mule deer, elks, and mountain lions.

Great Sand Dunes National Monument
Alamosa and Saguache Counties
Located at the edge of the San Luis Valley and the Sangre de Cristo Mountains about 40 miles northeast of Alamosa, the Great Sand Dunes are the tallest in North America, approaching heights of 800 feet. The dunes cover nearly 40 square miles. Although the environmental conditions are hot, dry, and windy, the dunes are partially anchored by patches of blowout grass, Indian ricegrass, scurfpea, and the prairie sunflower. The kangaroo rat lives in the dunes, as does the rare giant sand treader camel cricket and four other insect species that occur no other place in the world, such as the Great Sand Dunes tiger beetle. The dunes are flanked by grassland and pinyon-juniper woodland, which attracts numerous birds.

High Creek Fen
Park County
The Nature Conservancy owns this 1,147-acre property about 9 miles south of Fairplay. This ecosystem, the most biologically diverse fen in the southern Rocky Mountains, is centered around calcium-rich springs. Several endangered or rare plant species thrive in the fen, such as Porter feathergrass, hoary willow, little bullrush, green sedge, wood lily, and bladderwort. The wetland is also an important nesting area or stopover for a variety of birds. Colorado blue spruce grows on the drier uplands around the wetland, as does the rare mountain plover.

Hurricane Canyon Research Natural Area
El Paso County
Located about 3 miles northwest of Manitou Springs, this 528-acre RNA is part of Pike and San Isabel National Forests. Natural communities include undisturbed montane forest on high rocky slopes and riparian corridors. The site was designated an RNA because of the pristine old-growth forests of Douglas fir, ponderosa pine, and oakbrush; Engelmann spruce and Colorado blue spruce grow in the riparian corridors. The area escaped logging because of the extremely rugged terrain. It is an important habitat for big mammals, such as mountain lions and black bears.

Indian Peaks Wilderness Area
Boulder and Grand Counties
Located near Ward, this 73,400-acre property consists of subalpine-alpine conifer forest, meadow, lake, brook, and tundra natural communities. The unusual Parry's primrose can be found here. The alkaline Long Lake fen hosts a number of calcium-loving plants, such as rare sedges, rushes, sphagnum moss, and cottongrass.

Indian Springs Natural Area
Saguache County
This 640-acre tract protects Indian Spring, the largest natural cold spring in the San Luis Valley. Surrounded by warm-climate dune fields, its restricted drainage pattern has isolated its unique aquatic species. In addition to several rare plants found in the wetland margins, the natural area is a habitat for sever rare insects, including wasps,

butterflies, and an extremely rare species of tiger beetle.

La Garita Wilderness Area
Mineral County
Located north of Creede, this area comprises high mountain streams and rocky glacial slopes. Spruce-fir forests and meadows carry snowpack for most of the year. The meadows are lush with abundant wildflowers during spring. The exposed rocky cliffs are ideal nesting sites for golden eagles, white-throated swifts, and rock doves. Big horn sheep, mountain lions, mule deer, and elks are common. Some old-growth stands of conifers remain. Several rare or unusual species inhabit the area, including butterflies.

Lathrop State Park
Huerfano County
Three miles west of Walsenburg, this impressive state park protects high-quality examples of several natural communities, including mixed-grass prairie, wetland, Gambel oak woodland, rock face, and meadow around two reservoirs. Juniper and pinyon-juniper woodlands grow at the highest elevations, with some of the pinyon trees being nearly 500 years old. A diverse number of plant and animal species can be found here, including migrating shorebirds that visit during the spring and fall. The park is also an important mountain lion habitat.

McElmo Research Natural Area
Montezuma County
This site protects 443 acres along the canyons of the McElmo drainage. Natural communities include riparian corridor, cliff face, shrubland, and woodland plant assemblages. The area is well known for its high concentration of rare or endangered animals, including the endemic desert spiny lizard, leopard lizards, spotted night-snake, lesser earless lizard, and several toads.

Mishak Lakes Preserve
Saguache County
This 2,400-acre preserve encloses an entire shallow-water wetland system in the San Luis Valley. Located near Saguache, it consists of a series of numerous seasonal ponds that attract thousands of waterfowl and shorebirds. More rare species include Wilson's phalarope, the greater sandhill crane, the white-faced ibis, and the American avocet. It is one of the most critical habitats and stopover points for wetland birds in the state. Other rarities include the spiderflower, least chipmunk, Ord's kangaroo rat, and the endemic San Luis Valley sandhills skipper butterfly.

Mount Goliath Research Natural Area
Clear Creek County
A population of ancient bristlecone pine trees occurs on this 400-acre RNA 8 miles southwest of Idaho Springs. Located on the flank of Mount Evans in Arapaho and Roosevelt National Forests, this tract preserves the oldest trees in Colorado—some exceeding 1,500 years old. Other species in the area are Engelmann spruce, whiproot clover, and alpine kobresia. Elks and bighorn sheep also inhabit the site.

Mueller State Park and Wildlife Area
Teller County
This 12,103-acre site was donated to the state of Colorado by The Nature Conservancy. Located about 30 miles west of Colorado Springs, it comprises a variety of natural communities, including meadow, forested upland (fir-aspen-spruce), and rock face. It offers critical habitat for disappearing species, such as mountain lions, bighorn sheep, elk, songbirds, and a variety of raptors. Rare or unusual plant species also occur.

Narraguinnep Research Natural Area
Dolores County
Located about 17 miles northwest of Dolores, this 1,952-acre research natural area is part of San Juan National Forest. It was established to protect a section of Narraguinnep Canyon, an excellent undisturbed example of the canyon mesa country of southwestern Colorado. Natural communities include ponderosa pine forest, mixed pinyon pine–Utah juniper woodland, scrub oak forest, and rock-cliff plant assemblages. Aspen and Gambel oak grow in the canyon bottom, with lesser mountain snowberry and mountain mahogany. The canyon is an important habitat for black bears, elks, chorus frogs, and a variety of skinks, some of them rare or unusual. Several varieties of raptor nest and hunt in the area.

North Star Ranch
Pitkin County
Located on the Roaring Fork River, this former Nature Conservancy property protects 236

acres of critical high-elevation wetland. Natural communities include wet meadow and willow riparian shrublands. The wetland is an important ecosystem in itself and provides habitat for a variety of birds and animals, including elk and deer.

Paradise Park Research Natural Area
Grand County
A pristine portion of the upper Paradise Creek basin is preserved in this 5,760-acre RNA. Subalpine and alpine natural communities include virgin forests of Engelmann spruce-subalpine fir and windswept alpine meadows. The area is dotted with remarkably clear glacial lakes, and pure populations of the state-endangered Colorado River cutthroat trout can be found in Paradise Creek. It is an important habitat for a number of birds and large mammals, including the mountain lion.

Pawnee National Grassland
Weld County
Scattered across much of Weld County, these parcels totaling 193,000-acres are covered by short-grass prairie (blue grama and buffalo grass). Located on the ecotone between the High Plains and the Colorado Plateau, more than 400 plant species, and 354 wildlife species, have been identified here. Rocky buttes are excellent habitats for eagles, kestrels, prairie falcons, and turkey vultures. Rare plants, such as evening primrose, globemallow, pink vervain, and buckwheat, can be found in sheltered ravines. Sensitive wildlife species include mountain plovers, shrikes, Wilson's phalaropes, burrowing owls, swift fox, and the long-billed curlew. Seasonal ponds in natural depressions attract waterbirds, especially during migrations.

Phantom Canyon Preserve
Larimer County
Located on the North Fork of the Cache la Poudre River about 30 miles northwest of Fort Collins, this 1,700-acre tract is owned and managed by The Nature Conservancy. Several varieties of forest and wetland communities occur on the property. It is one of the last roadless canyons along the Rocky Mountain front. A globally rare member of the parsley family grows here. It is also an important habitat for other rare or endangered species, such as mountain lions and bald eagles. The canyon walls are important nesting sites for other raptors.

Rocky Mountain National Park
Boulder, Larimer, and Grand Counties
Part of the Front Range near Estes Park, this 266,000-acre park includes tundra, coniferous forest, subalpine meadows, streams, riparian habitats, and lakes. Montane forest between 5,500 and 9,000 feet consists of ponderosa pine and Douglas fir forest, with some aspen and lodgepole pine. Subalpine forests, between 9,000 and 11,500 feet, are dominated by Engelmann spruce–subalpine fir and some limber pine. Riparian habitats are lined by blue spruce, aspen, and narrowleaf cottonwoods. The tundra is windy and cold with stunted trees, low plants (including rare alpine phlox, pink moss campion, and yellow snow buttercups), white-tailed ptarmigan, and bighorn sheep. Nearly 1,000 plant species have been identified in the park, which is critical habitat for large mammals, such as bears, mountain lions, moose, elk, and sheep. The rare greenback cutthroat trout inhabits the local streams.

Roxborough State Park
Douglas County
Located about 10 miles west of Sedalia, this preserve encloses jagged sandstone outcrop, riparian corridor, rock face, dry grassland, wet meadow, cottonwood-box-elder riparian corridor, prairie, oak scrubland, and conifer-aspen forest communities. Straddling an ecotone between plains and mountains, its biodiversity reflects a mix of prairie and mountain species. Stunted forests of oak and mountain mahogany can be found on the higher windswept elevations. The rugged terrain is ideal habitat for golden eagles, black bears, and mountain lions. Meadows and prairies are resplendent with Indian paintbrush, golden banner, sunflower, and other prairie flowers during spring and summer.

Ryan Ranch
Hinsdale and Gunnison Counties
Ryan Ranch is located about 7 miles north of Lake City along the Gunnison River. It was purchased by The Nature Conservancy and the Bureau of Land Management. It is a critical riparian habitat for the rare narrowleaf cottonwood, Colorado blue spruce, and thinleaf alder trees.

Saddle Mountain Research Natural Area
Park County
Rare montane grasslands are preserved on this 485-acre RNA about 20 miles west of Cripple

Creek. In this site, located in Pike and San Isabel National Forests, Parry oatgrass grasslands are mixed with stands of Engelmann spruce, bristlecone pine, and aspen. Largely undisturbed by grazing and agriculture, the grasslands consist of Parry oatgrass-mountain muhly and Parry oatgrass–needleleaf sedge–American vetch–Arizona fescue associations. The grasslands are an important habitat for a variety of birds, insects, mammals, and reptiles.

San Miguel River Canyon Preserve
San Miguel County
This 279-acre preserve is located 4 miles from Placerville along a 2-mile stretch of the San Miguel River. It is critical riparian habitat for the globally rare narrowleaf cottonwood–Colorado blue spruce–thinleaf alder plant community. The understory is relatively free of exotic species, which makes the site that much more exceptional. The property is owned and managed by The Nature Conservancy.

South Fork San Miguel River Preserve
Dolores County
Located 10 miles from Telluride, this 67-acre property straddles a section of the South Fork of the San Miguel River. It protects the globally endangered narrowleaf cottonwood–Colorado blue spruce–black twinberry natural community along the river. The site is owned by The Nature Conservancy.

Specimen Mountain Research Natural Area
Larimer and Boulder Counties
Located on 9,056 acres in Rocky Mountain National Park, this site preserves a portion of the Continental Divide, including Specimen and Shipler Mountains. Special natural communities include virgin stands of Engelmann spruce–subalpine fir forest and undisturbed, highly diverse alpine meadows. It is a critical habitat for big mammals, such as black bears, Rocky Mountain bighorn sheep, elk, and mountain lions.

Tamarack Natural Area
Logan County
Excellent remnants of plains cottonwood riparian forest and sandhills prairie are protected on this 470-acre property along the South Platte River. Biodiversity is high. The prairie grasses are an important habitat for a variety of birds, insects, and mammals, including the rare greater prairie chicken. The river and its riparian corridor are an important migratory stopover for numerous waterfowl during spring and fall, including several threatened species.

Unaweep Seep
Mesa County
Located 8 miles east of Gateway, this BLM ACEC covers 55 acres of hillside marsh. The spring system is unusual in that it is not affected by flooding. Wet sedge marshes and seeps make up the marsh. A rare species of butterfly, the Nokomis fritillary, feeds on a species of violet that grows at the site. Unusual orchids can also be found here. As a wetland, it is an important water source for surrounding fauna.

West Creek Natural Area
Larimer County
Located in the Big Thompson River basin, this 9,043-acre state natural area protects part of the West Creek and Fox Creek drainage basins. The riparian corridor is edged by montane virgin forests of lodgepole and limber pine. Pure populations of the greenback cutthroat trout are plentiful in West Creek. The area is an important habitat for big mammals, such as elk, bighorn sheep, and mountain lions, as well as raptors such as the rare peregrine falcon.

Williams Creek Research Natural Area
Hinsdale County
Located about 20 air miles northwest of Pagosa Springs in San Juan National Forest, this 552-acre RNA encloses a large stand of virgin white fir. Forest communities include white fir–Douglas fir–forest fleabane, white fir–Douglas fir–Rocky Mountain maple, and Engelmann spruce–grouse whortleberry; other habitats include a small willow wetland and rock-cliff plant assemblages. The RNA is an important habitat for elk, mule deer, mountain lions, ground squirrels, meadow voles, and black bears. Raptors are attracted to the cliff faces. The towering pine trees show what the presettlement forests looked like prior to Anglo-European development.

Yampa River Preserve
Routt County
This 265-acre Nature Conservancy property is located about 19 miles west of Steamboat Springs. It is well known as one of the few sites in the world that contains the rare box

elder–narrowleaf cottonwood–red-osier dogwood natural community. It is also an important nesting habitat and migratory stopover for various birds, including songbirds, bald eagles, and the greater sandhill crane.

Yanks Gulch/Upper Greasewood Creek
Rio Blanco County
Located on 2,687 acres along Upper Greasewood Creek, this RNA protects excellent examples of mesic pinyon-juniper woodland, mesic sagebrush shrubland, and Western Slope grassland natural communities. Three rare plant species endemic to a unique geologic formation (the Green River Shale) grow here. Several rare or unusual reptiles can also be found in the gulch.

CONNECTICUT

Connecticut can be divided into the following physiographic regions: Northwest Highlands, Northern Uplands, Northern Hills, Southern Hills, and Coastal Hardwoods.

The Northwest Highlands occupy the northwestern corner of the state. A series of hills and plateaus, the area is covered by northern hardwood–hemlock–white pine forest. Dominant hardwood trees include sugar maple and beech, with lesser black cherry, hickory, and oak. Spruce bogs can be found in some of the ravines.

The Northwest Uplands, a northeast-southwest–trending zone also in the northwestern corner, is dominated by a transitional hardwood forest. Elevations are between 700 and 1,500 feet. Forest types include mesic northern red oak–basswood–white ash–black birch, with some beech and sugar maple, and red oak–white oak–black oak–hickory. A few acidic spruce bogs are scattered across the region.

Most of the northern half of the state forms the Northern Hills ecoregion (400–1,000 feet elevation). This series of hills and lowlands is covered by oak–hickory–hemlock–white pine forest. Other communities are grassland, red maple swamp, Atlantic white cedar swamp, black spruce bogs, and bottomland forest. Drier ridges support scarlet oak and chestnut oak woodlands.

A coastal upland comprises the Southern Hills ecoregion, which occupies most of the southern half of the state. Forest types include oak–hickory–tulip–yellow poplar–dogwood. Other communities are sand plains, grasslands, Atlantic white cedar swamp, and pitch pine barrens.

The Coastal Hardwoods zone is a narrow strip along the Atlantic coastline roughly 5 miles wide. Habitats include tidal flat, beach, tidal marsh, estuary, and rocky headland. The major forest type is oak–hickory–tulip poplar–black cherry–sassafras–hemlock. The understory is quite dense and choked with vines and shrubs. Several stands of old-growth forest can be found here.

Special Habitat Sites

Black Spruce Bog
Litchfield County
Located about 5 miles southeast of Cornwall in Mohawk Mountain State Park, this 36,000-acre tract encloses a black spruce bog. Rare in Connecticut, this kind of wetland supports black spruce, mountain laurel, hemlock, leatherleaf, sheep laurel, and a sphagnum floating mat. Other rare bog plants occur here.

Canaan Mountain Natural Area Preserve
Litchfield County
Located in the townships of Canaan and North Canaan east of Falls Village, this relatively undisturbed section of forest is fairly representative of presettlement times. It is one of the most rugged undeveloped forest tracts in the state. Natural communities include rock face, acidic basin swamp, dry acidic forest, and talus forest woodland. Many of these hardwood and pine trees are centuries old and quite large. A host of threatened, endangered, or special-concern species occur here, including a variety of sedges and ferns, stiff gentian, Canada violet, white mandarin, and the timber rattlesnake.

Cathedral Pines Preserve
Litchfield County
This Nature Conservancy property about 1 mile outside of Cornwall covers 42 acres. Although struck by a recent tornado, several stands of old-growth white pine and hemlock

that are more than 300 years old survived. The trees are more than 100 feet tall and nearly 4 feet at breast height.

Chapman Pond Preserve
Middlesex County
Located near East Haddam, this 60-acre freshwater pond is hydraulically connected to the Connecticut River, and the pond level fluctuates with the tides. Surrounded by mature hemlock forest and mixed hardwood (maple-cottonwood-ash) bottomland forest, it is an important wintering habitat for bald eagles. Twenty-five fish species thrive here, as do the rare tidewater arrowhead and Torrey's bulrush. The property is managed by The Nature Conservancy.

Devil's Den Preserve
Fairfield County
Located 3 miles southwest of Weston, this 1,500-acre Nature Conservancy property has ecological and archeological significance. Its granite ridges are covered by mature oak-hickory with swamp azalea and mountain laurel. It is an area of high biodiversity; rare or uncommon species here include the Louisiana waterthrush, pileated woodpecker, wood duck, and several amphibians.

Griswold Point
New London County
Located near Old Lyme, this area is a mile-long sandspit in the mouth of the Connecticut River with flanking tidal marshes. Owned by The Nature Conservancy, it is a critical nesting habitat for the rare piping plover and least tern during spring and summer. More than 200 bird species use the site as a migratory stopover to rest and feed.

Guilford Salt Meadows Sanctuary
New Haven County
Located near Guilford, this 200-acre Audubon Society property encloses one of the best remaining sections of salt marsh in the state. This productive marsh is an important feeding ground for tens of thousands of waterbirds, especially during spring and fall migrations.

Hammonasset Natural Area Preserve
New Haven County
Located within Hammonasset Beach State Park about 4 miles east of Madison, this 400-acre preserve along the Atlantic Ocean comprises high-quality salt marsh, beach, dune and swale, and upland forest. Several rare communities occur here. Rare birds nest along the beach and in the marsh, including the endangered piping plover.

Hopeville Pond Natural Area
New London
Located on 25 acres about 10 miles east of Hopeville, this property encloses a rare pitch pine–oak sand barren community. Several rare or unusual plants occur within this habitat.

Kitchel Wilderness Natural Area Preserve
Litchfield County
This proposed natural area consists of 580 acres within the Algonquin State Forest east of Colebrook. High-quality natural communities include mature acidic talus forest, northern hardwood forest, and ash-hickory forest. Dominant trees are red oak, sugar maple, beech, birch, and white pine; several stands of old-growth trees remain. Rare or special species here include squirrel corn, purple oats, and Virginia waterleaf.

Mount Higsby
Middlesex County
Owned and managed by The Nature Conservancy, this property 5 miles northeast of Meriden encloses rock ridge, cliff face, lowland forest, meadow, and mature to old-growth hemlock–mixed hardwood forest communities. Another forest subtype is stunted oak-hickory with grassy openings. The woodlands are an important habitat for the box turtle, a species that is rapidly disappearing. The rare bearberry and yellow corydalis plants also grow here.

Mount Misery Brook
New London County
This proposed 26-acre state natural area is located near Voluntown, in Pachaug State Forest. The tract encloses mature Atlantic white cedar swamp with a well-developed understory of great laurel. This kind of wetland is rapidly vanishing in the eastern United States. The swamp is relatively undisturbed, an important habitat for nesting birds, amphibians, reptiles, and fish. Several rare or endangered species occur here, including the green adder's-mouth and banded bog skimmer.

Pachaug Great Meadow
New London County
This proposed state natural area preserve covers about 600 acres of a wetland complex within Pachaug State Forest, northeast of Voluntown. Natural communities include the Pachaug River and its riparian corridor, wet sedge fen, Atlantic white cedar swamp, and wetland red maple–ash–hemlock forest. Some stands of timber are mature to old-growth. Rare or threatened species that depend on this habitat are the creeping snowberry, inkberry, beaked rush, Hessel's hairstreak, and banded bog skimmer.

Poquetanuck Cove Preserve
New London County
Located about a mile from Poquetanuck, this 200-acre property encloses an extensive brackish tidal marsh. Noted as a bird sanctuary with more than 100 recorded species, it is especially crowded with birds during the spring and fall migrations. The property is owned and managed by The Nature Conservancy.

Robbins Swamp Natural Area Preserve
Litchfield County
This proposed 785-acre preserve covers part of Robbins Swamp, the largest freshwater wetland complex in the state. Located along the Hollenbeck River, wetland communities include northern white cedar swamp, red maple–black ash seepage swamp, calcareous fen, and open marsh. Many rare or special species grow here, including foxtail sedge, chestnut sedge, white adder's-mouth, sweet coltsfoot, bur oak, northern white cedar, autumn willow, and the threatened blue-spotted salamander.

Sunny Valley Preserve
Litchfield County
Located west of New Milford, this Nature Conservancy property encloses high-quality forest, wetland, and meadow communities. The grassland is an important habitat for dwindling grassland bird species, mammals, and reptiles. A working dairy farm, the property is a good example of how development, industry, and preserved lands can coexist.

Turtle Creek
Middlesex County
Located near Essex, this property preserves a high-quality remnant of hemlock–mountain laurel–beech forest that covers glacial hills along the Connecticut River. The preserve is owned and managed by The Nature Conservancy.

Westwoods
Middlesex County
About 3 miles southeast of Guilford, this 1,000-acre property covers forested rocky hills and ravines. Dominant natural communities are stream, riparian corridor, bottomland forest swamp, and upland old-growth northern hardwood forest. Dense stands of mountain laurel grow on some of the ridges. Several of the hemlock trees form a towering overstory and have diameters up to 12 feet; rare rock communities cling to the granite outcrops. Ospreys, ducks, egrets, and peregrine falcons nest on the site.

Wharton Brook Natural Area
New Haven County
Located near Wallingford, this 21-acre property protects a pitch pine and oak sand barren natural community. This kind of community is one of the rarest in the eastern United States and harbors a host of rare or special plant, animal, and insect species.

White Memorial Foundation
Litchfield County
This 4,000-acre preserve of forests, wetlands, and meadows is about 2 miles southwest of Litchfield. Within the forest is a 30-acre stand of 200-year-old hemlock and white pine trees—the largest old-growth stand in state. The largest trees are nearly 4 feet at breast height.

Wier Preserve
Fairfield County
Located about 4 miles northeast of Wilton Center, this mature to old-growth oak-hickory forest includes dense stands of mountain laurel. Rock face communities support several rare lichens, mosses, and other clinging plants. The property is owned and managed by The Nature Conservancy.

DELAWARE

Delaware is covered mostly by forests and coastal-maritime communities. The terrain is flat and covered by Coastal Plain sediments. The southern third of the state is covered by remnants of southern yellow pine–loblolly pine forest that is occasionally mixed with oak-pine and oak-gum-cypress wetland forest. Marshy areas (red maple–ash–willow–cottonwood) are often interspersed with woodlands. The unique Cypress Swamp and several stands of Atlantic white cedar swamp occur in this part of the state.

Oak-hickory forest dominates the gentle hills and valleys in the western third of Delaware, along the border with Maryland. Lesser oak-pine and oak-gum-cypress swamp occur here as well. Oak-hickory is also abundant in the northern tip of the state, west of Wilmington.

Along its eastern edge, Delaware borders the Atlantic Ocean. Maritime communities include barrier island, tidal mudflat, beach strand, dune and swale, maritime forest, salt marsh, brackish marsh, freshwater marsh, brackish pond, Delmarva (Carolina) bay, and cordgrass-saltgrass salt meadow. Many of these communities are important nurseries for fish and shellfish and are feeding grounds for waterbirds. The coastal influences extend inland via creeks and streams, which are affected by ocean tides.

Special Habitat Sites

Blackbird State Forest
New Castle County
This property 5 miles northwest of Smyrna protects four unusual and rare Delmarva (Carolina) bays—shallow ponds in the Coastal Plain sediments. Most of these water bodies along the eastern coast have been drained and destroyed. Surrounded by thick forest, the bays support rare glade and swamp plants, such as the buttonbush. Several types of birds nest here, including the rare wood duck.

Bombay Hook National Wildlife Refuge
Kent County
Located about 15 miles northeast of Dover along Delaware Bay, this 16,000-acre wetland complex consists of beaches salt marshes, brackish marshes, and freshwater impoundments. More than 300 bird species have been identified here during spring and fall migrations, including Canada geese, snow geese, ducks, teals, plovers, and sandpipers. Bald eagles nest deep in the salt marsh. The beach is an important nesting habitat for northern diamondback terrapins.

Brandywine Creek State Park
New Castle County
This 850-acre park 3 miles north of Wilmington includes rolling hills covered with a unique old-growth forest that includes 200-year-old tulip trees taller than 100 feet. Some rare plants occur in the understory. Other communities are meadow, bottomland forest, and freshwater marsh, which harbors the rare Muhlenberg's bog turtle. Several varieties of nesting warblers, bald eagles, and peregrine falcons can also be seen.

Edward H. McCabe Preserve
Sussex County
Located near Milton, this Nature Conservancy preserve protects part of the riparian corridor of the Broadkill River, a tidal river. Within the dense mixed hardwood–conifer forest is the globally rare seaside aider plant. The corridor is an important habitat for birds and mammals.

Great Cypress Swamp
Sussex County
Only a remnant remains of this once-vast wetland complex about 20 miles south of Georgetown. The swamp was heavily used in the past for its cypress logs and its cedar trees for shingles. Though this site originally comprised more than 50,000 acres, only some 10,000 remain. Within its interior are some old-growth bald cypress trees that escaped the lumber saw. Dominant trees today are pine gum, swamp cottonwood, and red maple. Interesting or rare species include the rare bog frog, Swainson's warbler, barred owl, Boott's wood fern, redbay evergreen, bald eagle, pileated woodpecker, red-bellied water snake, swamp magnolia, sweetleaf, and exotic white fringed orchid. The wetland is an important wintering ground for birds traveling along the Atlantic Flyway. Cypress, loblolly, and white cedar are being replanted, and existing agricultural fields are being restored to wetlands.

Nanticoke Wildlife Area
Sussex County
Located on Broad Creek where it empties into the Nanticoke River 2 miles west of Bethel, this property encloses a freshwater tidal marsh that is connected hydraulically with the ocean tides. This improves the nutrient flow and makes it a very productive ecosystem. Dominant aquatic plants are spatterdock, pickerelweed, and large stands of wild rice up to 10 feet tall. The marsh is an important food source for waterbirds and a popular migratory stopover. The forest corridor includes loblolly, Virginia, and pond pines, Atlantic white cedar, and the endangered box huckleberry.

Norman G. Wilder Wildlife Area
Kent County
About 3 miles west of Viola, this tract protects a high-quality example of mature coastal-plain forest. Impressive specimens of white oak are taller than 100 feet. Chestnut oak and willow oak also grow on the sandy substrate. Delaware's best population of the rare club moss, a glacial relict, grows here.

Port Mahon Preserve
Kent County
Located just outside Port Mahon, this Nature Conservancy property includes marsh, wetland, and Delaware Bayshore beach communities. Its position under the Atlantic Flyway makes it is an important migratory stopover for thousands of birds during spring and fall, including waterbirds, bald eagles, peregrine falcons, and other raptors. The beach is a critical habitat for the springtime mating of the horseshoe crab.

Trussum Pond
Sussex County
Located near Laurel, this wetland-forest complex includes old-growth specimens of bald cypress and Atlantic white cedar. Some of the tallest trees are nearly 500 years old. The wetland is also an important habitat for a variety of birds, amphibians, and reptiles, some of them rare.

FLORIDA

More than 80 natural communities, an astounding number, have been identified in Florida. Community types take into account differences in vegetation structure and composition, topography, land form, substrate, soil moisture and chemistry, climate, and fire. All of these communities are threatened to some extent by fire suppression, lumbering, agriculture, overgrazing, plowing, and residential and commercial development.

Kautz (1998) divides Florida vegetation into two major types: upland and wetland. Upland communities include coastal strand, pine flatwood, dry prairie, scrub, sandhill, temperate hardwood forest, and rockland. Wetland communities are composed of extensive coastal salt marsh, freshwater marsh, cypress swamp, mixed hardwood swamp, bay swamp, mangrove swamp, and shrub swamp.

Most of the Atlantic Ocean and Gulf of Mexico coastlines are edged by coastal strand vegetation, which can be found on beaches, foredunes, dunes, and rocky coasts. Affected by wind and salt spray, coastal plants consist of low grasses, herbs, shrubs, and small trees. They include salt grass, beach elder, beach pennywort, cordgrass, sea oats, saw palmetto, saltbush, prickly pear, and sea grape. The coastal strand is the most rapidly disappearing community in Florida and exists only in small isolated stretches along the Florida coastline.

Dry prairies occur mostly in south-central Florida, west of Lake Okeechobee. The soil is generally dry. Prairie plant diversity tends to be high, including a variety of grasses, herbs, shrubs, sedges, and occasional clumps of trees. Prairies are a rapidly disappearing community upon which a number of grassland bird, mammal, and insect species depend.

Much of the pine flatwood community, very abundant in presettlement times, has been destroyed by development. It can be found today on flat sandy areas across Florida. The pine overstory (longleaf, slash, and pond pines) is underlain by shrubs, grasses, and forbs. Wetlands are typically interspersed with piney woods in these low-lying, poorly drained areas.

One of the most endangered communities is scrub, which occurs on former beach and lake ridges and is noted for its large number of narrowly endemic plants and animals. Not pleasing to the eye, scrub communities are often lost to development or damaged by off-road vehicle traffic and foot traffic, which leads to erosion. Typical scrub plants are myrtle oak, silk bay, scrub holly, Florida rosemary, and sand-live oak.

Sandhill communities can be found in the Panhandle and north-central portion of the state. Similar to savanna, this vegetation is dominated by widely spaced longleaf pine, various oaks, and grasslands consisting of wiregrass, milk peak, partridge pea, and other forbs.

Temperate hardwood forests can be found across the state, most commonly in the Panhandle and South Florida. They tend to be more mature than the flatwoods and are found on mesic slopes and hills. Depending on soil moisture content and slope aspect, forests can comprise white oak, mockernut hickory, basswood, white ash, spruce pine, American beech, and southern magnolia. Depending on how well drained the substrate is, temperate hardwoods can be further divided into xeric, mesic, and hydric categories.

South Florida rocklands occur on outcropping limestone ridges in the extreme southern reaches. They are covered by either tropical hardwood forest or an open forest of South Florida slash pine. Tropical hardwoods can be found in the Florida Keys; the very rare pine rocklands can be found in the Lower Florida Keys.

Coastal salt marshes as well as freshwater marshes are abundant in the state. Salt marshes vary from saline to brackish and are common in bays and estuaries along the coastline. Dominant plants include smooth cordgrass, glasswort, salt grass, marsh elder, and saltbush. Freshwater marshes are most common in south-central Florida; natural communities vary according to elevation, topography, and connection to stream systems. Diversity is high, with 200–300 plant species typically growing in the wetlands. A common type of Florida marsh is the sawgrass marsh.

Cypress swamps are one of the most visually impressive wetlands in the state and tend to develop in river floodplains and around lakes. The water is usually dark and acidic or tannic. Other important species are blackgum, sweetbay, buttonbush, and Virginia willow.

Occurring mostly in the northern part of the state, mixed hardwood swamp can be found as dense corridors along floodplains, blackwater rivers, and other topographic lows that channel runoff precipitates. The core swamp trees are blackgum, cypress, red maple, and water tupelo. Sand levees and slight uplands that are better drained support various oak, elm, and locust trees.

Bay, shrub, and mangrove swamps are other kinds of wetlands. Bay swamps are found most often in flatwoods and support broadleaf evergreen bay trees. Shrub swamps tend to be successional and have Carolina willow–buttonbush–red maple–black gum plant associations. Mangrove swamps, with distinctive tangled root systems, form in the intertidal zones of South Florida.

Special Habitat Sites

Anclote Keys State Park
Pasco and Pinellas Counties
This 287-acre barrier island complex includes beach strand, dune, and maritime hammock (forest) communities. The island is known as an important bird rookery. More than 200 birds have been identified, including laughing gulls, least terns, Wilson's plover, American oystercatchers, and black skimmers. This delicate coastal ecosystem is accessible only by boat.

Apalachicola Bluffs and Ravines Preserve
Liberty County
Located north of Bristol along the Apalachicola River, this 6,267-acre preserve protects a variety of natural features. Natural communities include upland pine forest, sandhill, river bluff, stream, spring, steephead ravine, and wetland. Several rare or threatened plants and animals can be found here. Two of the world's rarest conifer trees—northern relicts from the Ice Age—thrive in the deep, cool ravines. Bald eagles, Mississippi kites, vireos, warblers, alligators, gopher tortoise, Apalachicola dusky salamanders, and indigo snakes are some of the many rare or endangered animals that inhabit the woodlands. The preserve is owned and managed by The Nature Conservancy.

Apalachicola National Forest
Liberty, Gulf, Wakulla, Leon, Gadsden, and Franklin Counties
Located on about 500,000 acres, this national forest has one of the largest populations of red-cockaded woodpeckers in the United States. High-quality natural communities include magnolia bay, cypress swamp, longleaf pine forest, wiregrass prairie, and savanna. Some old-

growth longleaf pine trees remain. More than 40 rare or unusual plants can be found here, including toothache grass, blue-eyed grass, xyris, orchids, and violas. Wet areas in the savannas contain rare bog plants, such as pitcher plants, sundews, and butterworts. The endangered gopher tortoise and eastern indigo snake often live in grasslands in the savanna. The forest recesses are a critical habitat for the rare Florida black bear and flatwoods salamander.

Archie Carr National Wildlife Refuge
Brevard and Indian River Counties
This 28-acre strip of sand along the Atlantic coastline is a critical habitat for nesting loggerhead, green, and leatherback turtles, which are becoming rare. Natural communities include beach, dune, and maritime hammock (forest) with palmettos and prickly pear cactus. This refuge is one of the most important turtle habitats in the eastern United States.

Bear Lake Natural Area
Located along the Blackwater River, this wetland-bottomland forest complex includes some old-growth longleaf pine forest, cypress-tupelo swamp, and relict Atlantic white cedar swamp. These communities are important habitats for alligators, various waterbirds, red-cockaded woodpeckers, pink sundews, and the very rare red pitcher plant. Some of the cypress trees are old-growth and more than 500 years old.

Big Bend Coast Preserve
Taylor and Dixie Counties
Located along a stretch of Gulf of Mexico coast near Jena, this 67,000-acre wetland-slough-forest complex shelters a large number of natural communities, including ocean beach, tidal marsh, pond, river and stream, hardwood hammock, upland pine forest, swamp forest, and sawgrass slough. These diverse communities give rise to a rich biodiversity; hundreds of plant and animal species have been identified, many of them rare or unique. The snowy plover, peregrine falcon, eastern indigo snake, Homesassa shrew, salt marsh mink, bald eagle, West Indian manatee, green sea turtle, and Kemp's Ridley turtle are some of the rare or endangered species found here.

Big Cypress National Preserve
Collier and Dade Counties
This sprawling, 570,000-acre wetland-forest complex is located just northeast of Everglades City. Its natural communities include wet prairie, dry prairie, marsh, estuary mangrove swamp, slough, cypress swamp, island, and mixed hardwood hammock. Biodiversity is correspondingly high. A few giant virgin cypress trees that are nearly 1,000 years old remain in the cypress swamps. Rare or endangered species that live here include red-cockaded woodpeckers, Florida panthers, Cape Sable sparrows, bald eagles, herons, wood storks, mink, and alligators. The property has spiritual and ceremonial significance to the Miccosukee and Seminole tribes.

Biscayne National Park
Dade County
This 173,000-acre marine preserve just south of Miami protects a variety of rare communities, including coral reef, mature mangrove forest, open marine bay, and coral key. The remnant mangrove forest along the shoreline is an important buffer zone and habitat for nesting bird species. The bay is a critical breeding ground for aquatic organisms, such as fish and shellfish, which provide food for predators. The coral keys support rare plants, such as the rare gumbo-limbo tree, Jamaican dogwood, strangler fig, and many orchids. The rare zebra swallowtial and Schaus's swallowtail can also be found here. Numerous shipwrecks lie offshore near the coral reefs.

Blowing Rocks Preserve
Martin County
Located east of Jupiter, this 73-acre property protects a mile-long stretch of undeveloped Atlantic Ocean beach. Communities include limestone outcrop, sand dune, mangrove wetland, and tropical coastal hammock. Four threatened plant communities are found here. The preserve, owned and operated by The Nature Conservancy, is an important breeding habitat for West Indian manatees and three species of sea turtle. Thousands of birds nest here or visit during migration, including the endangered brown pelican, heron, and osprey.

Butterfield (Janet) Brooks Preserve
Hernando County
This 320-acre Nature Conservancy property is located a few miles northwest of Brooksville. Natural communities include old-growth longleaf pine sandhill, mixed hardwood hammock, pond, slough, swale wetland, and swamp forest. Several rare or endangered plant and ani-

mal species can be found here, including the indigo snake, gopher tortoise, and fox squirrel. The communities are important habitats for various bird species, including woodpeckers and waterbirds.

Caladesi Island State Park
Pinellas County
Located just outside Tampa, this 2,429-acre property includes a barrier island; its waters are an important feeding zone for the endangered manatee. The island is surprisingly undisturbed and has high-quality beach, dune, scrub, and maritime forest communities. Beach sands are anchored by saw palmetto and prickly pear cactus. Hammocks (forest pockets) contain extremely rare stands of virgin South Florida slash pine, an important habitat for red-cockaded woodpeckers. Ospreys and other raptors can also be found here. Endangered sea turtles nest on the beach, and millions of birds stop here to rest and feed during migrations.

Camp Helen
Bay County
This 183-acre property on the west side of Lake Powells covers a Pleistocene bluff. High-quality natural communities include beach dune, coastal grassland, scrub, mesic flatwood, and a maritime hammock (forest). Rare or uncommon species that depend on this habitat are the Godfrey's golden aster, gulf lupine, piping plover, and least tern.

Catfish Creek Preserve
Polk County
Located on 1,147 acres northeast of Lake Wales, this Nature Conservancy property protects some of Florida's rarest natural communities and species. These include seeps, seepage wetlands, pine flatwood, scrub forests, hydric hammocks, mesic hammocks, cypress swamps, lakes, and sand pine scrubs. Parts of this preserve are relatively undisturbed. Bald eagles, cutthroat grasses, scrub jays, Florida mice, gopher tortoises, and several globally rare endemic plants can be found among the rare or special species that thrive here.

Cedar Keys Scrub State Preserve
Levy County
Located on nearly 5,000 acres about 5 miles south of Gulf Hammock, this reserve protects vital sections of maritime scrub. Other natural communities include tidal marsh, hydric hammock, flatwood, scrub, and floodplain swamp. Rare or unusual species found here include the Florida mouse, indigo snake, Florida scrub jay, wood stork, striped newt, and one-toed amphiuma.

Corkscrew Swamp Sanctuary
Collier, Lee Counties
Owned by the Audubon Society, this 10,000-acre tract is about 10 miles south of Lehigh Acres. Part of the Everglades, its old-growth bald cypress forest is one of the largest in the world. The majestic trees are nearly 1,000 years old. Other communities include wet prairie, marsh, and bottomland pine forest. Several rare wood storks breed here. Unusual water plants are various orchids and yellow catopsis.

Crystal River National Wildlife Refuge
Citrus County
This 54-acre property just southwest of the town of Crystal River encloses nine islands in Kings Bay (Gulf of Mexico). This warm-water ecosystem is a critical wintering ground for the federally endangered West Indian manatee. The surrounding marshlands and shallow waters are important habitats for a variety of nesting and migrating birds, such as herons, coots, and laughing gulls. Rare or unusual fish include alligator gars. The property is owned by the U.S. Fish and Wildlife Service.

Cummer Sanctuary
Levy County
Located near Fowler Bluff along the Suwannee River, this 985-acre parcel is centered on mature bottomland pine-oak-apple hardwood forest, a rapidly vanishing natural community. Other communities are river swamp, scrub swamp, and cypress-tupelo swamp. Some large trees remain. The preserve is a critical habitat for a number of rare or endangered animals, including alligators, the Florida mouse, Bachman's warbler, bald eagle, gopher tortoise, and certain amphibians. Numerous waterbirds nest here or visit during migration. The property is owned by The Nature Conservancy.

Darling National Wildlife Refuge
Lee County
Located on 5,014 acres on Sanibel and Captiva Islands in the Gulf of Mexico, this property protects a barrier island ecosystem. Natural communities include mangrove swamp, bay, ocean

beach, and maritime grassland. The property is an important habitat for a number of rare or endangered species, including sea turtles, manatees, wood storks, brown pelicans, and numerous raptors, including the peregrine falcon. More than 200 different species of birds have been identified here. Alligators frequent the waters of the bays. The property is owned by the U.S. fish and Wildlife Service.

Disney Wilderness Preserve
Osceola and Polk Counties
Located about 22 miles southeast of Kissimmee, this 11,500-acre Nature Conservancy preserve encloses a variety of habitats. Natural communities include wetland, marsh, palmetto prairie, pine flatwood, and scrub. It contains one of the largest populations of bald eagles in the southeastern United States. Sandhill cranes, Florida scrub jays, and indigo snakes also live on the property.

Dry Tortugas National Park
Monroe County
One of the best-known and most pristine among national refuges that protect coral reef habitat, this park is located on a cluster of coral islands about 70 miles west of Key West. Constantly buffeted by the weather, these bleak islands are sparsely vegetated with sea grapes and mangroves. The park is an important nesting site and migratory stopover for a variety of songbirds, shorebirds, and waterfowl, including the uncommon frigate bird. Coral communities in the shallow, warm waters are vibrant with teeming underwater life.

Elgin Air Force Base
Walton, Okaloosa, and Santa Rosa Counties
This federal installation near Niceville encloses thousands of acres of undeveloped coastal barrier islands, beaches and dunes, and sandhills. Old-growth longleaf pine forest near Roberts Pond is a rare remnant of this once-vast forest ecosystem: It supports the second largest colony of reproducing red-cockaded woodpeckers in the nation. More than 100 species of endemic reptiles, amphibians, and plants can be found here.

Fakahatchee Strand State Preserve
Collier County
Located southeast of Fort Myers, this wetland complex is part of Big Cypress Swamp. Natural communities include sawgrass prairie, swamp, pine flatwood, cypress swamp, and wet meadow. Some towering cypress trees are nearly 1,000 years old. The property also supports the world's largest concentration of royal palms and the greatest variety of orchids in the United States, some of which are endemic to Florida. The primeval woods and wetlands are critical habitats for a variety of birds, including woodpeckers and songbirds as well as the rare endemic Florida panther (about 50 remain in the wild).

Florida Caverns State Park
Jackson County
Located 3 miles north of Marianna, this interesting cave system contains delicate microclimates that support albino cave salamanders, crayfish, and various isopods. The surface is covered by brilliant displays of wildflowers, some of which are unusual relicts from an ancient northern climate, including rue anemone and bloodroot. The rare Barbour's map turtle is endemic to the Chipola River, which flows through the cave system.

Florida Keys Managed Areas
Monroe County
Several properties totaling about 70,000 acres along the Florida Keys are managed for the conservation of critical habitats. The Key Largo State Botanical Site and Torchwood Hammock protect two of the largest tropical hardwood stands in the United States. High-quality pine rocklands are within the National Key Deer Refuge, which is a critical habitat for the endemic Key deer. The uncommon American crocodile is protected at Crocodile Lake National Wildlife Area. An abundance of rare plants and animals occurs at these sites, some of them endemic to the Keys.

Fred C. Babcock Wildlife Management Area
Charlotte County
Located on 69,000 acres in Charlotte County, this large tract is managed by the Florida Game and Fresh Water Fish Commission. It encloses a large tract of South Florida slash pine flatwoods with interspersed wetlands. It is a critical habitat for the red-cockaded woodpecker, the Florida black bear, and the crested caracara. Among the rare plants found here is the beautiful pawpaw.

Highlands Hammock State Park
Hardee and Highlands Counties
Located about 10 miles southwest of Avon Park, this 4,900-acre tract encloses rare old-growth hardwood hammocks and cypress swamps. Other high-quality natural communities are sand pine scrub, scrubby pine flatwood, bayhead swamp, and marsh. The hardwoods are more than 100 feet tall and centuries old. This habitat shelters the endemic Florida scrub jay and the rare red-cockaded woodpecker. Rare plants can also be found here.

Hobe Sound National Wildlife Refuge
Martin County
This 968-acre parcel is located along the Atlantic Ocean on Jupiter Island and, landward of Jupiter Island, near the community of Hobe Sound. Natural communities are ocean beach, mangrove swamp, sand pine scrub forest, and sand dune. These kinds of oceanfront communities are rapidly disappearing due to encroaching development. This property contains a variety of rare or endangered plant and animal species, including the federally endangered leatherback turtle, green sea turtle, pelicans, ospreys, scrub jays, indigo snake, manatees, and gopher tortoises. Its beaches are some of the most important sea turtle nesting sites in the United States.

Honorable Theodore Roosevelt Preserve
Duval County
This 640-acre parcel near the mouth of the Saint Johns River near Jacksonville is owned by the U.S. Forest Service. Natural communities here are oak-hickory maritime hammock, mixed hardwood forest, saltwater marsh, freshwater button brush–fetterbush and cordgrass–needle rush marsh, dune scrub, and mixed swamp forest. Hundreds of species of fish and shellfish reproduce in the marsh waters and are an important food source for waterbirds and raptors. Rare or endangered animals are the eastern diamondback rattlesnake, coral snake, vireos, certain skinks, herons, alligators, and the federally endangered wood stork. Ospreys also nest on the site.

Jonathan Dickinson State Park
Martin County
This undisturbed wetland-forest complex just northeast of Jupiter covers about 11,000 acres. Communities include mature sand pine scrub, oak scrub, pine flatwood, slash pine forest, mangrove swamp, and saw palmetto prairie. Unusual scrub plants include Florida rosemary, conradina, and gopher apple, an important food source for the rare gopher tortoise. Wetlands include cypress and mangrove trees and several rare orchids. The federally threatened red-cockaded woodpecker nests on the property.

Lake Wales Ridge Scrub Sites
Polk and Highlands Counties
These five tracts, comprising some 27,000 acres, preserve high-quality Lake Wales Ridge scrub sites. Located near the town of Lake Wales, it includes cutthroat grass seepage sites. Among the thick scrub vegetation are one of Florida's densest populations of rare and endemic plants and animals, including the Florida scrub jay and gopher tortoise.

Lignumvitae Key State Botanical Site
Monroe County
This 280-acre property midway along the Florida Keys protects a very rare, vanishing natural community: a lignumvitae forest. Located on Lignumvitae Key, these rare trees are more than 1,000 years old. Other rare plants and trees are the gumbo-limbo, mahogany, poisonwood, pigeon plum, and strangler fig. Numerous rare insects, spiders, and snails inhabit the forest. Several varieties of birds, some of them rare, nest or visit the property during migration.

Matanzas Pass Preserve
Lee County
This 42-acre preserve south of Fort Myers along the Gulf of Mexico is on the eastern end of Estero Island. Most of the preserve is tidal mangrove swamp, with lesser maritime live oak hammock on the sand ridges. Vegetation types include oak, saw palmetto, and myrsine trees. Rare plants include the nearly extinct gumbo-limbo tree, coral bean, geiger tree, sea hibiscus, and sabal palm. Estero Island is a critical habitat for a number of rare or endangered animals, such as the osprey, gopher tortoise, brown pelican, indigo snake, and southern bald eagle. Thousands of birds nest or visit during migration, including the yellow-

crowned night heron. The property is owned and maintained by The Nature Conservancy.

Merritt Island National Wildlife Refuge

Orange County

Just east of Titusville, this large refuge includes hardwood hammocks, pine upland woods, palmetto shrubs, brackish marshs, freshwater marshes, and ponds. Nearly 500 species of birds, mammals, fish, amphibians, and reptiles have been recorded here, many endemic and rare. Sabal palms edge the bays and marshes. Birds of special interest are southern bald eagles, brown pelicans, wood storks, mottled ducks, and several unusual herons and egrets. Hundreds of thousands of birds arrive during the fall and spring migrations. Twenty-one federally endangered or threatened wildlife species reside here, including the southern bald eagle, arctic peregrine falcon, piping plover, loggerhead turtle, Atlantic salt marsh snake, eastern indigo snake, least tern, and gopher tortoise.

Myakka River State Park

Sarasota County

About 15 miles northwest of North Port, this 8,500-acre park includes live oak forest, marshes, lakes, prairies, and riparian corridors. Parts of the park are nesting habitats for raptors, including ospreys and marsh harriers. The rare wood stork also lives here, as does of the largest alligator populations in the southern United States. Both dry and wet prairies are very diverse, with more than 250 plant species, including many colorful wildflowers, which are maintained through prescribed burns. Other rare species are the pine lily and also include several insects and butterflies.

Ocala National Forest

Marian, Lake, and Putnam Counties

This 382,000-acre forest just east of Ocala is the only semitropical national forest in the United States. Natural features include spring-fed pool, sinkhole, cypress-water oak bottomland, dry pine scrub, and maple swamp. One of best remaining stands of longleaf pine–wiregrass forest-savanna in the United States grows here. Its high biodiversity is reflected in the more than 300 species that inhabit the forest, including ospreys, Florida black bears, woodpeckers, Atlantic white cedars, and the very rare coonties, a member of the cycad family.

Ordway-Whittell Kissimmee Prairie Sanctuary

Okeechobee County

Located northwest of Okeechobee, this 7,500-acre Audubon Society property is located along the Kissimmee River. Its natural communities include wet and dry prairie, live oak and slash pine hammock, and freshwater marsh. These high-quality communities are inhabited by a variety of birds, mammals, and reptiles, including sandhill cranes, burrowing owls, alligators, rattlesnakes, and various raptors.

Paynes Prairie State Preserve

Alachua County

Located about 10 miles southeast of Gainesville, this 20,000-acre tract encloses one of the few large remnants of Florida's wet prairie. Wet prairie consists of cattails, sedges, and maiden cane developed on underlying limestone bedrock. Alligators, herons, egrets, ibis, and rare sandhill cranes inhabit or visit the property. Several varieties of rare plants species and wildflowers, including buckthorns, grow on the prairie.

Phipps (John S.) Preserve

Franklin County

This 40-acre tract southwest of Tallahassee along the Gulf of Mexico is owned by The Nature Conservancy. Also known as Alligator Point, this narrow peninsula–barrier island is windswept and sparsely vegetated. Natural communities include beach zone, sand dune, beach grassland, and stunted pine forest. It is an important nesting site for threatened least terns and snowy plovers and a migratory resting place for thousands of other birds. Monarch butterflies, during their migration to Mexico, stop here to feed on the diverse population of wildflowers.

Saint Joseph Peninsula State Park

Gulf County

This park encloses one of the oldest and least disturbed pine-scrub habitats in the United States. More than 200 years old, it consists of old-growth slash pines, live oaks, and sabal palms. Scrub oak and rosemary grow on the dunes. Other communities include tidal flat, coastal strand, and dune. Eagles and peregrine falcons are often seen, as the property is located along a raptor flyway. The scrub-oak community shelters rare plants, such as the endemic large-leaved jointweed and sand skinks. The park is an important habitat for a variety of bird species, including gulls and plovers; migrating monarch butterflies can also be seen.

Seminole Ranch
Orange, Seminole, Volusia, and Brevard Counties
This vast wetland complex, comprising some 26,380 acres, is located west of Titusville. Its natural communities include freshwater marsh, swamp forest, wetland, lake, and riparian corridor (Saint Johns River). Many of the communities are older-growth and remain relatively undisturbed. It is a critical nesting habitat and migratory stopover for thousands of birds, especially waterfowl and wading birds. Twelve endangered or threatened species can be found here, including the wood stork, osprey, bald eagle, and peregrine falcon. It is also an important corridor that connects wildlife areas and state forests.

Spruce Creek Preserve
Volusia County
This 150-acre tract, just inland from the Atlantic Ocean, is located north of Samsula. Owned by The Nature Conservancy, it encloses old-growth longleaf pine–wiregrass sandhills, wetland swales, magnolia–maple–cabbage palm river hammocks, and a blackwater creek with some tidal influences. These communities support a highly diverse population of endemic plants and animals. Rare or endangered species found here are gopher tortoises, alligators, certain turtles, indigo snakes, and other reptiles. The property is maintained with the help of prescribed burns.

Swisher Memorial Preserve
Putnam County
Owned by The Nature Conservancy, this 3,214-acre tract is located about halfway between Gainesville and Palatka. Natural communities include longleaf pine sandhills as well as wetland, pond, and wet prairie in swale. Although heavy logging occurred during the 1920s for the production of turpentine, some older-growth trees remain. The property is a critical habitat for numerous plants and animals, including the rare eastern indigo snake, Florida gopher frog, gopher tortoise, sandhill crane, bald eagle, and other raptors. Thousands of waterbirds congregate in the wetlands during spring and fall migrations.

Ten Thousand Islands
Dade County
Part of the immense Florida Everglades, this portion protects part of Mangrove Swamp, a rapidly vanishing ecosystem. The islands are covered in dense red mangrove vegetation shaped by wind, weather, and tide. Microenvironments within intertwined mangrove roots are important habitats for shellfish and fish. Rare bird species include bald eagles, peregrine falcons, brown pelicans, swallow-tailed kites, terns, and oystercatchers. Manatees inhabit the shallow waters between the islands.

Three Lakes Wildlife Management Area
Oceola County
This 60,000-acre wildlife management area near the Saint Johns River encloses several natural communities of high quality, including a rare xeric prairie. The forests, prairies, and riparian corridors are important habitats for a variety of rare or endangered species, including the Florida black bear, red-cockaded woodpecker, whooping crane, and several wildflowers and other plants, some endemic to Florida.

Tiger Creek Preserve
Polk County
This 4,500-acre preserve northeast of Frostproof encloses some of Forida's most unique wetlands. Habitats include a pristine blackwater stream (Tiger Creek), mixed hardwood swamp, hardwood hammock, sand pine scrub, oak scrub, sandhill, and mature pine flatwood. This preserve is a critical habitat for some of Florida's rarest species, including the federally threatened gopher tortoise, sand skink, Florida scrub jay, scrub lizard, pigmy fringe tree, large-flowered bonamia, and 13 other globally rare plants. The property is owned and managed by The Nature Conservancy.

Torreya State Park
Liberty County
Located 2 miles north of Rock Bluff, this 8,762-acre tract offers pine and hardwood forests. The Florida yew and torreya trees, endemic to this part of Florida, grow here. A variety of wildflowers grows among the groundcover. Rare relict plants, such as baneberry and leatherwood, from a colder northern climate survive among cool, moist microclimates in narrow gorges and ravines. The extremely rare torreya tree is a relict species from a genus that

existed millions of years ago; it has been nearly eliminated as a result of logging and disease. Older-growth beech and magnolia trees are also part of the hardwood forest. Species that depend on this habitat are the Apalachicola dusky salamander, one-toed amphiuma, and gopher tortoise.

Upper Suwannee River
Hamilton County
Located on 3,914 acres along the Suwannee River corridor northwest of Lake City, this preserve encloses blackwater rivers, pine flatwoods (bottomland forest), cypress–tupelo gum swamps, freshwater spring tributaries, rocky shoals, wetland sloughs, and underwater cave systems. These natural communities are home to an incredibly diverse group of plant, animal, and insect species, including nesting herons, endemic fish, the pallid cave crayfish, Florida cave amphipod, and Hobbs's cave amphipod. The extensive wetlands are critical nesting and migratory habitats for a variety of bird life. Some large hickory and oak trees hundreds of years old can be found along the river.

Waccasassa Bay State Preserve
Levy County
Located about 10 miles north of Yankeetown, this preserve encloses nearly 30 miles of coastline. Natural communities include hammock, savanna, tidal creek, tidal flat, extensive salt marsh, floodplain forest, pine flatwood, hydric and maritime hammock, and freshwater swamp. Various mammals, fish, shellfish, birds, and reptiles live in the park. Rare or unusual species include the cedar elm, pine-wood dainties, corkwood, manatee, bald eagle, gulf salt marsh snake, and Florida black bear.

Wakulla Springs State Park
Wakulla County
Located on 3,000 acres along the Wakulla River about 10 miles northwest of Saint Mark's, this property encloses one of the world's largest freshwater springs. Natural communities include spring basin, submerged limestone cave, extensive wetland, swamp forest, longleaf pine forest, upland beech-hickory-magnolia forest, and mixed pine flatwood. One of the state's largest basswood trees can be found here. The numerous rare plants and animals include egrets, herons, bald eagles, ospreys, purple gallinules, and a variety of waterbirds. Divers exploring the submerged cave systems have discovered rare catfish, crayfish, and eels that are specially adapted to the cave environment. The property is managed by the Florida Department of Natural Resources.

GEORGIA

Georgia contains a range of natural communities and vegetation types, divided into four larger natural divisions from northwest to southeast: the Appalachian Province, Blue Ridge Province, Piedmont Province, and Atlantic Plain. Elevations, stream erosion, ravines, and other conditions affect the spatial relationships of species and community types. Wetlands are present in all the provinces; swamps and marshes are more extensive in the Piedmont and Atlantic Plain terrains.

Located in the rugged country of northwestern Georgia, the Appalachian Province comprises a series of forested mountains, hills, and valleys. Vegetation types in this area are hickory-beech-oak forest, oak-hickory-chestnut forest, oak-pine forest, cedar glade, flatwood, and stunted shrub vegetation on rock cliffs; cave systems also exist.

The Blue Ridge Province, also in northwestern Georgia, is marked by mountainous terrain as well. Its forests are oak, chestnut, oak-chestnut-hickory, and shrub, with intergrowths of hemlock and white pine. More local and fragile ecosystems are attached to rock outcrops, cliffs, and canyon walls. Mountain laurel and rhododendron are abundant in the understory.

The Piedmont Province is a wide belt across central Georgia comprising rolling hills and valleys. Several varieties of forest types occur here, including evergreen heath, hardwood (black gum–beech–oak–hickory), flatwood, pine successional forest, and mixed pine-hardwood forest. The woodlands can be choked with a dense understory of shrubs, herbs, and vines.

The coastal Atlantic Plain covers southeastern Georgia. It has a variety of community types ranging from forest to dune, including deciduous-needleleaf pine forest, dwarf oak forest, lowland maritime forest, broadleaf evergreen forest, evergreen scrub forest, evergreen scrub–lichen, pine flatwood, loblolly–shortleaf pine, dune meadow, interdune, dune oak–evergreen, and rock outcrop.

Special Habitat Sites

Argent Swamp Natural Area
Effingham County
This 12,493-acre site north of Savannah is an excellent example of wooded swampland. Water tupelo, bald cypress, maple, and black gum trees are frequently flooded by the Savannah River. It is an important habitat for various animals and birds, including wood ducks, egrets, herons, and southern bald eagles.

Banks Lake Preserve
Lanier County
Located near Lakeland, this 3,153-acre Nature Conservancy property is part of a much larger blackwater lake–swamp complex called Grand Bay. Natural communities include pond-cypress forest, cypress–black gum forest, swamp, and other wetland communities. It is a critical habitat for nesting ospreys, alligators, herons, southern bald eagles, peregrine falcons, and the rare round-tailed muskrat. The rare green-fly orchid can also be found.

Big Duke's Pond
Jenkins County
This 472-acre wetland complex is located near Millen. It is a Carolina bay, a rapidly disappearing wetland type in the southeastern United States. Dominant natural communities are cypress savanna, pond–cypress swamp forest, slash pine forest, mixed pine-hardwood forest, swamp, and open water habitat. Several rare or unusual species are found here, including alligators, ferns, sedges, herons, wood storks, and amphibians. It is a critical nesting habitat to a large variety of birds. The property is owned and maintained by The Nature Conservancy.

Big Hammock Natural Area
Tattnall County
This 6,437-acre preserve south of Reidsville along the Altamaha River is one of the few undisturbed swamp forests in the state. Evergreen-hardwood forests are interspersed with cypress-gum swamps. The state-threatened Georgia plume shrub is found here. The wetlands are an important nesting habitat and migratory stopover for many birds, including the rare swallow-tailed kite, wood duck, white ibis, and various herons.

Black's Bluff
Floyd County
This 107-acre tract southwest of Rome protects an excellent example of mature oak-hickory forest. The plant life is diverse due to the rugged topography and north-facing slopes. Rare plants include large-flowered skullcap, limerock arrowwood, and Dutchman's breeches. The area is well known for its wildflowers during spring. It is owned and maintained by The Nature Conservancy.

Bond Swamp Refuge
Bibb and Twiggs Counties
Located just southeast of Macon, this 5,438-acre tract protects a diverse wetland complex along the Ocmulgee River. Natural communities include hardwood-pine upland forest, rocky area, bottomland forest, and swamp. Dominant trees are oak, black gum, pine, and sycamore. The wetlands are a critical nesting habitat and migratory stopover for many bird species.

Broxton Rocks
Coffee County
The Nature Conservancy has protected this 3,049-acre area near Broxton through land aquisition and registry agreements. Forested uplands and bottomlands, streams, wetlands, ravines, and cliff faces are some of the natural communities found here. Plant life is diverse, including the rare grit portulaca, silky creeping morning glory, Georgia plume, filmy fern, shoestring fern, and green-fly orchid. Turkey vultures nest in the forests; threatened animals include the indigo snake and gopher tortoise.

Buffalo Swamp Natural Area
McIntosh County
Primarily bottomland swamp along the Altamaha River, this 6,259-acre preserve south of Cox supports a number of rare and endangered species. Natural communities include gum-cypress swamp and forested sandhill ridges. Some of the more special species are the gopher tortoise, swallow-tailed kite, pitcher plant, and beak rush.

Camp Meeting Rock Preserve
Heard County
Located a few miles from Franklin, this 77-acre area protects a large rock exposure and its

highly specialized plant life. Some of these vary rare species include mosses, ferns, black-spored quillwort, pool sprite, longleaf sunflower, dwarf granite stonecrop, and stonecrop. The property is owned by The Nature Conservancy.

Cathead Creek Preserve
McIntosh County
This 752-acre Nature Conservancy preserve is located a few miles northwest of Darien. Natural communities include riverine bottomland forest, rare tidal forest, freshwater tidal marsh, and upland pine forest. Dominant trees (many of them old) are bald cypress, black gum, water tupelo, loblolly pine, willow oak, and dwarf palmetto. It is a critical habitat for a variety of plant and animal species, including green herons and great egrets.

Charles Harrold Preserve
Candler County
This 73-acre property near Metter is owned by The Nature Conservancy. The two major natural communities are scrub oak–longleaf pine sandhill habitat and mixed swamp flatwood. The rare Georgia plume, a deciduous shrub, the gopher tortoise, and various birds and amphibians can be found on the property.

Cooper Creek Scenic Area
Union County
This property protects part of the riparian corridor along Cooper Creek. It is best known for its rare stands of old-growth hardwood, hemlock, and white pine trees, many of them hundreds of years old and taller than 100 feet. The pine species are more typical of northern climates and are probable relicts from the Ice Age.

Crockford-Pigeon Mountain
Walker County
Located in the rugged mountains of northwestern Georgia, this 18,794-acre preserve protects the hardwood-pine forests on the slopes and flat summit of Pigeon Mountain. A network of more than 40 caves has developed in the limestone bedrock. The area is home to a variety of endangered plant and animal species, including the Ohio buckeye, hedge nettle, broadleaf puccoon, horse-balm, Alabama snow-wreath, green salamander, endemic Pigeon Mountain salamander, and various mosses and lichens.

Daniel Conservation Easement
Candler County
This 250-acre conservation easement near Metter is managed by The Nature Conservancy. Important communities include sand ridge, flowing stream, bottomland forest, and upland forest. The rare Georgia plume, longleaf pine–wiregrass habitat, silky camellia, green-fly orchid, Georgia feverbark, and gopher tortoise have all been discovered on the property.

Dry Creek Swamp Preserve
Worth County
Large trees in flatwood and cypress–black gum swamp communities are protected on this 20-acre tract near Sylvester. Owned by The Nature Conservancy, it contains populations of several rare plants, including the Cooley meadowrue, Turk's-cap lily, and wood poppy-mallow. The wetlands are also an important bird and animal habitat.

Egg and Little Egg Islands
McIntosh County
These islands form part of a fragile barrier island system along the Atlantic coast. The roughly 600 acres are covered with brush, gnarled trees, and dunes, edged by marshy cordgrass-needlerush shoreline. The islands are a very important nesting site and migratory stopover for thousands of shorebirds and waterfowl.

Hammock Conservation Easement
Thomas County
The Nature Conservancy established this 294-acre easement near Thomasville during the 1980s. It protects one of the few remaining undisturbed old-growth slope forests in the coastal Atlantic Plain. Natural communities include oak-hickory forest, beech-magnolia, and gum swamp. Its long existence makes it an excellent site to study how old-growth forests develop over time.

Heggie's Rock Preserve
Columbia County
Located near Appling along Benton Creek, this 101-acre property protects a pillar of granitic bedrock and its related communities. It was designated a National Natural Landmark in 1980. Unique plants survive in the rock depressions that periodically fill with water. Some of the rare species include stonecrop, pool sprite,

quillwort, cacti, reindeer moss, and broomstraw. A rare form of yucca plant can also be found. The site is managed by The Nature Conservancy.

Lewis Island Natural Area
McIntosh County
This 5,633-acre island tract is in the Altamaha River, 5 miles from Darien. It is an excellent remnant of bottomland hardwood forest. Other natural communities are virgin tidewater cypress and tupelo gum swamp. Some of these trees are 7 feet in diameter and more than 1,000 years old. The island is rimmed with biologically complex tidewater marshes. Rainbow snakes, spotted turtles, and swallow-tailed kites are some of the rare species found here.

Little Tybee and Cabbage Islands
Chatham County
These islands 18 miles east of Savannah are an important habitat for a number of rare species. Natural communities include tidal forest, creek, beach, dune, and marsh. Vegetation ranges from saltwater grasses to oak, pine, cedar, and saw palmetto. Rare species include the piping plover, Wilson's plover, American oystercatcher, and the threatened loggerhead sea turtle.

Lula Falls Area
Walker County
Located near Fort Oglethorpe, this 385-acre Nature Conservancy property protects unique communities. The pine-oak-hickory–covered hillsides are underlain by a series of caves and sinks. The subterranean water flows from the rock faces as springs. The rocks protect rare mosses, lichens, and the Virginia spiraea. A stand of mature to old-growth oak trees, some more than 100 years old, can be found on the steep talus slope.

Marshall Forest Preserve
Floyd County
Located just southeast of Rome, this old-growth pine-hardwood forest covers about 311 acres. Communities include pine-oak, chestnut oak, and mixed hardwood stands. Plant diversity is especially high here because northern and southern species overlap. The larger trees are hundreds of years old. The endangered large-flowered skullcap is also present. The site is owned by The Nature Conservancy.

Montezuma Bluffs Natural Area
Macon County
Located a few miles north of Montezuma, this 499-acre reserve overlooks the Flint River. Different natural communities on the site include floodplain forest, rocky slope forest, and upland beech-magnolia forest. The area is well known for its old-growth spruce, ash, gum, hickory, and oak trees. Rare plant species include the mountain catchfly, needle palm, ovate catchfly, slender sunflower everlasting, several species endemic to the calcium-rich soil, and the endangered relict trillium.

Oakbin Pond
Dooly County
This open cypress pond near Unadilla contains one of the country's healthiest populations of the endangered Canby's dropwort. The open nature of the pond enables the dropwort to flourish. The pond is surrounded by pine forest. Alligators have been sighted, and the waters attract a variety of waterbirds, including wood ducks and night herons. The property is managed by The Nature Conservancy.

Ogeechee River Natural Area
Bryan and Chatham Counties
Located near Richmond Hill, this 3,965-acre conservation easement protects an important wetland complex. Managed by The Nature Conservancy, the property encompasses blackwater swamp, marsh, and bottomland and upland forest communities, some of which are undisturbed and represent presettlement conditions. Dominant trees are cypress, gum, tupelo, maple, and pine. The wetlands are critical breeding habitats for rare wood storks, bald eagles, and shortnose sturgeons.

Ogeechee Swamp Natural Area
Chatham County
This 171-acre tract protects a small portion of blackwater wetland and cypress swamp near the mouth of the Ogeechee River. Communities include forested sand ridge and bottomland cypress-gum-tupelo woodlands. The wetland-woodland complex is a critical habitat for a number of birds, plants, and animals, including the threatened gopher tortoise, diamondback rattlesnake, and shortnose sturgeon.

Ohoopee Dunes Preserve
Emanuel County
This 267-acre segment of a complex sandhill ecosystem near Swainsboro was purchased by The Nature Conservancy in 1991. The sand dunes support scrub oak, evergreen shrubs, and lichen-moss communities. Although small in stature, some of the scrub oaks are nearly 200 years old. Rare species that live here are the eastern indigo snake, gopher tortoise, red-cockaded woodpecker, and numerous snakes. Rare plant species, such as rosemary and scarlet basil, have also been identified.

Ossabaw Island Heritage Preserve
Chatham County
This 25,000-acre preserve protects a complex barrier island ecosystem. Natural communities include oak-palmetto forest, tidal marsh, oak-pine upland forest, magnolia-holly forest, and red cedar–cabbage palm marshes. Rare or threatened species on the island include loggerhead sea turtles, alligators, eastern diamondback rattlesnakes that reach lengths of seven feet, painted buntings, piping plovers, and bald eagles.

Panola Mountain
Henry and Rockdale Counties
This 633-acre site about 20 miles southeast of Atlanta protects the nearly pristine natural communities on Panola Mountain, a granite dome. These include upland pine-hardwood forest and lichen-moss-fern communities on the rock faces. Rare or unusual plants include various lichen and moss species, spiderwort, Atamasco lily, dwarf granite stonecrop, and the Confederate daisy. It is also an important area for raptors such as red-tailed hawks.

Smith (George L.) State Park
Emanuel and Candler Counties
This 1,445-acre tract about 5 miles southeast of Twin City is dominated by river swamps and sand ridges. It contains one of the few remaining dwarf oak forests in the southeastern United States. The dry sandy soil supports a variety of rare lichens, yuccas, cacti, and gopher tortoises. Nearby wetlands contain tall cypress trees and are important bird and alligator habitats.

Stephen Foster State Park
Echols County
Located near Fargo on 80-acre Jones Island, this property encloses a portion of the Okefenokee Swamp, one of the largest wetland complexes in the United States. The landscape is dominated by swamp, bottomland forest, and wet prairie communities. Towering cypress trees, some of them 2,000 years old, line the water. A very diverse area, it is home to alligators, fish, snakes, amphibians, insects, and one of the largest black bear populations in the southern United States. More than 230 bird species, including sandhill cranes and the endangered red-cockaded woodpecker, have been recorded.

Tallulah Gorge State Park
Habersham County
Part of the eastern Blue Ridge Mountains, this park protects a portion of the Tallulah River gorge, a steeply incised canyon. The rugged edges are forested with Virginia and slash pine, shortleaf pine, rhododendrons, and the rare Carolina hemlock. The course of the river is lined with breathtaking displays of wildflowers. Several rare animal and plant species have been found here, including cliff-ledge species and amphibians.

HAWAII

The larger biotic regions in Hawaii are determined by elevation, precipitation, and vegetation type. Criteria include five elevation zones, three moisture categories (dry, mesic, wet), and vegetation structure (forest, shrubland, grassland, desert). The following biotic regions are recognized: alpine desert, subalpine forest-woodland-shrubland, montane dry and mesic forest and woodland, wet forest and woodland, lowland dry and mesic forest, woodland, and shrubland, lowland dry shrubland and grassland, and coastal communities.

Alpine deserts can be found on the highest mountain summits on Maui and Hawaii Islands. Typically very dry, the communities that survive here are subject to great temperature variations. Frost is common at night. Plant species include dwarf native shrubs, grasses, lichens, and mosses. Several rare and native species of insects and birds inhabit these windswept terrains.

Subalpine elevations are marked by distinctive forests, woodlands, and shrublands between 6,000 and 9,000 feet on Maui and Hawaii Islands. These slopes are more sheltered and receive more rainfall than the alpine desert, but they remain relatively dry. The vegetation communities are specially adapted to the dry soils and volcanic soil conditions. Some distinctive or endemic species here are tussock grass, lovegrass, *mamane* and *naio* trees, rare mints and spleenworts, yellow-faced bees, and rare forest birds.

Montane dry and mesic forests and woodlands can be found at lower elevations, such as the volcanic slopes of Hawaii, Kauai, and Maui Islands, transitions between the drier subalpine environment and the wet forest communities. Annual precipitation is less than 100 inches. The increased rainfall and gentler slopes have resulted in dense, junglelike forests of native trees, such as *Acacia koa, mamane, Chamaesyce,* and *Sapindus.* A large variety of endangered or rare insects, birds, and plants can be found here, including the Kamehameha and Blackburn butterflies and the Hawaii creeper. This ecoregion is especially threatened by the logging of *koa.*

Wet forests and woodlands dominate the windward slopes of Maui, Hawaii, Kauai, Oahu, Molokai, and Lanai Islands. Starting at sea level, they extend up the volcanic slopes to a maximum elevation of about 6,000 feet. Rainfall can be as much as 500 inches per year. The lush closed-canopy forests are dominated by *koa* and *'olapa;* the understories consist largely of dense tree ferns. Other communities include coastal lowland *hala* forest, *'ohi'a* and *'akala* shrubland, and bogs. Unique species found in these landscapes include the Hawaiian honeycreeper, lobelioids, mints, ferns, and the Maui parrotbill. Being very accessible, these forests and woodlands are being destroyed by development, construction, and agriculture.

Lowland dry and mesic forests and woodlands are most extensive on mature low islands such as Oahu. The natural communities that grow here are subject to hot temperatures and seasonal droughts. Grasslands and shrublands occupy the plains and gentle ridges—species include *pili* and *'a'ali'i.* Diospyros, Erythrina, Nestegis, and Pleomele survive in sheltered sites, such as gorges and ravines. Several rare or uncommon species live here, including several native birds and insects, as well as the Hawaiian hoary bat.

Lowland dry shrublands and grasslands are scattered across the islands, often in leeward lowlands and rain shadows. These specialized communities exist in hot and dry conditions on rocky soils. Remnants of native *pili* grassland, *'a'ali'i* shrubland, Hawaiian cotton, and *'akoko* dwarf shrubland can still be found. The heavy use of these areas by humans has eliminated many of these habitats and has resulted in the introduction of exotic species. Several rare native plants include members of the hibiscus and amaranth families.

The coastal communities that rim all the islands include cliff face, beach, and raised coral habitats. They are dominated by salt-tolerant grasses and shrubs and remnants of native sedges. Remnants of *hala* forest can also be found. The beaches are an important breeding habitat for the threatened green sea turtle and many rare seabirds.

Special Habitat Sites

'Ahihi-Kina'u Preserve

Maui Island

Located on the extreme southern corner of the island, southwest of Ulupalakua, this site is mostly covered by exposed volcanic rock with sporadic patches of vegetation. Located along the coast, its natural communities include tidal pool, low-salinity anchialine pool, lava tube, coastal dry shrubland, and coastal mesic boulder beach. Among the unique and rare species found here are Hawaiian shrimps and native cave animals that inhabit the specialized lava-tube environments.

Alaka'i Swamp

Kauai Island

This 16-square-mile area comprising swamp forest and muddy bogs is some of the wettest land on the planet. Its dense native wet forest and higher-elevation mesic forest support a great concentration of native birds, many of which are rare or endangered. The damp, chilly microclimate tends to keep out exotic species, which often carry diseases harmful to native species. Several rare bog plants can be found here, including the carnivorous *mikinalo.*

Hanawi
Maui Island
This 7,500-acre preserve west of Hana is on the flank of Mount Haleakala. Natural communities include lowland grassland and forest, montane grassland and forest, and a very rare subalpine grassland. Several important streams originate in these elevations and have lush riparian corridors. Various rare plant, animal, bird, and insect species inhabit the preserve.

Hawaiian Islands National Wildlife Refuge
Kauai Island
Ten groups of tiny barren islands northwest of the island are critical habitats for several endangered Pacific seabirds, land birds, and mammals. More than 15 million seabirds, including boobies, albatrosses, moddies, shearwaters, and sooty terns, nest or visit. The threatened and endemic Laysan duck and finch, Nihoa finch, miller bird, and endangered Hawaiian monk seals all find shelter in the coral reef, beach, dune, or inland scrub communities.

Hono O Na Pali
Kauai Island
Located along the northern shore of the island, this 3,150-acre state reserve is just south of Hanalei. Natural communities include streams and riparian corridors, beach, rock face, coastal shrubland, wet shrubland, grassland, and wet forest. This range of community types reflects changes in elevation and coastal influences, which has resulted in an extremely high biodiversity. Several rare plant and animal species can be found in these lush habitats, including the rare Hawaiian dark-rumped petrel and Newell's shearwater.

Honouliuli Preserve
Oahu Island
Located above Makakilo on Oahu Island, this 3,692-acre Nature Conservancy property protects part of the southeastern slopes of the Waianae Mountains. The dominant natural community is upland rain forest. More than 60 rare or endangered native forest birds, plants, and snails can be found here, such as the Hawaiian owl, the crimson-feathered honeycreeper, and yellow-green honeycreeper. This is the only location in the world where two mint species and two lobelia species are known to exist.

Ihiihilauakea Preserve
Oahu Island
This 30-acre property protects rare habitats along the dry southeastern coastline of Oahu near Hanauma Bay. The dominant natural community is a very rare vernal pool and its plant assemblage. The changing water conditions create a unique assemblage of plants, including several very rare ferns. The property is owned by The Nature Conservancy.

Ka'ena Point Natural Area Reserve
Oahu Island
This state natural area on 12 acres is at the westernmost point of the island. Its coastal natural communities include beach, dune complex, and dry shrubland. Tiny particles of coral and lava make up the beach sands. Several rare coastal plants can be found in the dune complexes and shrublands. The property is also an important habitat for the rare Laysan albatross, Hawaiian monk seals, other sea birds, and rare green sea turtles.

Kahauale'a Natural Reserve
Hawaii Island
Located on the slopes of Kilauea, the largest active crater in the world, this 16,726-acre tract is just outside the town of Volcano. Several varieties of natural communities can be found here, including montane wet forest, lowland wet forest, lowland mesic forest, grassland, and riparian corridor. These lush habitats shelter a variety of rare plant and animal species, including several endangered ferns.

Kamakou Preserve
Molokai Island
Part of Molokai Island, this 2,774-acre Nature Conservancy property protects a section of undisturbed high-elevation rain forest, an important source of water for the island. Hundreds of rare or endangered plants, animals, and insects can be found on the property, most of them endemic to the Hawaiian Islands. Some of these rare species are the Molokai thrush, Molokai creeper, and the rare Hawaiian owl. The property is located a few miles northeast of the city of Kaunakakai.

Kanaio Island
Maui Island
Located about 10 miles east of Keokea, this 876-acre tract protects communities that have developed on rough lava rocks. The site is

important because it represents a high-quality remnant of the native dryland forest that once covered a much larger area but is now being eliminated by development. Many of the plant assemblages within the forest are unique or rare, as are several of the individual species of plants, animals, birds, and insects that inhabit the area.

Kanepu'u
Lanai Island
This 590-acre Nature Conservancy property encloses high-quality remnants of the rare Hawaiian dryland forest that once covered much of Hawaii's lowland terrain. Located on a plateau on the western part of Lanai Island, the dryland forest shelters 50 rare endemic plant and animal species, such as the federally endangered Hawaiian gardenia, *nau,* and sandalwood.

Kapunakea
Maui Island
This 13,000-acre Nature Conservancy preserve on Maui northwest of Kahului protects dry lowland forest, summit rain forest, and their various communities, including rare montane bogs. The summit of the West Maui Mountains supplies the freshwater for the western half of the island. Nearly 30 rare plants can be found on the forested slopes, including the Maui bog violet and the nearly extinct *kauila* tree. Several rare or endangered birds and land snails inhabit the forests, such as the Hawaiian owl and West Maui silversword.

Kipahoehoe
Hawaii Island
Several varieties of rare natural communities can be found on this 5,583-acre state natural preserve on the lower-western slope of Mauna Loa, south of Captain Cook. A range of natural communities are represented, including lowland lava tube, lowland dry grassland, lowland dry forest, lowland mesic forest, and montane wet forest. Several rare or uncommon species occur here, many of them endemic to the Hawaiian Islands.

Ku'ia State Natural Area
Kauai Island
Located on 1,636 acres in the northwestern quarter of the island about 10 miles north of Kekaha, this natural area reserve protects some rapidly disappearing habitat. Within its borders are two rare natural communities: a mixed montane mesic forest, (*koa* and *'ohi'a*) and lowland mesic forest. Other plant assemblages include wet montane forest and lowland dry shrubland. Several varieties of rare and endangered plant, animal, and insect species can be found here.

Manuka Natural Area Reserve
Hawaii Island
Located on 25,550 acres on the southern tip of Hawaii Island, this tract encloses high-quality examples of lava anchialine pool, lowland mesic forest, lowland dry forest, wet montane forest, mesic montane kipuka forest, and subalpine forest and shrubland. A variety of rare or endangered plants and animals lives here, including the Hawaiian hoary bat.

Mauna Kea Ice Age Reserve
Hawaii Island
Located in the high-elevation alpine zone west of Hilo, this 3,894-acre nature reserve protects very rare alpine aeolian desert and an alpine lake. Mauna Kea, at 13,796 feet, is the highest point in the state. Alpine conditions are harsh, with strong winds and rapid changes in temperature. A unique group of rare plants and invertebrates inhabit the site.

Molokai Forest Reserve
Molokai Island
Covering nearly a third of the island, the Molokai Forest Preserve protects the upper heights of Mout Kamakou, north of Kamalo. Natural communities progress quickly from beach to upper-elevation forests toward the interior of the reserve. Although grazing has caused significant damage, the native vegetation at the upper elevations is relatively undisturbed and includes rare sandalwood trees, which were nearly eliminated by the shipbuilding industry. The deeper junglelike rain forests have some very large trees, as well as cooler microclimates among the narrow ravines. More than 200 Hawaiian species grow here, many of them endemic to the islands, forming critical habitats for several rare birds. An unusual upper mountain bog with dwarf trees occurs at the top of the reserve.

Moomomi Preserve
Molokai Island
Owned and managed by The Nature Conservancy, this 921-acre property on the north side of Molokai protects a section of ocean beach and its flanking dunes. More than

20 native Hawaiian plants grow on the dunes, including four rare or endangered ones. Unique plants include the heliotrope, beach *naupake,* and beach morning glory. It is an important nesting site for the endangered Hawaiian green sea turtle. The beaches and dunes may also attract the rare Hawaiian monk seal and Laysan albatross. Other unique birds, such as plovers and frigatebirds, visit the preserve.

Mount Ka'ala Natural Area Reserve
Oahu Island
Part of the central Leilehua Plateau about 5 miles south of Hale'iwa, this 1,100-acre reserve is located on the flanks of Mount Ka'ala, the highest peak on the island. Natural communities include cloud forest, steep rocky gulches, a fragile montane peat bog, and lower-elevation, drier foothills. Only patches of the lowland forests remain. Several varieties of rare plant, insect, and animal species, many of them endemic to the Hawaiian Islands, can be found here.

Oloku'i Natural Area Reserve
Molokai Island
Located just southeast of Kalaupapa, this 1,620-acre coastline tract encloses a high-elevation plateau, with steep slopes sweeping down to the sea. This is one of the most pristine wild areas left in the state and has not been invaded by many exotic or feral species. Natural communities include coastal dry grassland, lowland wet and mesic forest, montane wet and mesic forest, and a variety of unique cliff face plant assemblages. Among the rare or threatened plant and animal species found here are several snails.

Pahole Natural Area Reserve
Oahu Island
This lush 685-acre state reserve is located southeast of Dillingham Airfield. Part of the Waianae Range, its natural communities include dense rain forest, mesic forest, and dry grassland-shrubland. The preserve has an abundance of rare or endangered plant and animal species, including a rare dry forest, tree snails, and *'oha wai* (a lobeliad).

Pelekunu Preserve
Molokai Island
This 5,714-acre preserve is located in the Pelekunu Valley on Molokai's northern coast. It encloses ocean coastline and beach, dunes, tall sea cliffs, a rare pristine free-flowing stream, and dense lowland rain forest. The property is home to a great number of endemic plant, animal, bird, and insect species, many of them rare or endangered. The property is owned and managed by The Nature Conservancy.

Pu'u Ali'i Natural Area Reserve
Molokai Island
This rugged natural area reserve covers 1,330 mountainous acres just south of Kalaupapa. It is unique in that it is a high-elevation plateau. Communities include dense, junglelike wet forest, mixed fern, shrub montane cliff, wet shrubland, and stream with riparian corridor. This property protects an important watershed as well as numerous rare forest birds and mammals.

Pu'u Maka'ala Reserve
Hawaii Island
This sprawling, 12,106-acre tract is located on the eastern flank of Mauna Loa, a 13,680-foot volcano about 20 miles west of Mountain View. It was established to protect some of the highest-quality montane wet *'ohi'a* and *koa* forest remnants in Hawaii. Other communities include stream, riparian corridor, and montane wet grassland. Numerous rare plants, birds, and mammals inhabit the reserve.

Pu'u O 'Umi Preserve
Hawaii Island
The montane slopes and summits of the Kohala Mountains, including their abrupt coastal sea cliffs, are protected on this 10,412-acre reserve about 10 miles east of Kamuela. Natural communities include wet grassland, shrubland, wet forest, coastal dry shrubland-grassland, stream, lowland forest, and wet cliff face. Several rare montane bogs that support rare and beautiful plants can be found within the forests. The range of elevations and community types results in a high biodiversity. Many rare or endangered species can be found here, including the *koloa* (Hawaiian duck). The headwaters of several important streams also originate on the preserve.

Waikamoi
Maui Island
This 5,320-acre Nature Conservancy property protects part of the rugged eastern slope of Maui Island. Natural communities include high-elevation rain forest and subalpine shrub-

land. Hundreds of native plant, animal, and bird species live here. The forests (including some old-growth or virgin stands) are critical habitat for 12 bird species, seven of which are endangered, including the rare Maui parrotbill, Maui creeper, Hawaiian owl, and dark-rumped petrel. The property is also a major source of freshwater for Maui and its inhabitants.

Waikamoi Gulch
Maui Island
Rocky gorges and lush junglelike forests occupy this steep high-altitude gorge in Haleakala National Park. Biodiversity is quite high and includes numerous native species. Old-growth *'ohi'a* trees are nearly 300 years old. Several rare bird species can be found here that are endemic to Mount Haleakala. Other unique species are the parrotbill, an endangered honeycreeper, Hawaiin mint plants, and the world's only know tree geranium.

Waimanu Valley
Hawaii Island
Located north of Waimea along a coastal plateau, natural communities here include coastal cliff, pool, riparian habitat, lowland forest, estuary, tidal pool, and wetland. Native species include the bulrush *'aka'akai,* goby fish, mollusks, shrimp, and prawns. It is an important habitat for the black-crowned night heron, Hawaiian hawk, Hawaiian short-eared owl, and the endangered Hawaiian duck. The island's only native land mammal, the hoary bat, lives in the valley.

West Maui
Maui Island
This large natural area reserve (6,702 acres) west of Wailuku protects montane grassland, shrubland, wet *'ohi'a* forest, lake, and riparian corridor. The headwaters of several important streams originate within these communities and are important water sources. Several varieties of rare and endangered plant, animal, and bird species can be found in this preserve, some of them endemic to the Hawaiian Islands.

IDAHO

The major vegetation types in Idaho are sagebrush-grass, salt desert shrub, juniper woodland, mountain shrub, ponderosa pine, Douglas fir, grand fir–western redcedar–western hemlock, subalpine forest, and alpine. The distribution of these types is largely dependent on altitude, soil chemistry, and precipitation.

Most of the southern half of Idaho is covered in part by various sagebrush-grass communities. Stream drainages are lined with the native narrowleaf cottonwood, black cottonwood, coyote willow, and various smaller trees, such as box-elder and green ash. In the southwestern corner of the state is a small area of salt desert shrub dominated by greasewood and shadscale. Occasionally, juniper and mountain mahogany overlaps this zone.

Rocky Mountain juniper–pinyon pine–mountain mahogany woodlands occur sporadically in the sagebrush-grass zone at slightly higher elevations. They often give way to mountain shrub vegetation upslope—a dense forest of tough shrubs less than three feet tall. Species include mountain mahogany, bigtooth maple, chokecherry, bitterbrush, and mountain snowberry.

Well-developed ponderosa pine forests grow in west-central Idaho, from roughly Lewiston to Boise, and eastward along the Salmon River. They occur at elevations higher than the juniper-pinyon woodlands. As elevations increase, the ponderosa pine forest yields to the Douglas fir, grand fir–western redcedar–western hemlock, and subalpine limber pine–Engelmann spruce–subalpine fir–lodgepole pine forests, respectively. These cover the rugged mountains in the northern half of the state (subalpine forest can also be found in isolated high-elevation areas in southeastern Idaho).

At the highest elevations, alpine zone occurs on scattered ridge tops clustered in south-central Idaho. Located above timberline, this rocky environment is blasted by cold winds and temperatures. Alpine vegetation forms low, dense mats known as "cushion plants," or it may consist of a few twisted dwarf shrubs. Patches of meadows contain hardy grasses and colorful wildflowers.

Special Habitat Sites

Belvidere Creek
Valley County
A tributary of the Frank Church River in the No Return Wilderness Area, Belvidere Creek is protected by an RNA in Payette National Forest, near Krassel. Natural communities include pristine high-elevation subalpine fir forest, whitebark pine–subalpine fir forest, Douglas fir forest, meadow, and wetland. The endangered Rocky Mountain gray wolf is known to inhabit the area, as does the threatened bald eagle. The threatened Chinook salmon spawns in Big Creek.

Birch Creek
Lemhi County
This 1,160-acre Nature Conservancy preserve in Birch Creek Valley is located some 50 miles southeast of Leadore. More than 50 springs form the headwaters of the stream, which is one of the most pristine high-quality high-desert spring-creek systems in the United States. In addition to being an important habitat to a variety of bird and mammal species, the property supports one of the largest known populations of the rare alkali primrose.

Birds of Prey
Ada County
Owned and managed by The Nature Conservancy, this 358-acre sanctuary about 30 miles south of Boise protects a stretch of the Snake River canyon, including its riparian corridor and steep cliff faces. It is recognized as one of the country's most important raptor breeding sites and includes golden and bald eagles, falcons, owls, and various hawks.

Chilly Slough
Custer County
This 1,020-acre property is managed by The Nature Conservancy and five other agencies. Located about 20 miles northwest of Leslie, it protects an undisturbed high-quality spring-fed high-desert wetland. This type of natural community is very rare in the United States and is an important habitat for a variety of wildlife, including more than 130 bird species. Several rare plants, including rush aster and marsh felwort, grow here as well.

Cuddy Mountain
This proposed RNA is located in Payette National Forest. Natural communities include rare old-growth ponderosa pine forest, bluebunch wheatgrass steppe, Douglas fir forest, subalpine fir forest, and grand fir forest. Elevations range roughly from 3,700 feet to 7,000 feet. The old-growth trees are nearly 175 feet tall, 5 feet in diameter at breast height, and more than 300 years old. The grassy groundcover and steppe communities have been lightly grazed by cattle in the past.

Flat Ranch
Fremont County
Located about 10 miles north of Island Park, this 1,614-acre tract near Henry's Lake encloses important wetland and grassland habitats. The wetland is a vital fishery for native cutthroat trout and supports several rare plant species. An interesting aspect of this Nature Conservancy project is that it continues to operate as a cattle ranch to demonstrate that ranching and conservation can be compatible.

Formation Springs
Caribou County
This 115-acre property near Bancroft encloses a wetland system that includes thermal mineral pools, submerged cave openings, and lush riparian corridors. The system is an important water source for large mammals, such as elk, and a migratory stopover for various bird species. Several rare plant species also grow here.

Gamlin Lake
Bonner County
This 206-acre Nature Conservancy Project protects a sensitive wetland complex. Located about 30 miles north of Coeur d'Alene, this wetland harbors a variety of fish species, some of them rare, as well as rare or unusual plants. The lake and surrounding meadows and forest are important habitats for large mammals, such as moose, bears, and deer, and birds, such as ospreys.

Garden Creek Ranch
Nez Perce County
This 1,550-acre tract within Craig Mountain Wildlife Management Area south of Lewiston protects a section of the Snake River. Within this project area are forested slopes and rocky canyons that support a highly diverse biota,

including several rare plant species. Other important species found here are bighorn sheep, moose, black bears, ruffed grouse, raptors, and mountain bluebirds. Wildflowers are quite colorful during spring and summer.

Harriman State Park
Fremont County
This 16,000-acre park about 5 miles southwest of Island Park is in Targhee National Forest. Grasslands, meadows, wetlands, low- and high-elevation forest, and a 6-mile riparian corridor of Henry's Fork are the major community types. The park is an important habitat and watering hole for big game mammals, such as moose, elk, pronghorn, and black bears. Trumpeter swans, Canada geese, other waterfowl, sandhill cranes, and bald eagles are also present. Several rare and unusual plant and animal species have been recorded here.

Hells Canyon Wilderness Area
Adams and Idaho Counties
On the border with Oregon, this wilderness area consists of a high, isolated forest ridge between the Salmon and Snake Canyons; the ridge plunges nearly 7,000 feet. Natural communities include cliff face and high-elevation forest. For such a small, restricted area, it possesses very high biodiversity, mostly due to the extreme elevation changes. The canyon flatlands are arid, with a dry floor; the ridge tops are covered with grand fir forests and lush, wildflower-filled meadows. Intermediate elevations and plateaus support bunchgrass, prairie plants, and some old-growth Douglas fir and ponerosa pine woodland. It is an important habitat for bears, cougars, bighorn sheep, mountain goats, elk, lynx, and various raptors.

Heyburn State Park
Benewah County
Located along the shores of Coeur d'Alene Lake, this park is about 12 miles south of the city of Coeur d'Alene. Old-growth ponderosa pine and mature cedar-fir are the dominant forest types. Marshes, ponds, and sloughs fringe parts of the lake with lush ferns, shrubs, and mosses. The lake and wetlands are important bird habitats and support one of the largest nesting waterfowl populations in the state. The rare red-legged frog and an endemic salamander inhabit the marshes.

Hixon-Sharptail Preserve
Washington County
This 5,950-acre Nature Conservancy project protects a portion of grassland that is a critical nesting habitat for the rapidly diminishing Columbian sharptail grouse. Located about 10 miles north of Weiser, the preserve is part of a larger ACEC under BLM management. The sharptail grouse is known for its elaborate mating rituals and dancing grounds.

Idler's Rest
Latah County
Located a few miles east of Moscow, this 35-acre property, owned and managed by The Nature Conservancy, encloses a stand of old-growth grand fir–red fir–cedar forest. The impressive specimens are nearly 200 feet tall and 300–600 years old. It is an important habitat for species that inhabit old-growth forests. Several rare or unusual plant species also grow here.

Lost Basin Grassland
Adams County
Part of the Payette National Forest, this proposed RNA near Council encloses a very rare virgin Idaho fescue–bluebunch wheatgrass grassland. It has been isolated from grazing by cliffs and steep slopes and thus has maintained its presettlement character. It is an important habitat for a variety of small birds, mammals, insects, and reptiles that inhabit grasslands, including the rare savanna sparrow.

Moscow Mountain
Latah County
Located about 10 miles north of Moscow, this 295-acre Nature Conservancy project protects one of the best examples of virgin old-growth cedar forest in the United States. These impressive trees are more than 200 feet tall, up to 10 feet in breast-height diameter, and more than 1,000 years old. The groundcover consists of a lush display of ferns and wildflowers. This is a critical habitat for species that depend on old-growth forests.

Mount Harrison Research Natural Area
Cassia County
Part of the Sawtooth National Forest south of Burley, Mount Harrison is the highest point in the Albion Mountains. This 381-acre property encloses relatively undisturbed subalpine natural communities and a num-

ber of rare plants. These include cliff face, vernal pool, subalpine meadows, sagebrush-fescue grassland, and subalpine fir–limber fir forest. Several very rare plants can be found here; three are endemic to the Albion Mountains, and Christ's Indian paintbrush is found only on Mount Harrison. Rare sedges grow among the cliff walls. The property is also a habitat for a variety of birds and mammals, including the boreal owl, various raptors, and mountain lions.

Pondeosa State Park
Valley County
Located a few miles northeast of Lake Fork, this state park is known for its old-growth ponderosa pine forest. Other natural communities include cliff face, lake beach, and lodgepole pine, grand fir, white fir, Douglas fir, and Engelmann spruce forests. Marshy areas support tamarack stands, and riparian corridors are lined with black cottonwood trees. The largest trees approach 150 feet tall and are nearly 500 years old. The forest is an important habitat for a number of birds and mammals, some of them rare or of special concern.

Sawtooth Valley Peatlands
Custer County
This RNA in Sawtooth National Forest occupies part of Sawtooth Valley in central Idaho near Stanley. It was established to protect unique peatlands and related springs that harbor a number of rare or unusual species. Natural communities include peatland, fen, meadow, and upland lodgepole pine or western spruce–fir forest. A variety of rare plants grows here, including mosses, sedges, and the spoon-leaved sundew. The federally threatened Chinook salmon spawns in lower Mays Creek.

Selway-Bitterroot Wilderness Area
Idaho County
Located on 1.3 million rugged acres about 50 miles southwest of Missoula, Montana, this area protects the headwaters of Selway River. Natural communities include high-elevation lake, marsh, prairie, Douglas fir forest, lodgepole pine forest, Engelmann spruce and subalpine fir forest, and ponderosa pine woodland at intermediate elevations. Old-growth western redcedar and hemlock trees are hundreds of years old. The biologically diverse area is an important corridor for deer, elk, bears, moose, mountain lions, bighorn sheep, mountain goats, and wolves.

Silver Creek
Blaine County
Located near Carey, this 9,790-acre property protects one of the most pristine and sensitive trout streams in the country. More than 35 miles of Silver Creek Valley are protected by a joint venture among The Nature Conservancy, private citizens, and various agencies. Nearly 150 bird species inhabit the riparian corridor, including bald and golden eagles, sandhill cranes, and trumpeter swans. Large mammals, such as elk, bears, and mountain lions, also inhabit the site.

Snake River Birds of Prey National Conservation Area
Ada and Owyhee Counties
This portion of the Snake River corridor is about 25 miles south of Boise. It shelters one of world's densest concentrations of nesting raptors, including prairie falcons, golden eagles, and ospreys, which nest or perch on the high cliffs. More than 200 pairs of prairie falcons nest on the rock walls. The dry flats along the river are covered by sagebrush and winterfat and support small mammals like Townsend's ground squirrel.

South Fork of the Snake River
Bonneville County
Located about 40 miles east of Idaho Falls, this 4,714-acre Nature Conservancy project protects about 8 miles of riparian habitat along the South Fork of the Snake River. The cottonwood and fir forests along the river and rocky slopes are important nesting sites for various raptors, including bald eagles, peregrine falcons, and great gray owls. Other birds, such as the trumpeter swan, also nest here.

Strike Wildlife Management Area
Owyhee County
Located along a reservoir on the Snake River about 10 miles southwest of Mountain Home, this property is an important waterfowl wintering site. The desertlike terrain is covered by grass and sagebrush. Hundreds of bird species have been recorded, including American kestrels, raptors, bald eagles, sage grouse, turkey vultures, various owls, golden eagles, northern harriers, and bluebirds. Several rare

or unusual reptiles and amphibians also inhabit the area.

Targhee Creek Research Natural Area
Fremont County
This proposed RNA is located within Targhee National Forest a few miles from Island Park. Located on the Idaho-Montana border near West Yellowstone, Montana, it includes a variety of alpine and subalpine communities, including lake, meadow, and whitebark pine forest. Elevations range from about 7,400 feet to 10,400 feet. It is an important habitat for several threatened or endangered woodland species, such as the grizzly bear, gray wolf, and peregrine falcon. Bighorn sheep can also be seen during summer and fall.

Thousand Springs
Gooding County
This 400-acre preserve protects a 2-mile stretch of bottomland habitat along the Snake River. Located about 15 miles northwest of Twin Falls, the impressive series of springs and waterfalls that flow from the rocky basalt cliffs supplies some of the purest water in the state. The cliffs are an important raptor habitat, and the springs support some rare plant species. The property is owned and managed by The Nature Conservancy.

Trapper Creek
Cassia County
Located on a mesa in the South Hills of southern Idaho southeast of Twin Falls, this 453-acre RNA encloses high-quality woodland–dwarf-shrubland sagebrush–Utah juniper and pinyon-juniper communities. Elevations range from about 5,900 feet to 6,400 feet. The area has been slightly disturbed by past grazing of horses and cattle. The site is an important habitat for a variety of birds, including grassland sparrows and raptors such as the golden eagle. Mule deer and black-tailed jackrabbits support a mountain lion population.

ILLINOIS

Illinois is separated into 14 natural divisions. The Wisconsin Driftless division, in northwestern Illinois, is a rugged section of forest, savanna, and prairie developed on bedrock-derived soil. The Rock River Hill Country division consists of forested bluffs that border the Rock River in northern Illinois. The Northeastern Morainal division is defined by its development on glacial outwash around the edges of Lake Michigan. This terrain contains a number of bogs, peatlands, and Atlantic plant species. Tallgrass prairie once dominated the Grand Prairie division in north-central Illinois. Marshy backwaters, wetlands, and lush bottomland forests are characteristic of the Upper Mississippi River and Illinois River Bottomlands division. Closely associated with the bottomlands are the Illinois River and Mississippi River Sand Areas division, which contains sand dunes covered by forest, savanna, or prairie. The Forest-Prairie division comprises a mix of forest and prairie communities in the western part of the state. The Middle Mississippi Border division is located in rugged terrain along the Mississippi River, from Rock Island County to Saint Clair County, and contains unique cliff-dwelling plant species. The Southern Till Plain division consists of forested plains developed atop glacial outwash. Bottomland forest and marsh dominate the Wabash Border and Lower Mississippi River Bottomlands divisions. Rugged terrain covered by upland forests and hill prairies is typical of the Ozark division. The Shawnee Hills division occupies the extreme southern tip of Illinois and comprises forested sandstone and limestone hills and cliffs along the Mississippi River. Tupelo and bald cypress swamps dominate the Coastal Plain division in the southern part of the state.

Special Habitat Sites

Almond Marsh Preserve
Lake County
Located about 3 miles north of Libertyville, this 110-acre tract protects wetlands, oak woodlands, savannas, and a marsh-sedge meadow complex. The oak woods contain red, bur, and white oaks. The sedge meadow contains at least 10 different sedge species, including the rare beaked sedge and wheat sedge. Blue flag and marsh skullcap can also be found.

American Beech Woods
Clark County
This 20-acre tract 3 miles south of Marshall protects a presettlement-type old second-growth

forest. The dominant trees are American beech, tulip, and sugar maple. Ground plants include orchids and hepatica. Two rare plants that do not contain chlorophyll, beech drops and squaw root, also grow here.

Apple River Canyon State Park
Jo Daviess County
Located on 472 acres about 8 miles northwest of Stockton, this Nature Conservancy property encloses a stretch of the Apple River. Natural communities include the riparian corridor, mixed upland forest, cliff faces, floodplain forest, and dry prairie. Otters and more than 60 species of fish, some of them rare, live in the clean waters of the Apple River. The rare bird's-eye primrose, American stickseed, purple melic grass, and moschatel can be found in the cool, moist microenvironments among the cliff faces. Raptors can also been seen in the area.

Argyle Hollows
McDonough County
This rare barrens community, on 17 acres about 5 miles northeast of Colchester, contains stunted white and black oak and shagbark hickory. The herbaceous understory consists of typical herbaceous prairie plants, such as bluestem, white prairie clover, and poverty grass.

Ayers Sand Prairie Preserve
Carroll County
About 3 miles south of Savanna, this 115-acre tract includes a high-quality remnant of rolling dry sand prairie and sand dunes. There are nearly 150 different species of grasses, woody plants, and forbs. Dominant species include little bluestem, June grass, hairy grama grass, Carolina anemone, sandcress, and puccoon. Rare plants are the shaved sedge and Gray's umbrella sedge. The acreage is also home to the rare upland sandpiper, loggerhead shrike, grasshopper sparrow, and dickcissel, as well as several rare amphibians and reptiles.

Baber Woods
Edgar County
This 60-acre remnant of old-growth dry-mesic oak-hickory and maple forests is a few miles southwest of Grandview. The taller oaks and hickories are more than 100 years old. The presence of more than 210 plant species attests to its high biodiversity. It is also known for spectacular displays of ferns and wildflowers during spring and summer.

Baker's Lake
Cook County
Located just southeast of Barrington, this 210-acre complex of wetland and woodland communities contains a 112-acre lake. It is a protected rookery for the rare black-crowned night heron, great egret, double-crested cormorant, and great blue heron. It is also an important stopover for other waterfowl during spring and fall migrations.

Barrington Bog
Lake County
A graminoid bog, sedge meadow, and marsh occur on this 41-acre tract just north of North Barrington. A wide range of bog plants includes dwarf birch, buckbean, marsh cinquefoil, hoary willow, large cranberry, and marsh bellflower. The rare Baltimore checkerspot butterfly has also been found here.

Beall Woods
Wabash County
Located along the Wabash River about 8 miles southwest of Mount Carmel, this 329-acre tract contains excellent remnant old-growth bottomland and upland forest. Some of the largest and oldest trees in Illinois, including oak, hickory, pecan, and black gum, are found in this preserve. Pileated woodpeckers and rare vireos and warblers nest in the forest.

Belmont Prairie
Du Page County
This 10-acre parcel amid heavy urban development just north of Belmont is a high-quality remnant of original prairie. Prairie types are dry-mesic, wet-mesic, and wet. Representative species are porcupine grass, Canada blue-joint grass, cordgrass, and sedges. The area was protected from development by The Nature Conservancy during the early 1970s.

Berryville Shale Glade
Union County
This rare barren community occupies 40 acres about 3 miles west of Anna. The barrens are covered by little bluestem and prairie supurge, with scattered clumps of stunted oaks and hickory. Surviving barren communities are very rare in Illinois.

Black Hawk Forest Preserve
Rock Island County
This 106-acre high-quality upland forest in urban Rock Island is known as a bird habitat: More than 175 species have been seen in the old oak woods, including some disappearing species of warblers.

Bluff Springs Fen
Cook County
Located southeast of Elgin, this 91-acre tract contains dry gravel prairie, mesic blacksoil prairie, bur oak savanna, calcareous fen, wetland, and sedge meadow communities. Several threatened and endangered plants occur here. It is also an important breeding habitat for numerous birds, including woodcock, wood duck, and yellow-breasted chat. The property is owned and managed by The Nature Conservancy.

Bonnie's Prairie
Iroquois County
This 10-acre prairie remnant 3 miles north of Watseka contains wet sand prairie and wetlands. Prairie plants include cordgrass, fowl manna grass, water plaintain, pickerel weed, burreed, and yellow pond lily. Two very rare bee species that gather pollen only from the pickerel weed are found here. Other unusual plants are smartweed and blue flag.

Braidwood Dunes
Will County
This 259-acre tract about 2 miles southeast of Braidwood is one of the best remnants of savanna, mesic prairie, wet-mesic prairie, sedge meadow, and marsh complex in Illinois. Unusual savanna plants are the clammy false foxglove and huckleberry. Prickly pear cactus can be found on the dry dune ridges. Coastal Plain plants, rare relics from an ancient arm of the Atlantic Ocean, include yellow-eyed grass, grass pink, and tubercled orchid. Rare animals include the tiger salamander, western chorus frog, western slender glass lizard, and bullsnake.

Brookville Lutheran Cemetery
Carroll County
This remnant of original prairie is preserved within a small pioneer cemetery about 7 miles south of Lanark. Dominant plant species include Indian grass, little bluestem, big bluestem, sedges, New Jersey tea, and rosinweed.

Brownlee Cemetery Preserve
Mercer County
This small remnant of original prairie is preserved within a small pioneer cemetery about 6 miles northwest of Alexis. With the exception of the graves, the surface has not been disturbed. Native plants include Indian grass, bluestem, leadplant, coneflower, and rattlesnake master.

Burlington Prairie
Kane County
Although only 6 acres in size, this preserve protects remnant mesic and wet-mesic prairie, sedge meadow, and wetland. More than 215 native plants have been identified, including blue-eyed grass, yellow star grass, gentians, 11 species of sedge, and prairie blazingstar. Unusual aquatic plants, such as white water crowfoot, blue flag, and water smartweed, grow in the marsh. The site is also a habitat for the rare purplish copper butterfly.

Burton Cave Nature Preserve
Adams County
About 7 miles southeast of Quincy, this 85-acre tract protects dry-mesic upland forest, floodplain forest, and limestone cave natural communities. The area is known for its delicate cave habitat, which shelters federally and state-threatened bat species and blind, white invertebrates that never leave the cave.

Cache River Wetlands
Johnson and Pulaski Counties
Located about 10 miles south of Vienna, this wetland habitat along the blackwater Cache River has some of the oldest living trees in the United States. Natural communities include cypress swamp, high-quality upland oak-hickory-walnut forest, and swamp forest. Some of the huge cypress trees approach 2,000 years old. The plentiful holes in the cypress trees are excellent nesting sites for wood ducks. Other rare species that occur here are the bald eagle, heron, osprey, yellow-crowned night heron, rattlesnake, Cooper's hawk, orange-footed pearly mussel, dusky salamander, and bird-voiced tree frog. The property is owned and maintained by The Nature Conservancy.

Calamus Lake Nature Preserve
Macon County
This 120-acre wetland about 9 miles west of Harristown is an old oxbow of the Sangamon River and one of the few wetlands remaining

in central Illinois. It is an important migratory stopover for birds during spring and fall.

Carpenter Park Preserve
Sangamon County
This 322-acre tract just east of Springfield contains high-quality upland and floodplain forest, seeps, and stream communities. Among a heavily developed area, tall old-growth sycamore, maple, and cottonwood trees dominate the floodplain forest; oak, black cherry, and hickory are found on the uplands. These trees are more than 100 years old. Several rare birds nest here, including pileated woodpeckers, tanagers, and a variety of warblers.

Cedar Glen Preserve
Hancock County
Located along the Mississippi River, this hilly river-bluff country contains typical upland and floodplain forest communities reminiscent of presettlement conditions. The forests contain second-growth oak, hickory, maple, and basswood. Species diversity is high, and the preserve is a roosting site for wintering bald eagles. More than 400 eagles have been counted here. Nearly 30 different species of freshwater mussels can be found along the shoreline. The property is owned and managed by The Nature Conservancy.

Churchill Prairie
Du Page County
Located 2 miles southeast of Glendale Heights, this 65-acre tract protects remnant mesic and dry-mesic forest, savanna, sedge meadow, and prairie. More than 500 native plant and animal species have been documented, including foxglove, Indian plantain, stiff gentian, prairie sundrops, broad-leaved puccoon, and black-crowned night herons.

Colored Sands Bluff
Winnebago County
Located along the Sugar River, this 55-acre tract about 8 miles northeast of Durand contains sand bluff, savanna, and prairie communities. Rare species include a variety of snakes, Blanding's turtle, and the blue-spotted salamander.

Cotton Creek Marsh
McHenry County
Cotton Creek Marsh is an important wetland within the Fox River floodplain about 2 miles south of Holiday Hills. It also contains sedge meadow, wet prairie, and fen natural communities. The marsh is an important nesting site for sandhill cranes and a stopover for migrating birds during fall and spring.

Cretaceous Hills
Pope County
About 5 miles northeast of Metropolis, this 237-acre preserve protects a variety of forest communities and boggy seeps. The forests are dominated by white oak and hickory trees. Unusual plants are cinnamon fern, marsh fern, royal fern, and sphagnum moss.

Creve Coeur Nature Preserve
Tazewell County
Located on the western outskirts of Creve Coeur, this 23-acre plot contains an upland forest of oaks, sugar maples, and basswoods. A remnant of upland prairie is also present. It contains a variety of species, including little bluestem, Indian grass, sideoats grama, pale coneflower, and short green milkweed. More rare species are the Canada milk vetch and French grass.

Dean Hills Nature Preserve
Fayette County
Located in an isolated part of the county, this 75-acre tract protects representative high-quality mesic and dry-mesic upland forest communities. Dominant trees are large white oaks and hickories. This is an important habitat site for sensitive bird species, such as the red-eyed vireo, wood thrush, barred owl, tanager, and pileated woodpecker.

Detweiller Woods
Peoria County
Located a few miles northeast of Peoria Heights, this 246-acre woodland contains representative mesic and dry-mesic upland forest, cut by deep ravines. A variety of dominant trees includes red oak, sugar maple, basswood, hickory, pawpaw, and walnut. The rarer chinquapin oak is found atop the ridges. Remnant prairies are covered by native grasses like little and big bluestem. The forest is a critical habitat for songbirds, including the wood thrush, tanager, and red-eyed vireo.

Elizabeth Lake Nature Preserve
McHenry County
This diverse wetland occupies about 160 acres 2 miles northeast of Richmond. Natural com-

munities include bog, fen, marsh, sedge meadow, and dry-mesic savanna. The rare graminoid bog contains sphagnum moss, sedges, a calcareous floating mat of sedge peat, and marsh shield fern. Several threatened or endangered plant and animal species inhabit the wetland.

Exner Marsh
McHenry County
Located 5 miles east of Huntley, this 116-acre wetland complex is in the early stages of bog development, with scattered calcareous floating peat mats. Common bog plants are wire sedge, blue joint grass, and marsh shield fern. It is best known as a stopover for migrating birds, especially grebes and egrets.

Fairchild Cemetery Savanna
Vermilion County
This small pioneer cemetery about 6 miles northwest of Danville is one of the few remnants of original savanna remaining in Illinois. The savanna is dominated by big bluestem, Indian grass, coneflower, and prairie rose. It is also an important bird habitat.

Falks Bald Eagle Refuge
Rock Island County
This 173-acre tract about 2 miles northeast of Hampton is an important winter roosting site for bald eagles. The eagles hunt along the Mississippi River. The forest is primarily oak-maple-hickory.

Farm Trails
Lake County
This 20-acre parcel contains wetlands and remnants of fen and calcareous sedge meadow communities. Rare species are grass-of-Parnassus, Ohio goldenrod, small fringed gentian, and valerian. The marsh is also the habitat for the rare Baltimore checkerspot butterfly.

Fern Rocks Nature Preserve
Jackson County
This 170-acre tract about 3 miles northeast of Makanda is within Giant City State Park. Its most notable features are the shady, moist cliff faces and outcrops that shelter an amazing variety of ferns, mosses, and liverworts, including the rare Forbes's saxofrage, partridge-berry, French's shooting star, and small alumroot.

Ferson's Creek Fen
Kane County
Located about 2 miles north of Saint Charles, this 45-acre tract contains sedge meadows and wet floodplain forests. Fen species include fringed gentian, grass-of-Parnassus, Kalm's lobelia, marsh wild timothy, great Saint John's-wort, and northern fringed orchid. Wild rice also grows here. The preserve is an important nesting habitat for warblers, gnatcatchers, yellowthroats, and swamp sparrows. It is also a stopover for migrating birds during fall and spring.

Fogelpole Cave Nature Preserve
Monroe County
One of the largest and most pristine cave systems in Illinois, this 10,000-acre karst area is about 8 miles southwest of Waterloo. The cave system supports many species, including the federally endangered Indiana bat and invertebrate cave fauna. It is also the only location for a newly discovered species of cave beetle.

Forest Glen Seep Nature Preserve
Vermilion County
A complex of marsh, floodplain forest, and seep communities are found on this 8-acre tract above the Vermilion River, about 7 miles southeast of Westville. The area is known for large walnut and cottonwood trees, as well as rare orchids.

Fox River Fen
Kane County
This 12-acre parcel contains a wetland complex of fen, sedge meadow, and mesic prairie. Located 3 miles north of Elgin, it contains the rare green orchid, white camass, valarian, gentians, and Michigan lily. Rare wetland butterflies also inhabit the site.

Franklin Creek Nature Preserve
Lee County
Located about 3 miles west of Franklin Grove, this 96-acre tract is a mix of forests, wetlands, streams, and cliff faces. The maple-basswood forest is critical breeding habitat for interior birds such as pileated woodpeckers. Rare Canada yew, bulblet fern, cliffbrake, and downy arrowwood grow in crevices in the cliff faces.

Freeport Prairie Nature Preserve
Stephenson County
This 5-acre tract of virgin prairie is located on the outskirts of Freeport. A remnant of upland

prairie, it contains high-quality, undisturbed bluestem, prairie dropseed, sideoats grama, and Indian grass. The rare pasque flower and prairie smoke can also be found here.

Fults Hill Prairie Nature Preserve
Monroe County
Located along the Mississippi River about 25 miles southwest of Belleville, this 532-acre property protects a rare remnant of undisturbed loess hill prairie. Natural communities include mixed hardwood (oak-hickory) upland forest, savanna, hill prairie, cliff face, sinkhole pond, and limestone glade. Although the main grasses are big and include little bluestem and Indian grass, there are nearly 100 different plant species, some of them rare or unusual. The prairie is a critical habitat for grassland birds, animals, and insects, including the narrow-mouthed toad, coachwhip snake, and plains scorpion.

Funk's Grove Nature Preserve
McLean County
This 18-acre tract about 4 miles southwest of Shirley contains virgin mesic upland forest. The largest and oldest trees, the white and bur oaks, are probably much older than 200 years. They are gradually dying out and being replaced by sugar maples, a process called succession. Several varieties of wildflowers make a colorful display during spring.

Gensburg-Markham Prairie
Cook County
Located just west of Markham, this area of remnant wet-mesic and mesic prairie, dry mesic prairie, and sedge meadow communities covers about 110 acres. Rare plant species include grape fern, sundrop, sundew, and grass pink orchid. It is also an important nesting area for the rare Henslow's sparrow.

George Fell Nature Preserve
Ogle County
This 685-acre tract offers a variety of natural communities, including floodplain forest, upland forest, prairie, and sandstone cliffs. The cliff conditions permit the survival of rare relict boreal plants from the Ice Age, including white pine, bunchberry, hairy woodrush, round-leaved shinleaf, and wild sarsaparilla. Shining clubmoss and smooth white violet can also be found.

Glacial Park
McHenry County
Three hundred and thirty acres in size, this preserve 3 miles north of Wonder Lake includes diverse undisturbed glacial landforms, tallgrass savannas, dry prairie, bogs, and sedge meadows. Nine state-listed endangered or threatened plant species are present. Other unusual plants are Culver's root, shooting star, yellow star grass, leatherleaf, cinnamon fern, and rusty cottongrass.

Gleason Nature Preserve
Mason County
Located about 3 miles north of Topeka, this 110-acre tract protects sand dunes and dry sand prairie. The property includes a sand dune that reaches 60 feet. Rare species include the bladderpod, prickly pear cactus, porcupine grass, spiderwort, pocket gopher, six-lined racerunner, and regal fritillary butterfly.

Gooseberry Island Preserve
Kankakee County
Located in the Iroquois River, this 14-acre island escaped logging and is known for its old-growth maple and oak trees. Toothwort, wild ginger, green dragon, and starry Solomon's seal can also be found as groundcover.

Gooselake Prairie
Grundy County
Located about 10 miles east of Morris, this 1,537-acre tract is the largest remnant prairie remaining in Illinois. It also contains wetlands and seeps. In addition to containing indigenous prairie plants, it is home to the rare Henslow's sparrow, Virginia rail, least bittern, harriers, upland sandpipers, and several newly discovered moth species.

Grant Creek Prairie
Will County
More than 100 native plant species have been identified on this 78-acre tract of remnant wet and mesic prairie about 5 miles southeast of Channahon. Typical plants include big bluestem, switchgrass, Indian grass, and false wild indigo. Uncommon animals are the upland sandpiper, western chorus frog, and prairie vole.

Halesia Nature Preserve
Massac County
About 15 acres in size, this preserve is along the Ohio River 3 miles northwest of Metropolis. It contains a mix of floodplain and mesic upland forest. Dominant trees are oaks, hickories, and the occasional Kentucky coffee tree. The preserve is most noted for the best stand of silverbell trees in the state.

Hanover Bluff
Jo Daviess County
This 361-acre tract a few miles south of Hanover along the Mississippi River contains prairie, cliff, and dry-mesic and mesic upland forest communities. Prairie subtypes include sandhill prairie, dry dolomite prairie, dry-mesic prairie, and seep springs. More than 80 native prairie plant species have been identified. Several rare wildflowers occur in the oak-basswood forests. Rare animal species include the bald eagle, scarlet tanager, thrushes, the timber rattlesnake, and other snakes, frogs, and toads.

Harlem Hills Preserve
Winnebago County
This remnant gravel prairie on 53 acres about 2 miles north of Rockford contains a diverse number of prairie plants, including some state-endangered species. Rarer plants include the pink milkwort, downy yellow painted cup, Hill's thistle, and prairie buttercup.

Helm Woods
Kane County
This 80-acre tract is a well-preserved flatwood community with some very large and old white oak trees. Tall ash and elm trees also form part of the canopy. The wet areas contain the rare crowfoot fox sedge, buttonbush, and pagoda dogwood. The upland oak woods shelters rare long-leaved shinleaf, shooting star, wild geranium, and trillium.

Heron Pond/Little Black Slough
Johnson County
This 1,861-acre tract about 5 miles south of Vienna contains upland forest, swamp, bottomland forest, and limestone glade communities. Water tupelo and bald cypress in the swamps are some of the oldest trees in the state. It is critical habitat for a number of plant and animal species, including herons, vultures, red-shouldered hawks, barred owls, and songbirds. The property is owned and maintained by The Nature Conservancy.

Horseshoe Lake Preserve
Alexander County
This 492-acre tract about 5 miles southeast of Olive Branch is an oxbow separated from the Mississippi River and is a National Natural Landmark. The most notable features are a virgin beech-oak-elm forest and a bald cypress–tupelo community. Some of the trees are up to 4 feet in diameter at breast height. The preserve is host to a number of endangered or rare plant and animal species, including the green tree frog, mole salamander, raptors, warblers, and bald eagles.

Illinois Beach
Lake County
This 829-acre tract along Lake Michigan was designated a National Natural Landmark in 1980. Located just northeast of Waukegan, it protects sand dunes and swales deposited by a large glacial lake. The many communities include lakeshore, dune, sand prairie, savanna, fen, sedge meadow, marsh, and pond. Rare or unusual plants include bearberry, New Jersey tea, shrubby cinquefoil, prickly pear cactus, various sedges, Ohio goldenrod, fringed gentian, and Kalm's lobelia. The preserve is an important habitat for a variety of birds, especially during migration.

Indian Boundary Prairies
Cook County
Located a mile east of Markham, this Nature Conservancy property protects one of the best remaining remnants of virgin prairie in the Midwest. Less than 1 percent of Illinois's original prairie remains. The incredible diversity at this location includes nearly 500 species of mammals, birds, reptiles, and insects, some of which are rare, threatened, or endangered. Dominant grasses are prairie dropseed, Indian grass, little bluestem, big bluestem, and Indian paintbrush. The rare sundrop plant, Franklin's ground squirrel, prairie white fringed orchid, regal fritillary, dreamy dusky wing, various skippers, smooth green snake, yellow rail, and Henslow's sparrow can be found here.

Iroquois Woods Nature Preserve
Kankakee County
This 47-acre island in the Iroquois River 2 miles southwest of Aroma Park is a remnant of old-

growth mesic upland forest. The largest trees are red oak and basswood. The groundcover contains a variety of ferns, including some rarer species.

Jasper County Prairie Chicken Preserve
Jasper County
This 407-acre prairie habitat was acquired in 1959 to protect the greater prairie chicken. Located about 4 miles southwest of Newton, it protects a rare population of greater prairie chickens—fewer than 100 remain in the state. Other rare species include the Henslow's sparrow, grasshopper sparrow, upland sandpiper, northern harrier, and short-eared owl.

Jurgeson's Woods
Cook County
This 120-acre tract of woodland and prairie just north of Glenwood contains rare prairie plants, such as purple chokecherry, hardhack, huckleberry, low bush blueberry, Kalm's Saint John's-wort, sweetfern, grass pink orchid, and sundews. These natural communities are rare remnants of presettlement ecosystems.

Kankakee River Preserve
Kankakee County
This 139-ace preserve includes an island in the Kankakee River as well as some 15 miles of river shoreline. Located about 6 miles northwest of Kankakee, it contains oak upland forest, oak–basswood–sugar maple mesic forest, and bottomland forest of silver maple, cottonwood, and green ash. Cool, shaded, moist cliff communities along dolomite outcrops include the rare wild stonecrop and bulblet fern.

Kettle Moraine Preserve
McHenry County
This 242-acre nature preserve just southwest of Lilymoor contains remnant peatland, marsh, fen, and bog communities. Rare species include shrub leatherleaf, sphagnum moss, marsh fern, marsh marigold, willows, bladderwort, pitcher plant, and Blanding's turtle.

Kishwaukee Fen
McHenry County
Two miles south of Crystal Gardens, this 36-acre tract contains a wetland complex that includes a calcareous fen. Distinctive species include twig rush, sweet Indian plantain, turtlehead, grass-of-Parnassus, and Kalm's lobelia. Several varieties of sedges are also found. The larva for the rare Baltimore checkerspot lives on turtlehead plant, which makes this an important habitat for this butterfly.

La Rue Swamp
Union County
A National Natural Landmark, this 140-acre wetland about 3 miles north of Wolf Lake is one of the most biologically diverse areas in Illinois. Communities include mesic floodplain forest, shrub swamp, and open water habitats. Rare animal species include the golden mouse, green water snake, bluehead shiner, and bantam sunfish.

Lake Renwich Heron Rookery
Will County
Located about 2 miles south of Plainfield, this 320-acre preserve protects an island nesting site for great egrets, great blue herons, black-crowned night herons, and double-crested cormorants. Recent management of the area has focused on reducing the natural erosion of the island.

Lake-in-the-Hills Fen
McHenry County
This 207-acre preserve protects glacial landforms, wetlands, a variety of fens and floating mats, sedge meadows, and dry and mesic gravel prairies. The native prairie plants enclose rarer species, such as Hill's thistle, prairie smoke, prairie gentian, and leadplant. The calcium-rich water in the fens derives from the limestone gravel and is home to rare plant and insect species, such as the false asphodel and the blue bell, a tiny and rare damsel fly. Other uncommon species are shrubby cinquefoil and certain butterflies. Largely unchanged, this area closely resembles the presettlement landscape.

LeRoy Oakes Nature Preserve
Kane County
Located a few miles northwest of LeRoy Oakes, this 20-acre preserve includes a few acres of high-quality virgin prairie that has been little changed by encroaching development. It has a very diverse plant community, including the rare green milkweed, scurfy pea, and Richardson's sedge. Other plants include bird-foot violet, leadplant, false toadflax, downy gentian, and blazing star.

Liberty Prairie Nature Preserve
Lake County
This 47-acre tract near Libertyville contains a remnant of virgin prairie that has been only lightly grazed. Because it has never been plowed, it offers an incredible diversity of more than 150 plant species. Other natural communities include wetland, sedge meadow, and fen. Unique species include blue-eyed grass, yellow star grass, shooting star, blazing star, marsh phlox, grass-of-Parnassus, Riddell's goldenrod, sedges, and tufted loosestrife. The prairie is also home to six rare butterflies and the smooth green snake. The wetland is also an important nesting site for egrets and waterfowl.

Lockport Prairie
Will County
This very significant 254-acre parcel of prairie is on the floodplain of the Des Plaines River on the outskirts of Lockport. Located in an industrial area between the river, railroad tracks, a highway, and a state prison, it contains some very rare species. Habitats include floodplain prairie, wetland, and fen. Diverse plant species include the low calamint, slender sandwort, Ohio horsemint, the rare prairie satin grass, and the federally endangered leavy prairie clover, once thought to be extinct in Illinois.

Loda Cemetery Preserve
Iroquois County
This 4-acre cemetery is remnant original mesic blacksoil prairie. Located just outside Loda, it has never been cultivated or pastured. More than 130 native prairie plant species are found, including bluestem, Indian grass, rattlesnake master, compass plant, and wild quinine. Despite its small size, it is also a habitat for nesting prairie birds and butterflies, such as the rare grasshopper sparrow, fritillary, and skippers.

Lusk Creek Canyon
Pope County
Located within the Lusk Creek Wilderness Area, this 125-acre tract protects a steep sandstone canyon and related cliff communities. Rare plant species include arching dewberry, superb lily, ferns, and clubmosses. Timber rattlesnakes are also in the canyon.

Lyons Prairie and Marsh
Lake County
Remains of wooded upland, prairie, wet prairie, and marsh are protected on this 260-acre site northeast of Fox River Grove. Interesting species include big bluestem, cord grass, a variety of sedges, bluejoint grass, orchids, and Turk's cap lily. The wetland along the Fox River is an important stopover for birds migrating during spring and fall.

MacArthur Woods
Lake County
This largely undisturbed 446-acre tract of dry upland forest, mesic floodplain forest, and northern flatwood is representative of presettlement times. Located about 4 miles southeast of Libertyville, it is dominated by large white oak, silver maple, and bur oak trees. The preserve is an important breeding habitat for some rare birds, including raptors, pileated woodpeckers, and warbler species.

Main Street Prairie Preserve
McHenry County
A 114-acre tract along the Fox River north of Fox River Grove contains high-quality remnant dry prairie and restored mesic prairie. Uncommon plants include prairie buttercup, Hill's thistle, shooting stars, prairie cinquefoil, and gentians. It is also an important habitat for grassland birds, such as vespers and sedge wrens; bluebirds also nest on the property.

Marion County Prairie Chicken Sanctuary
Marion County
Located 2 miles south of Kinmundy, this 160-acre preserve protects prairie grasslands that are a critical habitat for the greater prairie chicken. It is also an important habitat for other grassland species, some of them rare or endangered, including certain plants and butterflies as well as the upland sandpiper, grasshopper sparrow, and northern harrier.

Massasauga Prairie Preserve
Warren County
About 5 miles northeast of Roseville, this 6-acre tract contains remnant wet and mesic prairie, as well as a colony of eastern massasauga rattlesnakes. Other snakes include the blue racer and western fox snake. Rare plant species include gentians and Turk's cap lily.

Matthiessen Dells
La Salle County
This 86-acre tract about 2 miles south of North Utica contains a mix of upland oak forest and

sandstone cliff communities. Cliff species include the rare white pine, Canada yew, and a variety of ferns. It is also an important habitat for gnatcatchers, tanagers, and other songbirds.

Meredosia Hill
Morgan County
This 30-acre tract of prairie about 5 miles east of Meredosia is rooted atop loess-covered hill-sides. Typical plant species are sideoats grama and little bluestem; other rare plants are pink milkwort and Hill's thistle.

Miller Swamp
Marion County
This spring-fed wetland covers about 35 acres 2 miles south of Helm. Common shrub swamp plants include buttonbush, water pepper, duck-weed, and a number of rare and sensitive ferns.

Miller-Anderson Woods
Bureau and Putnam Counties
This 268-acre tract is known for its old-growth oak-hickory upland forest community. Rare plants include the showy orchid, queen-of-the-prairie, and Schreber's aster. Peatlands occur in seep areas.

Munson Township Cemetery Prairie
Henry County
This 5-acre virgin black-soil prairie tract about 2 miles north of Cambridge has been mowed for hay in the past. Rare plant species include yellow star grass, Hill's thistle, and the one-flowered broomrape. The prairie is now managed with prescribed burns.

Nachusa Grasslands
Ogle County
This Nature Conservancy property covers 1,020 acres northwest of Rochelle. Natural communities include tallgrass prairie, oak savanna, dry prairie, tallgrass prairie, fen, sedge meadow, rocky meadow, marsh, and riparian corridor. Its large size makes it good habitat for threatened grassland birds, such as upland sandpipers, grasshopper and Henslow's sparrows, and dickcissels. Rare plants include the federally endangered prairie bush clover, Hill's thistle, kittentails, forked aster, prairie lion's tooth, and downy yellow painted cup.

Nelson Lake Marsh
Kane County
This 160-acre tract protects woodland, marsh, fen, and open water communities. Located about 4 miles southwest of Batavia, its rare occurrences include a floating peat mat, swamp milkweed, blue flag, mad-dog skull-cap, and clearweed. Rare fauna include four butterflies as well as the American bittern, Wilson's phalarope, black tern, various raptors, and the sandhill crane.

Norris Nature Preserve
Kane County
Located 3 miles west of Batavia, this 62-acre preserve encloses an excellent remnant of the "Big Woods" that once bordered the Fox River during presettlement times. Natural communities include oak–hickory–black cherry dry-mesic upland forest and willow-maple bottom-land forest. Some of the oak trees are nearly 200 years old and very tall. The woodland is a critical habitat for a number of nesting birds, including woodpeckers. Several rare or endangered plant and animal species have been found here.

North Dunes Nature Preserve
Lake County
Located just north of Zion along Lake Michigan, this 200-acre tract protects wetland, prairie, savanna, and beach habitats. Rare species that occur here are bugseed, bearberry, downy yellow painted cup, bog arrow grass, trailing juniper, Henslow's sparrow, Virginia rail, and upland sandpiper. It is also an important migratory stopover for raptors and waterfowl.

Oakwood Hills Fen
McHenry County
Located just north of Oakwood Hills and Silver Lake, this 12-acre property encloses undisturbed graminoid fen, calcareous seep, and sedge meadow. Rare plants include grass-of-Parnassus, fringed gentian, sedges, and the northern bog orchid. Peaty areas contain cotton sedge and bog valerian.

Olin Nature Preserve
Madison County
Old-growth upland forest, limestone bluff, active streams, and prairie communities exist on this 213-acre tract just west of Alton. The largest post, white, and rarer chinquapin oak

trees are found in steep-sided ravines. Bald eagles roost on the property during winter.

Palos Fen Nature Preserve
Cook County
This 48-acre site northwest of Palos Hills contains old-growth upland forest, graminoid fen, marsh, and oak savanna. The largest and oldest trees are oaks and shagbark hickory. Fen species include big and little bluestem, prairie dropseed, Ohio goldenrod, calamint, bog lobelia, and queen-of-the-prairie.

Park Memorial Woods
Putnam County
This 80-acre tract about 4 miles east of Henry is a remnant of virgin upland forest. The large hardwood trees are dominantly white and black oak, sugar maple, red oak, walnut, and Kentucky coffee. It is an important habitat for a number of bird species, including songbirds, such as warblers and thrushes.

Parklands Nature Preserve
Tazewell County
A large population of the rapidly vanishing heart-leaved plantain is protected within this 40-acre tract about 2 miles northeast of Mackinaw. The site encloses rubbly tributaries of the Mackinaw River that host the rare plant.

Pecatonica Bottoms
Winnebago County
Located about 10 miles southwest of Rockford, this 36-acre property includes floodplain forest, swamp, and oxbow lake communities along the Rock River. Rare animal species include pileated woodpeckers, bald eagles, and the blue-spotted salamander.

Pencumsaugan Creek
La Salle County
Just west of Utica, this 205-acre tract contains the abandoned Blackball Mine that is now an important bat habitat. Five bat species live in the mine, including the federally endangered Indiana bat. Other rare animal species are timber rattlesnakes and the mottled sculpin, a fish in Pencumsaugan Creek. Several endangered plant species occur in dolomite cliff, prairie, and floodplain forest communities within the preserve.

Pistakee Bog
Lake and McHenry Counties
Located about 4 miles east of McHenry, this 228-acre site protects bog, sedge meadow, marsh, floating mat, and open water communities. Species diversity is high, including rare orchids. The exotic glossy buckthorn and purple loosestrife are invading the preserve; the invaders are being controlled through chemical, mechanical, and biocontrol-insect methods.

Plum Grove
Winnebago County
Located in Rock Cut State Park east of Machesney Park, this 20-acre parcel contains old-growth upland forest. The largest white oak, red oak, and shagbark hickory trees may be as old as 200 years. The area is critical habitat to a number of flora and fauna, including the flying squirrel, woodpeckers, yellow-billed cuckoo, thrushes, and warblers.

Poplar Creek Prairie and Savanna
Cook County
This 600-acre preserve is located in suburban northwestern Chicago. Natural communities here include virgin pieces of hill prairie, tallgrass prairie, and rare black oak savanna. The disturbed sections are currently being restored with native prairie grasses. The virgin quality of the prairie remnants is reflected in the high plant diversity: More than 100 species have been identified, including some rare or endangered grasses and flowers. The preserve is an important habitat for grassland birds, insects, mammals, and reptiles, including endangered species such as Henslow's sparrow. The property is owned by The Nature Conservancy and is being managed through prescribed burns and by removing exotic species.

Reed-Turner Woodlands
Lake County
This 33-acre tract protects a remnant of presettlement prairie, savanna, and floodplain and upland forest communities. Located 2 miles northeast of Kildeer, it contains large oak and hickory trees and a variety of unique plant species. It is an important breeding habitat for waterfowl and herons.

Revis Hill
Mason County
This 417-acre prairie is the finest example of remnant loess hill prairie in the state. Located

about 10 miles east of Kilbourne, it contains most of the typical prairie vegetation. Some rare species also occur. Less common animal species include the western hognose snake, western slender glass lizard, prairie walking stick, and Ottoe skipper butterfly.

Ridgetop Hill Prairie Nature Preserve
Woodford County
Located on 16 acres about 10 miles southeast of Eureka, this Nature Conservancy property protects a rare remnant glacial drift hill prairie and surrounding oak-hickory dry-mesic upland forest. Bottomland maple-sycamore-elm-walnut can be found along a section of the Mackinaw River that crosses the property. Nearly 50 prairie grasses and forbs grow here, including little bluestem, purple prairie clover, and goldenrod. The grasslands are important habitat for grassland birds, insects, and reptiles, including nesting lark sparrows and grasshopper sparrows.

Rocky Branch Nature Preserve
Clark County
This 130-acre remnant of old-growth dry-mesic upland forest 2 miles northeast of Clarksville is dominated by oaks and hickories. It is noted for a sandstone cliff community that hosts rare species of mosses, liverworts, and lichens. It is also a critical habitat for the wood frog and long-tailed salamander.

Romeoville Prairie
Will County
Just east of Romeoville, this 108-acre tract comprises wet to mesic prairie, marsh, sedge meadow, fen, and wet forest communities. Typical prairie species include tufted hair grass, blue-joint grass, big bluestem, Indian grass, and prairie cordgrass. The fen contains a peat island. The preserve is an important habitat for nesting waterbirds and migratory birds.

Round Bluff
Johnson County
This 53-acre parcel about 2 miles south of Goreville is known for its upland forest, cliff face, and sandstone glade communities. The shaded rocky environments protect the delicate bottle gentian, chokecherry, rare ferns, bartonia, prickly pear cactus, rock pink, and blue curls. The forest is a critical nesting habitat for the yellow-billed cuckoo, eastern wood pewee, red-headed woodpecker, and a variety of songbirds.

Sagawau Canyon
Cook County
Located about 4 miles southwest of Hickory Hills, this 12-acre tract protects forested dolomite cliff faces. The cool, moist, shaded cliff conditions support rare cliff-dwelling species, such as bulblet fern, purple cliffbrake, walking fern, hairy rock cress, and ninebark.

Sand Ridge Nature Preserve
Cook County
Located in suburban Chicago between South Holland and Calumet City, this 70-acre tract protects glacial sand dunes and swales. Vegetation includes scattered black oak savannas and some prairie species, such as little bluestem, porcupine grass, June grass, cord grass, and the rarer tubercled orchid, grass pink rochid, chokeberry, and yellow-eyed grass.

Sand Ridge Savanna
Will County
This diverse sand prairie, wetland, and savanna complex occupies about 150 acres 2 miles west of Resthaven. It is unique in that coastal Atlantic Plain plants also occur here. Rare species include tubercled orchid, Saint John's-wort, narrow-leaved sundew, and savanna butterflies. The prairie and savanna are being restored through periodic burning, and former wetlands are being reclaimed to restore the natural hydrology of the area.

Section 8 Woods
Pulaski County
Located about 2 miles northwest of Karnak, this 326-acre nature preserve contains important wetland, pond, and bottomland communities. The most distinctive community is a cypress-tupelo swamp with very old, exceptionally large cypress trees. Buttonbush and Virginia willow can also be found near shallow water. The preserve is an important nesting area for a variety of birds, including songbirds, herons, woodpeckers, and black vultures.

Sentinel Nature Preserve
Carroll County
This 48-acre preserve located along the Mississippi River is 3 miles north of Savanna and part of Mississippi Palisades State Park.

Natural communities in the preserve are mesic upland maple–basswood–red oak forest, dry-mesic upland white oak forest, loess hill little bluestem–sideoats grama–rough blazing star prairie, caves and karst depression, and rocky dolomite cliff and bluff. The existence of nearly 600 plant species indicates high biodiversity. Several varieties of rare or threatened plant species can be found among the different communities, including American bugbane, ill-scented trillium, Canada violet, jeweled shooting star, purple prairie clover, bulblet fern, cliff goldenrod, and smooth cliffbrake. The property is managed by the Illinois Department of Conservation.

Severson Dells Nature Preserve
Winnebago County
About 4 miles southwest of Rockford, this 22-acre tract is best known for its 40-foot cliffs and related cliff communities. Rare species include bulblet fern, smooth cliffbrake, maiden hair fern, Canada yew, and white pine. Dolomite prairie also occurs on the site.

Skokie River Preserve
Lake County
Located just west of Lake Forest, this 100-acre tract contains woodlands, marsh, and virginal prairie. The prairie, never plowed, has been only lightly grazed. Thus, it contains a high diversity of plant species, including the rare compass plant, leadplant, prairie phlox, hoary puccoon, and Indian plantain.

Spring Bluff
Lake County
Located just northeast of Zion along Lake Michigan, this 274-acre tract protects a variety of natural communities, including sand dune and swale, prairie, savanna, marsh, and graminoid fen. Rare plants include royal fern, fringed puccoon, butterflyweed, white wild indigo, ragged fringed orchid, and button blazing star. The site is an important breeding site and migratory stopover for many bird species, including the rare Henslow's sparrow, king and Virginia rails, bitterns, and the upland sandpiper.

Spring Grove Fen
McHenry County
Located along Nippersink Creek, this 33-acre site encloses prairie, fen, and sedge meadow communities. Unusual fen species include cottongrass, dwarf birch, northern fringed orchid, and grass-of-Parnassus. Bog bedstraw and water parsnips also occur here.

Starved Rock Nature Preserve
La Salle County
Deep sandstone canyons are protected on this 582-acre site about 4 miles north of Oglesby. It is an important Native American site and was the site of a French fort during the 1700s. The sandstone cliffs host rare species, such as white pine, Canada yew, and northern white cedar. Very old oak trees occur in the upland forest areas.

Stubblefield Grove Nature Preserve
McLean County
About 3 miles northeast of McLean, this 30-acre tract contains old-growth forest with white and bur oaks that have diameters of 6 feet and are more than 100 feet tall; large maples are also present. The area is well known for its spectacular display of Virginia bluebell flowers during spring. It is also an important habitat for a variety of songbirds.

Thompson-Fulton Sand Prairie Preserve
Whiteside County
This 212-acre site about 5 miles north of Fulton contains remnant sand prairie. In addition to the typical plant species such as little bluestem, it also hosts prickly pear cactus and June grass. Rare animal species that inhabit the prairie are the western hognose snake, bullsnake, and six-lined racerunner.

Thorn Creek Woods
Will County
This 500-acre tract 2 miles southeast of Richton Park protects a variety of natural communities, including upland and bottomland forest, kettle, prairie, and the open water of Thorn Creek. White and red oak trees dominate the upland areas; maple and basswood cover the lower slopes. Prairie cordgrass and little bluestem can be found in the wet prairie remnants. Culver's root and swamp saxifrage also occur on the site. The wetland environment is an important breeding ground for the blue-spotted and spotted salamander.

Trout Park Nature Preserve
Kane County
This 26-acre tract contains old-growth woodlands as well as calcareous groundwater

springs and seeps in the lower ravines. The forest is dominated by oak trees, including the chinquapin oak. Rock elm can also be found. The rare arbor vitae plant occurs along springs and seeps, as does marsh marigold, bulbous cress, butternut, swamp saxifrage, and great Angelica. The tiny streams support seven species of caddis flies that are found nowhere else in the state.

Turner Lake Fen
Lake County
This large complex of wetland, peatland, fen, sedge meadow, and mesic savanna along the Fox River about 5 miles east of Spring Grove is one of the more important wetland sites in Illinois. About 100 acres total, it also protects interesting glacial landforms, calcareous springs, and marl deposits. Rare features include a calcareous floating mat. Delicate fen plants include pitcher plant, beaked spikerush, wild marsh timothy, dwarf birch, and numerous species of rushes and sedge. It is an important habitat for nesting waterbirds, such as the sandhill crane, and is also a popular migratory stopover for other birds.

Upper Embarrass Woods
Douglas County
This 65-acre tract about 6 miles east of Hindsboro contains old-growth oak forest. It also protects a variety of sensitive ferns. It is a critical nesting site for pileated woodpeckers, scarlet tanagers, and a variety of songbirds.

Volo Bog
Lake County
This 161-acre tract 2 miles northeast of Lily Moor is well designed for public visitation. A National Natural Landmark, it contains a classic bog zonation, complete with peat mat and acidic water, sedge meadows, and other wetlands. Specialized plants include shieldfern, bottlebrush sedge, sensitive fern, tamarack, and marsh cinquefoil. It is an important habitat for the little brown bat, beaver, and sandhill crane. Several varieties of migrating birds stop here during spring and fall.

Wadsworth Prairie
Lake County
This 176-acre preserve just east of Newport contains marsh and wet-mesic prairie communities. Dominant prairie grasses are big bluestem and cordgrass; rarer plants are tick trefoil and prairie dock. Rare wildlife species include Henslow's sparrow and the Virginia rail.

Wauconda Bog
Lake County
Located just east of Wauconda, this 67-acre preserve contains bog, marsh, and peatland communities. More than 100 plants have been documented, including the rare tamarack, various orchids, yellow birch, and winterberry holly. Sphagnum moss toussocks occur in the bog. It was designated a National Natural Landmark in 1974.

Wilmington Shrub Prairie
Will County
One of the few remnants of rare shrub prairie in Illinois, this 146-acre tract is located about 3 miles east of Braidwood. It is known for its herbaceous plants, including running pine. Drier prairie and savanna communities also exist on the property.

Windfall Prairie
Vermilion County
This delicate habitat covers 32 acres about 7 miles northwest of Hartshorn. A remnant of scenic prairie on a high bluff, it contains diverse flora. Tiny seeps support rare fen plants, such as marsh lousewort, Indian plantain, and grass-of-Parnassus.

Wolf Road Prairie
Cook County
This 44-acre tract just 2 miles west of Riverside contains virgin tallgrass prairie and oak savanna. Dominant species are big and little bluestem, prairie dropseed, and Indian grass. Rarer plants include blue aster, prairie phlox, obedient plant, rattlesnake master, Indian paintbrush, gentians, and Indian plantain. The western chorus frog and fox snake also inhabit the site.

INDIANA

Indiana comprises nine major natural regions: Northwestern Morainal, Grand Prairie, Northern Lakes, Central Till Plain, Southwestern Lowlands, Southern Bottomlands, Shawnee Hills, Highland Rim, and Bluegrass.

The Northwestern Morainal natural region occupies the northwestern corner of the state. It is developed on sandy moraines left by the last glacial Ice Age. Species diversity is high because of the merging of eastern deciduous forest (beech-maple or oak-hickory), tallgrass prairie, black oak–pine savanna, northern forest and wetlands, and beach–dune-and-swale vegetation.

Most of the northwestern quarter of Indiana is occupied by the Grand Prairie natural region. This is dominated by tallgrass prairie on glacial sediments (little bluestem–big bluestem–Indian grass–sideoats grama) with occasional pine-oak flatlands or savanna.

The northeastern sector of the state is called the Northern Lakes natural region. Several glacial freshwater lakes occur here, with marsh, swamp, and oak-hickory forest or beech-maple-tulip forest. Bogs are common.

The Central Till Plain natural region occupies the middle third of the state and was once heavily forested. Dominant forest types that remain are oak-maple, oak, elm-ash, and beech-maple-oak-hickory-basswood. Other specialized communities include fen, bog, cliff, and ravine vegetation. White pine–hemlock–Canada yew can be found on the north-facing slopes.

The southwestern corner of the state is the Southwestern Lowlands natural region. This generally flat-lying area with low hills is covered by sand barren, sand prairie, wet prairie, marsh, hickory-oak flatwood, and black ash swamp. River bottomlands that cut the area are characterized by pecan–swamp chestnut–red maple–green ash–bald cypress.

The southeastern corner of the state contains the Shawnee Hills, Highland Rim, and Bluegrass natural regions. Trending north-south, the rocky Shawnee Hills are covered by oak-hickory and mixed mesophytic forests. Some rare rock face and cave communities also occur here. The Highland Rim is dotted with caves and sinkholes. Natural communities include swamp forest, oak-hickory-beech-maple upland forest, and limestone glades. The Bluegrass natural region, in the extreme southeastern corner, comprises swamp forest, beech-maple-sweet gum-oak flatwood, oak-hickory upland forest, and lesser mixed mesophytic forest.

Special Habitat Sites

Anderson Falls
Bartholomew County
This 44-acre Nature Conservancy property about 5 miles east of Petersville encloses a rocky gorge, waterfalls, lowland beech-maple forest, and upland buckeye–oak–hickory forest. Some of the trees in the mature forests are quite large. Several rare or unusual plant species occur in the rock walls and ravines. Colorful wildflowers are abundant during spring.

Asherwood Nature Preserve
Wabash County
This 60-acre tract near Asherwood includes part of Asher Creek. The area is hilly and a good example of older mesic upland forest and dry-mesic upland forest. Dominant trees are beech, sugar maple, oak, and hickory. Seep springs on the property host some delicate and unusual plant species, including ferns and mosses.

Ashumbala Nature Preserve
Vanderburgh and Warrick Counties
Located along the Ohio River, this 63-acre remnant of floodplain forest near Evansville is one of the few remaining in the state. Large silver maple, cottonwood, and black willow trees dominate this natural community. Rarer species include green hawthorn, water purslane, clustered broomrape, and catbird grape. It is also a nesting habitat for a variety of bird species.

Babcock's Mill Gravel Wash Barrens
Crawford County
This 280-acre property south of Milltown is located along the Blue River. Natural communities include rare gravel wash prairie, mixed hardwood upland forest, and rocky cliff face. The prairie consists of prairie plants and grasses that are rooted in the sand and gravel bars along the river. Rare plants found in this environment include spreading rockcress, wild false indigo, and wallrue spleenwort. Several varieties of mosses, lichens, and ferns are

anchored to the rock faces along the river. The property is owned and managed by The Nature Conservancy.

Ball Wetlands
Kosciusko County
One of the largest wetland complexes remaining in Indiana, this 221-acre tract east of Leesburg is located between Tippecanoe Lake and James Lake. It protects open marsh, tall shrub swamp, red maple swamp forest, and oak-covered upland-island natural communities. Several rare or threatened plant species occur in the marsh, such as narrow-leaved bladderwort, shinleaf, cuckoo-flower, and bog buckbean. It is also an important nesting site and migration stop for many bird species, mostly waterfowl. The property is owned and managed by The Nature Conservancy.

Baltzell-Lenhart Woods
Adams County
This 37-acre tract about 4 miles south of Decatur contains typical mesic and wet-mesic flatwood communities. The mesic forest is dominated by large oaks and hickories; the wet-mesic community contains bur oak, swamp oak, sycamore, and cottonwood.

Barker Woods
La Porte County
Owned by The Nature Conservancy, this 31-acre tract about 1 mile north of Michigan City preserves an excellent example of wet-mesic forest. It is unusual in that white and yellow birch grow in peaty areas, far south of their typical terrain. The forest is dominated by pin oak and red maple, with lesser tulip, black gum, and beech. Several varieties of ferns and the rare Canada mayflower, Canada cinquefoil, and Indian cumber-root can be found in the moist depressions. Other rare or endangered plants are various sedges, cow wheat, and wintergreen.

Barnes Nature Preserve
Newton County
Four tracts totaling 249 acres protect oak savanna, sand forest, flatwood, prairie, and sedge meadow natural communities. Located about 3 miles west of Morocco, these communities are remnants of presettlement sand country. Rare plant species include the cream wild indigo. The area is managed with prescribed burns to maintain the savanna-prairie character.

Beaver Lake Preserve
Newton County
This 640-acre tract about 4 miles northwest of Enos is an area of drained wetland that has converted to wet prairie. Plant species include little bluestem and switchgrass. Bluejoint grass and sedges occur in the wetter areas. The site is a prairie chicken refuge and an important habitat for other prairie-wetland species.

Bell-Croft Woods
Jay County
Located on 40 acres in eastern Jay County, this property protects wetlands and a rare stand of old-growth oak-hickory forest that is a remnant of the larger "Big Woods" that covered eastern Indiana during presettlement times. Some of the larger trees are hundreds of years old and nearly 3 feet in breast-height diameter. The woodlands are an important habitat for interior forest birds and migrating songbirds; rare frog species may exist in the small ponds. The property is owned and managed by The Nature Conservancy.

Bendix Woods
Saint Joseph County
Located 10 miles southwest of South Bend, this 27-acre preserve shelters rare old-growth forest. The largest trees are beech and sugar maple, with lesser red elm and basswood. Ground cover consists of jewelweed, Solomon's seal, wild ginger, running euonymus, blue cohosh, and numerous ferns, some of them rare.

Berns-Meyer Woods
Pulaski County
A remnant of the "Big Woods" that once covered eastern Indiana, this 20-acre parcel about 3 miles east of Pulaski contains old-growth forest. The large trees are dominantly white oak, red oak, maple, and hickory, many nearly 100 feet tall and hundreds of years old. An abundance of colorful wildflowers appears during spring and summer. The forest is an important nesting habitat for a variety of birds, including the pileated woodpecker. The property is owned and managed by The Nature Conservancy.

Beverly Shores Dunes Area
Porter County
Located on 7 acres along the Lake Michigan shoreline, this Nature Conservancy property protects an undisturbed section of high sand

dunes. Adjacent to Indiana Dunes National Lakeshore, it contains a variety of dune plants, some of them relicts, rare, or endangered.

Bicentennial Woods
Allen County
This 80-acre tract near Fort Wayne is a virginal remnant of the original forest that existed during the 1700s. The acreage is dominated by towering oak trees that are more than 200 years old. The understory is ablaze with wildflowers during spring, including rue anemone and goldenseal.

Biesecker Prairie
Lake County
Located on 34 acres 2 miles south of Saint John, this property shelters a rare remnant of the larger prairie that once covered western Indiana. The dominant natural community is a mesic medium-grass prairie atop rolling hills. Little bluestem, Indian grass, prairie dock, and rattlesnake master are abundant. Several rare or endangered plants, such as flax, can be found here. The property is owned and managed by The Nature Conservancy.

Big Spring Preserve
Washington County
This 10-acre tract about 8 miles south of Salem shelters a spring flowing from a cave that is surrounded by old-growth forest. The larger trees are dominantly sugar maple, poplar, hickory, and various oaks.

Big Walnut
Putnam County
Located on 1,552 acres in scenic, rugged Big Walnut Creek Valley, this Nature Conservancy tract protects rare virgin stands of timber. Virgin eastern hemlock and Canada yew trees, relicts from the Ice Age, thrive on the cooler north-facing slopes. Some trees within the old-growth climax beech–maple forest are among the tallest and oldest in the state. The largest hemlock in Indiana is found on this property. A variety of bird and animal life inhabits the preserve, including some rare or endangered species. The property adjoins Hall Woods Nature Preserve.

Binkley Bog
Steuben County
This 10-acre property protects part one of the best remaining bogs in the state. Marsh, bog, and lowland forest communities can be found here. The bog supports a number of rare or endangered plant species, including wintergreen and the cuckoo-flower. The property is owned and managed by The Nature Conservancy. Because of the extreme fragility of this ecosystem, it is not open to the public.

Bloomfield Barrens
Spencer County
Rare barren communities are protected on this 794-acre Nature Conservancy property near Lincoln City. Rare natural communities include dry, grassy barren, stunted post oak flatwood, Little Pigeon Creek, bottomland forest, wetland, and oxbow pond. Nearly a dozen rare, threatened, or endangered plant species grow here, including the spider lily.

Blue Bluff
Morgan County
Located above the West Fork of the White River, this 32-acre parcel atop a steep, rocky bluff encloses upland beech-maple and oak-hickory forest. Biodiversity is high in this tension zone between the two forest types and includes the rare purple-flowering raspberry and several rare ferns. Owned and managed by The Nature Conservancy, the property is not open to the public at this time.

Bluffs of Beaver Bend
Martin County
This 210-acre property outside the town of Shoals protects a rugged hillside along the East Fork of the White River. Natural communities include sandstone cliff faces, mature and undisturbed upland oak forest, bottomland forest, and riparian wetlands. The rock cliff faces support a variety of rare plants, including lobed spleenwort, walking fern, rock tripe, stonecrop, rock geranium, and wild hydrangea. The property is owned and managed by The Nature Conservancy.

Boone Creek Barrens
Perry County
Located on 144 acres a few miles east of Oriole along Boone Creek, this Nature Conservancy property protects a rare barrens ecosystem developed on rocky, rugged terrain. Natural communities include stunted oak tree upland forest and grassy barren openings. Rare plants among the barrens include Canada lily, large-leaved phlox, waxy-leaved aster, Maryland but-

terfly pea, and little-headed nut-rush. The preserve is an important habitat for various reptiles, mammals, and birds, including raptors.

Brennan Marsh
Marshall County
Located on 65 acres a few miles west of Plymouth, this Nature Conservancy property protects a deepwater marsh that is part of the Menominee Wetland Conservation Area. Natural communities include open water, marsh, wet prairie, and vegetated sand dune. Marsh species include sedges, smartweed, burr reed, and spikeweed. Little bluestem and asters are the prairie plants. The wetlands are an important nesting site and migratory stopover for songbirds, shorebirds, and waterfowl.

Brock-Sampson Ridge
Floyd County
This 549-acre property about 7 miles south of New Albany protects a rare siltstone glade habitat. Natural communities include outstanding remnants of a siltstone glade and rugged upland Virginia pine–oak forest. The woodlands are an important nesting habitat for a variety of birds. Among the rare or unusual plants found here are rattlesnake weed and blazing star. The property is owned and maintained by The Nature Conservancy.

Browning Hill
Brown County
Located near Story, this 60-acre property is part of some of the most rugged land in the state. An old-growth mixed hardwood upland forest occurs on the steep rocky slopes of Browning Hill. These large trees are excellent examples of the forest that once covered the southern portion of the state. The property is an important habitat for a variety of mammals, reptiles, insects, and birds, including owls and vultures. The property is owned and maintained by The Nature Conservancy.

Bryan Nature Preserve
Clinton County
This 29-acre tract of old-growth forest is surrounded for miles by farmland. Located about 7 miles northwest of Frankfort, the preserve protects old-growth white oak trees that are hundreds of years old and up to 4 feet in diameter. Pawpaw, buttonbush, and winterberry surround a small pond on the property. Unusual wildflowers include yellow trout lily and wild crane's-bill.

Buffalo Flat
Dubois County
Important bottomland forest, swamp forest, wetlands, and open water ponds are located on this 375-acre Nature Conservancy parcel a few miles northeast of Jasper. The pumpkin ash–black willow–swamp cottonwood forest community is one of the rarest in the world. Flora and fauna are diverse, including the rare or endangered American snowbell, Virginia willow, red-shouldered hawk, yellow-crowned night heron, and several snake species.

Calvert and Porter Woods
Montgomery County
This 40-acre forest tract contains one of the best stands of old-growth timber in Indiana. Natural communities are upland oak-hickory forest, pin oak–maple–elm bottomland forest, and beech-maple lowland forest. Some wetlands occur seasonally. The tallest trees are nearly 3 feet in breast-height diameter and hundreds of years old. The woodlands are an important breeding habitat for a number of interior forest birds, including woodpeckers and great blue herons. The property is owned and maintained by The Nature Conservancy.

Carnes Mill
Crawford County
Located along the Little Blue River a few miles southeast of Grantsburg, this 73-acre Nature Conservancy property shelters highly diverse flora and fauna in a rugged terrain. Natural communities are the Little Blue River, its riparian borders, sandstone cliff faces, wetlands, and old-growth upland hardwood forest. Some oak, hickory, and beech trees are hundreds of years old. The moist, cool microenvironments along rock edges support a number of ferns, lichens, and mosses; other rare plants include mountain laurel, hairy mint, and skullcap. The wildflowers are abundant during spring.

Cedar Bluffs
Monroe County
This 23-acre tract southwest of Bloomington encloses rocky cliff faces, red cedar forest, mixed upland forest, and Clear Creek. The red cedar forest is a rare community in Indiana; the rare black-seeded rice grass can be found along the edges of the canyon. The display of

wildflowers during spring is colorful. The property is owned and maintained by The Nature Conservancy.

Cedar Bluffs Preserve
Monroe County
Located on 23 acres south of Bloomington, this tract protects rocky hills covered by red cedar and lesser oak trees. The less common chinquapin oak is found on shaded talus slopes. Wahoo is the dominant shrub plant. Unique herbaceous plants in this rocky terrain include white boneset, pussytoes, and various lichens.

Chapman Lakes Nature Preserve
Kosciusko County
Located just west of Big and Little Chapman Lakes and about 5 miles northeast of Warsaw, this wetland is a complex mix of cattail-bullrush, open water, and marl-beach prairie natural communities. Covering 147 acres, the preserve is an important nesting habitat for various bird species, including bitterns, rails, and herons; it is also a migratory stopover for birds during spring and fall. The rare bog plant species include pitcher plant, few-flowered spikerush, and cranberry, which are found on the southern half of the preserve.

Chelsea Flatwoods
Jefferson County
This 238-acre property a few miles west of Madison protects a largely undisturbed remnant of bluegrass till plain flatwood, a rare natural community in the state. The larger hardwood trees that escaped cutting are quite old and tall. Rare or unusual plants found here include the blunt-lobe grape fern, tree clubmoss, and running ground pine. The woodland is an important habitat for a variety of nesting birds, including songbirds and woodpeckers. The property is owned by The Nature Conservancy.

Ciurus Park
Jasper County
Located about 2 miles southwest of Demotte, this 40-acre parcel contains sandy upland forest communities. It is most noted for having two rare plant species: bristly sarsaparilla and cream wild indigo. Wildflowers, including rough blazing star, are abundant during spring and summer.

Clifty Canyon Preserve
Jefferson County
This 179-acre tract just southwest of Madison encloses steep canyons and ravines along Clifty Creek. A flowing stream tumbles through a series of waterfalls. Mature mesic forests are found on the lower slopes, and oak-hickory forests occur on the upper slopes. The cool, shaded cliff environment is host to a number of fragile, and several rare, species of mosses, lichens, and ferns.

Cline Lake Fen
La Grange County
Part of a large wetland complex, this 64-acre Nature Conservancy property is just northeast of Plato. Natural communities include stream, marsh, swamp, and fen. The fen is undisturbed and supports 15 rare or threatened species, including the spotted turtle, northern leopard frog, swamp metalmark, smooth gooseberry, and purple avens. The wetlands are an important nesting habitat and migratory stopover for a variety of bird life.

Conley Wildlife Refuge
Elkhart County
Located on 100 acres southeast of Elkhart, this Nature Conservancy tract encloses unique glacial deposits, ponds, sand prairie, and upland oak-hickory forest. Among the rare plant species can be found a strong population of pipsissewa. Many birds nest on the property.

Conrad Savanna
Newton County
This 313-acre tract about 2 miles south of Lake Village preserves a remnant of oak savanna reminiscent of presettlement times. White and black oaks dominate the clusters of trees. The sand prairie is home to a variety of typical prairie plants, including little bluestem, junegrass, Pennsylvania sedge, cleft phlox, milkweed, leadplant, and New Jersey tea. Goat's rue and hairy puccoon can also be found.

Cressmoor Prairie
Lake County
This 36-acre parcel is one of three high-quality remnants of blacksoil prairie in the state. Located near Crown Point, it has a remarkable biodiversity—more than 180 native prairie species have been identified, including a number of rare or threatened plants. The property is owned and maintained by The Nature Conservancy.

Crooked Lake Preserve
Whitley and Noble Counties
This 100-acre preserve in Thorn Creek Township protects more than 3,000 feet of undeveloped shoreline along one of the Indiana's quality lakes. Beech and sugar maple trees dominate the woodlands; important wetlands and calcareous seeps with unusual plant species occur along the shore. Natural communities include marl beach, meadow, cattail marsh, and spring-fed Crooked Lake—one of the deepest and most pristine in the state. Trout and the state-endangered cisco fish inhabit the lake.

Dauby Hollow
Perry County
Located a few miles southwest of Branchville, this 120-acre tract is part of a rugged and scenic area in extreme southern Indiana. Natural communities include sandstone cliffs, streams, waterfalls, and upland hardwood forest. The damp and cooler microclimates of the canyons and streams support rare plants, such as mountain laurel, rock clubmoss, and French's shooting star.

Deep River
Lake County
Located on 39 acres just east of Ainsworth, this preserve encloses a mature upland oak-hickory forest and lowland floodplain forest along Deep River. In places, the wilderness seems impenetrable, an excellent example of what presettlement bottomlands were like along this river. Several rare species occur here, including the purple fringed orchid. It is an important habitat for a variety of mammals, birds, and reptiles.

Dewey Hickman Nature Preserve
Harrison County
This 125-acre parcel a few miles west of Corydon preserves well-developed examples of dry-mesic upland forest, cliff face, mesic upland forest, and mesic floodplain forest communities along Indian Creek. This waterway becomes a subterranean tributary to the Blue River as it flows through cave systems. Several rare or endangered species can also be found here. The property is owned and managed by The Nature Conservancy.

Dogwood Nature Preserve
Ripley County
Located about 3 miles east of Versailles in Versailles State Park, this 20-acre parcel is all that remains of virgin old-growth hardwood forest. The largest trees are beech, maple, walnut, basswood, and hickory and are probably hundreds of years old. Flowering dogwood is also abundant.

Donaldson Cave/Donaldson's Woods
Lawrence County
Located about a mile southeast of Spring Mill Village on two tracts of land totaling 73 acres, these preserves enclose a cave with a flowing stream and virgin old-growth forest. Considered to be one of the best stands of original forest left in Indiana, it is a mix of beech-maple and oak-hickory types. The cave is also a bat habitat.

Douglas Woods
De Kalb and Steuben Counties
A rare remnant of old-growth "Big Woods" is preserved on this 200-acre property a few miles east of Hamilton. Some of the taller oak, maple, and beech trees are hundreds of years old. Fish Creek runs through the property and is home to several rare species of fish and mussels. The wildflowers are colorful during spring. The property is owned and maintained by The Nature Conservancy.

Dune Nature Preserve
Porter County
This 1,530-acre part of Indiana Dunes State Park near Gary protects a series of sand dunes along Lake Michigan's shoreline. The sand is anchored by beachgrass, bearberry, sumac, juniper, and jackpine. The backdune area is forested with oak, sassafras, blueberry, and white pine. Wetlands are also part of the preserve.

Efroymson Woods
Shelby County
An excellent remnant of old-growth maple-beech forest occurs on this 57-acre Nature Conservancy property in northeastern Indiana. The older oak, beech, maple, and tulip trees are nearly 100 feet tall and centuries old.

Elkhart Bog
Elkhart County
This wetland complex is located on 154 acres north of Elkhart. Classified as a circumneutral bog, its waters have a neutral pH. Several varieties of rare or unusual plant and animal species can be found here, including green adder's tongue, bog rosemary, cottongrass, rose pogo-

nia, orchids, and the massasauga rattlesnake. The property is owned and managed by the Indiana Department of Natural Resources.

Fall Creek Gorge
Warren County
Located about 3 miles northwest of Attica, this 37-acre tract encloses a steep sandstone gorge and surrounding upland forest. Fall Creek has carved a unique series of round potholes in the sandstone bedrock. Rare plant species include snow trillium, grass-of-Parnassus, white pine, forked aster, and a native bush honeysuckle. Liverworts, mosses, and lichens also thrive in the cool, shaded cliff environment.

Falling Timber Preserve
Ripley County
This 202-acre area is located within Versailles State Park near the town of Versailles. The preserve contains mesic forest, dry-mesic upland forest, riparian forest, and limestone glade natural communities. Rare plant species include the threatened white gentian and wavy-leaf aster.

Fawn River Fen
La Grange County
Located on 122 acres, this fen-wetland complex is a rare ecosystem in the state. The fen is fed by calcium- and magnesium-rich groundwater. Pristine Fawn River flows through the preserve. Several rare or unique plant species thrive in the fen, including certain sedges and the globally threatened kittentail wildflower. The property is owned by The Nature Conservancy.

Fern Cliff Preserve
Putnam County
Owned by The Nature Conservancy, this 115-acre property about 5 miles west of Greencastle encloses mesic upland forest and sandstone cliff natural communities. The property is unique in that it has a high diversity of ferns, liverworts, and mosses (some of them rare) that thrive in the shaded, moist cliff environment.

Fish Creek Fen
La Porte County
Located on 49 acres near La Porte, this Nature Conservancy property protects a fen wetland, a rare ecosystem in the state. The alkaline-rich chemistry of the soil and groundwater supports a variety of rare or unique plants, including arrow grass, valerian, and white lady's slipper. It is also an important bird habitat.

Flint Barrens
Harrison County
This 56-acre property protects the last remaining flint barrens in Indiana: a grassy expanse of prairielike grasses developed on a silica-rich, cherty soil. The barren has never been plowed and is surrounded by upland oak forest. It hosts a diverse array of prairie species, some of them rare or unusual.

Fox Island Nature Preserve
Allen County
This area of wetland occupies about 270 acres near Fort Wayne, including sphagnum-covered quaking bogs. Dry sand dunes are covered by black and white oak trees. The wetland and open water communities are important nesting sites and migratory stopovers for nearly 200 bird species. Rare species found here are trailing arbutus, bracted orchid, and Blanding's turtle.

German Methodist Cemetery Prairie
Lake County
This single-acre plot is the best remnant of virgin blacksoil prairie dropseed prairie in Indiana. Because it has never been plowed, it has retained its high natural diversity—more than 70 native prairie plants have been identified, including the rare cream wild indigo and several asters. The wildflowers are beautiful from spring to fall, with vibrant displays of shooting stars, prairie dock, compass plant, asters, and blue gentians. The property is owned by The Nature Conservancy.

Gibson Woods
Lake County
Just north of Hammond, this 120-acre tract shelters a remnant of the very rare ridge-and-swale dune environment formed thousands of years ago. It also has one of the longest undissected dune ridges in the state. Natural communities found here are dry-mesic savanna (black oak and bracken fern), mesic prairie (big bluestem), and wet-mesic forest (pin oak). Rare plant and animal species found here include speckled alder, dune goldenrod, orchids, paper birch, Kalm's Saint John's-wort, golden sedge, plains pocket gopher, red-shouldered hawk, blue winged teal, and Franklin's ground squirrel.

Gilford Sand Prairie
Lake County
Located on 18 acres in a rapidly developing area outside Gary, this property encloses an excellent remnant of sand prairie. Nearly 100 prairie species have been found here, including the rare, threatened, or endangered zig-zag bladderwort, Muhlenberg's nut-rush, round-headed rush, few-flowered spikerush, and cream wild indigo. The dominant prairie grass is bluestem.

Goose Pond Cypress Slough
Posey County
Owned by The Nature Conservancy, this 60-acre tract preserves part of a 3-mile cypress slough that runs parallel to the Ohio River. About 18 miles west of Evansville, the preserve protects one of the state's few remaining cypress stands. Some of the cypress trees are quite old. The wetland is a critical habitat for nesting and migrating birds.

Grand Kankakee Marsh
Lake County
Located south of Lowell, this 940-acre wetland complex borders the Kankakee River. This property is one of the few remnants of a 750,000-acre swamp and marsh. The many wetland habitats are critical habitats for birds, animals, and fish. It is an important resting and feeding stop for thousands of waterbirds during spring and fall migrations.

Gray Estate Cypress Slough
Posey County
Located near the confluence of the Ohio and Wabash Rivers, this 520-acre parcel southwest of Mount Vernon protects an extensive wetland complex. Natural communities include oxbow lake and slough, bald cypress swamp, and buttonbush swamp. In addition to harboring a variety of rare or unusual plant and animal species, the wetlands are a critical nesting habitat and migratory stopover for thousands of birds. The property is owned and managed by The Nature Conservancy.

Green's Bluff Nature Preserve
Owen County
This 260-acre property encloses steep, rocky slopes that border Raccoon Creek. These rock-and-cliff communities support a variety of rare species that require shade and moisture, such as mountain spleenwort, hay-scented fern, goldenseal, spotted wintergreen, and eastern hemlock. The uplands are forested with oak and hickory. Older-growth beech and maple trees grow in the steep, rugged ravines. The property is owned by The Nature Conservancy.

Guthrie Woods
Jennings County
This 61-acre Nature Conservancy property near North Vernon protects a relatively undisturbed flatwood natural community. Mature to older-growth stands of large beech and oak trees on steep slopes and in ravines contain some very large trees. Wildflowers are colorful in the wet areas during spring. The rare dwarf ginseng and sweet white violet are among some of the rare or endangered plants found here. The property is owned and managed by The Nature Conservancy.

Hall Woods Preserve
Putnam County
Two tracts totaling 385 acres about a mile east of Bainbridge preserve old-growth mesic upland forest, running stream, and floodplain forest communities. Very old white oak trees dominate the mesic uplands, with lesser hickory and maple. The property includes one of the finest stream segments in the state.

Hammer Wetlands
Noble County
This 115-acre tract just west of Waldron Lake contains wetland, open water, cattail marsh, shrub marsh, and floodplain forest natural communities. Large cottonwood and sycamore trees grow in the floodplain forest. The wetlands are an important nesting site and migratory stopover for a variety of bird species.

Harrison Hoke Memorial Woods
Knox County
One of the last old-growth hickory-oak remnants from the "Big Woods" that once covered much of the state is preserved on this 60-acre property a few miles east of Oaktown. Although selective timbering has occurred, there are still stands of old trees. It is an important habitat to a variety of plant, animal, and bird species, including woodpeckers and other forest-interior birds.

Hemlock Bluff Preserve
Jackson County
This 44-acre property just south of Leesville is located along a steep north-facing bank of

Guthrie Creek. The property is known for its stands of old hemlock trees, remnants from the last glacial period. The state's largest hemlock tree (3 feet in breast-height diameter) grows here.

Hemmer Woods
Gibson County
A National Natural Landmark, this old-growth forest 3 miles east of Buckskin covers 73 acres. This remnant is typical of the oak-hickory forest that covered southwestern Indiana during presettlement times. There are also large specimens of ash, sassafras, and pignut hickory.

Hitz-Rhodehamel Woods
Brown County
Located on 263 acres about 4 miles north of Nashville, this property is owned and managed by The Nature Conservancy. Natural communities include rock outcrop, cliff face, shady ravine, and upland chestnut oak forest. Several rare or endangered species can be found here, including certain mosses, lichens, ferns, and the whorled pogonia. It is an important habitat for a variety of birds.

Hoosier Prairie
Lake County
Located about 3 miles west of Griffith, this 548-acre tract preserves a large remnant of the presettlement prairie that once covered northwestern Indiana. It includes oak savanna, mesic prairie, wet prairie, sedge meadow, and marsh natural communities; plant and animal biodiversity is high as a result of this variety. More than 350 plant species have been identified. Several rare or endangered plant, animal, and insect species inhabit the prairie, which is managed through prescribed burning.

Hoot Woods
Owen County
This 82-acre property is one of the largest tracts of virgin forest remaining in Indiana. Among the beech-maple climax forest, some trees are 120 feet high and centuries old. Because of its undisturbed nature, it is an excellent place to study the dynamic relationships in a climax forest. The property is managed by The Nature Conservancy.

Indian Bitter Preserve
Washington County
Located east of Plattsburg, this 35-acre preserve protects mature mesic forest, mostly maple-beech-oak-hickory. This site is best known for containing one of the few stands of cucumber magnolia in the state.

Indian Creek Woods
Harrison County
This 150-acre property southwest of Corydon shelters special riparian habitats along Indian Creek. Natural communities include rock bluff, mature hardwood upland forest, rocky knob, and minor wetland. Several varieties of rare plants occur here, including certain ferns, mosses, and lichens on the rock faces. The black stemmed spleenwort, extremely rare in Indiana, is also found on the preserve.

Jordan Seeps
Vigo County
This 29-acre property a few miles southeast of Terre Haute protects an acidic seep and its surrounding wetland vegetation. A rare wetland community for Indiana, this seep reflects near-presettlement condition. Several rare or unusual plant species that are adapted to the acidic water and soil can be found here. The wetland is also an important bird and amphibian habitat.

Kankakee Fen
Saint Joseph County
This 13-acre tract encloses a rare fen community in the Kankakee River valley. As in other fens, a number of rare or unusual plants can be found here, including prairie dock, cotton-grass, and marsh valerian. It is an important habitat for waterfowl and wading birds, such as herons and egrets.

Kankakee Mallow
Elkhart County
This undisclosed small parcel was discovered to contain the globally rare Kankakee mallow plant, related to hollyhock. This mallow species has been found in only four places worldwide. The property is in an industrialized setting, and the habitat is endangered; the plants are being removed and will be replanted in a more suitable natural site. The Nature Conservancy is managing the transplantation.

Kieweg Woods
Vigo County
This excellent and rare remnant of old-growth beech-maple forest is located on 65 acres near Terre Haute. The trees are more than 100 feet tall, reach nearly 3 feet in breast-height diame-

ter, and are hundreds of years old. The understory is highly diverse and contains many species of wildflowers. The property is a research area for Indiana State University.

Knobstone Glades
Jackson County
This 60-acre preserve about 4 miles south of Brownstown consists of three tracts that enclose rugged, rocky knobstone glades, rare natural communities for this state. Stunted chestnut and oak trees surround the glade openings, which contain a variety of shrubs and herbs, including blueberries and prairie plants.

Koontz Lake Preserve
Starke County
Located on 148 acres just northeast of Koontz Lake, this preserve encloses mesic upland forest, fen, black oak savanna, prairie, open water, marsh, and shrub bog communities. Unusual, more northern plant species, such as tamarack, leatherleaf, gold thread, and wintergreen, are found in the wetlands.

Laketon Bog
Wabash County
Just north of Laketon, this 32-acre property contains a peat-filled bog, a rare natural community in Indiana. Other communities are shrub swamp, marsh, and floodplain forest. Unusual plant species in the bog are tamaracks (relics from the last glacial period), ninebark, shrubby cinquefoil, and dwarf birch. Several varieties of wildflowers can be seen here, including the rare yellow lady's slipper. The bog is accessed by a boardwalk.

Lakewood Dune Forest
Lake County
This 8-acre property just south of Lake Michigan encloses a rare high dune forest community. High sand dunes are forested by oak trees and various grasses. Rare plants found growing in the sand are arctic bearberry, round-leaved dogwood, prostrate juniper, and marrum grass. The property is owned and maintained by The Nature Conservancy.

Laughery Bluff
Ripley County
This 81-acre preserve is located within Versailles State Park, a few miles from Versailles. Old-growth trees survive on the rugged slopes along Laughery Creek. Some beech trees are more than 3 feet in breast-height diameter and probably 200 years old. Other large specimens are tulip, maple, and black walnut. Groundcover includes bloodroot and white baneberry.

Leatherleaf Bog
Kosciusko County
This 23-acre Nature Conservancy tract protects one of the rare bog communities in Indiana. Other communities on the property include swamp and lowland mixed forest. Rare bog plants include cranberry, bladderwort, buckbean, bog rosemary, and carnivorous plants. The wetland is also an important nesting habitat and migratory stopover for waterfowl and wading birds.

Leavenworth Barrens Preserve
Crawford County
Located a mile north of Leavenworth, this 762-acre tract protects excellent examples of relatively undisturbed barren communities typical of presettlement in the Shawnee Hills natural region. Barrens consist of scrubby oak trees and open areas of prairie grasses, such as Indian grass, little bluestem, prairie dock, blazing star, rattlesnake master, and hoary puccoon. The barrens are surrounded by upland oak-hickory forests.

Lieber Nature Preserve
Elkhart County
Located north of Bristol, this 135-acre property encloses Pipewort Pond as well as bog, mudflat, wetland, shrub swamp, and mixed lowland forest communities. An abundance of rare coastal Atlantic Plain species still grows here, including cranberry, small bullrush, brown-fruited sedge, long-beaked bald rush, and umbrella sedge. The property is a critical habitat for numerous horebirds, waterfowl, forest birds, and amphibians.

Little Bluestem Prairie
Vigo County
This 9-acre tract just north of Terre Haute protects the only surviving piece of original prairie in Vigo County. Plants growing in the steep, sandy soil include Indian grass, little bluestem, hoary puccoon, big bluestem, and round-headed bush-clover. Controlled burns will eliminate the incursion of woody plants and shrubs.

Little Grass Lake
Steuben County
Part of a larger wetland complex, this 118-acre tract comprises a shallow lake that often has mudflats during summers. Natural communities include open water pond, marsh, floating peat mat, and bottomland forest. Several rare or unusual water plants can be found here, including the tall-beaked rush. Thousands of wading birds and water fowl use the wetland as a nesting site or migratory stopover. The property is owned and managed by The Nature Conservancy.

Liverpool Preserve
Lake County
Located within the town of Liverpool, this 20-acre tract contains savanna, prairie, sedge meadow, and marsh communities. Although the marsh and meadow have been disturbed by sand mining, they have been revegetated with rare, moisture-loving plants. These include carnivorous sundews, bluecurles, bladderworts, black-fruited spikerush, and club mosses.

Lookout Point Prairie
Tippecanoe County
Located on 15 acres, this remnant of gravel hill prairie is the largest in the state. Developed on a river outwash terrace, it contains a number of rare or unusual prairie plants, such as western wallflower, rare asters, fringed puccoon, slender sandwort, and prairie satin grass. It is an important habitat for grassland birds, small mammals, and reptiles. Raptors frequently hunt over the prairie. The closed property is owned by The Nature Conservancy.

Loon Lake Preserve
Steuben County
This 99-acre preserve, 2 miles northwest of Angola, is an important wetland that rims the southern shore of Loon Lake. It has a variety of different wetland communities, supporting more than 12 rare, threatened, or endangered plant species, including purple bladderwort, Robbin's spike rush, cuckoo-flower, pitcher plant, cottongrass, rose pogonia, and tamarack.

Mallard Roost
Noble County
This 145-acre tract along the floodplain of the South Branch of the Elkhart River about 10 miles west of Albion is a high-quality remnant of what was once one of the largest wetlands in the state. Natural communities include marsh, bog, swamp, fen, and willow-cottonwood bottomland forest. It is a critical habitat for a number of rare or endangered plant and animal species, including the badger, start-nosed mole, mountain holly, Loesel's twayblade, herons, rails, egrets, and the sandhill crane. Thousands of birds land here to rest and feed during spring and fall migrations.

Manitou Islands Area
Fulton County
Located at the south end of Lake Manitou southeast of Rochester, this 298-acre wetland is accessible only by boat. It is surrounded by old-growth oak-hickory upland forests; lower islands in the marsh are covered with large tamarack trees. It is a critical nesting habitat for a number of birds, such as marsh wrens, rails, ducks, herons, and swamp sparrows. Thousands of birds stop here to rest and feed during spring and fall migrations.

Marsh Lake Preserve
Steuben County
Located just north and east of Marsh Lake, this 103-acre wetland protects cattail marsh, sedge marsh, muck flat, calcareous fen, and shrub swamp natural communities. Unusual or rare plant species include tamarack, yellow birch, dwarf birch, Canada mayflower, and starflower. The wetland is an important site for nesting and migrating birds, especially waterfowl. Rare butterfly species and other insects have also been found here. The property is owned and managed by The Nature Conservancy.

Mayme Hinton Glade
Harrison County
Located on 40 acres near the town of Elizabeth, this Nature Conservancy property shelters a globally rare limestone glade amid high-quality upland mesic forest. Several rare or unusual plants occur here, including angle pod, little-headed nut-rush, and orange coneflower. The property is still being inventoried; visitation is by permission only.

McClue Nature Preserve
Steuben County
This 80-acre tract a few miles southwest of Fremont encloses some of the best old-growth forest in the state. The trees, mostly tulip and white oak, are hundreds of years old. The for-

est is also a critical nesting habitat for several rare bird species, including the American redstart warbler and the cerulean warbler.

McKenny Nature Preserve
Steuben County
Located about 45 miles northwest of Fort Wayne, this 80-acre tract is owned and managed by The Nature Conservancy. Its natural communities include mudflats, muck flats, open water pond, sedge-rush marsh, buttonbush swamp, lowland forest, and upland oak forest. Boglike portions of the wetlands have floating mats of sphagnum moss. Among the diverse plant life are a number of rare species, including the tall-beaked rush. The wetlands are a critical habitat for nesting and migrating waterbirds.

Merry Lea Preserve
Noble and Whitley Counties
Located a few miles southwest of Wolflake, this 880-acre tract of forest and wetland contains excellent examples of natural communities, including upland oak-hickory forest, black walnut forest, buttonbush swamp, red maple swamp forest, the shoreline of High Lake, and cattail marsh. This diversity supports an array of rare plants, such as running ground pine, pink lady's slipper, bristly tamarack, carnivorous pitcher plant, orchids, water arum, and whorled pogonia. It is also an important nesting and migratory stopover for birds.

Mongoquinong Nature Preserve
La Grange County
This 65-acre property about 2 miles west of Mongo contains a fen community that is kept wet by flowing alkaline groundwater. Fens are rare in Indiana. Unusual fen plants include a variety of sedges, blazing stars, gentians, and shrubby cinquefoil.

Nine Penny Branch Preserve
Clark County
This 121-acre preserve about 4 miles north of Charlestown contains old-growth mesic upland forest. The tall trees are dominated by beech, oak, and tulip poplar. Nine Penny Run flows through a ravine cut in the limestone bedrock and contains interesting herbaceous plants.

Nipsco Savanna
Jasper County
High-quality black oak savanna, black oak savanna flat, and prairie grow on this 840-acre tract about a mile northeast of Wheatfield. All of these communities are now rare in the state, and several rare or unusual plants can be found. The tract is also an important habitat for grassland birds, insects, and reptiles. Wildflowers are abundant during spring and summer.

North Branch Elkhart Wetlands
Elkhart County
Located on 50 acres along the North Branch of the Elkhart River southeast of Elkhart, this property protects a diverse wetland area along the floodplain. Natural communities include bottomland forest, cattail-willow marsh, sedge meadow, and swamp. Several rare plant species can be found. It is also an important habitat for the rare massasauga rattlesnake and star-nosed mole, two rapidly disappearing species. Open water within the wetlands attracts thousands of waterfowl, especially during spring and fall migrations.

Ober Savanna
Starke County
Located just west of Ober, this 59-acre parcel is owned by The Nature Conservancy and contains a high-quality remnant of black oak savanna. Clusters of large black oaks grow among prairie grasses and wildflowers such as little and big bluestem, Indian grass, porcupine grass, June grass, and puccoons. The rare prickly pear, yellow wild indigo, globally endangered fame flower, crea, and wild indigo are found here, as are some rare, prairie-specific butterflies.

Ogle Hollow
Brown County
Located near Nashville in Brown County State Park, this 41-acre tract sits among forested hill country dissected by ravines. The cool, moist north-facing slope creates an excellent habitat for the rare yellowwood tree and a number of unusual ferns, including Christmas fern, maidenhair fern, and narrowleaf spleenwort. This is one of only a few locations in Indiana where the yellowwood tree grows.

Olin Lake Nature Preserve
La Grange County
This 269-acre property about 3 miles northeast of Wolcottville encloses the largest lake in Indiana possessing an undeveloped shoreline. Natural communities include open water, marsh, swamp forest, and upland forest. Unusual trees along the swampy shoreline

include green ash and tamarack. A remnant of old-growth forest is marked by large maple, beech, and red oak trees that are more than 100 years old. Rare plants include cisco, early ladies' tress, and alder buckthorn. The property is managed by the Indiana Department of Natural Resources.

Orangeville Rise
Orange County
Located in an area of caves and sinkholes about 7 miles west of Orleans, this 3-acre property is located where a disappearing tributary of the Lost River reappears at the base of a low cliff face. Inward from the opening the hidden cave system is home to rare cave species such as the blind cave fish. The site has been designated a National Natural Landmark and is owned and managed by The Nature Conservancy.

Pine Hills Preserve
Montgomery County
Located about 5 miles north of Waveland, this 480-acre parcel is covered by evergreens and upland forest. Old-growth white pine, hemlock, and Canada yew are rare evergreens, relicts from the Ice Age that grow in the deeper ravines. The maple, oak, and beech trees of the uplands were heavily logged during the 1800s, but virgin stands remain. Abundant ferns and several rare plants grow along cliff faces and rocky ravines and canyons.

Pipewort Pond
Elkhart County
Located a few miles north of Bristol and just south of Round Lake and Indiana Lake, this 135-acre parcel protects wetland, open water, peaty muck flat, and sand flat communities. It is an important nesting and migrational habitat for birds, especially herons, ducks, and shorebirds. It also shelters rare relict plants from the coastal Atlantic Plain, such as umbrella sedge, Robbin's spike rush, green-keeled cottongrass, and pipewort.

Portland Arch
Fountain County
Located about 5 miles south of Attica, this 293-acre parcel protects a high-walled canyon through which Bear Creek flows. The unique rock-cliff communities include the rare white pine, wintergreen, and Canada blueberry. Rare ferns, mosses, and lichens grow on the rock walls and include bulblet, hayscented fern, and marginal shield fern. Liverworts, walking fern, and Forbes's saxifrage can also be found. The site is also a National Natural Landmark.

Potawatomi Preserve
Steuben County
Located in Pokagon State Park about 6 miles northwest of Angola, this 256-acre property protects one of the few remaining remnants of the presettlement Northeastern Morainal Lake natural region. Largely a lowland, it contains open water, cattail marsh, sedge meadow, and swamp communities. Rarer vegetation includes tamarack and yellow birch in the swamps. The surrounding uplands are covered by oak-hickory forest. The wetlands are also an important nesting habitat for birds.

Ritter Nature Preserve
Martin County
Located on 79 acres, this property protects one of the three acidic seep springs that remain in the state. The groundwater and soil in this rare community are slightly acidic, giving rise to a number of unique plant species. Other communities include sandstone cliff faces, wet bottomland forest, midslope beech-maple forest, and oak-hickory upland forest. Some of the rare plants found here are various mosses, lichens, and ferns. The permission-only property is owned by The Nature Conservancy.

Rocky Hollow–Falls Canyon
Parke County
Part of Turkey Run State Park about 3 miles north of Marshall, this 1,609-tract preserves old-growth forests and rocky canyon–cliff natural communities. Many rare or unusual ferns, hydrangeas, and hemlock trees cling to the sandstone cliffs. Old-growth forest contains tall stands of beech-maple and oak-hickory trees that are probably hundreds of years old.

Ropchan Memorial Preserve
Steuben County
This 77-acre tract about 2 miles east of Orland is well known for its diversity of habitats and landforms. The hilly terrain comprises moraines, ridges, kettle holes, and swamps, all the result of glaciation. The swampy areas are dominated by interesting or rare plant species, such as tamarack, yellow birch, white pine, winterberry, mountain holly, and cinnamon fern. Part of the lowlands are peaty with sphagnum moss accumulations.

Ropchan Wildlife Refuge
Steuben County
This 247-area just south of Fremont contains upland forest, mix mesophytic forest, lowland swamp forest, and wetland natural communities. It is an important bird habitat. Rare plant species include tamarack, yellow birch, and shinleaf. Rare animals include star-nosed moles, Blanding's turtles, and massasauga rattlesnakes.

Round Lake Nature Preserve
Starke County
This 140-acre property about 6 miles east of North Judson encloses the entire shoreline of Round Lake and its adjacent wetlands. A floating sedge marsh rims the lake and hosts several rare plant species. It is also an important bird habitat.

Saalaman Hollow
Perry County
Owned by The Nature Conservancy, this 100-acre tract about 16 miles north of Tell City protects a section of steep-walled, rocky streambed that has waterfalls, gorges, and outcrop exposures. It is best known for its rare sandstone cliff communities, which include rare ferns, mosses, liverworts, shining clubmoss, mountain laurel, farkleberry, Virginia saxifrage, and French's shooting star.

Salamander Swale
Henry County
Located in the middle of a suburban residential area, this 6-acre parcel of woodland and swamp is unique in that it supports a variety of rare or threatened salamander species. Some of these are the four-toed salamander, silvery salamander, and the ravine salamander. A globally endangered species of sedge can also be found growing in the wetland. The property is owned and managed by The Nature Conservancy.

Sandhills Savanna
Jasper County
This 240-acre property southeast of Demotte encloses a high-quality remnant of what was once a vast black oak sand savanna. The rolling hills are the product of sand ridges and swales deposited during the Ice Age. The savanna contains a diverse mix of prairie plants and sedges, including rare species, such as silky aster and cream wild indigo. It is also an important habitat for grassland birds, insects, and reptiles. Prescribed burns are an effective management tool in maintaining the savanna.

Section 6 Flatwoods
Posey County
Located in the extreme southwestern tip of Indiana, this 437-acre property southwest of Mount Vernon encloses a high-quality remnant of rare flatwoods. Other communities include cypress slough, bottomland forest, swamp, and oxbow wetland along the Wabash River. The flatwood community is globally rare today because so much of it has been converted to farmland. This property has a high biodiversity, including a number of rare or threatened species, such as the Carolina spider lily and buffalo clover. It is an important habitat for a variety of mammals and birds, including forest interior species such as woodpeckers. The property is owned and operated by The Nature Conservancy.

Shoup–Parsons Swamp Woods
Elkhart County
Located on 17 acres southeast of Elkhart along the Elkhart River, this 17-acre parcel encloses a number of communities, including bottomland hawthorn-locust-cherry-tulip forest, upland forest, marsh, shrub swamp, and riparian habitat. Several varieties of rare or unusual plants can be found in this wetland complex, including cardinal flower, bottle gentian, sphagnum moss, orchids, swamp grass, and the flowering rush. It is also a critical habitat for a number of nesting and migrating bird species.

Shrader-Weaver Preserve
Fayette County
This 96-acre tract 2 miles southeast of Bentonville is a National Natural Landmark. It contains a pioneer building and a 30-acre area of old-growth upland forest. The forest is dominated by huge specimens of beech, maple, and walnut that are probably hundreds of years old. Virginia creeper and spicebush can be found in the shrub layer. The property is well known for its colorful display of wildflowers during spring.

Spicer Lake
Saint Joseph County
This 167-acre wetland complex is centered on one of Indiana's few remaining pristine kettle lakes. Located north of New Carlisle, the property encloses open water, peatlands, loosestrife-lily swamp marsh, buttonbush-winterberry

shrub swamp, and red maple swamp forest. Parts of the lake and swamp are covered by a solid mat of floating sphagnum moss. Several rare or unusual plants, some of them relicts from an ancient boreal ecosystem, thrive here. The wetlands are an important nesting habitat and migratory stopover for thousands of waterbirds. A boardwalk allows easy access.

Spinn Prairie Nature Preserve
White County
This 29-acre parcel about 2 miles north of Reynolds is owned by The Nature Conservancy. It is one of the largest remnants of original tallgrass prairie remaining in Indiana. Different communities include wet prairie (cordgrass), mesic prairie (big bluestem and Indian grass), and oak savanna. The mesic prairie also contains blazing star, wild white indigo, rattlesnake master, and downy sunflower. It is an important habitat for grassland birds, insects, and reptiles.

Springfield Fen
La Porte County
Located near Springville, this 45-acre tract contains wetlands, prairie, and a high-quality example of prairie fen. Fens are rare in the state; this example contains a variety of prairie plant species, including some that are rare, threatened, or endangered in Indiana.

Steele Nature Preserve
Brown County
This 92-acre tract a few miles southeast of Belmont contains examples of high-quality mesic and dry-mesic upland forest communities along a high ridge. Some of the oak, hickory, and maple trees are quite large. Whorled pogonia, a rare orchid, grows here among the wildflowers.

Stoner Hill Glade
Harrison County
Located on 461 hilly acres in Harrison County, this Nature Conservancy preserve shelters four rare limestone glades. These open grassy areas are surrounded by mixed hardwood upland forest, which conceals them from view. The glades are vegetated with a diverse number of prairie grasses and forbs, including some rare or unusual species.

Stoutsburg Savanna Preserve
Jasper County
This 236-acre property near Wheatfield encloses rolling sand ridges covered by good examples of sand prairie and black oak savanna. The savanna shrub contains chokecherry, sumac, and blueberries; little bluestem, puccoons, coreopsis, and blazing stars can also be found. Rare plants are the silky aster, cream wild indigo, and sweet fern.

Swamp Angel Nature Preserve
Noble County
An important wetland complex covers this 94-acre parcel in Noble County. Natural communities include lake, fen, bog, swamp forest, shrub swamp, and upland oak forest. Several rare plant and animal species can be found here, especially the spotted turtle and massasauga rattlesnake. Wildflowers are colorful and abundant during spring. The property is owned and managed by The Nature Conservancy.

Tamarack Bog
La Grange County
This 170-acre tract about 2 miles southeast of Mongo contains a rare relict boreal swamp forest. Located along Pigeon Creek, it is the largest remaining tamarack swamp in Indiana. Rare plant species that survive in this cool microclimate are tamarack, speckled alder, dwarf birch, Canada mayflower, and starflower. The wetland and open water communities are important nesting habitats for birds. Rare animal species, including the massasauga rattlesnake, can also be found here.

Tamarack Lake
Steuben County
This lake-wetland complex on an 80-acre tract in Steuben County is surrounded by dense swamp vegetation. Open water, swamp, shrub swamp, wet meadow, and bottomland forest are the dominant natural communities. Several relict boreal plants from the colder climate of the Ice Age survive, including tamaracks and other evergreens. Other rare or unusual species include the massasauga rattlesnake, spotted turtle, sandhill crane, cuckoo-flower, alderleaf buckthorn, and insectivorous pitcher plants. The property is owned and managed by The Nature Conservancy.

Teeple Glade
Harrison County
Located on 82 acres in a mature oak-hickory hardwood forest, this high-quality grassy limestone glade is only one of four remaining in Indiana. It is vegetated with a variety of grass-

es and other plants, including the rare axe-shaped Saint John's-wort, false aloe, heart-leafed meadow parsnip, gaura, angle-pod, and fringed greenbrier. This prairielike expanse is an important habitat for grassland birds, insects, and reptiles. Raptors can also be seen here. The property is owned and managed by The Nature Conservancy.

Tefft Savanna
Jasper County
This 480-acre tract west of Medaryville is a highly diverse savanna complex built on sand dunes. Prairie openings are interspersed through the savanna, which is dominated by black oak, pin oak, blueberry, huckleberry, and a variety of herbs. The wetter areas between the dunes support sedge meadows, wet prairie, and marsh communities. Several varieties of rare animal and plant species inhabit the site, including relict plants from the coastal Atlantic Plain.

Thornhill Farm
Huntington County
This abandoned farm on 77 acres southwest of Fort Wayne is now an important bird habitat. Natural communities include a small stream, man-made ponds, old fields, and young pine and deciduous forests. Although the communities have been radically disturbed or are young and regrowing, more than 135 bird species nest or feed on the site. Many are rare or endangered, including the least bittern, great horned owl, sandpipers, herons, various warblers, savanna sparrow, and northern waterthrush. It is one of the few disturbed properties in the eastern United States that has taken on special wildlife values to become a very important resource.

Three-Way Sedge Sinkhole Swamp
Harrison County
This 10-acre wetland is one of a handful of pristine sinkhole swamps remaining in Indiana. It occupies a shallow surface depression that resulted from the collapse of an underground cave. The sinkhole swamp is a globally rare community. Several rare plants grow here, including several boreal relicts from the Ice Age, as well as log sedge and bur reed. It is an important habitat for amphibians and water-loving birds. The property is owned and managed by The Nature Conservancy.

Trail Creek Fen
La Porte County
Located on 38 acres in La Porte County, this property shelters a rare raised graminoid fen. The diverse vegetation grows on a mineral-rich substrate and groundwater. Nearly a dozen of the more than 200 plants that grow here are rare or endangered, including the insectivorous pitcher plant and tamarack, a relict from an ancient boreal climate. The property is owned and managed by The Nature Conservancy.

Tribbett Woods
Jennings County
This is one of four remaining old-growth flatwood communities in southeastern Indiana. Located on 33 acres a few miles northwest of Paris Crossing, the old-growth specimens of oak and beech are nearly 150 feet tall, 3 feet in breast-height diameter, and hundreds of years old. The property is a critical habitat for interior forest birds such as pileated woodpeckers.

Twin Swamps
Posey County
This 598-acre parcel is located in extreme southwestern Posey County. It contains cottonwood-cypress swamp, overcup oak swamp, and wet flatwood. Only a few remnants of once-extensive bald cypress swamp remain in the state. Visitors can see some very large cypress, cottonwood, and overcup oak trees. It provides critical habitat for a variety of birds, including the blue-gray gnatcatcher and the prothonotary warbler. Rare plants are the spider lily and state-endangered featherfoil.

Versailles Flatwoods Preserve
Ripley County
Located about 2 miles northeast of Versailles in Versailles State Park, this 45-acre tract is an excellent example of a relatively undisturbed mesic flatwood community. Very old American beech and black gum trees grow in the wetter sections. The preserve is also home to the state-rare redstem white violet.

Wabash Lowlands Ecosystem
Posey County
Located in the extreme southwestern corner of the state southwest of Mount Vernon, this 1,268-acre property protects an extensive wetland-forest complex. Natural communities include cypress swamp, buttonbush swamp,

and post oak flatwood. The undisturbed flatwoods are considered globally rare in that there are fewer than 20 lying in their accepted range. The cypress trees, and some of the flatwood trees, are old-growth and quite large. The wetlands are an important nesting and migratory habitat for numerous waterbirds. The property is owned and managed by The Nature Conservancy.

Wawasee Wetlands Preserve
Kosciusko County
This 17-acre tract just north of Sunrise Beach on Lake Wawasee contains open water, cattail marsh, and sedge meadow natural communities. Leatherleaf, buttonbush, and wild gooseberry form a shrub layer along the upland edges. It is an important nesting habitat for a number of bird species, as well as a feeding and resting stop for thousands of migrating birds during spring and fall.

Wening-Sherritt Seep Springs
Dubois County
This 75-acre tract encloses a rare acidic seep spring and surrounding wetland forest. The plant assemblage includes a number of rare boreal relicts from the Ice Age. Rare plants include sphagnum moss, various ferns, Virginia willow, orchids, and the federally endangered bog bluegrass. The property is owned by The Nature Conservancy.

Wesselman Woods
Vanderburgh County
Located just east of Evansville, this 197-acre nature center preserves an exceptional remnant of old-growth lowland forest. The larger specimens of sweet gum, tulip, and oak are probably hundreds of years old and dominate the high canopy. The basal area of 187 square feet per acre is the highest known of any woodland in the state.

Yellow Birch Ravine
Crawford County
This 441-acre tract just south of Taswell is unique as the only place in Indiana where mountain laurel, yellow birch, and hemlock grow together. These rare species survive in the cooler, moist ravines that dissect the hilly upland oak-hickory forest. The birch and hemlock stands are relicts from the last glacial age, when the climate was colder. Indiana's largest yellow birch tree can be found here. Rare plant or animal species include the American bittern, cottongrass, showy lady's slipper, orchids, and the rare yellow lady's slipper.

IOWA

Iowa is divided into the following major natural regions: Paleozoic Plateau, Iowan Surface, Southern Iowa Drift Plain, Des Moines Lobe, Northwest Iowa Plains, and Loess Hills.

Located in the extreme northeastern corner of the state, the Paleozoic Plateau has rugged hills and deep valleys. Natural communities include cliff face, algific talus slope, springs and seeps, hill prairie, oak-juniper glades, upland xeric oak-hickory forest, and mesic black walnut–red oak–hard maple–basswood forest. Many of Iowa's rarest plants are associated with this plateau.

The Iowan Surface region covers about one-seventh of the state in the northeast. Major communities include wetlands, tallgrass prairie, loess-capped ridges (pahas), fens, and sandy areas anchored by various grasses and forbs. Developed on calcium-rich soils groundwater, the biologically diverse fens shelter a number of rare or unique species.

The north-central portion of Iowa is occupied by the Des Moines Lobe. Its flat surface is dotted with pothole lakes, tallgrass prairie, remnant marsh, savannas, peatlands, a sphagnum bog with rare plants, and several fens containing several rare shrubs.

The northwestern corner of the state is covered by the Northwest Iowa Plains. Treeless and flat, this dry region was once part of a vast tallgrass prairie. Although most of this area has been converted to farms, prairie remnants remain. Unusual rock communities have also developed on abundant quartzite outcrops in the extreme northwestern corner.

The Southern Iowa Drift Plain covers most of the southern half of Iowa. It was once heavily forested, and significant stands remain. Communities include prairie, rock face, algific talus, riparian corridor, and various xeric and mesic forest types, including oak-hickory and other mixed hardwoods.

The Loess Hills form a thin north-south strip along the western edge of the state that parallels the floodplain of the Missouri River. The sediments in these dry hills were originally deposited by wind. Some of Iowa's rarest and most diverse prairie remnants are hidden in these hills.

Special Habitat Sites

A. F. Miller
Bremer County
This 6-acre tract near Sumner protects a high-quality sedge meadow–shrub swamp in an old oxbow of the Little Wapsipinic River. Among the plant species are some rare sedges and ferns. The wetland is also an important bird habitat.

Anderson Prairie
Emmet County
Located about 2 miles northwest of Estherville, this 200-acre site contains excellent remnants of original prairie. Natural communities include dry prairie, wet prairie, oak savanna, and floodplain forest. The prairies are dominated by big and little bluestem, sideoats grama, and Indian grass. Parts of the prairie have never been plowed. Large bur oaks are present on the savanna. Some rare prairie plant and animal species are found here.

Behrens Pond and Woods
Linn County
This 30-acre tract about 2 miles northwest of Toddville contains woodland, prairie, and wetland communities. The Nature Conservancy bought it to protect critical wetland habitat for the blue-spotted salamander and other rare amphibians that live in the area.

Berry Woods
Warren County
This 42-acre tract about 3 miles north of Indianola preserves an excellent example of old-growth upland forest. It is reminiscent of the forest that covered this part of Iowa during presettlement times. Logging was limited. The largest trees—red oak, white oak, and basswood—are more than 200 years old.

Bixby Woods
Clayton County
Located on 184 acres about 3 miles north of Edgewood, this preserve includes rock-cliff, karst, and upland forest communities. Because of the north-facing bluffs, ice remains under the talus slope as late as September. This helps maintain a rare, cool microclimate that can support fragile and delicate species of plants (ferns, mosses, lichens) and animals such as snails. Many of the plant species are relicts from the last glaciation.

Bluffton Fir Stand
Winneshiek County
Located on 94 acres 8 miles northwest of Decorah, this preserve along the rocky bluffs of the Upper Iowa River contains the largest known relict stand of balsam fir in Iowa. The north-facing slopes create cool and shady conditions that support relict species, including other evergreens as well as mosses and ferns.

Brushy Creek Canyon
Fayette County
Located about 2 miles northwest of Arlington, this 217-acre tract preserves a section of wooded, rocky gorge, which plunges up to 150 feet and contains a variety of habitats, including dry forest, mesic forest, and cliff communities. The cliff species include rare ferns, forbs, and shrubs. Forest trees are oak, hickory, sugar maple, and basswood.

Cameron Woods
Scott County
This 36-acre tract a mile west of Maysville encloses an old-growth upland forest. Large, old specimens of red oak and hickory grow on the property, which has been only slightly disturbed since presettlement times. It is also well known for spectacular displays of wildflowers during spring and summer.

Cayler Prairie
Dickinson County
This National Natural Landmark covers 160 acres about 8 miles southwest of Montgomery. It contains remnant wet prairie (big bluestem and Indian grass), dry prairie (little bluestem), and marsh (slough grass and sedge) communities. Several rare prairie-specific butterflies have been found here. The marsh is also an important bird habitat.

Cedar Hills Sand Prairie
Black Hawk County
This 35-acre tract 4 miles northwest of Cedar Falls is owned by The Nature Conservancy. It

protects an excellent unplowed remnant of prairie that has developed on old sand dunes. The dry sand prairie is marked by big and little bluestem, switchgrass, prairie dropseed, and purple lovegrass. A sedge meadow and fen are also located on the site.

Cheever Lake
Emmet County
Two miles southwest of Estherville, this 366-acre tract is one of the few remnants of a much larger wetland that once covered much of northern Iowa. Typical plant species include cattail bullrush, water lily, and bladderwort. It is an important nesting habitat and migratory stopover for a number of bird species.

Clay Prairie
Butler County
Located 9 miles north of Parkersburg, this unplowed remnant prairie is preserved within a small pioneer cemetery. Typical plant species include little bluestem, Indian grass, shooting star, prairie smoke, and golden alexander.

Crossman Prairie
Howard County
Owned by The Nature Conservancy, this 10-acre parcel 6 miles northeast of Riceville protects remnants of virgin prairie. Dominant prairie plant species include big and little bluestem, prairie dropseed, Indian grass, sedges, and blue joint grass.

Derald Dinesen Prairie
Shelby County
This 20-acre tract 4 miles northeast of Harlan contains one of the best remaining remnants of tallgrass prairie in western Iowa. Plant species include slough grass, sedges, and big and little bluestem. Much of this prairie has never been plowed.

Doolittle Prairie
Story County
Located on 25 acres southwest of Story City, this preserve contains pothole wetlands, remnant mesic prairie, and remnant wet prairie. Prairie species are little and big bluestem, Indian grass, prairie dropseed, slough grass, and sedges. The relatively small wetlands attract a variety of bird life.

Fallen Rock
Hardin County
This 122-acre tract just southwest of Steamboat Rock encloses steep bluff faces and sandstone cliffs along the Iowa River. The north-facing slopes provide a cooler microclimate that allows rare relict northern species, such as white pine, yellow birch, and leatherwood, to survive here. It is also on an important bird migration route, especially for raptors.

Five-Ridge Prairie
Plymouth County
Located on 789 acres 10 miles northwest of Sioux City, this property protects good examples of remnant dry prairie and upland woods. The woods are dominated by bur oak; the dry prairie is covered with little bluestem and sideoats grama. Several rare prairie butterflies have been identified here.

Freda Haffner Preserve
Dickinson County
This 110-acre parcel, located about 4 miles northwest of Milford, contains remnants of dry and mesic prairie as well as kettlehole wetlands. Prairie species include little and big bluestem, hairy grama, switchgrass, muhly grass, and sedges. Birds are attracted to the kettleholes. The property is owned by The Nature Conservancy.

Gitchie Manitou
Lyon County
This 91-acre tract near Granite is along the Big Sioux River. It contains quartzite rock exposures, prairie, and woodland communities. The woodlands contain bur oak; the remnant prairie contains big and little bluestem, sideoats grama, prairie dropseed, Indian grass, and porcupine grass. Unusual lichen and moss species can be found among the rock outcrops.

Hanging Bog
Linn County
Owned by The Nature Conservancy, this 16-acre tract 2 miles northwest of Covington contains excellent examples of upland sugar maple–basswood–red oak forest and bog communities. The bog is permanently wet from seeping calcareous groundwater and contains abundant examples of marsh marigold.

Hayden Prairie
Howard County
This 240-acre parcel about 6 miles north of Saratoga is the largest tallgrass blacksoil prairie remnant in the state. It is dominated by big and little bluestem and Indian grass. It is an important habitat for a number of prairie plant and animal species, including snakes and voles.

Hoffman Prairie
Cerro Gordo County
This 36-acre tract of remnant virgin prairie 3 miles west of Clear Lake is owned by The Nature Conservancy. Wetland communities are also on the site. The prairie is dominated by big bluestem and Indian grass. Several rare plant species and butterfly species have been identified here.

Kalsow Prairie
Pocahontas County
Located 4 miles northwest of Manson, this 160-acre prairie contains remnants of mesic prairie (big bluestem and prairie dropseed), wet prairie (Indian grass, switchgrass, and slough grass), and pothole wetlands. Although smaller in size, the wetlands attract a variety of bird species.

Kish-Ke-Kosh Prairie
Jasper County
Located 2 miles south of Reasnor, this 17-acre tract comprises remnant original prairie. It is covered by big and little bluestem, Indian grass, and prairie dropseed. The wildflowers are spectacular during spring and summer. Rare butterfly species have also been identified here.

Liska-Stanek Prairie
Webster County
Located 3 miles south of Fort Dodge, this 20-acre tract contains remnant original prairie. Plant species include big and little bluestem, needlegrass, slough grass, blue joint grass, sedges, and a larger population of the rarer compass plant. Pothole wetlands on the property attract a variety of birds.

Malanaphy Springs
Winneshiek County
Located 2 miles northwest of Decorah, this 66-acre tract contains upland forest, cliff, and algific talus communities. The north-facing slopes and long-lasting ice under the talus create a cooler microclimate in which rarer northern relict species can thrive. These include ferns, lichen, moss, and evergreens. The talus community is very delicate and should not be walked on. The spring is surrounded by mature maple-basswood upland forest.

Manikowski Prairie
Clinton County
This 40-acre tract near Elwood contains remnant original dry prairie dominated by little bluestem. Several rare or unusual forbs also grow here.

Marietta Sand Prairie
Marshall County
Original wet prairie, dry prairie, and wet woodland communities are protected in this 17-acre tract about 2 miles northwest of Marietta. Prairie plants are reed grass, lovegrass, big and little bluestem, sedges, and sensitive ferns. The woodlands are dominated by large specimens of elm, box elder, and aspen.

Mericle Woods
Tama County
Located 3 miles northwest of Toledo, this 132-acre tract preserves an excellent example of mature upland oak-sugar maple-walnut woods. Patches of old-growth trees include white oaks that are more than 200 years old. It is also an important nesting habitat for birds.

Merrit Forest
Clayton County
About 5 miles south of Guttenburg, this 20-acre tract encloses old-growth mesic upland forest. Trees include oak, maple, basswood, hickory, and walnut. Many of the oak trees are older than 200 years.

Mossy Glen
Clayton County
Located about 4 miles north of Edgewood, this tract contains rocky bluff, cliff, and upland forest communities. The north-facing cliff faces create a cooler microclimate in which a number of rarer northern plant species, such as mosses, ferns, lichens, and evergreens, can thrive.

Mount Talbot
Woodbury and Plymouth Counties
Lowland forest, prairie, and rock communities are protected within this 90-acre tract just

northwest of Sioux City. The valleys are forested with bur oak; the prairie is covered with little and big bluestem, sideoats grama, and Indian grass. The prairie is developed on loess soils. More than 50 butterfly species, some of them rare or endangered, inhabit the site.

Nestor Stiles
Cherokee County
This unplowed mesic tallgrass prairie remnant covers 10 acres 2 miles south of Aurelia. Prairie species include big bluestem, porcupine grass, and prairie dropseed. Wildflowers include the more unusual Jerusalem artichoke.

Pecan Grove
Muscatine County
Located 2 miles southwest of Muscatine, this 23-acre parcel is rare in that it contains the northernmost population of pecan trees in Iowa. It is found in the Mississippi River floodplain, along with large specimens of box elder and cottonwood trees.

Pilot Knob
Hancock and Winnebago Counties
Located 3 miles east of Forest City, this 230-acre tract contains oak-hickory upland forest and wetland communities. The wetlands occur as open water ponds or marshes in kettles. A bog is also present with a floating sphagnum mat. Rare plant species are found in the bog. The wetlands are an important bird habitat.

Roberts Creek
Clayton County
This 13-acre tract in central Clayton County encloses steep, rocky, north-facing cliffs and bluffs, including algific talus slopes that preserve ice well into the summer months. These conditions create a cooler microclimate in which rarer, more northern plant species can survive, including mosses, lichens, ferns, birch, and evergreens.

Rock Creek Island
Cedar County
This 35-acre tract just west of Rochester protects a large Cedar River island, one of the few stable river islands in the state. Covered by floodplain woods that may be old-growth, the island displays some very large and old cottonwood, maple, and sycamore trees.

Roggman Boreal Slopes
Clayton County
Located on 20 acres in Clayton County, this preserve is owned by The Nature Conservancy. It contains north-facing slopes and bluffs that have algific talus slopes. These slopes preserve underlying ice well into the summer months. These conditions create a cool microclimate in which northern relict plants can survive, such as ferns, lichens, moss, and trees, including white pine or birch. The talus slopes are very delicate.

Rolling Thunder Prairie
Warren County
This 123-acre remnant original prairie 3 miles east of New Virginia is an excellent example of dry and mesic tallgrass prairie. Plant species include needlegrass, junegrass, big bluestem, and Indian grass. It is rare in that it has never been plowed and retains the plant and animal diversity of the original prairie.

Savage Memorial Woods
Henry County
This 12-acre tract 5 miles from Mount Pleasant is rolling upland oak-hickory forest that contains old-growth trees. Huge white oak trees are probably older than 200 years.

Searryl's Cave
Jones County
This cave is on a 43-acre tract a few miles southeast of Cascade. Located along the North Fork of the Maquoketa River, the cave system is surrounded by mature upland oak-hickory forest. The cave is an important habitat for a number of bat species.

Sheeder Prairie
Guthrie County
Located on 25 acres about 10 miles west of Guthrie Center, this tract is an excellent remnant of original tallgrass prairie. Dominant species are little bluestem, big bluestem, and prairie dropseed. It is well known for spectacular blooms of wildflowers during spring and summer.

Silver Lake Fen
Dickinson County
This unusual fen community is located on 10 acres 8 miles east of Harris. It is a water-saturated peat deposit complete with mineralized mats, called tufa, and floating moss mats.

Uncommon fen species include a variety of sedges, lobelia, and the sawtooth sunflower.

Steele Prairie
Cherokee County
This 200-acre tract a mile west of Larrabee is one of Iowa's largest remnants of original tallgrass prairie. Other communities include sedge meadow and riverine marsh. Prairie species are big bluestem, porcupine grass, panic grass, sedges, blue joint grass, and cordgrass. The wetlands attract a variety of bird life.

Stinson Prairie
Kossuth County
Located 5 miles southwest of Algona, this 32-acre property contains remnant mesic prairie, pothole wetland, and sedge meadow communities. Dominant prairie species include needlegrass, little bluestem, big bluestem, sideoats grama, and Indian grass. Slough grass and a variety of sedges occur in the wet areas. The pothole wetlands attract a number of different bird species.

Strasser Woods
Polk County
Located in Des Moines, this 40-acre tract contains old-growth upland forest communities. Although some logging has occurred, there are patches of large red and white oak trees that may be as old as 200 years.

White Pine Hollow
Dubuque County
Located 2 miles west of Luxemburg, this 712-acre preserve is a National Natural Landmark. It protects a grove of very old white pine trees that has grown below north-facing cliffs for hundreds of years. The white pine is a relict species more common in northern climates; it was established in Iowa during the last glaciation and is now rare for this state. It grows here only because of the cooler microclimate created by the north-facing cliffs.

Williams Prairie
Johnson County
This 20-acre tract about 6 miles north of Oxford protects unplowed remnant wet prairie near the floodplain of the Iowa River. Dominant species include big bluestem, Kentucky bluegrass, blue joint grass, and various sedges. Rare butterflies have been identified here.

Woodman Hollow
Webster County
This 63-acre tract is well known for its sandstone cliff–bluff plant communities. The cooler, moist, shaded cliffs and rock faces support one of the most diverse populations of ferns, mosses, and liverworts in Iowa, including some rare or unusual species. The rocky valley is surrounded by mature oak-hickory uplands.

KANSAS

Kansas is divided into the following major natural divisions: Ozark Plateau, Osage Plains, Dissected Till Plains, Arkansas River Lowland, Dissected High Plains, and the High Plains.

Located in the extreme southeastern corner of the state, the Ozark Plateau consists of rocky, mountainous terrain covered by oak-hickory forest.

The Osage Plains cover most of the eastern third of the state. Topography is flat to gently rolling atop shale, sandstone, or limestone substrates. Vegetation types include oak-bluestem cross timbers, post oak or blackjack oak savanna, and big bluestem–little bluestem–Indian grass–switchgrass tallgrass prairie. Tall cottonwood-sycamore-willow trees are restricted to riparian corridors.

Located in the northeastern corner of the state, the Dissected Till Plains are noted for their glacially derived soil, including loess hills. The surface is flat to gently rolling, with some steeper bluffs along the Missouri River. Vegetation types include oak-hickory forest and tallgrass prairie openings. Cottonwood, elm, and willow trees form lush riparian corridors along streams and rivers.

The Arkansas River Lowland forms an east-west–trending band in the western half of the state. The soils in the valley of the Arkansas River are mostly sandy and covered by sand prairie or sandsage prairie dominated by big bluestem, little bluestem, switchgrass, sandhill sage, and sandreed.

The Dissected High Plains are found in north-central Kansas. This upland area consists of gentle hills, ridges, and ravines atop a base of sandstone and limestone. The dominant vegetation is bluestem–sideoats grama–blue grama prairie, with occasional stands of red cedar.

Rolling prairie covers the sweeping High Plains in the western third of the state. An expanse of few hills and trees, the landscape is covered by bluestem-grama tallgrass prairie that changes toward the west to blue grama–buffalo grass short-grass prairie to the west. Other plants in the short-grass prairie include saltgrass, buckwheat, and broomweed.

Special Habitat Sites

Baker Wetlands
Douglas County
A National Natural Landmark, this 573-acre site is an important habitat for the sensitive northern crawfish frog. Located just south of Lawrence, it is part of Wakarusa River bottomlands. The site is covered by prairie, wetland, and riparian woodland. Prairie species include cordgrass, grama grass, switchgrass, big bluestem, and compass plant. It is home to a variety of salamanders and frogs, some of them rare. It is an important nesting habitat for a variety of birds and a migratory stopover during spring and fall.

Big Hill Reservoir/Wildlife Area
Montgomery County
This 2,560-acre tract near Cherryvale is an excellent example of oak forest and bluestem prairie. There is a high diversity of forest and prairie flora and fauna, including turkeys, flying squirrels, songbirds, and armadillos. The prairie is well known for its brilliant display of wildflowers. The wetlands are an important migratory stop for thousands of birds during spring and fall.

Byron Walker Wildlife Area
Kingman County
This 4,530-acre tract about 7 miles west of Kingman is well known for its high biodiversity. It shelters a mix of woodland, prairie, wetland, and flowing-water communities. The river environment is along the South Fork of the Ninnescah River and is an important nesting and migratory habitat for a variety of bird species.

Cheyenne Bottoms
Barton County
Located about 10 miles southeast of Hoisington, where Deception and Blood Creeks join to form the upper stretch of Little Cheyenne Creek, this 20,000-acre wetland is the largest marsh in the interior United States. Natural communities include marshes, mixed-grass prairies, wheatgrass-saltgrass, open water, and mudflats. It is one of the most important shorebird migration points in the Western Hemisphere. About 45 percent of all North American shorebirds stop here during spring migration. It is a critical habitat for the threatened or endangered whooping crane, bald eagle, peregrine falcon, and least tern. Cheyenne Bottoms is threatened by the increased use of water resources in the area and has occasionally dried up in drier seasons. Groundwater irrigation began in 1940. Restoration plans are under way. Part of Cheyenne Bottoms is owned by The Nature Conservancy.

Cimarron National Grassland
Morton County
This huge, 108,175-acre tract of U.S. government land about 10 miles south of Richfield largely comprises rolling sandsage prairie. The prairie is dominated by gray-green sand sagebrush that stands about 3 feet tall. A variety of wildflowers blooms during spring and summer. Unique wildlife includes herds of elk and pronghorn. Rare species consist of several prairie plants, mountain chickadees, scrub jays, steller's jays, roadrunners, brushtits, and western tanagers. A portion of the historic, rutted Santa Fe Trail also crosses the property.

Dingus Natural Area
Linn County
Purchased by The Nature Conservancy and managed by the Kansas Ornithological Society, this tract protects eastern deciduous woodlands, oak-hickory uplands, cliff communities, and wetlands. The lush forest understory shelters numerous ferns, including the rare purple cliffbrake. The wetlands shelter great blue heron rookeries; the natural area is also an important wild turkey nesting habitat.

El Dorado Reservoir
Butler County
Located on 16,000 acres about 5 miles northeast of El Dorado, this preserve encloses rock face, open water, woodland, prairie, and wetland communities. The dominantly bluestem tallgrass prairie contains a rich diversity of prairie plants and animals, including the greater prairie chicken. It is also on an impor-

tant bird migration route. Bald eagles nest here during winter.

Elk City Reservoir
Montgomery County
This 13,000-acre preserve about 3 miles west of Independence includes a large reservoir, rolling prairie, upland forests, and unique rock formations. Thousands of snow geese stop here as they migrate from November through March. Other visitors are neotropical birds, such as buntings, tanagers, warblers, and pileated woodpeckers. It is also important winter habitat for the bald eagle.

Flint Hills
Lyon County
This natural division is a 40 mile-wide stretch of tallgrass prairie—the largest in the world. Centered on the towns of Cassoday and Lapland, this rolling prairie is largely privately owned. Big bluestem, little bluestem, Indian grass, and switchgrass can be viewed from a number of roads. It is an important habitat for coyotes and raptors, including hawks, falcons, and kestrels. The area also contains one of the largest populations of prairie chickens in the world, and their ritualistic mating behavior can be viewed during March. Pronghorns, rare in Kansas, also live here.

Flint Hills Wildlife Refuge
Coffey County
Located on the Neosho River at the upper end of John Redmond Reservoir, this 18,500-acre tract preserves wetlands, croplands, and tallgrass prairie. The bird life is highly diverse, including rarer species of woodpeckers, flycatchers, sparrows, turkey vultures, and warblers. It is an important stopover for monarch butterflies that migrate to Mexico in September. Bald eagles also winter in the area. The threatened Neosho madtom, a small endemic fish, lives in the Neosho River.

Fort Riley Military Reservation
Riley County
This 101,000-acre section of U.S. Army land about 2 miles west of Ogden offers the highest diversity of wildlife in north-central Kansas, including elk. Natural communities include wetland, woodland, and prairie. Bald eagles are also frequently sighted.

Gypsum Hills
Barber County
Located near Medicine Lodge, this area of rolling countryside, buttes, and mesas is an important wintering habitat for mountain bluebirds. Thousands of bats, including some rare or threatened species, inhabit cave systems in the area. The prairie grasses are dominated by little bluestem and sideoats grama. Groves of cedar trees attract mountain bluebirds, cedar waxwings, and Townsend's solitaires. Other interesting species are armadillos, roadrunners, black-tailed prairie dogs, cliff swallows, tarantulas, and the Texas horned lizard. Most of the Gypsum Hills are privately owned.

Jamestone Wildlife Area
Republic and Cloud Counties
One of the largest wetland complexes in Kansas, this 3,230-acre site about 3 miles southwest of Kackley protects a variety of natural communities, including marsh, open water, prairie, and woodland. It is well known as a critical nesting habitat for various bird species and is an important stopover for thousands of migrating birds, mostly waterfowl.

John Redmond Reservoir
Coffey County
This 10,870-acre area is located in the broad floodplain of the Neosho River, about a mile west of Strawn. Part of the Central Flyway, it is an important stopover for migrating birds. Bluestem prairie also grows here. Rarer bird species include bald eagles, rough-legged hawks, snow geese, and various herons.

Kirwin National Wildlife Refuge
Phillips County
Located in the Smoky Hills about 10 miles southeast of Glade, this 11,000-acre tract is located on the Central Flyway. It is therefore one of the major feeding stops for migrating waterfowl during spring and fall. It encloses lake, riparian woodland, and mixed-grass prairie communities. Unusual species include black-tailed prairie dogs, Franklin's gulls, golden eagles, bald eagles, and kestrals. The prairie is managed through controlled burning.

Konza Prairie
Riley and Geary Counties
Formerly part of a cattle ranch, this 8,600-acre tract about 7 miles south of Manhattan was purchased by The Nature Conservancy during

the 1970s. It protects virgin tallgrass prairie and limestone-capped hills covered in mature oak-walnut-hickory-hackberry forest. Other communities include shrub and claypan. The woodlands as well as big and little bluestem–Indian grass–switchgrass prairie support a rich variety of plant and animal species, including a large population of prairie chickens. A buffalo herd also grazes on the prairie.

La Cygne Lake and Wildlife Area
Linn County
This 4,600-acre tract about 5 miles east of La Cygne is an important habitat for birds, including bald eagles, ducks, grebes, mergansers, and herons. This lake and wetland is surrounded by oak-hickory woodlands. It is also a stopover for migrating birds during spring and fall.

Marais des Cygnes Wildlife Area
Linn County
Located about 5 miles north of Pleasanton, the translated name for this 7,235-acre preserve is "Marsh of the Swans." Its extensive wetlands host up to 150,000 birds during fall and spring migrations, including trumpeter swans, herons, white pelicans, and cormorants. The prehistoric-looking paddlefish also lives in its waters. Southern flying squirrels can be found in the surrounding woodlands.

Maxwell Wildlife Area
McPherson County
Located about 6 miles north of Canton, this area, about 4 square miles, is one of the best to view large North American mammals in the wild—especially bison and elk—from nearby roadways or an observation tower. The area is mixed woodland and prairie. The rare regal fritillary butterfly and rare catclaw brier are also found here.

McPherson Valley Wetlands
McPherson County
Once one of the largest wetlands in Kansas, McPherson Valley is now mostly drained and agriculturally developed. Located about 3 miles east of Inman, this 1,310-acre preserve protects the remaining Lake Inman, Clear Pond, and a few other small marshes. A program to buy adjacent property and convert it to wetlands began in 1989. It is an important nesting habitat for a variety of waterbirds and is a stopover for thousands of migrating birds during spring and fall.

Neosho Wildlife Area
Neosho County
Located just east of Saint Paul in the Neosho River valley, this 3,240-acre tract protects marshes and riparian woodlands. Longnosed and shortnosed gars are found in its waters. Tens of thousands of migrating birds stop here during spring and fall to rest and feed. Bald eagles nest from November to March. It is also critical habitat for nesting wood ducks and pileated woodpeckers. Southern flying squirrels are found in the woodlands.

Prairie Center
Johnson County
Just 6 miles west of Olathe, the Prairie Center protects remnant tallgrass prairie. Dominant prairie species include big and little bluestem, switchgrass, Indian grass, and a variety of wildflowers. The endangered Meade's milkweed can also be found here.

Pratt Sand Hills and Texas Lake Wildlife Areas
Harvey County
This 5,797-acre tract about 5 miles northwest of Cullison is a large expanse of unique sand prairie habitat. It contains a diverse array of prairie plants and animals and is the easternmost range for the lesser prairie chicken. Other unusual species include kangaroo rats and Swainson's hawks.

Quivira National Wildlife Refuge
Stafford County
This large marsh complex along Rattlesnake River covers some 850 acres 15 miles northeast of Stafford. It contains abundant and diverse wildlife; more than 270 species of birds inhabit the marsh. Hundreds of thousands of birds visit during fall and spring migrations, including pelicans, sandhill cranes, sandpipers, and whooping cranes. It is also home to a variety of raptors, including bald eagles.

Sand Hills State Park
Reno County
This rolling sand prairie covers 1,123 acres about 10 miles northeast of Hutchinson. The hills are formed by wind-deposited sands from the Ice Age Arkansas River. The sand has been stabilized by sand reed, big and little bluestem, dropseed, and switchgrass. It contains a high diversity of prairie plants and animals, including some rare or special-concern species.

Schermerhorn Park
Cherokee County
Located just south of Galena, this 24-acre park protects upland forest, wetland, stream, and cave communities. The hillsides are covered with mature oaks, bitternut, and shagbark hickories. Many threatened and endangers species live here, including the Ozark minnow, black redhorse, and greenside darter. Schermerhorn Cave contains a number of rare cave fauna, including the brilliant orange cave salamander and a variety of protected bat species (gray myotis, eastern pipistrelle, big brown, and red). Pileated woodpeckers and scarlet tanagers also frequent the park.

Scott State Park
Scott County
Located along Lake Scott in Ladder Creek Canyon 13 miles north of Scott City, this 1,120-acre site was first recognized as important by Native Americans who lived there during the 1600s. Large springs distinguish the area. This is the only place in the world where the Scott riffle beetle can be found; it lives in the springs that feed the lake. The park is also an important rock wren nesting site. Managed elk and bison herds can also be seen.

Slate Creek Wildlife Area
Sumner County
This 667-acre tract about 8 miles southwest of Oxford protects woodland, cropland, prairie, wetland, native prairie, and mudflat natural communities. Its mature woodlands host a variety of neotropical songbirds whose habitats are rapidly disappearing. Great blue herons nest in the area, which is also an important stop for migrating birds, especially waterfowl.

Weston Bend Bottoms
Leavenworth County
This 5,600-acre tract just east of Leavenworth is owned by the U.S. Army. It is located below limestone bluffs along the Missouri River. One of the best examples of riparian bottomland in Kansas, its woodlands contain cottonwood, sycamore, hackberry, and walnut. Pileated woodpeckers and nesting great blue herons frequent the site.

Woodson Wildlife area
Woodson County
Part of the Chautauqua Hills, this 2,850-acre tract 10 miles southwest of Yates Center protects upland prairie and oak woodland communities. Dominant prairie plant species include big and little bluestem, sideoats grama, and Indian grass. The cool, wooded ravines shelter delicate species of fern and moss. It is an important habitat for many species, including butterflies, herons, and woodpeckers. The mating rituals of greater prairie chickens can be observed from blinds during spring.

Yates Center Reservoir
Woodson County
This 500-acre site about 5 miles southwest of Yates contains a reservoir surrounded by upland native grasses, including tallgrass prairie. The reservoir is an important site for migrating birds during spring and fall. The prairie vegetation shelters sensitive plant and animal species, such as the rare prairie vole.

KENTUCKY

The Bluegrass State is divided into the following major natural divisions: Bluegrass, Knobs, Appalachian Plateau–Cumberland Mountains, Mississippian Plateau, Shawness Hills, Mississippi Embayment, and Mississippi Alluvial Basin.

Located in north-central Kentucky, the Bluegrass Region has developed atop calcareous, phosphate-rich soil and bedrock. Native forest communities include black maple, bitternut hickory, Kentucky coffee, black walnut, beech, and various oaks and ashes. The English bluegrass and white clover that dominate pastures in the area today were introduced by Europeans more than 200 years ago.

The Knobs comprises a narrow band that borders the Bluegrass region. The terrain consists of rolling hills and valleys, which are covered by grassy plains, moist valleys, second-growth mixed pine-hardwood forest, rocky glades, and dry grassy cliff tops.

The northeast-trending eastern quarter of the state consists of the mountainous, forested Appalachian Plateau and the Cumberland Mountains. The terrain is largely covered by mesophytic second-growth mixed pine-hardwood forests. Some grassy prairie openings can also be found. The rugged cliffs and rock faces shelter a number of rare plant and animal species.

A major portion of central-southwestern Kentucky is made up of the Mississippian Plateau, a flat region underlain by limestone and calcareous shale. Karst topography is common. Major community types include mixed forests and grassy prairies. A few remnant tracts of old-growth forest remain.

Most of the western third of the state is known as the Shawnee Hills, comprising numerous loess hills formed by blowing silt during the last glacial era. An extensive wetland during presettlement times, it is now a patchwork of agricultural fields, segmented wetlands, dry upland forest, and mesophytic forest.

The Mississippi Embayment, in the extreme southwestern corner of Kentucky, is the upper tip of the Gulf Coastal Plain. Also once an extensive wetland complex, it is now a mosaic of farms, wetland remnants, grassy prairies, bottomland forest (cypress and overcup oak) along the major rivers, and dry upland oak-hickory forest on limestone hills and loess bluffs. Other communities include cypress–water tupelo swamps.

Bottomland forests and wetlands once covered most of the Mississippi Alluvial Basin, an area that was once frequently flooded by the Ohio and Mississippi Rivers.

Special Habitat Sites

Aimee Rosenfield Memorial Preserve
Livingston County
This 59-acre Nature Conservancy tract near Smithland protects dry upland forest and grassy glades on a rocky substrate. The glades are a rare natural community that support a number of rare prairie plants.

Audubon State Park Nature Preserve
Henderson County
This property boasts one of the highest biodiversities in Kentucky. Although it is a popular place to visit, its uniqueness as a special area has not been compromised. Mature to old-growth mixed hardwood forest (beech-maple-basswood or oak-maple-tulip), with several streams and riparian corridors, dominate the landscape. Being an old-growth forest, the trees are hundreds of years old and more than 100 feet tall. More than 300 colorful wildflowers bloom during spring. The preserve is an important habitat for a variety of forest mammals and birds, including numerous warblers, tanagers, owls, and woodpeckers.

Axe Lake
Ballard County
This 146-acre preserve east of Barlow protects an important wetland complex centered on Axe Lake. The dominant natural community is an undisturbed cypress-tupelo swamp containing many large trees hundreds of years old. The wetland is an important nesting site for great blue herons, wood ducks, and the great egret. Thousands of waterbirds, including some rare or endangered species, descend on the wetlands during spring and fall migrations.

Bad Branch Nature Preserve
Letcher County
Located about 2 miles southeast of Whitesburg, this 1,199-acre Nature Conservancy property shelters a gorge and waterfall amid a mixed pine-hardwood forest. Natural communities include rock face, open woodland, and mature hemlock-tulip-poplar-birch-beech forest. The cooler microclimate in the wet gorge supports colder-temperature relict plants, such as Fraser's sedge, American burnet, Brook-saxifrage, rose pogonia orchid, and Michaus's saxifrage. Rare or unusual animals here include the arrow darter, various shrews, and the woodland jumping mouse. It is also the home of Kentucky's only know nesting pair of common ravens.

Ballard County Wildlife Management Area
Ballard County
Located on 8,373 acres about 30 miles west of Paducah along the Ohio River, this tract shelters a wetland–bottomland forest complex. Natural communities include pond, marsh, cypress swamp, and bottomland forest. Lotus flowers are abundant along the pool and wetland margins. The property is an important nesting habitat for waterbirds, including herons, and provides wintering habitat for bald and golden eagles. Other rare or threatened species include the orange-footed pimpleback mussel, spotted sunfish, evening bat, and hooded merganser duck.

Bat Cave/Cascade Caverns State Nature Preserve
Carter County
Part of Carter Caves State Resort Park about 5 miles south of Carter, this 146-acre property

encloses a cave system that is inhabited by the federally endangered Indiana bat. Also on the property are two rare plants: the Canadian yew and mountain maple, relicts from a cooler microclimate in the geologic past. About 30,000 bats winter in the cave and can be seen spiraling outward in a funnel at dusk. The surrounding mature mixed pine-hardwood forest is an important habitat for a variety of bird species.

Baumberger Barrens Preserve
Grayson County
This 435-acre property 10 miles northwest of Leitchfield is owned by The Nature Conservancy. It encloses dry limestone barrens and surrounding upland mixed deciduous-pine forest. Rare, calcium-loving prairie plants are found in the grassy barrens.

Blanton Forest State Nature Preserve
Harlan County
This 1,075-acre tract on Pine Mountain northeast of Baxter is part of the largest old-growth forest in the state. Tree types include sugar maple, oak, hemlock, beech, magnolia, and tulip poplar. The largest specimens are more than 100 feet tall, up to 5 feet in diameter at breast height, and centuries old. It is an important habitat for bird and animal species that thrive in old-growth forest, including interior forest birds.

Boone County Cliffs
Boone County
Located about 7 miles southwest of Burlington, this 72-acre Nature Conservancy property protects rugged cliff faces and their specialized flora. The dominant mesophytic forest is composed of mature sugar maple–basswood–beech–white oak–slippery elm. Plant diversity is high; more than 300 different ground species have been recorded, including several rare or unusual ferns, mosses, wildflowers, and lichens. The riparian corridor along Middle Creek shelters the rare red-backed salamander and dusky salamander. Nearly 100 bird species nest in the cliffs.

Buffalo Trace Preserve
Fleming County
This 15-acre property south of Fairview is owned and managed by The Nature Conservancy. The preserve was established to protect high-quality remnant prairie and a population of federally endangered Short's goldenrod.

Canoe Creek Preserve
Henderson County
Located on 78 acres just south of Henderson, this property protects a unique forested riparian habitat along Canoe Creek. Within the shady corridor is a population of leaf cup, a globally rare plant species. The property is owned and managed by The Nature Conservancy.

Cascade Cave
Carter County
This 18-acre property protects the entrance to Cascade Cave. The cave is an important breeding site for a large population of Indiana myotis, a federally endangered bat. No tours are allowed during hibernation. A huge cloud of bats exits the cave at dusk to feed on insects during the summer. Two unusual plants that grow here—mountain maple and Canadian yew—are relicts from a cooler glacial climate.

Cave Hollow Preserve
Lee County
This 168-acre Nature Conservancy property along Big Sinking Creek is a critical bat habitat. Located east of Crystal, the cave system supports ten of the dozen bat species found in Kentucky, including two federally endangered species. The property is owned and managed by The Nature Conservancy.

Clyde Buckley Wildlife Sanctuary
Franklin County
Located on 275 acres, this Audubon Society property is 10 miles south of Frankfort and encloses a transition forest between northern and summer climates. Some old-growth stands of Ohio buckeye, sugar maple, hickory, and tulip poplar approach 200 years old. Nearly 100 different species of wildflowers bloom during spring, including the unusual yellow corydalis, bloodroot, and dwarf larkspur. The rare pileated woodpecker, various songbirds, and owls are some of the more than 100 bird species that nest in the sanctuary. Other species of interest include the Tennessee cave salamander, northern dusky salamander, Fowler's toad, and Blanchard's cricket frog.

Cumberland Falls State Park
McCreary and Whitley Counties
Located about 5 miles northeast of Honeybee, this 1,294-acre park encloses well-developed riparian habitats along the Marsh River. Mature

mixed hardwood forest of hickory, oak, beech is the major forest type. The sandstone cliffs and waterfalls shelter a number of endangered plant and animal species, including the Cumberland bean pearly mussel, green salamander, brook saxifrage, riverweed, elktoe mussel, and kidneyshell mussel. Several rare or unusual ferns and lichens can be found on the cliff faces.

Cypress Creek Swamp Preserve
Marshall County
This 378-acre wetland complex is a few miles from Palma. Natural communities include wet bottomland forest and cypress-tupelo swamp. The wetlands are a nesting habitat for the great blue heron and several other rare wetland species. The property is owned and operated by The Nature Conservancy.

Dinsmore Woods
Boone County
This 107-acre property just south of Petersburg is an excellent example of undisturbed mesophytic maple-oak-ash forest along the Ohio River. This tract has never been logged. A population of federally endangered running buffalo clover grows along the river. The property is owned and managed by The Nature Conservancy.

Eastview Barrens Preserve
Hardin County
This 122-acre property about 20 miles southwest of Elizabethtown is owned by The Nature Conservancy. Natural communities include sandstone barren, limestone glade, and dry oak forest. Rare species include prairie gentian, frostweed, long-haired hawkweed, spikemoss, and several rare insects. Bats are sheltered in small caves and cracks in the limestone rock. The area is managed with prescribed burns.

Flat Rock Barrens
Simpson County
Located north of Franklin, this 65-acre property has large, flat expanses of limestone bedrock, which forms openings in surrounding oak-hickory woods. These limestone glades support seven rare plants, including the limestone fame flower, that love the sunny, dry, calcium-rich soil.

Hi Lewis Pine Barrens
Harlan County
Located on 164 acres, this state nature preserve encloses high-quality pine-oak woodland on Pine Mountain near Baxter. Parts of the forest contain areas of pine barrens, a very rare community. These open grassy areas are dominated by little bluestem and Indian grass and support rare plants, such as frostweed, screwstem, and a healthy population of yellow wild indigo. The many rock faces harbor a variety of lichens and mosses.

Hunter Bluff Preserve
Caldwell County
This 34-acre property 10 miles northeast of Princeton protects cliff face and old-growth western mesophytic forest communities. Several rare rock plants can be found here. The property is owned and managed by The Nature Conservancy.

Jessamine Creek Gorge Preserve
Jessamine County
Located near Nicholasville, this 70-acre property includes a rocky gorge with many specialized rock plants. Other communities are riparian, aquatic, and upland mixed hardwood forest. Biodiversity is high; seven rare plants and two rare bat species live in the rock niches.

Jim Beam Preserve
Jessamine and Garrard Counties
Located near Camp Nelson, this 110-acre Nature Conservancy property is located in the Kentucky River Palisades. These steep limestone cliffs along the Kentucky River are forested with oak, hickory, maple, elm, and buckeye trees. The rocky cliff-top sites support some rare plants, such as water stickwort, Svenson's wildrye, mountain clover, and cleft phlox. The site is also a critical habitat and flyway for rare bat species that live in the limestone cliffs and caves.

Jim Scudder Nature Preserve
Hardin County
Located on 58 acres, this state nature preserve protects a high-quality limestone glade and prairie amid a mature mixed pine-hardwood forest. These communities are extremely rare in Kentucky. Several rare plant species can be found in the dry glades, including silky aster and small white lady's slipper. It is an important habitat for several birds and insects.

Kaler Bottoms Wildlife Management Area
Graves County
This 2,150-acre property 8 miles northeast of Mayfield is situated along the Mississippi

Flyway. Natural communities include cypress swamp and bottomland hardwood forest. Nesting herons and vultures frequent the wetlands, which are an important migratory stopover for numerous bird species.

Kentucky River Palisades
Jessamine County
Located about 10 miles south of Nicholasville, this 356-acre state nature preserve protects a section of the 220-foot limestone cliffs along the Kentucky River. The tops of the cliffs are covered by mature upland mixed hardwood forest (oak-hickory-beech). Several rare plants can be found on the thin calcareous soils at the cliffs' edge; Eggleston's violet and starry cleft phlox thrive in this alkaline habitat.

Kingdom Come State Park Nature Preserve
Letcher and Harlan Counties
This 225-acre site is located on Pine Mountain in Kingdom Come State Park. It protects Line Fork Cave, an important nesting and nursing site for the federally endangered Indiana bat. This cave system shelters Kentucky's third largest Indiana bat colony.

Land Between the Lakes
Trigg and Lyon Counties
The center of this 170,000-acre peninsula between two reservoirs is near the town of Golden Pond, west of Hopkinsville. It was designated an international biosphere reserve in 1991 because of its biodiversity and its high-quality, representative natural communities. More than 250 bird species have been recorded, including the bald eagle. Nearly 900 vascular plant species can be found in the mature to old-growth mixed hardwood forest and intermittent wetlands. Several varieties of rare or unusual groundcover plants grow here, including drooping sedge and spangle grass. The woodlands are critical habitats for the endangered red wolf and interior forest birds.

Lilley Cornett Woods
Letcher County
Located west of Kings Creek, this remnant tract of virgin forest includes more than 100 species of trees and nearly 500 different flowering plants. The largest maple-oak-beech-basswood trees are more than 100 feet tall and hundreds of years old. Narrow streams, creeks, and seeps are home to the extremely rare Wehrle's salamander.

Lloyd Wildlife Preserve
Grant County
Located just outside of Crittenden, this state-managed wildlife management area contains a 20-acre stand of virgin "Big Woods" forest. Typical of the forest that covered this part of Kentucky before settlement, the towering walnut, hickory, and red and white oak trees are more than 150 feet tall and centuries old. The old-growth stand is an important habitat for several rare or special-concern species, such as the pileated woodpecker.

Logan County Glade State Nature Preserve
Logan County
This 41-acre property encloses a rare limestone glade near Russellville. Exposed on a rocky southwest-facing slope in a mixed hardwood forest, the glade community is dominated by little bluestem, sideoats grama, and other typical prairie grasses. Rare plants found here that thrive in the dry, calcium-rich soils are limestone fame flower, Carolina larkspur, and glade violet.

Mammoth Cave National Park
Edmonson County
Located about 30 miles northeast of Bowling Green, this famous site is enclosed within a 53,000-acre park. More than 100 rare cave-dwelling species make Mammoth Cave one of the most biologically diverse cave systems in the world. The high diversity continues at the surface, where nearly 1,000 different plant species grown in a mature, lush setting along the Green River. Relict northern trees, such as hemlock and yellow birch, can be found in the cool, narrow sandstone ravines and gorges. Old-growth yellow poplar–sugar maple–American beech forests remain, the ancient trees reaching heights of nearly 150 feet. An important nesting habitat for interior forest birds, these woods shelter the rare pileated woodpecker and several rare songbirds, including the cerulean warbler.

Mantle Rock Preserve
Livingston County
Located 2 miles north of Joy, this 19-acre Nature Conservancy property is located in Kentucky's Shawnee Hills. The highlight of the natural

communities is the fragile sandstone glade, which is scattered throughout the upland pine-hardwood forest. Developed on sandstone outcrops and derivative rocky soil, the glades support a variety of rare specialized plants in the acidic, dry soil. Some of these include June grass, prickly pear cactus, hairy lipfern, little bluestem, pinweed, and poverty grass. The wildflowers are vibrant during spring. The preserve is also a significant historical site, marking a location where displaced Cherokee tribespeople took shelter during their 1838–1839 march to Oklahoma along the Trail of Tears.

Metropolis Lake
McCracken County
Located 8 miles northwest of Paducah, this 123-acre property surrounds Metropolis Lake, one of the few remaining undeveloped natural lakes in the Ohio River floodplain. This unique wetland-water ecosystem is rimmed with tall bald cypress and swamp tupelo trees. Within the pond are four rare species of fish and a water snake.

Natural Bridge State Park
Powell County
This rugged, rocky, 2,000-acre property south of Slade is well known for its natural arch. It is also a very pristine part of the cliff country along the Cumberland Plateau. The slopes are covered by mature mixed hardwood forests. Rocky outcrops and caves are important habitats for the federally endangered Virginia big-eared bat, Rafinesque's big-eared bat, and the rare or threatened sweet pinesap, grass pink orchid, and small yellow lady's slipper.

Pine Creek Barrens
Bullitt County
Located northwest of Highgrove, this 112-acre Nature Conservancy property protects a rare limestone glade (limestone barren) community. The diverse species include purple coneflower, Great Plains ladies' tresses, and Indian grass. Several rare or unusual insects and birds can also be seen here.

Pine Mountain State Resort Park
Bell County
Located a few miles from Pineville, this state park includes about 800 acres of sparsely forested sandstone outcropping that protects the very rare pale corydalis and pale blue gentian. Other portions of the forest enclose about 90 acres of old-growth hemlock forest. These trees are more than 300 years old. The rugged topography is dominated by mixed pine-hardwood forests with occasional meadows, glades, and steep rock cliffs. It is an important habitat for a number of bird species, including raptors.

Red River Gorge Geological Area
Wolfe County
Part of Daniel Boone National Forest, this 26,000-acre area about 10 miles northeast of Slade protects sheer cliffs and gorges of sedimentary rock, including a series of natural geological arches. The site is heavily forested in mixed pine-hardwood with lesser stands of mature hemlock and bigleaf magnolia trees. In addition to its well-known geological features, more than 500 species of plants have been identified here, including several rare rock-clinging species—the cranefly orchid, New York fern, smilax, bloodroot, fire pink, grass pink, and mistflower. It is also an important habitat for several endangered bat species and raptors.

Sally Brown Nature Preserve
Mercer County
Located within the Kentucky River Palisades, this 500-acre Nature Conservancy property is located in a big bend of the Kentucky River near the small town of High Bridge. Excellent examples of beech–tulip–white oak, oak-hickory, and blue ash–chinquapin oak–sugar maple forest cover the steep limestone bluffs. More than 400 species of plants can be found within the Palisades area, some of them rare or unusual, such as snow trillium, water stitchwort, and the cleft flox. The federally endangered gray bat and Indiana bat may also live in the limestone cliffs.

Sunset Barrens
Warren County
Located 5 miles north of Anna, this 100-acre property includes sandstone barren, sandstone glade, flowing stream, and upland forest communities. Rare quillwort grows along an ephemeral stream. This property is owned and managed by The Nature Conservancy.

Thompson Creek Glade
Larue County
This 99-acre Nature Conservancy property is 5 miles west of New Haven. A limestone-hill glade along Thompson Creek supports several rare plants in the calcium-rich soil, including fame flower and several prairie grasses.

Woodburn Glade Preserve
Warren County
Located on 20 acres a few miles south of Rockfield, this Nature Conservancy property preserves a rare glade community. It is the only known location in the state for the rare Gattinger's lobelia.

LOUISIANA

Six major vegetation types are recognized in Louisiana: loblolly, longleaf pine, bottomland hardwood and upland hardwood forests, prairie, and marsh wetlands.

Loblolly forest comprises loblolly and shortleaf pine, a variety of oaks, hickory, dogwood, and black gum. The largest acreages of this forest type occur in the northwestern quarter of the state, excellent habitat for large mammals such as bears.

Restricted to southwestern Louisiana and along the border with Florida, the longleaf pine forest is a less common forest habitat. Tree types include slash pine, longleaf pine, and scrub oak. Longleaf pine forest occurs on dry, sandy soil, and is an excellent habitat for birds, hawks, and owls.

Corridors of bottomland hardwood forest can be found on the floodplains of major streams and rivers. Dominant species types are cypress and tupelo gum, red maple, bitter pecan, green ash, black willow, and honey locust. This habitat has been greatly reduced by industrialization.

Upland hardwood forest has been greatly reduced by logging and development. The largest single tract of this excellent wildlife habitat is in West Feliciana Parish along the Mississippi River. Trees include a variety of oaks and hickories, magnolia, and beech. The forest hosts a number of endemic plant and animal species.

Vast tracts of prairie once covered southwestern Louisiana. Most of this land has been converted to agriculture, ranching, and other development. There are a few patches of remnant virgin prairie with native grass communities. The prairie is an ideal habitat for specialized species of insects, birds, and small mammals.

Marshes and wetlands are an important wildlife habitat. More than 3 million acres of coastal delta and subdelta marshes line the coast—important wintering grounds for millions of birds. Freshwater marshes also occur along streams and rivers; swamps can be found in low-lying areas dominated by cypress, tupelo, and gum trees. All wetlands are important habitats for birds, mammals, and unique species such as alligators.

Special Habitat Sites

Abita Creek Preserve
Saint Tammany Parish
This 794-acre nature preserve near Folsom protects a large wetland complex. Natural communities on the site include longleaf flatwood, longleaf savanna, bayhead, cypress swamp, riparian, and flowing stream. The property is relatively undisturbed and an important bird, reptile, and amphibian habitat. More than 20 rare or threatened plants have been identified, including the federally endangered Louisiana quillwort. The Nature Conservancy owns and maintains this property.

Acadiana Park
Lafayette Parish
Located within the city of Lafayette, this 150-acre park contains excellent mature to old-growth bottomland hardwood forest. Some of the larger oak, pecan, and sycamore trees are more than 100 years old and dominate the upper canopy. The woodland is an important nesting habitat for a variety of songbirds and woodpeckers. Rare or unusual amphibians and reptiles spotted here include green and gray tree frogs, the leopard frog, Gulf Coast toad, green anole, certain skinks, and the Texas rat snake.

Alexander State Forest Wildlife Management Area
Rapides Parish
This 9,020-acre preserve is located 2 miles east of Woodworth. Natural communities include upland pine forest, upland pine-hardwood for-

est, bottomland hardwood forest, and riparian habitat along streams; wetlands and swamps are also present. The combination of wetlands and forest types results in a high biodiversity of flora and fauna, including reptiles and amphibians. Several varieties of songbirds, such as tanagers and pine warblers, and the endangered red-cockaded woodpecker breed in the area.

Attakapas Wildlife Management Area
Saint Mary Parish
This 26,300-acre wetland complex is located about 5 miles east of Baldwin. As it is connected hydraulically to the Atlantic Ocean, water levels are constantly changing from tidal action and flooding. Natural communities include open water lake and bayou, sand levee, mudflat, shallow pool, swamp, and bottomland hardwood (ash–black willow–oak) forest. The sand ridges are drier and support maple, oak, beech, and hackberry trees. Some stands of mature trees are very old and tall; mature cypress and tupelo trees dominate swampy areas. Hundreds of plant and animal species can be found in this diverse ecosystem, including the rare or unusual bald eagle, Mississippi kite, swamp rabbit, nutria, mink, otter, and alligator. Wildflowers are abundant during spring.

Avery Island Jungle Gardens
Saint Martin Parish
Located on 250 acres about 10 miles southwest of New Iberia, this nature preserve encloses a remnant of mature to older-growth bottomland hardwood forest and extensive swamplands. Avery Island is unique in that it is perched atop an emergent salt dome. The tall oak trees are hundreds of years old. The wetlands are an important nesting habitat for thousands of cranes and herons and are a migratory stopover for birds during spring and fall. Other unique species include the armadillo and alligator. The wildflowers are spectacular during spring.

Baptiste Collette Islands
Plaquemines Parish
Located near the mouth of the Mississippi River, these five river islands near Venice are a prominent nesting site and migratory stopover for millions of birds. The islands are covered with grasses, shrubs, and trees. Some of the unusual, rare, or endangered species that live here are the peregrine falcon, Wilson's plover, black skimmer, merlin, pelican, and several rare terns.

Bayou Bodcau
Bossier Parish
Located just west of Alden Bridge, this 752-acre parcel protects virgin cypress-tupelo swamp and thick bottomland hardwood forest. The cypress trees are very large and hundreds of years old. The site is an important wintering habitat for thousands of migrating waterbirds. The property is a joint venture between The Nature Conservancy and Ducks Unlimited.

Ben's Creek Wildlife Management Area
Washington Parish
Located along Ben's Creek, this 13,856-acre nature preserve contains stream, bog, seep, longleaf pine forest, and lesser bottomland hardwood forest communities. Rare carnivorous plants, such as pitcher plants and sundews, grow in the bogs. The management area also shelters the state's largest population of the rare gopher tortoise. The woodlands are an important nesting habitat and migratory stopover for a variety of bird species.

Big Lake Wildlife Management Area
Winn Parish
This 19,212-acre preserve is located about 15 miles east of Winnsboro. Natural communities include mature bottomland hardwood forest, swamp, slough, and canebrake. This property contains one of the largest contiguous tracts of bottomland hardwood forest remaining along the upper delta of the Mississippi River. The combination of forest and wetlands results in high biodiversity. Several rare or unusual plant and animal species can be found here. Large mammals, such as bears and deer, as well as waterbirds, raptors, vireos, pileated woodpeckers, and numerous warblers and thrushes inhabit the property.

Biloxi Wildlife Mangement Area
Saint Bernard Parish
Located about 40 miles east of New Orleans, this 39,583-acre property encloses brackish to saline marsh in a number of sloughs and bayous. The dominant vegetation is wiregrass, hog cane, salt grass, glasswort, and black rush. This extensive wetland complex is an important habitat for nesting and migratory waterfowl. The endangered brown pelican and bald eagle

have been sighted here. Alligators are abundant.

Blue Hole Lake Recreation Area
Vernon Parish
Located about 4 miles northwest of Cravens, this 109,855-acre nature sanctuary contains a diverse array of natural communities, including longleaf pine–hardwood forest, longleaf pine forest, swamps, streams, oak-dominated flats, and rare sphagnum bogs. Special plant species include older-growth oak and pine trees and rare bog plants, such as sphagnum moss and the carnivorous pitcher plant and sundew. The variety of fauna includes raptors, songbirds, roadrunners, lark sparrows, and the threatened red-cockaded woodpecker.

Boeuf Wildlife Management Area
Caldwell Parish
Located about 10 miles southeast of Columbia, this 48,595-acre tract encloses bottomland forest, canebrake, upland forest, and swamp. The drier uplands are covered by oaks, sweet gums, and cedar elms; the bottomlands are dominantly water locusts, bitter pecans, and overcup oaks. Wetlands-sloughs are dominated by older-growth cypress and tupelo trees. Several varieties of unusual or rare plant and animal species can be found here, including raptors, snowy egrets, orioles, and swamp rabbits. The property is also an important nesting habitat for wood ducks and is a migratory stopover for thousands of birds.

Briarwood Nature Preserve
Winn Parish
Located south of Saline, this property contains wetlands, bogs, and longleaf pine forests. Some old-growth stands of longleaf pine are more than 300 years old. Large bald cypresses, bay magnolias, and gallberry hollies can also be seen. Rare plants, including various orchids, azaleas, pitcher plants, and bog plants, grow in the wetlands.

Cameron Prairie
Cameron Parish
Nearly 25,000 acres in size, this large wetland-prairie complex about 11 miles south of Holmwood is located between Grand and Calcasieu Lakes. Natural communities include extensive freshwater, brackish, and saltwater marsh, coastal prairie, and abandoned agricultural fields that are being reclaimed to prairie. The native prairie grasses are critical habitat for a number of small mammals, grassland birds, reptiles, and insects. The marshes display an incredible diversity of resident and migratory bird life, including ibis, egret, dowitcher, roseate spoonbill, and rail, in addition to millions of shorebirds during spring and fall. Other interesting species include alligator, mink, nutria, and several rare frog species.

Catahoula Lake
La Salle County
This 26,000-acre wetland complex is one of the most diverse in Louisiana. Located about 25 miles northeast of Alexandria, this backwater lake is connected to the Little and Black Rivers and varies from a large, shallow, open water lake to a smaller pool surrounded by marsh grass, mudflat, and swamp. The unique mudflat vegetation is a critical food source for shorebirds and waterfowl. Flanking cypress swamps are graced by tall trees hundreds of years old. The swamp-wetland complex is an important habitat for a variety of bird, animal, and plant species, some of them rare and unusual. Spotted sandpipers, bald eagles, and ospreys are some of the rarer or threatened birds that inhabit the wetlands.

Charter Oak Preserve
Saint Tammany Parish
This Nature Conservancy preserve is located about 15 miles east of Covington. Its 160 acres enclose an old-growth bayhead swamp. The trees are very tall and hundreds of years old. The site has a high diversity of evergreen species, including the only known Louisiana location of the buckwheat tree.

Chicot State Park
Evangeline Parish
Located on more than 6,400 acres 6 miles north of Ville Platte, this state park includes bottomland forest, riparian corridor, wetland, and upland forest habitats. Upland forests include pine-oak-hickory, oak-hickory, and loblolly pine–oak–hickory–sweet gum–dogwood–beech assemblages. A few old-growth hardwood stands remain. The property is an important habitat for a number of flora and fauna, several of which are rare or endangered. The woodlands and Chicot Lake are an important spring and fall migratory stopover for a variety of bird life, including kinglets, vireos, thrushes, warblers, and various waterfowl.

Cunningham Brake
Natchitoches Parish
This 2,000-acre parcel is 4 miles south of Flora in Kisatchie National Forest. It is one of the most important wildlife habitats in Louisiana. An exceptional remnant of a brake-back-swamp, its junglelike environment shelters a number of rare or unusual species, including amphibians and reptiles. Natural communities include bottomland hardwood forest, yellow pine forest, and bald cypress–water tupelo forest. The diversity of bird life is outstanding and includes raptors and rare or endangered species, such as pileated and red-bellied woodpeckers, the southern red-backed salamander, Cooper's hawk, javeline crayfish, and hispid pocket mouse. The low-lying brake is sheltered by mature stands of oak-hickory-pine upland forest. Poisonous cottonmouth and copperhead snakes are common.

Cypress-Black Bayou
Bossier Parish
Located about 10 miles northeast of Bossier City, this 100-acre preserve encloses a remnant of mature mixed pine-hardwood forest with some wetland areas. A variety of seasonal bird life can be found here, including woodpeckers and some rare songbirds, such as the prairie warbler. Ospreys and bald eagles may also be seen in the winter.

Cypress Island Preserve
Saint Martin Parish
This 2,800-acre property east of Lafayette encloses high-quality bottomland hardwood forest, old-growth oak upland forest, and virgin cypress swamp. Some of the larger cypress trees are probably 500-plus years old. An important breeding habitat for a variety of birds, mammals, insects, and reptiles (some of them rare or endangered), the wetlands are one of the largest rookeries in the United States for herons, egrets, and ibis.

D'Arbonne Wildlife Refuge
Union and Ouachita Parishes
This 17,421-acre tract is located northwest of West Monroe. A mix of wetlands, bottomland forest, and upland forest covers the property. The drier, sandier upland areas support white and red oak, hickory, dogwood, and loblolly pine forest (including several mature stands). The frequently flooded bottomlands are forested with willow and overcup oak, cypress, and water elm. Biodiversity is high and includes a variety of mammals, amphibians, reptiles, birds, insects, and plants. Several rare or unusual species can be found here. An important migratory stopover during spring and fall, the refuge hosts a large population of ducks, egrets, yellowlegs, sandpipers, and, occasionally, bald eagles.

Fort Polk Wildlife Management Area
Vernon Parish
This 110,000-acre tract is located about 10 miles southeast of Leesville. The rolling hills and lowlands are covered by upland pine forest, bottomland forest, stream, bog, and wetland. The dominant trees are longleaf pine, various oaks, beech, and magnolia. The wet portions are wooded with cow oak, willow oak, sweet gum, beech, and blackgum. This is an important habitat for woodland mammals and birds, including bears, mink, and bobcats. The endangered red-cockaded woodpecker nests here. Rare plants, such as pitcher plant, sundew, bladderwort, and sphagnum moss, can be found in the bogs.

Fourchon Beach
Lafourche Parish
This 10-mile strip of beach preserves the transition zone between sand dune–grassland and extensive salt marsh. Located about 8 miles southeast of Leeville, this property hosts a large number of breeding and migratory shorebirds and waterfowl. The salt marsh is an important habitat for shellfish and is a food source for the shorebirds. Bird species include a variety of gulls and terns, willets, sanderlings, ruddy turnstones, herons, brown pelicans, rails, and the rare piping and Wilson's plovers.

Frederick's Swamp
Saint Martin Parish
This 76-acre Nature Conservancy property is located just south of Arnaudville and west of Bayou Teche. This is one of the few areas that escaped complete logging during the 1930s. Natural communities include cypress–tupelo gum swamp, upland live oak forest, and bottomland hardwood forest. Many of the cypress trees and some of the oaks are old-growth and hundreds of years old. Frederick's Swamp is an important habitat for an abundance of wildlife, including rare or threatened species.

Jacobs (Walter B.) Memorial Park
Caddo Parish
Located about 3 miles west of Blanchard, this 160-acre nature park preserves an excellent example of mixed hardwood-pine forest. Some larger, older species of oak trees can be found here. It is a well-known habitat for songbirds, including waterthrushes, a number of warblers, and scarlet tanagers. It is also an important wintering stopover for a variety of bird species.

Kisatchie National Forest Cloud Crossing Area
Winn Parish
Located about 15 miles west of Dodson, this 6,300-acre section of national forest contains a highly diverse mix of natural communities, including wetland, bottomland hardwood forest, upland mixed hardwood-pine forest, and stands of the more unusual yellow pine. Some of the trees are mature to older-growth. Species found here include raptors, pileated woodpeckers, vireos and warblers, herons, alligators, frogs, the eastern narrow-mouthed toad, the Woodhouse toad, and several unusual or threatened salamanders.

Lake Cocodrie Preserve
Evangeline and Rapides Parishes
An excellent example of mixed pine-hardwood forest is preserved on this 160-acre tract within Lake Cocodrie. The diverse plant population includes the state-rare crested coral-root orchid. The property is owned and maintained by The Nature Conservancy.

Lake D'Arbonne State Park
Union Parish
Located on 550 acres about 5 miles west of Farmerville, this preserve encloses an excellent remnant of mature loblolly pine–shortleaf pine–white oak–red oak–hickory–black gum forest. Wetland communities occur along the shore of Lake D'Arbonne. A variety of bird and animal life can be found here, including raptors, woodpeckers, some rare warblers and thrushes, egrets, bald eagles, and armadillos.

Lake Fausse Point State Park
Iberia Parish
This 6,100-acre tract 20 miles northeast of New Iberia is part of the Atchafalaya River basin, the largest riverine swamp in the United States. Some portions are pristine and contain mature to old-growth bald cypress trees. Bald eagles, ospreys, and rare swallow-tailed kites nest in the trees. Rare songbirds and pileated woodpeckers also frequent the forest. Other interesting species are black bears, alligators, swamp rabbits, and the Louisiana iris.

Lake Martin
Saint Martin Parish
This 780-acre lake-wetland complex is an important nesting habitat for a variety of birds. Freshwater marshes also occur along streams and rivers; swamps can be found in low-lying areas that are dominated by old-growth mixed hardwood–loblolly pine forest, longleaf pine flatwood savanna–wetland, bayhead forest, riparian corridor, and bottomland forest. Biodiversity is high, and nearly 20 rare or unusual plant species can be found here. The forests are an important bird habitat, especially for the red-cockaded woodpecker.

Little Pecan Island
Cameron Parish
Located about 4 miles southeast of Lake Charles, this 1,800-acre island includes old-growth chenier forest, oak-hackberry forest, and high-quality freshwater wetlands. This site has the only chenier habitat that is fully protected in the state: Larger trees are more than 100 years old, probably much older. Tens of thousands of birds use the island as a migratory stopover during spring and fall.

Loggy Bayou Wildlife Management Area
Bossier Parish
This 4,211-acre tract about 20 miles southeast of Bossier City encloses one of the last remaining tracts of bottomland hardwood forest in northwestern Louisiana. Its large size makes it an especially important habitat for a number of plant and animal species, including amphibians and reptiles. The area is frequently swampy from flooding. Some of the species seen here are a variety of herons and owls, warblers, and swamp sparrows. Several rare or unusual plant and animal species can also be found. Wildflowers are abundant during spring.

Longleaf Scenic Area
Vernon Parish
This 265-acre tract in Kisatchie National Forest is located about 12 miles west of Cravens. It encloses a rare old-growth tract of longleaf

pine forest. The rolling hills and bottomlands are covered by longleaf pine forest, with some stands of hardwoods. In addition to supporting an abundance of woodland mammals, the forest is also a nesting habitat for the threatened red-cockaded woodpecker and several raptors. Several rare amphibians exist along streams that cross the area.

Louisiana State Arboretum
Evangeline Parish
Located on 600 acres about 8 miles north of Ville Platte, this tract is a rare old-growth upland beech-magnolia climax forest. Many of the beech trees are hundreds of years old. The open climax forest is an important breeding habitat for a number of birds, including raptors, woodpeckers, and songbirds. As a wintering area, the arboretum hosts kinglets, various woodpeckers, warblers, and several rare or unusual secies. Several varieties of warblers, including the hooded warbler, can be seen during spring.

Mary Ann Brown Preserve
West Feliciana Parish
Located 7 miles east of Saint Francisville, this 110-acre Nature Conservancy property shelters high-quality hardwood–loblolly pine forest. Flowing streams and riparian habitats are also found within the hilly woodlands. Old-growth beech and magnolia trees are hundreds of years old. Several rare or threatened plants have been discovered here.

North Toledo Bend State Park
Sabine Parish
Located about 10 miles southwest of Zwolle, this 185,000-acre state park is centered on Toledo Bend Reservoir. The lake is edged by wetland communities and bottomland forest; mature mixed hardwood forest and pine forest cover the upland areas. Some stands of longleaf pine approach 200 years old and are critical habitats for the endangered red-cockaded woodpecker. Several rare plant and animal species have been identified in the area.

Pearl River Wildlife Management Area
Saint Tammany Parish
This 35,000-acre site near Slidell is an excellent example of mature to older-growth bottomland hardwood forest; pine-hardwood upland forest, swamp, cypress-tupelo swamp, canebrake, and slough natural communities can also be found. The property represents one of the best examples of what presettlement Louisiana resembled. Large specimens of hickory, beech, elm, magnolia, live oak, Nuttall oak, and bitter pecan can be found across the tract. Stately cypress and tupelo trees shade the swamps and sloughs. This is one of the most biologically diverse areas in the state. Rare or unusual species found here include swamp rabbits, otters, mink, bald eagles, ospreys, swallow-tailed kites, hooded and Swainson's warblers, rattlesnakes, and the barking tree frog.

Peason Ridge Wildlife Management Area
Sabine, Natchitoches, and Vernon Parishes
Located about 20 miles north of Leesville, this 33,488-acre wildlife management area is located on U.S. Army property. The major natural communities include high-quality loblolly pine forest, longleaf pine–oak forest, sandjack oak forest, riparian corridor, and wetland. Biodiversity is high, and several rare or unusual plant species can be found here, including the endangered red-cockaded woodpecker.

Persimmon Gully
Calcasieu Parish
Located about 25 miles northwest of Lake Charles, this 160-acre property encloses high-quality remnants of wet longleaf pine savanna and rare saline barren. The tall longleaf pines are hundreds of years old. Wildflowers abound during spring and summer. Several rare, unusual, or threatened species have been found here, including silveus dropseed, chaetoppa daisy, blazing star, Bachman's sparrow, and Henslow's sparrow. The property is owned and maintained by The Nature Conservancy.

Pomme de Terre Wildlife Management Area
Avoyelles Parish
Located about 6 miles east of Moreauville, this 7,084-acre tract encloses an excellent example of poorly drained, swampy, bottomland hardwood forest. Stands of oak-ash-maple-locust occupy the sandy ridges that separate lakes, swamps, and sloughs. Some of the cypress trees that edge the sloughs are more than 100 years old. The complex is an important breeding habitat for unique species, such as alligators, angingas, pileated woodpeckers, prothonotary warblers, vireos, and yellow-crowned night herons.

Pushepatapa Creek Preserve
Washington Parish
This 22-acre parcel along Pushepatapa Creek is the only land in the state that protects the state-rare mountain laurel shrub. The mature woodlands are home to a diverse population of plant and animal species, especially birds, some of them rare or endangered. The property is owned and maintained by The Nature Conservancy.

Rutherford Beach
Cameron Parish
This well-preserved 8-mile stretch of beach just west of Oak Grove is covered by fragile, shallowly rooted grasses and other plants. The shrubby vegetation is an excellent habitat for a variety of mammals and birds, including coyotes and the unusual ash-throated flycatcher and burrowing owl. Mudflats and saltwater marshes harbor shellfish populations. The shoreline is an important migratory stopover and feeding site for millions of waterfowl and shorebirds, including some rare species, during spring and fall.

Sabine Island Wildlife Management Area
Calcasieu Parish
Part of the Sabine River system, this 8,103-acre wetland complex about 5 miles from Vinton contains bottomland hardwood forest, slough, and brackish cypress-tupelo swamp natural communities. The fluctuating water levels from tidal action create a series of different floral assemblages. Biodiversity is high in this remote property, including a variety of mammals, alligators and other reptiles, and numerous birds, including the rare bald eagle, osprey, and several woodpeckers.

Saline Bayou
Winn Parish
This 6,390-acre roadless area in Kisatchie National Forest is located about 2 miles northeast of Goldonna. Excellent examples of loblolly pine, shortleaf pine, mixed pine-hardwood, bald cypress–tupelo gum, and sweet gum–Nuttall oak forests are the dominant natural communities. Saline Bayou was designated a National Wild and Scenic River in 1986. Several forest stands are old-growth and hundreds of years old. Sensitive species that occur in this undisturbed section of the national forest are the southern red-backed salamander, Cooper's hawk, red-cockaded woodpecker, hispid pocket mouse, javeline crayfish, and several reptiles.

Sam Houston Jones State Park
Calcasieu Parish
One of Louisiana's larger state parks, this 1,087-acre preserve is located about 10 miles northwest of Lake Charles on Beckwith River. Natural communities include upland pine-hardwood forest, bottomland forest, cypress swamp, canebrake, and lagoon. The abundance of water and wetlands attracts a variety of nesting and migratory bird life, including some rare or unusual species. Woodpeckers, owls, raptors, and the prothonotary warbler can be found here. Reptiles include alligators. Some of the larger cypress and oak specimens are hundreds of years old.

Sandy Hollow Wildlife Management Area
Tangipahoa Parish
This 3,617-acre preserve is located 10 miles northeast of Amite. The property consists of riparian habitat, some stands of bottomland hardwood forest, and relatively young longleaf pine–hardwood forest. The woodlands are an important nesting habitat for a number of bird species, including the eastern bluebird, Bachman's sparrow, and various raptors. Perhaps its greatest value is as a habitat for the gopher tortoise, an extremely rare reptile in Louisiana. The pygmy rattlesnake, another rapidly diminishing species, is also protected here.

Schoolhouse Springs Preserve
Jackson Parish
This 30-acre tract protects a rare spring-fed swamp; six globally rare aquatic insects have been discovered in its waters. The property is owned and maintained by The Nature Conservancy.

Sherburne Wildlife Management Area
Point Coupee and Iberville Parishes
This 43,000-acre tract in the Atchafalaya National Wildlife Refuge is about 5 miles southeast of Krotz Springs. It encloses one of the better remnants of mature bottomland hardwood forest in the state. Stands of tall hardwoods are interspersed in vast wetlands, sloughs, and bayous; bamboo brakes are another unique natural community. Stately sycamore, cypress, tupelo, and sweet gum

trees dominate the swampy terrain; the drier, sandier elevations are covered by oak, pine, and, occasionally, beech and maple. Although the area is traversed by roads and power lines, it remains a critical habitat for a number of rare or unusual species, including ospreys, eagles, pileated woodpeckers, Mississippi kites, ruby-throated hummingbirds, wood ducks, and the rare wood stork. Rattlesnakes can be found in the brakes.

Sicily Island Hills Wildlife Management Area
Catahoula Parish
This 7,424-acre property 6 miles west of Sicily Island is an isolated rocky upland covered by upland hardwood forest. Trees include loblolly pine, shortleaf pine, hickories, oak, white oak, magnolia, and beech. Some of the larger specimens are more than 100 years old. The mixed pine-hardwood forest is an important nesting and migratory habitat for a number of birds, including the American redstart, the rare pileated and red-cockaded woodpeckers, warblers, and vireos. Several varieties of animal and reptile life can be found in the forest, including armadillos and one of the state's largest populations of the nationally declining timber rattlesnake.

Spring Bayou Wildlife Management Area
Avoyelles Parish
This 12,078-acre wildlife management area about 3 miles southeast of Marksville is a bottomland hardwood-swamp ecosystem. Flooding typically keeps about 40 percent of the area underwater, forming an intricate system of densely vegetated, junglelike sloughs, bayous, canebrakes, lakes, and lowland forests. Water-loving trees found here are overcup and Nuttall oaks, cypress, black willow, and bitter pecan. The abundance of water and the intricate mix of woodland and swamp have resulted in high biodiversity. Several rare plant and animal species can be found here. The sloughs and bayous are an important habitat for alligators, other reptiles and amphibians, and nesting herons.

Thistlethwaite Wildlife Management Area
Saint Landry Parish
This 11,000-acre tract is located about 3 miles northeast of Washington. It is an excellent example of mature mixed bottomland hardwood forest, interspersed with wetlands and sloughs along the Atchafalaya River. Some large specimens of oak, hickory, cypress, tuypelo, and sweet gum can be found, especially among the wetlands. Biodiversity is high and includes several rare or unusual plant and animal species, such as wood ducks and ospreys. It is an important breeding habitat and migratory stopover for a variety of bird life.

Three Rivers Wildlife Management Area
Concordia Parish
Located east of Big Bend, this 23,222-acre property is located at the confluence of the Mississippi, Red, and Old Rivers. The mix of natural communities includes open water, swamp, slough, brake, bottomland hardwood, and a few islands of pine-hardwood forest. Dominant species include older-growth sycamore, oak, sweet pecan, and cottonwood trees. Much of the forest is developed on a series of sand levees that were formed by river action. Plant and animal life is highly diverse, and the tract is a critical habitat for some rarer species, such as alligators, certain raptors, and Mississippi kites. The bottomlands are an important stopover for thousands of other birds during spring and fall migrations.

Tunica Hills Wildlife Management Area
West Felicia Parish
This 2,944-acre property about 15 miles northwest of Saint Francisville protects a large expanse of mature to older-growth upland hardwood forest. The steep loess hills create shadowed ravines and glades that host several rare or unique plant and animal species, including the Louisiana black bear. Some oak and beech trees are more than 200 years old and form a majestic upper canopy. The junglelike brakes in the bottoms shelter timber rattlesnakes. The worm-eating warbler, Cooper's hawk, and other raptors nest in the area. Tunica Hills is also the only known Louisiana habitat for the rare Webster's salamander.

Union Wildlife Management Area
Union Parish
Managed by the Lousiana Department of Wildlife and Fish, this 13,400-acre parcel 4 miles northwest of Marion encloses upland pine-hardwood forest, bottomland forest, marsh, and riparian habitat. Although the forest is logged for timber, it has become an important habitat for a number of bird species,

including nesting wood ducks, herons, warblers, and a breeding colony of red-cockaded woodpeckers. Beaver lodges can be seen along Tick Creek. Several rare or unusual groundcover plant species also grow here.

White Kitchen Preserve
Saint Tammany Parish
Located along the West Pearl River, this 586-acre Nature Conservancy property is located about 20 miles southeast of Covington. It shelters mature to older-growth cypress swamp, old-growth loblolly pine–hardwood forest, and freshwater marsh communities. An important bird habitat, the property is also a nesting ground for bald eagles, angingas, herons, egrets, and other waterbirds.

Wisner Wildlife Management Area
Lafourche Parish
Located about 5 miles east of Leeville, this 21,621-acre wildlife management area protects a wetland complex of bayou, channel, and subdelta saline marsh. Vegetation is dominated by oyster grass and salt grass; small stands of oak trees also occur. The wetlands are an important breeding ground for shellfish and various fish species—and thus is an important food source for many waterbirds. Among the many birds that nest here, or visit during migrations, are the rare peregrine falcon and brown pelican.

MAINE

Maine has the following major vegetation types: maritime forest, mixed conifer-hardwood northern forest, oak-hickory forest, pine barren, fen, bog, lake and pond, and coastal dune-marsh-headland.

Maritime forest occurs in a narrow belt along the Atlantic coast, including spruce-fir, pitch pine, and jack pine woodlands. Tree growth is affected by wind and salt spray.

Almost the entire northern half of Maine is covered by various mixed conifer northern forests, interspersed with wetlands and bogs. Forest types include pine-hemlock-spruce, jack pine woodland, northern white cedar, and red pine. Numerous seeps and vernal pools also are present. Mixed hardwood-conifer forest, especially oak-pine and oak-hickory forest, dominate the southern half of the state.

Pine barrens are typically found on sandy outwash plains. Trees are mostly pitch pine, jack pine, or red pine. Heath shrubs, such as blueberry, and grassy openings often underlie the pine barren canopy. Pine barrens are found in eastern and southwestern Maine.

Both patterned and unpatterned fen complexes occur in the state, mostly to the north. Vegetation is typically a sedge-grass-reed-moss assemblage. Chemistry can vary from acidic to neutral and can include certain peatlands and Atlantic white cedar swamp. Coastal plateau, domed, level, eccentric, and kettlehole bogs can be found across the state, but most are concentrated in the east-central portion.

Lakes and ponds occur within all the vegetation types. Lakes are classified as monomictic (i.e., the water mixes from top to bottom once a year) or dimictic (twice a year). Water bodies are further classified according to acidity and shoreline fringe vegetation. A few rare alkaline hardwater ponds occur in Aroostook County.

Coastal ecoregions include freshwater tidal marsh, brackish tidal marsh, tidal creek, salt marsh, beach strand, sand dune, pitch pine dune forest, brackish pond, rocky coastal headland, and tidal flat. These occur along the Atlantic coast and harbor a number of rare plant, animal, and bird species.

Special Habitat Sites

Baxter State Park
Piscataquis County
Located about 20 miles west of Patton, this 210,000-acre property includes forested mountains, wetlands, ponds, and lakes. Forty peaks occur within its borders, including Katahdin, the tallest mountain in Maine. Blue fringed gentians, yellow avens, and pink moccasins are some of the abundant summer flowers; are bog plants include the pitcher plant. Pine-spruce forest and alpine communities grow on the upper slopes with blueberry heaths and moss-fern assemblages on windswept ridges. The park is an important habitat for large mammals, including bears and moose.

Bigelow Preserve
Somerset County
Located about 10 miles east of Stratton, this 40,000-acre preserve borders Flagstaff Lake. The mountainous terrain is covered by upland

spruce-pine forest. Lowland conifer–red maple–American beech forest supports a large population of relatively rare spruce grouse.

Borestone Mountain Sanctuary
Piscataquis County
Located near Monson, this 1,600-acre Audubon Society sanctuary encloses northern boreal hardwood and conifer forest, riparian, and wetland communities. A portion of the forest at the upper elevation contains virgin timber reminiscent of presettlement vegetation. Several species, some of them rare or endangered, such as the pileated woodpecker, depend on the old-growth forest.

Damariscove Island
Sagadahoc County
Located east of Boothbay, this offshore island is accessible only by boat. Its marsh, beach, dune, and inland forest communities support more than 300 different plant species. The island was also the first permanent European settlement on the Atlantic coast; it is owned and maintained by The Nature Conservancy.

Fernald's Neck Preserve
Waldo County
On Megunticook Lake, south of Lincolnville, this 320-acre preserve includes 18,000 feet of shoreline. Natural communities are white pine–red pine–hemlock and oak-beech-birch forest, rock cliff, and an expansive bog with leatherleaf, sweet gale, rhodora, pitcher plants, blue flag iris, and several rare sedges. The wetlands are an important nesting site for waterbirds, blackbirds, woodcocks, and Canada geese. Lupine is among the wildflowers that grow in the meadows.

Great Wass Island Preserve
Washington County
This 1,540-acre Nature Conservancy property south of Jonesboro is along the Atlantic Ocean. The windswept, rocky headlands and coastal strand support rare beach-head iris, marsh felwort, bird's-eye primrose, roseroot stonecrop, pearlwort, and seabeach sandwort. The interior has the state's largest stand of stunted jack pine. Coastal raised bogs and peatlands host rare plants, including carnivorous plants, baked-apple berry, and dragon's mouth orchid. Eiders, osprey, bald eagles, spruce grouse, and palm warblers nest on the property.

Kennebunk Plains
York County
A few miles from Kennebunk, this 15,000-acre property encloses coastal sand-plain grassland, pitch pine–scrub oak woodland, and inland white pine–white oak forest. The very rare grasshopper sparrow inhabits the dry grasslands. Rare grassland plants include northern blazing star, toothed white-topped aster, and upright bindweed. About 90 different birds breed here, including New England's largest population of grasshopper sparrows.

Maine Coastal Island Sanctuary
Knox County
Located offshore in the Atlantic Ocean, these 14 rocky islands, owned by the Audubon Society, total about 463 acres. The endangered puffin has been gradually reintroduced to these islands since 1973. A permanent puffin population has been established; it returns every year to breed. The islands also host nesting colonies of the endangered roseate tern as well as ospreys, eiders, and great horned owls.

Mattawamkeag Wilderness Park
Penobscot County
Within this park 20 miles southeast of Millinocket is old-growth forest along the Mattawamkeag River. More than 1,000 acres of old-growth pine-spruce-hemlock forest guard the riverbank, some of the giant trees being more than 300 years old.

Mount Desert Island
Hancock County
Located in Acadia National Park about 20 miles southeast of Ellsworth, this island has granite cliffs and mixed conifer forest with more than 400 species of plants and flowers. Natural communities include rocky headland, coastal forest, old-growth red spruce–white cedar–balsam fir forest, and coastal strand. More than 130 bird species nest here, including rare boreal chickadees and ospreys. Seals congregate on the northwestern shore of the island.

Rachel Carson Salt Pond
Lincoln County
This Nature Conservancy preserve is located along the Atlantic coast about 9 miles north of New Harbor. The property protects a recurring salt pond that is exposed when the tides retreat.

Saco Heath Preserve
York County
This 865-acre Nature Conservancy property shelters a rare bog community. Called a coalesced domed bog, it contains a raise island surrounded by peatland. Rare or unusual bog plants include Atlantic white cedar, sheep laurel, leatherleaf, and the carnivorous pitcher plant.

Waterboro Barrens
York County
Located a few miles northeast of Waterboro, this Nature Conservancy property protects a rare pine barren ecosystem. Natural communities include wetland, dune, and pine–scrub oak forest. It is one of only 20 major pine barrens systems in the world, and it supports rare or unusual flora and fauna.

Wells Reserve
York County
Located just south of Wells, this estuarine research area is located at the confluence of the Little, Merriland, and Webhannet Rivers along the Atlantic coast. Natural communities include barrier beach, beach, salt marsh, and brackish estuarine marsh. The estuary is an important nursery for fish, shellfish, and crustaceans. It is also an important bird habitat, especially during spring and fall migrations. Special or threatened species seen here include least terns, peregrine falcons, piping plovers, bald eagles, several orchids, and blue flag iris.

MARYLAND

Maryland is segmented into three northeast-trending zones, from east to west: coastal zone (Coastal Plain), midland zone, and mountain zone.

The coastal zone comprises roughly the eastern third of the state; its interior border runs roughly from the northeastern corner of the state to Washington, D.C. This flat-lying terrain encloses Chesapeake Bay, its estuaries, and a mosaic of different forest and wetland communities. Dominant forest types include loblolly–Virginia pine, oak-pine, and oak-hickory. Forest type is generally transitional from pine and mixed pine-hardwood along the coast to oak-hickory on the western side of Chesapeake Bay. Important wetlands are salt marsh, cypress swamp, and Atlantic white cedar swamp.

The midland zone occupies the central portion of the state; its western boundary runs northeast-southwest through the center of Allegany County. Equivalent to the Piedmont districts found in other southern states, this zone consists of rolling hills and valleys and is largely forested with oak-hickory, especially in the western half. Other forest communities include oak-pine and mixed mesophytic. Unique habitats in the midland zone include shale barrens and several bogs.

Hemlock, white pine, and northern hardwood forest dominate in the mountain zone, located in the extreme northwestern corner of the state. Meadows and grassy balds occur occasionally along the rocky ridges. Dominant trees include basswood, tulip, maple, birch, beech, pine, and oak. Several boreal-type bogs with unique plants can be found among these hills and ridges.

Special Habitat Sites

Battle Creek Cypress Swamp
Calvert County
Located about 4 miles south of Prince Frederick, this 100-acre property protects a bald cypress swamp and upland Virginia pine–tulip forest. The swamp is at least 10,000 years old. The impressive old-growth cypress trees are more than 100 feet tall and 2 feet in diameter at breast height. Several rare or unusual species inhabit the wetlands, including a variety of rare or endangered warblers. The spring wildflowers include violets and pink lady's slippers.

Black Bottom Ponds
Kent County
This property contains four Coastal Plain ponds (Delmarva bays or Carolina bays). Other natural communities within this wetland complex are maple-gum swamp forest, pepperbush-greenbrier forested scrub swamp, and buttonbush marsh. The pond environments support a number of rare or endangered species, including featherfoil, Harper's fimbristylis, clustered bluets, hornrush, and giant

sedge. Rare animals include the eastern tiger salamander, barking tree frog, and carpenter frog. It is also an important habitat for forest interior breeding birds.

Cedarville State Forest
Prince Georges and Charles Counties
About 5 miles southeast of Cedarville, this site protects a portion of Zekiah Swamp, at 70,000-acres the largest wetland in the state. Pond, stream, wetland, bog, and bottomland Virginia pine–sycamore–willow–oak–tulip poplar forest are the major natural communities. A great array of bog plants grows here, including the carnivorous sundew and pitcher plant. It is also one of the largest blue heron rookeries on the eastern U.S. coast.

Choptank Wetlands
Talbot and Caroline Counties
Located just south of Kingston Landing, this 656-acre property protects an extensive brackish tidal marsh along the Choptank River. The marsh is dominated by switchgrass, arrow arum, cattails, and phragmites. It is a critical habit for nesting as well as migrating birds, including snipe, northern harrier, osprey, bittern, and heron. The Nature Conservancy owns and maintains the preserve.

Coon Ridge Shale Barrens
Washington County
This preserve, part of Indian Springs Wildlife Management Area, protects a series of rare shale barren communities along Licking Creek. Some rare species found in this rare ecosystem are the rusy woodsia fern, Kate's-mountain clover, downy heuchera, and shale-barren primrose. The barrens are surrounded by hardwood and oak-pine forests.

Crabtree Cave and John Friend Cave
Garrett County
Totaling 58 acres, these two Nature Conservancy properties in western Maryland protect the entrances to caves with very fragile ecosystems. Cave species have adapted to continual darkness and a poor nutrient supply. John Friend Cave, near Hoyes, contains three nationally rare subterranean crustaceans and five bat species, including the federally endangered Indiana bat. Crabtree Cave, near Luke, shelters rare aquatic invertebrates, such as Franz's cave amphipod and isopod and the springtail. It is the largest bat habitat in Maryland; the eastern pipstrel and Keen's bat live here.

Cranesville Swamp
Garrett County (Maryland) and Preston County (West Virginia)
Located on the border with West Virginia, this 560-acre wetland was designated a National Natural Landmark by the National Park Service in 1965. Natural communities are larch–red spruce peatland bog, shrub swamp, and mixed hardwood upland forest. The wetland is a relict boreal swamp that can survive in the cooler microclimate created by frost pockets in the rugged terrain. Rare or unusual species found here include the larch (tamarack) conifer, cottongrass, the carnivorous sundew, and creeping snowberry. The preserve is also an important nesting or breeding habitat for the rare northern water shrew, Nashville warbler, saw-whet owl, and golden-crowned kinglet. The property is owned and maintained by The Nature Conservancy.

Cypress Swamp
Calvert County
Located a few miles south of Prince Frederick, this 100-acre nature sanctuary protects a rare bald cypress swamp at the northernmost tip of its range. The old-growth cypress trees are more than 100 feet tall and up to 1,000 years old. Colorful wildflowers, such as the cardinal flower, add color during spring and summer. It is owned and managed by The Nature Conservancy.

Delmarva Bays
Caroline, Dorchester, Wicomico, and Queen Annes Counties
Known as Delmarva bays, these six Nature Conservancy properties total 593 acres and include rare, undisturbed, seasonally flooded freshwater wetlands. They are marked by abundant grasses and sedges, such as Walter's sedge and twig-rush, in a meadowlike opening. The wetlands have been in existence for more than 15,000 years. Different wetland communities include shrub (buttonbush-pepperbush) swamps and gum-maple flatwoods (swampy forest). Waterbirds, including herons and egrets, nest on the property. More than 20 rare or threatened plants grow in the bays, including Canby's dropwort, pondspice, Harper's sedge, red-root, and horned-rush. The rare carpenter frog can also be found here.

Eastern Neck National Wildlife Refuge
Kent County
This 2,285-acre reserve is at the mouth of Chester River, 10 miles south of Rock Hall. The protected island is a wintering ground for tundra swans and resident mute swans. Communities include brackish marsh, lowland forest, grassland, tidal mudflat, saltwater marsh, and brackish pond. The rare Delmarva fox squirrel, peregrine falcon, piping plover, and bald eagle live in the refuge, which supports more than 250 bird species.

Finzel Swamp
Garrett and Allegany Counties
Dense forests and wetlands are protected on this 333-acre property a few miles north of Frostburg. Natural communities include swamp, wetland forest, upland hardwood forest, and meadow. The swamp contains rare relict tamarack, wild calla, Canadian burnet, cottongrass, and bog fern. Rare fauna include the woodland jumping mouse, smoky shrew, bobcat, alder flycatcher, Virginia rail, sedge wren, and various warblers. Purple finches also frequent the site. It is also an important migratory stopover for thousands of birds during spring and fall. The property is owned and maintained by The Nature Conservancy.

Fort Hill
Allegany County
Located east of Danville, this 155-acre property protects limestone glade, rocky outcrop–cliff face, and old-growth limestone forest. The lime-rich soil and rock supports a host of unusual or rare plants, such as purple harebell, rock sandwort, toothache tree, cliff stonecrop, tall larkspur, and mountain pimpernel. Although the forest trees are stunted and scattered, they are old-growth remnants of what was once an extensive forest. The open conditions attract raptors, such as the broad-winged hawk. The preserve is owned and operated by The Nature Conservancy.

Glades
Garrett County
Located east of McHenry, this 592-acre property encloses a relict northern cranberry bog in rugged western Maryland. Natural communities include open peatland, sphagnum-polytrichum moss bog, and surrounding upland forest. The cooler microclimate supports rare relict boreal plants, such as small cranberry, carnivorous bladderwort, and bog clubmoss. The rare bog copper butterfly—the larva feed only on cranberry bushes—inhabits the glades. State-rare black bears and bobcats inhabit the wetland, as do the rare Nashville warbler and alder flycatcher. The property is owned and managed by The Nature Conservancy.

Green Ridge State Forest
Allegany County
Located about 20 miles southeast of Cumberland along the Potomac River, this forest encloses a series of rare shale barrens plant communities on its south-facing slopes. The microclimate is hot and very dry, and a group of specialized plants has adapted to the harsh conditions, including evening primrose, shale ragwort, pussytoes, and the very rare Kate's mountain clover. Virginia pine and hardwood trees that cling to the barrens are short and stunted. The uncommon five-lined skink, timber rattlesnake, wood turtles, and Carolina wrens also inhabit the barrens community.

Hammel Glade
Garrett County
Located about 10 miles northeast of Oakland, this 206-acre preserve shelters a pristine piece of rugged western Maryland. Natural communities include peatland, shrub swamp, conifer (white pine–hemlock–red spruce) forest swamp, and sedge glades. The relict boreal wetlands support a number of rare species, including Jacob's ladder, buckbean, goldthread, and five other plants. Rare fauna include the northern watershrew, bald eagle, sharp-shinned hawk, and alder flycatcher. Large red spruce trees in the swamp represent what are probably the state's finest examples. The property is owned and maintained by The Nature Conservancy.

Hellen Creek
Calvert County
A relict stand of Canadian hemlock, very rare in the Coastal Plain of the eastern United States, is protected on this 67-acre property about 3 miles from Lusby. Hemlock thrived here about 10,000 years ago, at the end of the last glaciation. Other trees are mature red maple, beech, hickory, and oak. Rare or unusual species include the deer mouse and climbing fern. The upland woods and flanking marsh are an important habit for numerous birds, including the rare pileated woodpecker

and bald eagle. The property is owned and maintained by The Nature Conservancy.

Janes Island State Park
Somerset County
Located at the mouth of Crisfield Harbor on the eastern shore a few miles north of Crisfield, this 3,000-acre park includes islands, pristine salt marsh, sand beach, and tidal creek. Although heavily used in the past, it is now a sanctuary for the diamondback terrapin, seaside sparrow, and marsh wren. Loblolly pine–wax myrtle stands exist in the interior.

Little Catoctin Highland Glades
Frederick County
Located on 38 acres about 10 miles northeast of Myersville, this Nature Conservancy property protects a rugged portion of the Catoctin Highlands. Natural communities include mature maple-oak-hickory-basswood upland forest, seepage slope rich in ferns and moss, brook, sandstone outcrop, glade, and cliff face. Microenvironments on the rock faces support the rare and delicate mountain sandwort. Wildflowers are abundant and colorful during spring.

Merkle Wildlife Area
Prince Georges County
Located on 1,600 acres about 4 miles east of Croom, this Department of Natural Resources property shelters wetlands along the upper section of Patuxent River. Natural communities include tidal creek, bottomland forest, pond, upland forest, and marsh. The wetlands attract thousands of nesting and migratory birds, including Canada geese, herons, ospreys, wood ducks, and the federally threatened bald eagle. Bluebirds and purple martins can also be found here. An uncommon plant, the pawpaw, grows in the understory, and its fruit is an important food source for a variety of small woodland mammals and rodents.

Nanjemoy Creek Great Blue Heron Sanctuary
Charles County
One of the largest great blue heron rookeries along the eastern U.S. coast occurs on this 288-acre parcel just south of Nanjemoy. The treetops contain up to 1,000 heron nests. Natural communities include mature beech–tulip–Virginia pine upland forest and wetlands. Rare plants include the overcup oak, basket oak, several orchids, and Virginia wild ginger. The property is owned and maintained by The Nature Conservancy.

Nassawango Creek
Worcester and Wicomico Counties
Located on 3,300 acres about 13 miles south of Salisbury, this Nature Conservancy property protects an important wetland complex along Nassawango Creek. Natural communities include rare bald cypress forest and swamp, rare Atlantic white cedar swamp, black gum–maple–ash riparian forest, and mixed upland forest. The cypress trees are old-growth and more than 100 years old. More than 50 rare, threatened, or endangered species have been found here, including a variety of orchids, the golden prothonotary warbler, and other warblers. The property is an important nesting site and migratory stopover for a variety of forest birds and waterfowl.

Oldtown Shale Barren
Allegany County
Located near Oldtown, this 16-acre Nature Conservancy property protects a rare shale barren ecosystem in western Maryland. Natural communities include shale barren and mature upland mixed hardwood forest. The barrens host a variety of rare or unusual species specially adapted to the shale substrate, including evening primrose, shale ragwort, Kate's mountain clover, and three-flowered melic grass. The hot, dry microclimate attracts rare five-lined skinks and copperhead snakes. Warblers, vireos, and Carolina wrens can be found in the woodlands.

Otwell Woods
Talbot County
This 94-acre property about 7 miles east of Oxford on Chesapeake Bay protects an excellent remnant of the mixed hardwood–loblolly pine forest that covered much of the state during presettlement times. Old-growth beech, oak, and hickory trees on the western half of the preserve are probably hundreds of years old. The understory is American holly–dogwood–sassafras–pawpaw. The salt marsh along the shoreline contains cordgrass and cattail. Both the forest and marsh habitats are excellent habitats for resident as well as migratory birds, including songbirds, ducks, herons, kingfishers, and ospreys.

Pilot Serpentinite Barren
Cecil County
Located about 2 miles west of Oakwood, this 32-acre property protects the rare natural communities that have adapted to growing on serpentinite bedrock and serpentinite-derived soil. Serpentinite contains high amounts of iron, magnesium, nickel, and chromium. The communities include grassland and stunted forest trees. Rare vegetation includes fame flower, prairie dropseed, and serpentinite aster. This fragile ecosystem is owned and managed by The Nature Conservancy.

Pocomoke River State Forest and Park
Worcester County
This 12,000-acre preserve about 10 miles northeast of Pocomoke City protects an extensive, relatively undisturbed piece of the Eastern Shore. The dominant natural community is Pocomoke River Swamp, a riverine wetland containing towering old bald cypress trees and stands of rare Atlantic white cedar. Other communities include freshwater marsh, as well as pine barrens on the raised sandy ridges. It is a critical habitat for a variety of nesting and migratory birds, including bald eagles and rare warblers. Some experts believe this swamp is a northern extension of the Great Dismal Swamp in Virginia. The property is owned and managed by the Maryland Department of Natural Resources.

Potomac Islands
Montgomery County
Owned by the Nature Conservancy, these four islands—Bear Island, Conn Island, the Cyclades, and Tenfoot Island—are in the Potomac River near Poolesville. They total 176 acres. Natural communities include very mature floodplain forest, wetland pool, patches of dry forest, and savanna. The climatic extremes and flooding have created a remarkably diverse collection of flora and fauna; more than 100 rare, endangered, or threatened species have been found here, including the bald eagle, yellow nailwort, and starry false Solomon's seal. The old-growth trees on the islands are some of the largest in Maryland.

Ranck Woods
Talbot County
Located west of Unionville, this 43-acre Nature Conservancy tract protects an older remnant of the beech–oak–maple–loblolly pine forest that once covered most of the state. The large trees are more than 100 years old and a critical habitat for the globally rare Elmarva fox squirrel. The ground is ablaze with wildflowers, including the showy pink lady's slipper, during spring.

Robinson Neck Preserve
Dorchester County
This 920-acre Nature Conservancy property near Chesapeake Bay is about 20 miles west of Cambridge. It protects old-growth upland loblolly pine forest, brackish tidal marsh, and freshwater marsh natural communities. Marsh plants include salt marsh cordgrass, needlerush, and Olney threesquare bulrush. The high biodiversity is reflected by the number of rare or endangered plants and animals on the property, such as the Delmarva fox squirrel, bald eagle, osprey, and other species. The wetlands are an important nesting habitat for the black duck and attract thousands of migrating birds, including warblers, during spring and fall.

Selinger Marsh
Allegany County
Located about 4 miles southeast of Flintstone, this 85-acre property protects a marsh that occupies a former oxbow of Town Creek. Natural communities include sedge marsh, swamp, spring, and hilly mixed hardwood-pine upland forest. The wetlands are host to a variety of rare or endangered species, including the fringe-tipped gentian, upland chorus frog, Jefferson salamander, and several sedges. The forest and wetlands are important breeding and migratory habitats for numerous birds, including herons. The property is owned and managed by The Nature Conservancy.

Sideling Hill
Washington and Allegany Counties
This 1,500-acre preserve is located in the central Appalachians. Natural communities along Sideling Hill Creek are riparian habitat, rare shale barren, and mature upland forest. Rare species found in the fragile shale barren ecosystem are the Olympia marble butterfly, Allegheny plum, Kate's mountain clover, tiger beetle, and evening primrose. The stream has never been disturbed and is one of the most pristine waterways in the state; it contains a diverse population of mussels and has the world's second largest population of the endangered Harperella plant.

Soldiers Delight Preserve
Baltimore County
Developed on ultramafic, serpentinized bedrock, this unique ecosystem is marked by rare plants and stunted oak trees because of the underlying soil chemistry. The 1,800-acre tract just outside Owings Mills supports 39 rare or endangered species. Natural communities include prairie grassland, stunted Virginia pine forest, and stunted oak savanna. Some of the rare species include vanilla-scented hollygrass, tufted hairgrass, sandplain gerardia, serpentinite aster, fringed gentian, fame flower, salamanders, Edwards hairstreak butterfly, leafhoppers, and beetles.

Sugarloaf Mountain
Frederick and Montgomery Counties
Located just north of Washington, D.C., this 3,000-acre property protects unbroken oak forest along the rocky slopes of Sugarloaf Mountain. The fact that more than 500 plant species grow here, including some rare and unusual ones, underscores its high biodiversity. The preserve is home to a variety of forest mammals and birds, including the rare pileated woodpecker and red-shouldered hawk. The rocky bluffs and cliffs of the mountain attract raptors.

Third Haven Woods
Talbot County
This 393-acre Nature Conservancy property about 25 miles east of Oxford protects a relatively undisturbed hardwood forest reminiscent of the "Big Woods" that once covered most of Maryland. Specimens of oak, beech, and hickory are hundreds of years old and up to 4 feet in diameter at breast height. Rare or endangered species found here include the Delmarva fox squirrel. The wetlands contain a variety of flora and fauna, including skinks, turtles, and mud minnows. Both the forest and wetlands are important nesting habitats for a number of bird species, including warblers and woodpeckers.

Wolf Swamp
Garrett County
Located southeast of Grantsville, this 26-acre Nature Conservancy project encloses a relict boreal swamp. Natural communities include peatland, wetland, and old-growth white pine–hemlock–red spruce forest. Some of Maryland's tallest trees occur here. The swamp has a floating sphagnum moss mat and is home to nine rare plant or animal species, including wild calla, mountain rice, clintonia, pale green orchis, mountain chorus frog, and golden-crowned kinglets. Several varieties of warblers, thrushes, and vireos nest in the forest.

Wye Island Natural Area
Talbot County
Located along the Wye River a few miles south of Wye Mills, this 30-acre tract protects a stand of old-growth oak-hickory-gum forest. The largest trees are 300–400 years old. Ospreys nest along the shoreline.

MASSACHUSETTS

Massachusetts is covered by the following major natural communities: wetland, Coastal Plain pond, coastal-tidal ecosystem, floodplain forest, pitch pine-scrub oak barren, sandplain grassland, and spruce-fir forest.

Wetlands include acidic peatland, Atlantic white cedar swamp, calcareous fen, and calcareous seepage swamp. Acidic peatlands, found throughout the state, are boggy accumulations of peat dominated by moss, shrubs, and conifers. Atlantic white cedar swamps are an East Coast habitat that is vanishing rapidly. They are dominated by dense growths of Atlantic white cedar with lesser maple and birch. These swamps are most common south and southeast of Boston. Calcareous fens are peaty wetlands that thrive on calcareous groundwater and usually have a diverse assemblage of plant species. Most of them have been impacted by development; typically they are found in the southwestern corner of the state. Calcareous seepage swamps are forested wetlands fed by alkaline groundwater. Trees include red and white cedar, maple, birch, tamarack, hemlock, and white pine. Most of these occur in Berkshire County.

Coastal plain ponds (also called Delmarva bays or Carolina bays) have formed within depressions in glacially deposited sediments along the Atlantic coast. Typically shallow and sandy, Coastal Plain ponds contain a variety of aquatic plants. Their specialized habitats nurture a variety of globally restricted plants, including blue-joint and boneset. Such ponds are threatened by development, human visitation, and a lowered water table from the pumping of groundwater.

Coastal-tidal ecosystems include estuaries, salt marshes, and freshwater marshes. Estuaries form where freshwater from rivers mixes with ocean waters along the coast. They are dominated by eelgrass, pickerelweed, and salt reed grass. Vegetation patterns in saltwater marshes are determined by the height of tides; salt marsh cordgrass, salt marsh hay, and spike grass are the more common species. Freshwater tidal marshes are slightly brackish and dominated by emergent vegetation, cattails, sweet flag, or wild rice. Water levels are still affected by the tides. Coastal-tidal ecosystems can be found in rivers that empty into the ocean, but they have been greatly reduced by development along the rivers and contaminated by industrial discharge.

Floodplain forests are located along the state's rivers and streams. Tall, scattered trees with a somewhat closed canopy are distinctive. Silver maple, black willow, sycamore, cottonwood, and elm are the more dominant species. The largest tracts of floodplain forest can be found along the Merrimack, Connecticut, and Housatonic Rivers. Dams that have impacted flood patterns and logging have greatly reduced the acreage of floodplain forest.

Pitch pine–scrub oak barrens are an open community developed on Coastal Plain or glacial outwash sand. They tend to be low in height and very dense. Although plant and animal life are not very diverse, the barrens are a critical habitat for number of rare or endangered insect species, such as moths and butterflies. Barrens communities are mostly found in the southeastern part of the state.

Sandplain grasslands and heathlands can be found along the Atlantic coast and some river terraces. Open and treeless, the gentle sandhill plains are covered with prairie grasses, such as bluestem, bush clover, and wild indigo. Heathlands are more shrubby; bearberry, wintergreen, huckleberry, and chokecherry dominate.

The spruce-fir, or boreal, forest is found at higher elevations in the state. The forest floor is typically mossy and damp. Dominant species are red spruce, balsam fir, birch, and hemlock. These forests have been reduced by acid rain as well as mountain development—roads, parking lots, and ski slopes.

Special Habitat Sites

Atlantic White Cedar Swamp
Barnstable County
Located west of South Wellfleet on Cape Cod, this property protects a rare remnant of dense Atlantic white cedar swamp. It is a critical nesting habitat for the northern parula warbler.

Barnum Street Swamp
Berkshire County
Located near Sheffield, this 168-acre property encloses a rare alkaline wetland-fen complex. The calcium-rich groundwater and soil support rare and unusual orchids, sedges, and grasses that are specialized to this unique chemistry. The wetland is an important nesting and site and migratory stopover for numerous birds. A highly diverse wildflower population blooms during spring. The property is owned and managed by The Nature Conservancy.

Bartholomew's Cobble
Berkshire County
This area is located among a series of limestone outcroppings about 2 miles northwest of Sheffield. Both alkaline and acidic conditions support nearly 700 species, including 53 types of ferns, some endemic only to this location. Natural communities include the Housatonic River, floodplain forest, riverine forest, swamps, and rock outcrop. Black vultures and wood ducks nest here. More than 500 wildflowers bloom during spring, such as the great blue lobelia, cinquefoils, and the New England aster.

Beaver Brook
Middlesex County
Located near Boxborough, this 82-acre Nature Conservancy property contains forests and wetlands that cover hilly glacial terrain. The wetlands are an important nesting site for herons, owls, hawks, and other raptors. Several rare plants have also been found here.

Black Pond Nature Preserve
Plymouth County
This 101-acre Nature Conservancy property 3 miles southwest Norwell encloses virginal wetland and mature upland forest. Wetland communities consist of an acidic bog and Atlantic white cedar swamp, two rare communities in the eastern United States. The bog hosts a sphagnum moss floating mat, rare orchids such as rose pogonia, carnivorous pitcher plants

and sundews, several orchids, and white cedar and red maple swamps.

Black Rock
Berkshire County
Located near Sheffield, this 77-acre preserve protects a rugged pitch pine–oak forest developed on steep, rocky outcrops and cliffs. The rock faces are specialized microenvironments for rare plants, such as mountain spleenwort, as well as certain mosses and lichens. The property is owned and managed by The Nature Conservancy.

Boat Meadow
Barnstable County
Part of the Cape Cod National Seashore, this 126-acre preserve near Eastham protects salt marsh and sandy-grassy uplands. In addition to some rare plants, the northern diamondback terrapin is found here. The Nature Conservancy is the owner of this property.

Charles W. Ward Reservation
Essex County
Located outside Andover, this property encloses a northern spruce-tamarack quaking bog. Vegetation includes black spruce, tamarack, sundews and other carnivorous plants, and unusual sedges.

Cold River Old-Growth Forest
Berkshire County
Old-growth hemlock–yellow birch–sugar maple grows here along Cold River, 10 miles southwest of Florida. Some of the trees are 300–400 years old.

Daniel Webster Wildlife Sanctuary
Plymouth County
This former farm just north of Marshfield has been converted to grasslands and wetlands. It is an important habitat for northern harriers, peregrine falcons, bald eagles, swallows, sandpipers, plovers, herons, and monarch butterflies.

Duxbury Beach
Plymouth County
This property is one of the least disturbed barrier beaches in Massachusetts. Located 10 miles northeast of Plymouth, its sand beach and dunes are nesting sites for rare least terns and piping plovers.

Greene Swamp
Franklin County
Overlooking the Connecticut River valley, this 68-acre Nature Conservancy property near Sunderland protects forested hills, spring-fed swampy lowlands, and riparian habitats. Rare rock-clinging plants, such as green rock-cress, can be found on the bedrock outcrops.

Halfway Pond
Plymouth County
This preserve near Plymouth covers an undeveloped, 12-acre island in the middle of a Coastal Plain pond. The island is covered by old-growth tupelo-hemlock-beech-holly forest, an unusual mix. Some of the trees are hundreds of years old. The property is owned and managed by The Nature Conservancy.

Hawley Bog
Franklin County
A rare undisturbed kettle bog can be found on this 65-acre property near Hawley. It contains floating sphagnum moss mats and rare plants, such as bladderworts, rose pogonias, and carnivorous pitcher plants. The bog is surrounded by mature hemlock–mountain laurel woodlands. The property, owned and managed by The Nature Conservancy, is also an important bird sanctuary.

High Ledges Wildlife Sanctuary
Franklin County
Located 5 miles northwest of Shelburne on the slopes of Massamett Mountain, this 400-acre site preserves high-quality northern hardwood and hemlock forest. Ravens nest among the rock outcrops. More than 500 plant species grow here, including pink, yellow, and the rare white lady's slipper.

Homer-Watcha
Dukes County
Located on Martha's Vineyard near West Tisbury, this 142-acre property is managed by The Nature Conservancy through a conservation restriction. The acreage protects the rapidly disappearing maritime grassland and maritime heathland natural communities, rare or threatened plant and insect species, and the threatened northern harrier and other raptors.

Hyannis Ponds
Barnstable County
These properties on Cape Cod near Hyannis protect ten Coastal Plain ponds, a rapidly disappearing ecosystem along the Eastern Seaboard. More than 20 rare or state-listed plant species, including the Plymouth gentian and slender marsh pink, grow along the shores of these special ponds. The globally rare green comet dragonfly can also be found here. The wetlands are also an important bird habitat.

Ipswich River Wildlife Sanctuary
Essex County
Located along the Ipswich River 3 miles east of Topsfield, this 2,800-acre riparian strip includes meadow, swamp, and bottomland silver maple forest. It is a vital bird habitat; residents include thrushes, woodpeckers, warblers, wood ducks, egrets, marsh wrens, and peregrine falcons.

Katama Plains
Dukes County
Located near Edgartown on Martha's Vineyard, this 192-acre parcel protects the largest remnant of virgin sandplain prairie remaining on the island. This maritime grassland is home to a number of rare and specialized plant, animal, and insect species. The preserve is owned and managed by The Nature Conservancy.

Massachusetts Audubon at Arcadia
Hampshire County
Located near Easthampton, this 550-acre tract includes lowland orchards, old-growth river-bottom forest, ponds, and wetlands. Nesting wood ducks and rare butterflies and ferns can be found here.

Miacomet Moors
Nantucket County
Located on Nantucket Island, this Nature Conservancy property is part of a protected section of undeveloped maritime (coastal) heathland, a rare natural community. Rare plant and insect species also occur here.

Mount Greylock State Reservation
Berkshire County
Located on 11,000 acres about 10 miles northwest of Adams, this forested hill country shelters 46 rare, threatened, or endangered Massachusetts species. Natural communities include rare stunted balsam fir, red spruce, birch, and northern mountain ash (boreal forest). Wildflowers, such as various violets, painted trilliums, and the unusual hobblebush, grow in abundance. Rare warblers, pileated woodpeckers, barred owls, and ravens nest in the rocks and trees. There are several stands of old-growth hardwoods and conifers, including the red spruce and unusual mountain azalea.

Mount Holyoke
Hampshire County
This 1,000-foot mountain 15 miles southeast of Hadley has southern slopes cloaked in oak-hickory hardwood forest. Northern slopes are covered with hardwood–hemlock–white pine. Up to 10,000 raptors, which are attracted to the thermal updrafts along the mountain, can be seen during fall. The rare timber rattlesnake lives in the rocky ridges.

Myles Standish State Forest
Plymouth County
Glacial kettle ponds and an important coastal pinch pine barrens community are located on this 15,000-acre property. Rare species include Plymouth gentian, Maryland meadow beauty, and the endemic Plymouth redbelly turtle. The pitch pine barren is one of the best examples on the East Coast.

Newhall Woods Preserve
Dukes County
Located on Martha's Vineyard near West Tisbury, this 512-acre property contains rare pitch pine barrens, scrub oak barrens, upland oak forest, riparian habitat along Mill Brook, and red maple swamps. Rare plant and animal species can be found here. The land is managed by The Nature Conservancy through a conservation restriction.

Plum Island
Essex County
This barrier beach complex includes beach strand, sand dune, salt marsh, and maritime forest. Located 10 miles east of Ipswich, it is a critical nesting habitat for the rare least tern and piping plover. Migratory Canada geese, teals, falcons, and swallows frequently visit.

Quabbin Reservoir
Worcester, Hampshire, and Franklin Counties
This "accidental" wilderness is the result of a flooding of the Swift River. Almost 60,000 acres of contiguous forest habitat surround this large

lake, which is 40 square miles. Its recesses offer the sizeable ranges needed by the black bear, bobcat, and mountain lion, all of which have been seen here. Bald eagles and pileated woodpeckers nest in the woods.

Ram Pasture
Nantucket Island
Located on 634 acres, this property encloses coastal grassland, pond, barrier beach, low woodland, swamp, and coastal heath natural communities. Sometimes called the easternmost natural prairie in the United States, its grasses include little bluestem, New England blazing star, sandplain flax, New England silvery aster, and sandplain blue-eyed grass. Rare piping plovers nest in the grasses. The declining short-eared owl and northern harrier feed on the meadow voles that live here.

Ravenswood Park
Essex County
Located on 500 acres near Gloucester, this property is known for its old-growth hemlock–white pine forest and rare sweet bay–magnolia swamp forest. A diverse array of rare or unusual wildflowers grows here.

Reed Brook Preserve
Berkshire County
This 100-acre property in the Berkshires near the town of Florida is covered by mature pine-hardwood forest. A large exposure of serpentinite bedrock supports a very rare, specialized plant community that includes maidenhair spleenwort and mouse-ear chickweed. The Nature Conservancy owns the property.

Robinson-Gilmore Preserve
Plymouth County
This 40-acre parcel near Rochester is owned by The Nature Conservancy. It contains old-growth hemlock forest, vernal pool, Coastal Plain pond, marsh, and swamp. The hemlock trees are hundreds of years old. The wetlands are an important nesting site and migratory stopover for waterbirds.

Sandy Neck
Barnstable County
Located near Barnstable, this 12-acre Nature Conservancy property on a section of barrier beach shelters sand dunes and a coastal thicket harbor. This sheltered wetland is home to a number of rare plants, including the thread-leaved sundew. It is also an important bird sanctuary.

Sandy Neck Reservation
Barnstable County
Located on Cape Cod Bay north of Barnstable, this reservation includes 7 miles of beach, dune, salt marsh, tidal creek, and interior oak-beech interdunal woodland. It also encloses 3,000 acres of Great Marsh, one of the oldest, largest, most complete saltwater marsh systems in the eastern United States. Breeding populations of the rare and threatened diamondback terrapin, least tern, and piping plover live in the dunes and swales. The productive saltwater marsh is a tremendous food source for tens of thousands of seabirds.

Schenob Brook Fen
Berkshire County
Located near Sheffield in the southern Berkshire Mountains, this 450-acre preserve encloses one of the most important wetlands in New England. Natural communities include rare pitch pine barren, floodplain forest, riparian habitat along Housatonic River, calcareous fen, swamp, and pond. The property has the greatest number of rare species in the northeastern United States, including small yellow lady's slippers, American bitterns, northern harriers, and hooded ladies' tresses. Bears, bobcats, and other woodland mammals inhabit the property, as do a variety of interior forest birds and waterfowl. The property is owned and maintained by The Nature Conservancy.

Stacy Mountain
Franklin County
This 140-acre tract near Gill protects the rugged beauty of the Connecticut River's French Mountain Gorge. Communities include mixed hardwood-pine upland forest, wetland, vernal pool, cliff face, and riparian habitat. Several state-threatened plants and the rare Jefferson salamander can be found here.

Wachusett Mountain State Reservation
Worcester County
Located about 7 miles north of Princeton, this granite monadnock rises above the surrounding countryside. The higher elevations support relict northern spruce–hemlock forest. Old-growth northern red oaks are found on the lower slopes. The rock cliffs and thermals

attract up to 10,000 nesting or migrating raptors annually.

Waquoit Bay National Estuarine Research Preserve
Barnstable County
Located about 5 miles south of Waquoit on the southern shore of Cape Cod, this 2,250-acre property is an important bird and wildlife habitat. Natural communities include barrier beach, tidal flat, submerged eelgrass bed, sand beach and dune, salt marsh, brackish pond, freshwater stream and estuary, upland pitch pine–oak forest, and pine barren. Estuaries are an important spawning and nursing ground for sealife; they are also a rich food source for birds. Rare species include sandplain gerardia, an extremely rare coastal plant found in the sandplain. Other important species are the federally endangered roseate tern, least tern, and the federally threatened piping plover, which nests on the sand beach.

Wellfleet Bay Wildlife Sanctuary
Barnstable County
Located on 1,000 acres about 4 miles southeast of Wellfleet, this property includes pitch pine barren, oak-hickory forest, salt marsh, barrier beach, dune, tidal flat, coastal heathland, sandplain grassland, brackish marsh, and pond communities. More than 260 birds have been recorded, as have several rare turtles and the Fowler's toad. Among the 16 state-listed plants found here are the butterflyweed, broom crowberry, and wild lupine. Diamondback terrapins nest in the dunes.

MICHIGAN

The southern half of the Lower Peninsula of Michigan consists of rolling hills and flat lake plains. This area is heavily developed, and most of the original oak savanna, oak-hickory forest, beech–sugar maple forest, and wet prairie have been converted to agricultural purposes. Only remnant patches of forest remain.

The northern half of the Lower Peninsula is more reminiscent of the continuous forest that covered this area during presettlement times. Dominant communities are northern hardwood forest, white pine–red pine forest, hardwood-conifer swamp, conifer swamp, and jack pine–northern pin oak barren. Logging has removed much of the white pine forest and eastern hemlock forest.

The Upper Peninsula possesses rugged bedrock exposures, low hills and mountain ranges, and interspersed wetlands. The eastern half is heavily forested with northern hardwoods (maple and beech), jack pine barren, white pine–red pine, hardwood-conifer swamp, conifer swamp, bog, and muskeg. Rare grassland Alvar communities that cover flat surfaces of limestone and dolomite also occur here. The west half contains more rugged bedrock hills and forests of maple–eastern hemlock–basswood–birch–white pine, white pine–balsam fir–northern white cedar, and white pine, as well as some wetlands, bogs, and tamarack–black spruce swamps. The wetlands are narrower because of the irregular topography and narrower channels.

Special Habitat Sites

Algonac State Park
Saint Clair County
Located just north of Algonac, this 1,023-acre park encloses remnants of virgin Michigan prairie. The prairie segments have a diverse population of more than 300 plants, many of them rare or threatened. The prairie is surrounded by mature mixed hardwood forest.

Allegan State Game Area
Allegan County
This 608-acre parcel in Allegan County preserves the following natural communities: oak savanna, marshland, tamarack bog, red maple swamp, white cedar swamp forest, and lowland (oak-hemlock-birch-ash forest). The variety of communities results in a highly diverse population of flora and fauna, including the rare wild turkey, Canada warbler, and spotted turtle.

Au Sable Preserve
Crawford County
This 360-acre property about 2 miles south of Frederic is part of the Au Sable State Forest. Natural communities include stream, white cedar swamp, and upland pine-aspen forest. Jack pine tracts are critical nesting sites for the endangered Kirtland warbler. Rare plants that grow in the forest openings are Hill's thistle and Allegheny plum.

Baker Sanctuary
Calhoun County
Located on 871 acres a few miles from Battle Creek, this sanctuary was one of the first ones established to protect the graceful sandhill crane. Mostly swamp and lowland forest, it is an important nesting site and migratory stopover for sandhill cranes as well as many other birds. It is also well known for prominent displays of wildflowers during spring and summer.

Barry State Game Area
Barry County
This 45-acre tract in Barry County protects sand barren, prairie fen, and prairie communities. Plant species include bluestem, coreopsis, yellow coneflower, cordgrass, bluejoint, and shrubby cinquefoil. Rare bird species include the Henslow's sparrow and grasshopper sparrow.

Beaver Dam Nature Sanctuary
Mackinac County
This 40-acre parcel preserves a wetland complex developed on glacial sediments. Communities include aspen-tamarack forest and balsam fir forest. Portions of the swamp have never been logged. The understory contains spiraea, dogwood, willow, and sweet gail. This property is a critical nesting habitat for a number of rare or unusual birds, including Wilson's snipe, mergansers, flycatchers, nuthatches, and warblers.

Besser Natural Area
Presque Isle
Located about 6 miles southeast of Presque Isle, this 137-acre tract includes shoreline (Lake Huron), wetland, and upland forest natural communities. The main attraction is a stand of virgin white pine that is nearly 300 years old and about 150 feet tall. Species of special concern include the rare Pitcher's thistle.

Betsy Lake Natural Area
Chippewa County
This extensive bog-marsh-riparian swamp complex covers about 14,919 acres in Chippewa County. The communities are separated by the crests and swales of ancient sand dunes from the last glacial period. The bogs demonstrate some zonation, including sphagnum moss floating mats. Unique or rare bog plants include a variety of sedges, beak-rush, the carnivorous sundew and pitcher plant, and leatherleaf. The wetlands are an important habitat for a number of bird and aquatic species.

Big Stone–Cecil Bay
Emmet County
Part of Wilderness State Park, this 1,360-acre tract on a point in Lake Michigan preserves riparian, swamp, and lowland forest communities. White cedar, black spruce, and tamarack are the common swamp trees. Large specimens of white pine, beech, and sugar maple grow on sandy dune ridges. This property is an important habitat for a variety of rare or endangered species, such as Wilson's snipe, a number of warblers, the northern flying squirrel, and the water shrew.

Colonial Point Forest
Cheboygan County
Located about 5 miles east of Brutus, this peninsula in Burt Lake contains old-growth mixed forest. Specimens of red oak, maple, beech, and white pine are nearly 200 years old and about 150 feet tall. Some rare plants have been identified in the groundcover; wildflowers are prominent during spring and summer.

Cranbrook Nature Area
Oakland County
This 50-acre tract near Bloomfield Hills preserves a rare remnant of the virgin forest that once covered southern Michigan. Natural communities include mixed hardwood upland forest and wetland. Oak and shagbark hickory trees that exceed 3 feet in breast-height diameter, are 150 feet tall, and are 200 years of age are found in the virgin timber stands. The forest is an important habitat for interior forest birds, such as woodpeckers, and songbirds.

Crane Island Natural Area
Emmet County
Part of Wilderness State Park, this 230-acre parcel is an offshore island in Lake Michigan. The upland interior is covered by a mixed hardwood-conifer forest; the lowland edges support fragile marsh, low dune, and lagoon communities. Unusual plant species include Indian paintbrush, ragwort, and bird's-eye primrose. The island is a critical nesting habitat for a number of rare or endangered bird species, including the piping plover, spotted sandpiper, and American bittern. It is also an important migratory stopover for a variety of shorebirds.

Dead Stream Swamp
Missaukee and Roscommon Counties
A few miles west of Houghton Lake, this 30-square-mile wetland complex is one of the largest in the state. The swamp has been only slightly affected by development and logging. The wetlands are dominated by thick white cedar, balsam fir, and black spruce. Pockets of bottomland forest contain maple and birch. It is an important habitat for a number of rare or threatened animal species, including bald eagles, upland sandpipers, osprey, and black terns.

Dowagiac Woods
Cass County
Located about 5 miles west of Dowagiac, this 220-acre tract of land is one of the largest virgin-soil forests remaining in the state. It has highly diverse plant and animal populations. Prairie, meadow, wetland, and mixed hardwood forest are the principal natural communities. The forest is dominated by large, old oak, beech, and Ohio buckeye trees. Groundcover contains a variety of species, including trilliums, ferns, orchids, and the rare marsh blue grass and blue-eyed Mary. Rare animals include the spotted turtle. The woods are an important nesting habitat for birds, including woodpeckers, tanagers, and buntings.

Erie Marsh Preserve
Monroe County
This 2,618-acre stretch of marsh along Lake Erie near Monroe is one of the few remaining remnants of a huge postglacial wetland complex. Natural communities include woodland, wet meadow, marsh, swamp, and open water. It is an important nesting or migratory stopover sites for hundreds of bird species, including the rare Forester's tern, great black-backed gull, Icelandic gull, great blue heron, coots, egrets, and sandhill cranes. The wetlands also harbor rare plant species, including swamp rose mallow and American lotus.

Escanaba-Gwinn Area
Marquette County
This 420-acre tract within Escanaba River State Forest encloses a section of relatively undisturbed land. Natural communities include conifer swamp, Great Lakes marsh, mesic northern forest, dry northern forest, oak barren, and cliff face. Biodiversity is high. Rare species found here include the lake sturgeon, pine marten, dwarf milkweed, and Vasey's rush.

Estivant Pines Nature Sanctuary
Keweenaw County
Located in northern Michigan, this 200-acre tract is dominated by a rare old-growth stand of northern hardwood forest. Large specimens of sugar maple and white pine that are nearly 200 feet tall dominate the canopy. One of the white pine trees is more than 500 years old. Oak and balsam fir also occur on the property. The property is known as an important nesting habitat for pileated woodpeckers, warblers, owls, and raptors.

Fish Point Wildlife Area
Tuscola County
Located about 6 miles northwest of Unionville, this wetland-lake beach environment is home to hundreds of bird species. Located on the shore of Lake Huron, it is an important migratory stopover for thousands of birds during spring and fall, including ducks, geese, tundra swans, and sandhill cranes. The area is an important nesting habitat for a number of songbirds.

Fitzgerald County Park
Eaton County
This 80-acre park about 2 miles west of Grand Ledge preserves a narrow sandstone gorge and its habitats, including forested slope, wetland, and shaded cliff–dwelling communities. Because of the colder temperatures and shade, rare ferns, mosses, lichens, and liverworts cling to the rock faces; colder climate relict evergreens such as eastern hemlock can also be found.

Five Lakes Muskegon Nature Sanctuary
Muskegon County
This 53-acre preserve near Muskegon protects a very diverse wetland complex and some prairie. It is an important habitat for a variety of birds, including songbirds, great blue herons, hawks, and other raptors. It has numerous rare relict sedges, rushes, and grasses that are typically found only along the Atlantic coastline.

Goose Creek Grasslands
Lenawee County
This 69-acre property just south of Cement City is a virgin remnant of a much larger, nearly extinct prairie ecosystem that covered much of

the Lower Peninsula. Natural communities include wet and dry prairie, marsh, and calcareous fen. The prairie grasses are home to variety of grassland birds, including bobolinks, wrens, and meadowlarks. Plant life is diverse with rare or threatened species, including orchids.

Grand Mere State Park
Berrien County
Located 2 miles south of Stevensville along Lake Michigan, this state park protects wetland, marsh, sand dune, and lowland forest natural communities. It is an excellent example of natural succession from coastal sand dunes to forests and wetlands. Several rare plant species, including orchids and ferns, have been identified on the site. It is an important habitat for a variety of bird species, including Bell's vireo, rare warblers, loons, herons, gulls, and raptors. It also attracts large populations of birds during spring and fall migrations.

Grass Bay Preserve
Cheboygan County
This Nature Conservancy property is located about 5 miles south of Cheboygan. It is an excellent example of a complete Great Lakes interdunal wetland complex. Natural communities include marsh, flat, wetland, beach pool, and sand dune–interdunal environments. More than 300 plant species have been identified here, including 11 rare and threatened species, such as Pitcher's thistle, dwarf lake iris, Lake Huron tansy, and Houghton's goldenrod. Carniverous plants can also be found, such as the bladderwort, sundew, and pitcher plant. The wildflowers are thick and colorful during summer.

Grass River Natural Area
Antrim County
Located 4 miles west of Mancelona, this 965-acre preserve protects wetlands along the Grass River. Relatively pristine and undisturbed natural communities include meadow, river marsh, cedar swamp, lowland forest, and upland forest. Older trees can be found in the cedar swamp. Plant life is diverse with more than 300 species that have been identified, some of them rare or threatened. The area is well known for its diverse display of wildflowers. The rare bald eagle, osprey, and whistling and mute swans frequent the preserve.

Haehnle (Phyllis) Memorial Sanctuary
Jackson County
This property encloses a wetland complex about 10 miles northwest of Jackson. It is an important nesting habitat and migratory stopover for hundreds of bird species, including the rare sandhill crane, yellow rail, prairie chicken, black tern, and great blue heron.

Hartwick Pines State Park
Crawford County
This 49-acre tract of virgin presettlement forest is protected within Hartwick Pines State Park about 5 miles northeast of Grayling. Giant white pine trees more than 300 years old and nearly 200 feet high dominate the canopy. The forest is also an important nesting habitat for various bird species.

Haven Hill Natural Area
Oakland County
Located near Highland, this 721-acre property protects a variety of wetland and forest communities. Upland sand hills are covered by oak-hickory forest. Swamp forests include tamarack, elm, ash, basswood, white cedar, and yellow birch. Older-growth maple-beech stands occupy higher islands within the swamp. The land has been undisturbed for nearly a century; the larger trees are probably old-growth. The preserve is an important habitat for a variety of birds and animals, including the rare mink, cerulean warbler, ovenbird, owls, and turkey vultures.

Holly Recreational Area
Oakland County
This 7470-acre park near Holly contains a variety of natural communities, including streams, lakes, wetlands, and upland and lowland forests. The forests contain stands of mature mixed hardwoods, such as maple and beech. Near 200 bird species have been identified on the property, including the rare or threatened northern harrier, caspian tern, American avocet, osprey, and cerulean warbler.

Huron Mountain Club Nature Area
Marquette County
Much of this rugged terrain on 17,700 acres near Marquette is virgin timber that was only selectively logged. Tall species of hemlock, white pine, and sugar maple are frequently seen. Sandstone bluffs that edge Lake Superior host specialized boreal communities, including

hemlock and balsam fir. It is an important habitat for a number of large mammals and birds, including the rare black-bearded woodpecker and bald eagle.

Interlochen State Park
Grand Traverse County
This 186-acre park about 15 miles southwest of Traverse City protects lake, lowland forest, and upland forest communities. It encloses stands of virgin white pine and maple-oak trees that are more than 200 years old. It is an important bird habitat.

Jonathan Woods
Lapeer County
Owned by The Nature Conservancy, this 144-acre property south of Lapeer encloses glacial hills and valleys covered by upland forest, pothole lake, swamp, and bog natural communities. Old-growth sections of forest contain tall specimens of maple, oak, and beech. Rare plant species have also been identified.

Kitchell Dune Preserve
Ottawa County
Located about 3 miles northwest of Grand Haven, this 52-acre preserve is owned and operated by The Nature Conservancy. It encloses typical natural communities that grade from open dunes to shrubland to oak-maple forest. Unusual or rare plants here include sand cherries and Pitcher's thistle.

Kope Kon/Adaline Kershaw Preserve
Branch County
Located about 9 miles south of Coldwater, this 56-acre tract preserves virgin woodlands representative of presettlement Michigan. The upland forest trees include the uncommon Kentucky coffee tree, tulip tree, and sassafras tree. Old-growth oak trees hundreds of years old dominate the canopy. Lowland hardwood forests include red maple. Several rare plants have been found in the forest understory. Open water ponds and wetlands are important habitats for birds, including herons and wood ducks.

Lefglen Nature Sanctuary
Jackson County
Located about 5 miles south of Grass Lake, this 210-acre sanctuary protects a unique section of Michigan where northern and southern habitats overlap. Natural communities includes wetland, meadow, prairie, and upland and bottomland forest. Several rare plants and wildflowers have been identified, including orchids. The overlap of northern and southern plant species makes this one of the most diverse areas in the state. Rare bird species found here include nesting sandhill cranes.

Maple River State Game Area
Gratiot County
Located about 8 miles north of Saint John's along the Maple River, this 1,170-acre property encloses one of the largest wetland complexes in southern Michigan. Natural communities include bottomland and upland hardwood forest, meadow, marsh, and swamp. It is an important habitat to various bird species, including egrets, swans, sandhill cranes, geese, great blue herons, sandpipers, and wild turkeys. Raptors, including ospreys, can also be found here.

Maxton Plains Preserve
Chippewa County
This 180-acre Nature Conservancy property near the town of Maxton Plains is located on Drummond Island. The tract is an outstanding example of bedrock grassland, or alvar, community, extremely rare in the eastern United States. Rare species on the prairie are prairie dropseed, prairie smoke, and several sedges. It is also an important habitat for insects, birds, and small mammals that prefer grassland habitats.

Mud Lake Bog
Washtenaw County
This 200-acre property encloses a bog lake in a glacial kettle depression. The surrounding uplands are covered by remnants of an oak-hickory forest. The bog is mostly marsh and hardwood or tamarack swamp. The bog features a classic floating sphagnum–sedge mat and supports a variety of unusual or rare plant species.

Newton Woods
Cass County
This 40-acre tract near Dowagiac protects a remnant of old-growth hardwood forest that once covered southern Michigan. Impressively large specimens of white oak, black oak, maple, beech, and tulip trees are nearly 200 years old and more than 100 feet tall.

Northshore Natural Area
Mackinac County
Located on Bois Blanc Island, this 817-acre tract is located along the Lake Huron shoreline. Natural communities include lake beach, grassland, and mixed pine forest. Several stands of old-growth hemlock and white pine trees are hundreds of years old and more than 100 feet tall. Rare or unusual species found here are the dwarf lake iris and American bald eagle.

Pigeon River State Forest
Otsego County
Located about 10 miles east of Vanderbilt, this 93,000-acre tract along the Pigeon River protects a variety of natural communities, such as wetland, swamp, river, and upland and bottomland forest. It is an important habitat to a variety of bird and animal species and includes Michigan's only elk herd.

Platte Springs Pathway
Benzie County
Located 2 miles southeast of Honor, this rugged area along Platte River contains old-growth stands of maple and beech. Maple trees are nearly 3 feet in diameter at breast height and more than 200 years old. The streams that flow into the Platte contain trout and salmon.

Point Betsie Dunes Preserve
Benzie County
Located 4 miles north of Frankfort, this 71-acre Nature Conservancy property protects a pristine sand dune–swale complex along Lake Michigan. Natural communities include sand and cobble beach, sand dune, interdunal wetland, and a backing balsam fir forest. The dunes are anchored by Pitcher's thistle and Lake Huron locust, two species endemic to Great Lakes shorelines. The endangered piping plover may also nest along the beach.

Point Detour
Delta County
This 96-acre tract is the result of an effort between The Nature Conservancy and Michigan to protect Point Detour in Lake Michigan southeast of Fairport. Natural communities include sand and cobble beach, sand dune, and a backing forest of white cedar, spruce, and fir. At the edge of the woodland is one of the world's largest populations of the dwarf lake iris—a state-threatened wildflower—and the rare Lake Huron tansy. The tract is also an important nesting site for a variety of warblers and nuthatches.

Pointe Mouillee State Game Area
Monroe County
Located at the mouth of the Huron River along Lake Erie, this property protects about 4 miles of shoreline marsh. A few miles from Rockwood, it encloses a variety of wetland subcommunities. It is a critical habitat for nesting birds and a stopover for migrating shorebirds and waterfowl during spring and fall.

Porcupine Mountains Wilderness Area
Ontonagon County
This 42,246-acre rugged tract of rocky hills along Lake Superior just west of White Pine is famous as the most extensive virgin hardwood forest in east-central America. It is covered by large specimens of basswood, hemlock, maple, and yellow birch. The cliff communities are dominated by bearberry, blueberry, and dwarf pine. This mix of forests, lakes, and wetlands is an important habitat for just about every bird, animal, and plant species in this climate. This is the largest and most pristine hemlock-hardwood forest in the United States west of New York.

Port Crescent State Park
Huron County
Located along Lake Huron's Saginaw Bay, this 665-acre tract about 2 miles west of Port Austin protects sand dune–lake beach and wetland communities. The ridges and swales in the dunes support a variety of relatively rare plant species. It is also an important habitat for several rare bird species, including the piping plover.

Price Nature Center
Saginaw County
Located on 173 acres about 5 miles southeast of Saginaw City, this preserve encloses wetlands and an old-growth hardwood forest. Tall beech, maple, and oak trees are nearly 300 years old. Woodcocks nest along the stream that runs through the property.

Proud Lake Nature Study Area
Oakland County
Located within Proud Lake State Recreation Area about 10 miles southwest of Pontiac, this 105-acre parcel encloses a very fragile floating

bog ecosystem. The bog fills a kettle depression formed during the last Ice Age. The bog is covered by a thick mat of floating sphagnum moss. A highly diverse population of wetland plants, many of them rare or unusual, occurs here.

Rare Fern Plant Preserve
Mackinac County
This 25-acre property is well known for its diverse number of plant and moss species. Located along a rocky escarpment of dolomite near Mackinaw City, the delicate species grow in cool, moist, shaded crevasses and gullies. Unusual species include the holly fern, maidenhair fern, and maidenhair spleenwort. It is also an important habitat for the snowy owl, pileated woodpecker, and raptors.

Roscommon Red Pines
Roscommon County
Located on 160 acres in Houghton Lake State Forest, about 10 miles east of Roscommon, this extremely rare community of virgin red pine forest is one of the last in the state. Individual trees are more than 100 feet tall and nearly 30 inches in diameter at breast height. Large white pine trees also occur. It is an important habitat for a number of bird species, some of them rare or unusual.

Saugatuck Dunes State Park
Allegan County
Located along the Lake Michigan shoreline, this secluded state park about 2 miles north of Saugatuck is well known for its delicate interdunal wetlands, sandhills, and pine-hardwood forest. Specialized communities are home to a variety of rare or endangered plant and animal species, including spotted wintergreen, ginseng, Pitcher's thistle, and the common tern. Some sand dunes are nearly 200 feet tall. The dunes are an important stopover for migrating birds during spring and fall.

Shiawassee National Wildlife Refuge
Saginaw County
This 9,000-acre preserve about 5 miles southwest of Saginaw is a giant wetland complex where six rivers converge. Natural communities include open water swamp, marsh, bottomland hardwood forest, and upland hardwood forest. It is famous for its abundant bird life and is a critical nesting habitat for more than 100 species, including bald eagles, great blue herons, and a variety of songbirds. During the spring and fall it is filled with thousands of migrating geese, ducks, swans, and terns.

Skegemog Lake Wildlife Area
Kalkaska County
This 2,700-acre wetland complex along the southeastern shore of Lake Skegemog is located a few miles south of Rapid City. Natural communities include marsh, quaking bog, cedar swamp, open water, and lowland forest. Sections of the wetland complex and forest contain old-growth trees that date to the 18th century. It is home to a variety of rare or endangered species, including the massasauga rattlesnake, bald eagle, bobcat, and certain turtles.

Snake Island–Mudland Natural Area
Mackinac County
Located on eastern Bois Blanc Island in Lake Huron, this 244-acre natural area contains cobble beach, dune and swale, cedar swamp, and mixed coniferous forest communities. A variety of rare and unusual plants occurs here, including the endemic dwarf lake iris, pitcher plant, sundew, prairie Indian plantain, and Canada yew. Bald eagles and ospreys nest and hunt on the property.

South Manitou Island Natural Area
Leelanau County
Located on 80 acres on South Manitou Island in Lake Michigan west of Leland, this National Park Service property encloses a grove of ancient white cedar trees. These virgin specimens are more than 100 feet tall, up to 6 feet in diameter at breast height, and more than 500 years old. The largest white cedar in the United States grows here. Rare ferns found on the site include northern holly fern, walking fern, and green spleenwort.

Strangmoor Bog Research Natural Area
Schoolcraft County
This 640-acre tract within Seney National Wildlife Refuge about 7 miles southwest of Seney protects a rare ecosystem of string bogs, or strangmoor, developed between ancient glacial sand dunes and sand bars. The interdunal wetlands are channeled between the ridges, forming a streaky pattern of birch-cinquefoil-willow communities; the higher areas are covered by red and white pine and tamarack. It is an important habitat for a vari-

ety of animals and birds, including some rare or endangered species.

Sturgeon Bay Natural Area
Emmet County
Located in Wilderness State Park along the Lake Michigan shoreline, this unique natural area protects stretches of beach and bluff along the Straits of Mackinac. Natural communities include coastal wetland, gravel beach, wooded dune-and-swale complexes, and northern hardwood forest. Rare or endangered species that inhabit the site include the federally threatened Houghton's goldenrod and the piping plover.

Sylvania Recreational Area
Gogebic County
This 5,072-acre tract near Watersmeet preserves a large portion of unlogged virgin forest communities. Large specimens of sugar maple, hemlock, white pine, and red pine occur throughout the property. The forests also enclose sphagnum bogs that show excellent zonations, including floating mats. Rare and endangered bog plant species grow within the bog community.

Tahquamenon Falls State Park
Chippewa County
Located about 10 miles southwest of Paradise, this 35,000-acre state property shelters large waterfalls, gorges, old-growth hardwood forest, wetlands, muskeg swamp, and bogs. Plant and animal life are diverse. The old-growth forest consists of stands of beech and maple that are more than 300 years old and almost 200 feet tall. It is an important nesting habitat for a variety of birds, including songbirds.

Thompson's Harbor Natural Area
Presque Isle County
Located about 14 miles southeast of Rogers City, this 307-acre tract encloses a beach-dune-swale complex along Lake Huron. Natural communities include wetland and mixed hardwood-pine forest. The area is renowned for containing one of the world's largest populations of dwarf lake iris, a federally threatened species. Several other rare plants have also been discovered here.

Thorne Swift Nature Preserve
Emmet County
Located about 4 miles northwest of Harbor Springs, this 30-acre preserve protects a variety of natural communities, including cedar swamp, evergreen lowland forest, beach (Lake Michigan), and dune. Parts of the cedar swamp have never been logged. Faunal successions are well preserved across the beach and dune environments. There includes a number of rare plant and animal species, such as certain orchids, Pitcher's thistle, Lake Huron tansy, cerulean warblers, and pileated woodpeckers.

Timberland Swamp
Oakland County
Located about 5 miles west of Clarkston, this 245-acre preserve is relatively pristine and untouched. It is part of an 800-acre wetland that serves as the headwaters of the Huron and Shiawassee Rivers. The hardwood forest includes virgin stands of tall oak trees nearly 300 years old. Rare salamanders and turtles inhabit the swamp. Rare nesting birds include the cerulean warbler, Cooper's hawk, and pileated woodpecker. The area is well known for its diverse display of wildflowers during spring.

Tobico Marsh
Bay County
Roughly half of this 2,000-acre site about 5 miles northeast of Bay City is high-quality wetland, which is surrounded by mature hardwood forest. The property is a critical nesting habitat or migratory stopover for nearly 200 bird species, including the rare ruddy duck, yellow-headed blackbird, and redhead duck. Rare or unusual herbs, wildflowers, and other plants can be found in the marsh or the forest undergrowth.

Toumey Woodlot
Ingham County
This 14-acre tract contains a rare remnant of virgin beech–sugar maple forest. Tree specimens are more than 300 years old and nearly 3 feet in diameter at breast height. The wildflowers are colorful during spring and summer.

Two-Hearted Watershed–McMahon Lake Preserve
Luce County
Located on 2,880 acres about 2 miles east of Pine Stump Junction, this property protects part of the headwaters of Two Hearted River, a pristine wilderness watershed. The various forest and wetland communities here support a variety of rare or unusual plant and animal species, including the moose, pine marten, var-

ious sedges and rushes, and the state-threatened bog valerian. The property is owned and managed by The Nature Conservancy.

Vorhees Brothers Wildlife Sanctuary
Calhoun County
This hilly, 40-acre tract contains virgin old-growth maple-oak-cherry stands that are more than 200 years old. Large hickory and tulip trees also occur here. It is an important nesting habitat for interior forest birds, such as woodpeckers, warblers, and other songbirds.

Warren Woods Natural Area
Berrien County
Located about 4 miles southwest of Sawyer, this 200-acre Natural National Landmark is exclusively virgin beech-maple forest. It has not been touched by development in any way. The huge tree specimens are hundreds of years old, 125 feet tall, and 5 feet in diameter at breast height. It is an important nesting site for interior forest birds, such as thrushes, flycatchers, and the rare cerulean warbler and pileated woodpecker. The wildflowers are spectacular in April and May.

Waugoshance Point Preserve
Emmet County
This 250-acre tract in Wilderness State Park is located along the shore of Lake Michigan. Its unusual and delicate shoreline flora include Kalm's lobelia, Indian paintbrush, grass-of-Parnassus, and bird's-eye primrose. Twig-rush and arrowgrass cover the landward dunes. Interdunal wetlands are dominated by white cedar and white spruce. It is an important habitat for a number of rare animals and birds, including the northern flying squirrel, water shrew, spotted sandpiper, bald eagle, piping plover, and other raptors.

White Pigeon River Sanctuary
Saint Joseph County
Located along the White Pigeon River about 4 miles west of White Pigeon, this 58-acre property includes wetlands, bottomland forest, flowing streams, and steep-sided ravines. These natural communities are home to a variety of animal and bird life, including the rare prothonotary and cerulean warblers, buntings, and great blue herons. The massasauga rattlesnake and rare turtles can also be found here. Unusual plant species are lizard's tail and buttonbush.

Woodruff Creek
Livingston County
This 15-acre preserve covers a rare fen community as well as riparian communities. The fen is restricted to a low area and hosts a number of rare plant species, such as swamp valerian, sedge, pitcher plant, prairie-dock, and shrubby cinquefoil. The fen is surrounded by oak-hickory woodlands.

MINNESOTA

Minnesota contains four natural divisions: aspen parkland, prairie grassland, deciduous woods, and coniferous forest.

Aspen parkland occupies a small, flat region of northwestern Minnesota. It is developed on soils deposited by glacial Lake Agassiz. It is characterized by rolling lowlands, brush prairie, wet-mesic brush prairie, and shrubby wet meadow communities. During presettlement times it burned regularly, which controlled aspen intrusions. Postsettlement fire suppression has reduced the native tallgrass and wet prairie and encouraged succession to woodland.

Aspen parkland is essentially a transition between prairie and coniferous forest. Oak savanna and sand prairie grow on sand dunes and beach ridges. Wet areas contain willows, hazel, chokecherry, bur oak, and aspen. Prairie grasses include bluestem, prairie dropseed, mat muhly, prairie cordgrass, reed grass, bluejoint grass, and sedges. Rare northern gentian and prairie fringed orchid can also be found. Dominant animal species include moose, white-tailed deer, grouse, sandhill cranes, and yellow rails.

One-third of Minnesota was once covered by prairie grasslands of the Great Plains. Presettlement prairie was virtually treeless and covered by little bluestem, Indian grass, cordgrass, and needle grass. Wildfires reduced the encroachment of trees and brush. Of the original 18 million acres of prairie, fewer than 150,000 acres remain, found mostly in western and southern Minnesota. Several varieties of prairie subtypes have been identified: dry lime, dry sand, mesic blacksoil, wet blacksoil, saline, and savannas. These are essentially defined by the soil content and moisture level. Species include little bluestem, sideoats grama, sand reed, cordgrass, bluejoint grass, sedges, salt-grass, glasswort, and sea blite. Grasses on the blacksoil prairies can grow to 8 feet. Several varieties of bird, mammal, and reptile species live in prairie grasslands.

Deciduous woods occupy central Minnesota. Presettlement forests were largely maple-basswood; only a small portion remains today. Other species include sugar maple, oak, ironwood, elm, hickory, butternut, birch, and aspen. Within the woods are wetland communities and patches of prairie. Woods communities are further subdivided according to dry, moderate, and wet soil conditions. Dry forests are dominated by oak trees; maple and basswood grow on moderate soils; and willows, ash, birch, and cottonwood grow on wet soil (swamps and floodplains).

Coniferous forest occupies the northern third of Minnesota. The ground is rocky and boulder-strewn, with shallow soil cover. Vegetation in this ecosystem is tolerant of extremely cold temperatures and limited sunlight. Subtypes include wet and dry forests of spruce, fir, cedar, maple, basswood, birch, and oak; pristine peatlands that contain a number of rare animal, bird, and plant species; and bogs and fens. Wetland communities within swamps, bogs, and fens are abundant.

Special Habitat Sites

Agassiz Dunes
Norman and Polk Counties
Located about a mile south of Fertile, this 417-acre tract of grassland prairie has formed atop ancient dunes. Communities include oak savanna and dry prairie dominated by big and little bluestem. Other species are Indian rice-grass and skeleton weed. The prairie is home to some rare plant and animal species, including the upland sandpiper and lark meadow.

Black Lake Bog
Pine County
Located about 30 miles northeast of Sandstone, this 1,400-acre tract contains a mix of swamp, meadow, fen, and peat bog communities. Rare pitcher plants and round-leaved sundews have been identified. This terrain is a favorite of the timber wolf.

Black Dog Prairie
Dakota County
About 3 miles north of Burnsville, this 129-acre tract supports excellent segments of mesic prairie and calcareous fen communities. The rare calcareous fen species are adapted to calcium-rich groundwater, including sedges, shrubby cinquefoil, fringed gentian, and grass-of-Parnassus. Other rare plants are sticky false asphodel, hairlike beak-rush, whorled nut-rush, marsh arrowgrass, and valerian, blooming into spectacular spring colors.

Blazing Star Prairie
Clay County
Located about 20 miles northeast of Moorehead, this 160-acre prairie is owned and managed by The Nature Conservancy. The prairie it protects reflects the presettlement landscape. Natural communities include wet, gravel, and mesic prairies. About 120 plants grow here, including the dominant sideoats grama and needlegrass. Rare species that depend on this habitat are the regal fritillary, Assiniboia skipper, Dakota skipper, Poweshiek skipper, and prairie chicken.

Blue Devil Valley
Yellow Medicine County
Granitic bedrock and boulders dominate this 30-acre site of grassland prairie a mile southwest of Granite Falls. Dry prairie species include brittle cactus, bluestem, hare bell, and pasque flowers. A few bur oaks are also present. This unusual natural community supports a large population of the rare five-lined skink.

Bluestem Prairie
Clay County
A remnant of original prairie, this 1,300-acre tract 11 miles east of Moorehead is one of the finest prairie sites in the United States. Prairie types include mesic tallgrass, wet blacksoil, and calcareous fen. A variety of rare species occurs, including sticky false asphodel, alkali cordgrass, sedges, butterflies, the prairie vole, plains pocket mouse, northern grasshopper mouse, upland sandpiper, and Henslow's sparrow. Its high biodiversity is reflected by more than 300 plant species.

Bonanza Prairie
Big Stone County
Located about 14 miles northwest of Ortonville, this 85-acre tract of original, undisturbed dry prairie supports a variety of plant communities, including oak savanna. Little and big bluestem, purple coneflower, and grama grasses are abundant. Rare species that grow

here are the prairie moonwort and Missouri milk vetch. Coneflowers, purple prairie clover, little bluestem, Indian grass, and blazing star are plentiful.

Boot Lake
Anoka County
This 592-acre property about 7 miles northeast of Bethel consists of deciduous woods containing oak forest, aspen thicket, white pine stand, and wetland communities. Some of the wetlands are covered with floating mats. Rare plant species are the water willow, sea-beach needle grass, and long-bearded hawkweed. Sandhill cranes and Blanding's turtles also inhabit the site.

Botany Bog
Itasca County
Wetlands and bogs occur on this 127-acre site about 10 miles north of Grand Rapids. Typical plants are bog laurel, rosemary, joe pye-weed, and swamp milkweed; rare species are the four-angled water lily, humped bladderwort, olivacious spike-rush, and yellow-eyed grass.

Burntside Islands
Saint Louis County
These two islands total 64 acres and are located in Burntside Lake, about 8 miles west of Ely. The islands are covered by very rare, virgin, old-growth coniferous forest, including 300-year-old red pines.

Butterwort Cliffs
Cook County
Located within coniferous forest, this 53-acre parcel protects a strip of wet, rocky shore along Lake Superior some 8 miles northeast of Grand Marais. The cold microclimate of the cliff faces has allowed the growth of a rare assemblage of arctic-alpine plants, such as the butterwort and northern eyebright. Butterwort is an insect-trapping plant that grows in fragile mats.

Caldwell Brook Cedar Swamp
Koochiching County
This 80-acre tract about 25 miles southwest of Big Falls protects the threatened white cedar tree, which does not regrow after being logged. This cedar swamp is an excellent remnant of virgin northern white cedar forest and is a special habitat for a number of boreal plants.

Cannon River Trout Lily
Rice County
This 95-acre tract of deciduous woods about 2 miles northwest of Cannon City is a remnant of the original maple-basswood forest. This site sheltered the dwarf trout lily, a federally endangered plant that is found only in Minnesota. More than a quarter of the total known plants in existence grows on this property.

Chamberlain Woods
Le Sueur County
Floodplain forest, oak savanna, and upland deciduous forest along the Minnesota River are protected on this 254-acre property a mile north of Ottawa. The area is known for its tall, very old cottonwood, basswood, and elm trees along the river.

Chippewa Prairie
Lac qui Parle County
Located on 1,102 acres above the Minnesota River valley 3 miles north of Milan, this Nature Conservancy preserve encloses a remnant of blacksoil prairie. Six-foot-tall big bluestem, blazing star, sideoats grama, prairie dropseed, and purple coneflowers dominate the landscape. It is a critical habitat for the short-eared owl.

Clinton Prairie
Big Stone County
This 160-acre tract 6 miles west of Clinton is an important remnant of blacksoil prairie. It offers a diverse range of flora and fauna, including the rare Poweshiek skipper butterfly, upland sandpiper, and chestnut-collared longspur.

Compass Prairie
Nobles County
This 20-acre remnant of mesic blacksoil prairie about 5 miles west of Worthington is host to compass plants, locoweed, a variety of grasses, and the rare Poweshiek skipper butterfly.

Cottonwood River Prairie
Brown County
The federally endangered prairie bush clover is found on this 125-acre remnant prairie about 5 miles southwest of Springfield. Vegetation includes purple coneflower, lead plant, puccoon, and blazing stars. Rare prairie butterflies also abound.

East Rat Root River Peatland
Koochiching County
This 2,700-acre expanse of peat bog and wetland is located about 5 miles east of Ericsburg. Plant species include ericaceous shrubs, sedges, mosses, sundews, pitcher plants, and orchids.

Falls Creek
Washington County
Located 3 miles north of Copas, this 136-acre tract is covered by remnant deciduous forest. Stands of virgin hardwoods and white pines are on the property. The rare Louisiana water thrush and kitten-tails are also found here.

Felton Prairie
Clay County
Mesic blacksoil prairie, gravel prairie, and wet blacksoil prairie cover this 410-acre site about 2 miles south of Felton. The various communities support a variety of rare or threatened species, including several skipper butterflies, the upland sandpiper, marbled godwit, Sprague's pipit, Baird's sparrow, and chestnut-collared longspur. Gentians and orchids also bloom during summer. The site is an important nesting area for the greater prairie chicken.

Frenchman's Bluff
Norman County
This mix of remnant tall- and short-grass prairie covers 52 acres 4 miles east of Syre. Grasses include big bluestem, grama, needle grass, thread grass, and prairie dropseed. Rare species seen include skipper butterflies, the northern grasshopper mouse, Nuttall ground rose, plains reed grass, and the tiny moonwort fern.

Greenwater Lake
Becker County
This rare, sensitive, spring-fed lake and its protective, surrounding coniferous woodlands occupy 815 acres about 5 miles west of Ponsford. The lake contains only one kind of algae that is rarely found in other lakes. This suggests the lake is chemically stratified and that its waters do not circulate. The shoreline is undeveloped and lined with birch and aspen trees. The rare Appalachian brown, mustard white, and pine elfin butterflies have been discovered in the woodlands.

Gustafson's Camp
Lake of the Woods County
Located about 25 miles northwest of Waskish, these 185 acres are marked by tall stands of old-growth red and white pine. White cedar also occurs in nearby uplands. These trees escaped the widespread logging during the 1800s and 1900s. Rare ground species can also be found.

Helen Allison Savanna
Anoka County
One of the better examples of oak savanna in Minnesota, this 80-acre tract 4 miles west of Martin Lake contains oak savanna, dry upland prairie, and sedge marsh communities. Savanna plants include a variety of oak trees, chokecherry, willow, quaking aspen, lead plant, steeple bush, aster, and goldenrod. Rare species are sea-beach needle grass, hairy panic grass, Hayden's sedge, blue-joint grass, long-bearded hawkweed, evening primrose, and tall nut-rush. It is also known for its spectacular blooms of wildflowers during late summer.

Hole-in-Bog Peatland
Cass County
Located 2 miles south of Bena, this 1,482-acre peatland is the state's best example of a basin-filled raised bog. It is an important setting for peatland research and contains a peatland lake. The bog is a habitat for a diverse range of plant and animal species, some of them rare or threatened.

Hole-in-the-Mountain Prairie
Lincoln County
Located just south of Lake Benton, this 775-acre property is owned and managed by The Nature Conservancy. It protects a highly diverse remnant prairie ecosystem. Nearly 300 plants grow here, including more than 200 different wildflowers. Unusual plants include soft goldenrod and slender milk vetch. The grasses are an important habitat for insects, mammals, and birds that depend on prairie habitat, including 25 rare or endangered butterfly species.

Hovland Woods
Cook County
Hovland Woods is a 280-acre coniferous forest preserve about 4 miles northwest of Hovland. Natural communities include northern hardwood-conifer forest, upland white cedar forest,

spruce-fir forest, and Great Lakes pine forest. Nearly a third of the property consists of bog, marsh, and beaver flowage. The property is noted for its old-growth virgin forest, especially maple, balsam fir, white pine, and white spruce, some examples being more than 200 years old. The tract is also home to the eastern gray wolf and marten; rare plant species are Carolina spring beauty and large-leaved sandwort.

Hythecker Prairie
Dodge County
This 40-acre property is a rare remnant of original tallgrass prairie. Located about 2 miles southwest of Claremont, it is home to a number of rare plant species, including valerian, rattlesnake master, and compass plant.

Iron Horse Prairie
Dodge County
This 35-acre wedge of mesic tallgrass prairie is the largest remaining original piece in Minnesota. It is marked by a high diversity of wildflowers and other plant species, including the rare Sullivant's milkweed, Indian plantain, wild quinine, valerian, saxifrage, rattlesnake master, and blazing star.

Iron Springs Bog
Clearwater County
Located about 20 miles north of Park Rapids, this 215-acre tract contains bog, boreal forest, pine forest, and conifer swamp forest habitats. Iron-rich springs give the water a red color. The conifer swamp hosts many of Minnesota's rarest species, including coral root and ladies' tresses orchids.

Itasca Wilderness
Clearwater County
Coniferous forest covers this 1,580-acre tract of land along Lake Itasca about 30 miles northwest of Park Rapids. It is noted for its large virgin stands of white and red pine that are up to 300 years old. Rare species that inhabit the forest are the bald eagle, northern goshawk, bog adder's mouth, and matricary grapefern.

Kawishiwi Triangle
Lake County
Located about 8 miles southeast of Ely, this 80-acre tract of coniferous forest contains old-growth white pine trees that are about 150 years old. Other old trees include birch, balsam, maple, and aspen. The forest is the habitat for numerous animal and plant species, including rare lichens.

Kettle River
Pine County
This mix of fen, bog, floodplain forest, upland forest, and coniferous forest covers 801 acres about 3 miles east of Hinckley. Plant and animal species are diverse. Rare occurrences include bog bluegrass and nesting Louisiana waterthrush.

King's and Queen's Bluff
Winona County
Located on high bluffs along the Mississippi River, a mix of deciduous forest communities occurs on this 178-acre site 12 miles southeast of Winona. In addition to the old trees, more than 15 species of rare or endangered plants have been discovered here.

Lake Bronson Parkland
Kittson County
This 80-acre tract of aspen parkland is about 7 miles east of the town of Lake Bronson. Located along the South Branch of Two Rivers, it supports a mixture of prairie, wetlands, and aspen groves. Big bluestem, prairie cordgrass, northern reed grass, and a variety of sedges dominate the prairie. Shrubs include willow, bog birch, and shrubby cinquefoil. This is one of the few good sites of aspen parkland that remain.

Lost 40
Itasca County
This 114-acre site is an excellent remnant of black spruce and tamarack bog, red pine forest, and spruce-fir forest. Located about 11 miles east of Alvwood, it contains virgin old-growth red and white pine more than 300 years old. Rare ground species include fringed polygala and bluebead lily.

Lost Lake Peatland
Saint Louis County
This extensive peatland north of Virginia covers almost 3,000 acres. It is a transition zone between two peatland systems and supports a number of different peatland subcommunities. Rare plants include coast sedge and American bog rush. Some of the nearby birch trees are nearly 250 years old.

Lost River Peatland
Koochiching County
This 12,000-acre peatland 15 miles west of Big Falls contains a number of peatland subcommunities, including tear-drop island, fen, and patterned water tract. It also boasts a number of rare plant species, including a type of moss previously considered to be extinct in North America. Other important plants are sundews, rushes, marsh arrowgrass, and sticky false asphodel.

Lost Valley Prairie
Washington County
Located 3 miles south of Afton, this 200-acre tract contains a variety of prairie grasses in bluff prairie, dry swale, and lowland communities. Amid the dominant bluestem and grama species are rare rock sandwort and Hill's thistle. Wildflowers include prairie smoke, puccoon, yellow star grass, and beardtongue.

Lutsen Natural Area
Cook County
Extensive virgin hardwood-conifer forest covers 720 acres in this area about 2 miles northwest of Lutsen. The larger maples and conifers are 200 to 300 years old. Unusual flowers include the Carolina spring beauty and moschatel.

Mary Schmidt Crawford Woods
Wright County
Located 4 miles east of Buffalo, this 120-acre tract between Green Mountain Lake and Schmidt Lake contains a virgin remnant of the original maple-basswood deciduous forest. Trillium and false rue bloom during late spring among the giant trees.

Malmberg Prairie
Polk County
This 80-acre piece of virgin mesic and wet prairie is about 10 miles west of Crookston. Dominant plant species include big and little bluestem, mat muhly, and cordgrass. Rare or threatened species include the white-tailed jackrabbit, sharp-tailed sparrow, short-eared owl, and prairie vole. Several varieties of wildflowers, including various gentians, bloom during spring and summer.

Margherita Preserve Audubon Prairie
Clay County
This 480-acre Nature Conservancy property is about 24 miles southeast of Moorhead. Natural communities include remnant mesic and wet prairie, meadow, and aspen woodland. Dominant plants include big and little bluestem, Indian grass, purple prairie clover, and heath aster. The site is an important habitat for the greater prairie chicken, marbled godwit, savanna sparrow, grasshopper sparrow, and various prairie-dependent butterflies.

McGregor Marsh
Aitkin County
This 400-acre tract of wetland about 2 miles southeast of McGregor is a special habitat for the rare yellow rail and sharp-tailed sparrow.

Moose Mountain
Saint Louis County
This 55-acre tract about 5 miles north of Duluth is an excellent remnant of old-growth hardwood forest. The larger trees are sugar maple, basswood, and yellow birch. Rare plant species include white baneberry and moschatel.

Mulligan Lake Peatland
Beltrami County
This expansive patterned peatland around Mulligan Lake has water tracks, fens, tear-drop islands, and flark patterns. Subcommunities include tamarack and conifer swamps with bog birch and feather moss. Rare species include sandhill cranes, Wilson's phalaropes, yellow rails, sundews, dragon's mouth, and four-angled water lilies.

Myrtle Lake Peatland
Koochiching County
This 22,630-acre peatland is the most studied peatland in the United States. It is noted for its massive raised bog, large water track, and ribbed fen. It also includes a 3-foot-high waterfall. Numerous rare plants inhabit this site, including sundews, sedges, and mountain yellow-eyed grass. Animal species include moose, white-tailed deer, eastern timber wolf, and bog copper butterflies. Myrtle Lake Peatland is a National Natural Landmark.

Nerstrand Big Wood State Park
Rice County
Located on 1,500 acres about 5 miles northeast of Faribault, this old-growth maple-basswood forest supports a variety of rare and threatened plants and animals, including cerulean warblers and endangered dwarf trout lilies, the only truly endemic species to southeastern Minnesota.

North Black River Peatland
Koochiching County
This 1,220-acre peatland 8 miles west of Loman offers a variety of features, including an ovoid island, fens, and water tracks. Amid the typical plant species are the rare linear-leaved sundew, English sundew, and American bog rush.

Ordway Prairie
Pope County
Located on 582 acres near Brooten, this Nature Conservancy property contains diverse prairie habitats, including pond, marsh, wet prairie, dry gravel prairie, and calcareous fen. Rare species include false asphodel, bog birch, and hairlike beak-rush. The rare Hill's thistle and several endangered prairie butterflies can be found in the tall bluestem grasses.

Osmundson Prairie
Faribault County
This 6-acre remnant of mesic blacksoil prairie is all that survives from intensive agriculture in the area. Located about 3 miles southeast of Kiester, the rolling bluestem and grama grasses on the hills enclose the rare rattlesnake master and tuberous Indian plantain.

Ottertail Prairie
Otter Tail County
This 320 acres represents oft-flooded wet prairie. Located about 8 miles southwest of Fergus Falls, the property contains a variety of sedges, grasses, and forbs. Rare species include the downy gentian, upland sandpiper, and marbled godwit.

Partch Woods
Stearns County
Located about 5 miles north of Saint Joseph, this 120-acre site is a an excellent example of climax maple-basswood deciduous woods with representative plant communities.

Pembina Trail Preserve
Polk County
Several varieties of prairie communities are preserved on this 2,320-acre ridge-and-swale tract about 10 miles west of Mentor. The most common is mesic tallgrass prairie with lesser meadow, marsh and shrub swamp, and calcareous fen. Moose and sandhill cranes are commonly seen. Rare plant and animal species are hairlike beak-rush, marsh arrowgrass, sticky false asphodel, upland sandpiper, yellow rail, marbled godwit, Wilson's phalarope, and skipper butterflies.

Pennington Bog
Beltrami County
This property is a virgin piece of coniferous forest 108 acres in size. It is located about 16 miles east of Bemidji. Several varieties of unusual plant species grow beneath the tall canopy of cedar, fir, and spruce.

Plover Prairie
Lac qui Parle County
This 650-acre property is owned and managed by The Nature Conservancy. Natural communities include remnants of native blacksoil prairie, wet prairie, and wetland depression. The site is an important habitat for the small white lady's slipper, upland sandpiper, marbled godwit, Carolina foxtail, slender milk vetch, burrowing owl, and Wilson's phalarope.

Prairie Coteau
Pipestone County
Rare dry mixed-grass hill prairie and dry sand-gravel prairie survive on this 329-acre tract about 2 miles northeast of Holland. Common wildflowers are puccoon, prairie smoke, prairie phlox, and blazing star; more than 200 wildflower species attract 40 rare butterflies, including the Dakota and Ottoe skippers. Little bluestem, sideoats grama, and prairie dropseed also occur.

Prairie Creek Woods
Rice County
This 87-acre tract of prairie about 2 miles northwest of Nerstrand is one of the few sites in Minnesota that contains the dwarf trout lily—a federally threatened species endemic to Minnesota. It is believed to be a relict plant from the last glaciation, about 13,000 years ago. It has not been successfully raised in greenhouses, as it requires natural habitat to thrive.

Purvis Lake–Ober Foundation
Saint Louis County
This 140-acre tract of old-growth forest is located about 10 miles west of Ely. It is most known for its population of timber wolves and very old white and red pines. Several varieties of natural communities, including rare plant species, are found within the forest, bogs, and lakes.

Racine Prairie
Mower County
This 6-acre remnant of mesic prairie is especially rare because almost all of Minnesota's mesic prairie has been converted to farmland. Located about a mile north of Racine, it hosts rare plant species, including wild quinine, white wild indigo, rattlesnake master, and compass plant.

Red Lake Peatland
Beltrami, Koochiching, and Lake of the Woods Counties
At nearly 83,000 acres, this peatland is the largest and most complex in the United States. It contains every imaginable peatland feature. The different peatland subcommunities are complexly interwoven, and this is an important area for research. Some of it has been designated a National Natural Landmark. Rare animal species are the eastern timber wolf, short-eared owl, yellow rail, Wilson's phalarope, and sandhill crane. Rare plants include rushes, sundews, and mountain yellow-eyed grass.

Richard M. Elliott Property
Wilkin County
Remnant tallgrass and wet prairie communities exist on this 320-acre tract about 7 miles northwest of Rothsay. Sedges and willows occur in wetland areas. Rare species include upland sandpiper, prairie vole, northern gentian, glasswort, and alkali grass. It is also an important breeding ground for the greater prairie chicken.

Roscoe Prairie
Stearns County
Virgin blacksoil prairie and wet swales occupy this 56-acre site about 2 miles west of Roscoe. Plant and animal species are diverse. Dominant plants are dropseed, Indian grass, cordgrass, sedge, and big bluestem. Rare species include the Dakota skipper, thirteen-lined ground squirrel, white-tailed jackrabbits, Hill's thistle, orchids, spiderwort, purple prairie clover, and upland sandpipers.

Rothsay Prairie Wildlife Management Area
Wilkin County
This 4,000-acre tract 4 miles west of Rothsay comprises wet prairie, wet meadow, marsh, and calcareous fen communities. The big bluestem–sedge prairie was never plowed because it was too wet. Wet meadows have various sedges, rush, and prairie cordgrass. Important bird species that rely on this habitat are prairie falcons, yellow rails, and the rare greater prairie chicken.

Rush Lake Island
Chisago County
This 21-acre wooded island in Rush Lake about 7 miles west of Rush City is Minnesota's largest great blue heron rookery. Egrets have also started nesting on the island. Access is restricted during spring and summer to protect the young broods.

Rushford Sand Barrens
Fillmore County
About 3 miles southwest of Rushford, this 230-acre property protects a mix of rare natural communities with deciduous woods, including oak and jackpine savannas. Species of concern include the fame flower, clasping milkweed, sea-beach needlegrass, long-bearded hawkweed, toadflax, compass plant, Leonard's skullcap, witch hazel, and the six-lined racerunner lizard.

Saint Croix Savanna
Washington County
Oak savanna and dry prairie occupy this 84-acre site a mile south of Bayport. Savanna grasses include hairy and sideoats grama, prairie dropseed, and needle-and-thread. Wildflowers are varied and abundant. Rarer species include gentian, hoary vervain, kitten tails, James's polanisia, Illinois ticktrefoil, and pinweed.

Sand Lake Peatland
Lake County
Located about 12 miles southeast of Babbitt, this 4,500-acre peatland is unique for its bed of glacial outwash. It is of special scientific value because it is still expanding, compared to other peatlands that have stabilized. Rare species on the site include the eastern timber wolf, coast sedge, dragon's mouth, and the American bog rush.

Sandpiper Prairie
Norman County
Located about 7 miles southwest of Twin Valley, this 160-acre tract preserves remnant pieces of dry, wet, and dry-mesic prairies. Rare species include purple prairie clover, bunch speargrass, Culver's root, and alkali cordgrass.

Santee Prairie
Mahnomen County
A remnant blacksoil prairie and wetland complex occupies this 442-acre property about 6 miles northeast of Mahnomen. In addition to a variety of plants, numerous rare bird species, including the greater prairie chicken, marbled godwit, sandhill crane, Wilson's phalarope, and upland sandpiper, can be found.

Schaeffer Prairie
McLeod County
About 8 miles west of Glencoe, this remnant prairie is owned by The Nature Conservancy. The main prairie grasses are cordgrass and reed grass. Mima mounds occur near the wetlands and are important habitats for toads. The area is highly diverse, with nearly 300 plant species represented. Grassland birds, including the goldfinch, bobolink, and meadowlark, frequent the prairie. Rare plants include Sullivant's milkweed, Riddell's goldenrod, and Nuttall bur-reed.

South Black River Peatland
Koochiching County
At nearly 6,000 acres, this virgin peatland complex about 25 miles northwest of Big Falls shows excellently developed internal water tracks. Amid the typical peatland vegetation are groups of rare linear-leaved sundew and English sundew plants.

Spring Beauty Northern Hardwoods
Cook County
Located about 4 miles northwest of Hovland, this 400-acre tract of hardwoods forest contains rare old-growth trees. The largest trees are sugar maples; large white pine, white cedar, and yellow birch can also be found. Rare plant species include the Chilean sweet cicely, blunt-fruited sweet cicely, and the Carolina spring beauty.

Susie Island
Cook County
This 126-acre Nature Conservancy preserve is on Susie Island in Lake Superior, near Grand Portage. The rocky island contains lakeshore, sphagnum bog, and old-growth black spruce–balsam fir–northern white cedar forest in the interior. Rare relict plants from a colder climate include Norwegian witlow grass, northern eyebright, pearlwort, Arctic lupine, slender hairgrass, rock cranberry, and purple crowberry.

Swedes Forest
Yellow Medicine and Redwood Counties
This 202-acre rocky prairie site about 7 miles south of Sacred Heart contains prairie, savanna, and oak woodland communities. Granite boulders and outcrops shelter the rare five-lined skink. Undisturbed ponds and wetlands are also on the property.

Tauer Prairie
Brown County
Although this area was farmed, the low-lying portions were never drained, and virgin pieces of mesic and wet-mesic prairie remain. Covering 80 acres about 8 miles south of New Ulm, a rich variety of prairie grasses and wildflowers shelter two rare species: tuberous Indian plantain and rattlesnake master.

Townsend Woods
Rice County
About 5 miles northwest of Morristown, this 73-acre tract contains a rare virgin remnant of deciduous forest. Large trees include sugar maple, oaks, and basswood.

Twin Valley Prairie
Norman County
Remnants of wet to dry prairie occur on this 240-acre property about 7 miles southwest of Twin Valley. Dominant plants are sedges, cottongrass, big bluestem, prairie cordgrass, blazing star, and golden alexander. Moose, white-tailed deer, and jackrabbits are commonly seen. It is also an important habitat and breeding ground for the greater prairie chicken. Other rare species include the Dakota skipper, sandhill crane, upland sandpiper, and prairie vole.

Two Rivers Aspen Prairie Parkland
Roseau County
This 1,333-acre tract about 2 miles northeast of Pelan contains one of the best examples of parkland remaining in northwestern Minnesota. Communities include wet-mesic brush prairie, wet meadow, and aspen woodland. Grasses include big bluestem, Indian grass, prairie dropseed, and bluejoint grass; sedges dot the meadows. Several varieties of prairie flowers bloom during spring and summer. Plant and animal species are diverse and include the rare northern gentian and nesting sandhill cranes.

Uncas Dunes
Sherburne County
About 4 miles west of Zimmerman, this 745-acre tract contains a mix of prairie, sand dune, scrub-oak savanna, fen, oak forest, and wetland communities. Rare species include the Uncas skipper butterfly, sea-beach needlegrass, and annual skeleton weed.

Weaver Dunes
Wabasha County
Located about 10 miles south of Kellogg, this 697-acre Nature Conservancy property protects wetlands and sand dunes. Rare or unusual species on the property include Blanding's turtles, butterflies, the rough-seeded fame flower, clasping milkweed, cliff goldenrod, upland sandpipers, and peregrine falcons. The surrounding woodlands are home to songbirds and shrikes.

Western Prairie South
Wilkin County
A mix of sedge meadow, tallgrass prairie, and wet prairie exists on this 320-acre tract about 6 miles northwest of Rothsay. Certain portions contain salty groundwater. Tallgrass prairie is marked by big and little bluestem and switchgrass; the mesic prairie carries gayfeather and goldenrods. Plant and animal species are diverse, including rare skipper butterflies, upland sandpiper, marbled godwit, Wilson's phalarope, sharp-tailed sparrow, Nuttall alkali grass, and the regal fritillary.

Whitewater Wildlife Management Area
Winona and Olmsted Counties
This 28,000-acre wildlife management area sits along the Whitewater River. Natural communities include riparian corridor, cliff, oak forest, meadow, prairie, and dry oak savanna. Algific talus slopes with permanent ice and rare plant microcommunities occur along the river. The dry oak savanna supports the state-endangered sea-beech needlegrass, Ottoe skipper, rough-seeded fame flower, goat's rue, golden and bald eagles, and the federally endangered Karner blue butterfly. More than 200 bird species have been recorded here.

Wolsfeld Woods
Hennepin County
This 221-acre tract of old-growth deciduous woods is about 2 miles north of Long Lake. There are very few remnants of representative deciduous woods remaining in the state. The large trees on the rolling hills include red oak, ironwood, butternut, maple, elm, and basswood.

Yellow Bank Hills
Lac qui Parle County
This 80-acre tract beside Pegg Lake about a mile southeast of Nassau is a prairie developed on a glacial ridge of sand and gravel. The soil is coarse and well drained. In addition to the typical grasses and flowers for upland prairie, the following rare plants also occur: Missouri milk vetch, Nuttall violet, cutleaf ironplant, soft goldenrod, and prairie vole. There are also numerous butterflies.

MISSISSIPPI

Mississippi is divided into five natural regions: oak-hickory-pine forest, oak-hickory forest, southern floodplain forest, southern mixed forest, and prairie belt.

Oak-hickory-pine forest is the biggest natural region, covering almost the entire eastern half of the state. Dominant tree species vary from area to area. Once-extensive pine forests are being succeeded by hardwood forests in many parts of Mississippi. Other trees are pignut, bitternut, loblolly pine, and mockernut. Magnolias and sweet gum can be found in the wetter areas. Dogwood is common in the understory.

Most of the western third of the state is covered by a belt of oak-hickory forest that runs north and south, marking drier soil conditions. Several oak and hickory species can be found, such as black, white, red, and post oak, shagbark and mockernut and black hickory, poplar, and hackberry. Ironwood and dogwood are common in the understory.

Along the Louisiana border is a belt of southern floodplain forest. Topography is generally low and periodically to permanently flooded. These wet conditions support a dense forest of bald cypress, tupelo gum, pecan, swamp chestnut oak, and willow oak. Understory plants are vines, red haw, and cabbage palm.

The prairie belt consists of two narrow zones that cross the oak-hickory-pine forest terrain from northwest to southeast. The blacksoil supports red cedar and a variety of oak trees. Open, gladelike areas are covered with thick prairie grasses like cornflower, Cherokee sedge, big bluestem, and prairie clover.

The very diverse southern mixed forest extends from the beaches of the Gulf of Mexico inland for about 100 miles. The moist, warm climate has created a very dense, lush forest mix of pine, sweet gum, beech, oak, cucumber, magnolia, Florida maple, ironwood, red bay, dogwood, and holly. Open sandy areas are dominated by stands of pine.

Special Habitat Sites

Bienville Pines Scenic Area
Scott County
The largest tract of old-growth shortleaf and loblolly pine forest in Mississippi is located on this 189-acre tract. These trees are at least 200 years old and are important habitats for the federally endangered red-cockaded woodpecker.

Buttercup Flats
Hancock County
Located on 70 acres about 7 miles east of Picayune, this Nature Conservancy preserve encloses a rare Coastal Plain bog. Other natural communities are longleaf pine forest, longleaf pine savanna, meadow, relict pond cypress, upland oak-pine forest, mixed harwood forest, and riparian habitat along Catahoula Creek. The bog supports a number of rare or unusual plants, including orchids and the carnivorous pitcher plant. Colorful displays of wildflowers peak during May and early June.

Clark Creek Natural Area
Wilkinson County
Several varieties of natural communities are preserved on this 2,000-acre site about 15 miles west of Woodville. The rugged hills are forested with old-growth beech, magnolia, and pine; nearly 50 waterfalls can be found on the steep, rocky slopes. Some of the beech trees are hundreds of years old. Rare or endangered species found here include the southern red belly dace fish, Louisiana black bear, land snails, and the Carolina magnolia vine. Colorful wildflowers are abundant in April.

Dahomey National Wildlife Refuge
Bolivar County
This 9,500-acre Nature Conservancy–U.S. Fish and Wildlife Service property is one of the largest remnants of bottomland hardwood forest and shrub swamp in Louisiana. Located about 20 miles northeast of Greenville, the property is reminiscent of what the "Big Woods" wilderness looked like prior to European settlement. Nuttall oak, cherry bark oak, hickory, pecan, elm, and beech are hundreds of years old. Rich flora and fauna are present, including some rare or threatened species. The woodlands are an important nesting and migratory habitat for a variety of birds. It is likely the federally endangered pond spicebush is present on the preserve.

Dead Dog Bog
Greene County
This preserve about 6 miles north of Stateline encloses 360 acres of rare Coastal Plain bog and surrounding woodlands. Bogs are one of the most rapidly disappearing natural communities, as they are usually small and easily drained and filled. Eight rare or endangered species thrive in this fragile environment, including Wherry's pitcher plant, the lavender burrowing crayfish, and Drummond's yellow-eyed grass. The property is owned by The Nature Conservancy.

Donivan Slough
Lee County
Located 20 miles north of Tupelo, this property consists of mature to old-growth swamp bottomland forest. Some of the water oaks, sycamores, sweet gums, beech, and bald cypress trees are hundreds of years old. Trilliums are abundant during spring.

Grand Bay Savanna
Jackson County
Located about 7 miles east of Pascagoula and about 4 miles inland from the Gulf of Mexico, this 6,000-acre property protects part of an important wetland complex. Natural communities include marsh island, shell midden, salt marsh, cypress savanna, pine savanna, and brackish marsh. It is the largest and most pristine wet savanna in the United States and is a critical groundwater recharge site. The diversity of bird and aquatic life is astounding. The savanna is breeding habitat for at least 20 rare or endangered plant and animal species and is being considered as a reintroduction site for the federally threatened bald eagle. The property is managed by the U.S. Fish and Wildlife Service.

Hillside National Wildlife Refuge
Holmes and Yazoo Counties
Located on 22 acres about 15 miles southeast of Silver City, this Nature Conservancy property near Eden protects part of the important wetlands along the Yazoo River. The property adjoins the larger Hillside National Wildlife Refuge. The diverse wetland communities are important nesting, feeding, and migratory sites for thousands of waterfowl along the Mississippi Flyway. The refuge also supports a healthy alligator population, which nests during March and April. The surrounding bottomland forest contains some very large old-growth specimens of cypress, sycamore, tupelo, and cottonwood trees.

McIntyre Scatters
Leflore, Carroll, and Grenada Counties
A high-quality remnant of the presettlement "Big Woods" survives on this 808-acre Nature Conservancy tract about 15 miles north of Greenwood. The relatively undisturbed old-growth bottomland hardwood forest supports very diverse flora and fauna, including the rare Mississippi map turtle, paddlefish, and hoary bat. Colorful wildflowers are abundant during April and May.

Panther Swamp
Yazoo County
Located on 28,600 acres about 10 miles west of Yazoo City, this tract of dense mixed woodland and wetland is an important habitat for bobcats, alligators, and black bears. Old-growth stands of tall bald cypress remain.

Pascagoula River Wildlife Management Area
George County
Located southwest of Lucedale, this 40,000-acre preserve protects the most pristine natural habitat in the state. Natural communities include the Pascagoula River, oxbow lake, wetland, swamp forest, bottomland hardwood forest, and mixed upland forest. The swamps contain old-growth cypress and tupelo trees. The wetlands support rare or unusual plant and animal species, including rainbow trout, gopher tortoise, yellow-blotched sawback turtle, swallow-tailed kites, green-fly orchids, Florida flame azaleas, and silky camellias.

Shipland Wildlife Management Area
Issaquena County
This 3,642-acre Nature Conservancy property along the Mississippi River is about 45 miles northwest of Vicksburg. The property contains mature bottomland hardwood forest and wetlands in the river's floodplain. A great variety of flora and fauna can be found on the property; some of the species are rare or unusual, such as the scissor-tail flycatcher. The forest and wetlands are important bird habitats for nesting species as well as migratory visitors.

Sweetbay Bogs
Stone County
Located on 194 acres about 8 miles west of Wiggins, this property encloses a rare hillside seepage bog and surrounding longleaf pine forest. The bog has existed for thousands of years and has a floating sphagnum moss mat. Like most bogs, Sweetbay is host to a number of rare or endangered plant species, such as large beak-rush, Harper's yellow-eyed grass, coast sedge, bog spice bush, grass-of-Parnassus, certain orchids, sundew, southern butterwort, and pitcher plant. Bog spice bush is known to exist in fewer than 20 locations worldwide. The wildflowers peak during May and early June.

Tishomingo State Park
Tishomingo County
This property is located about 25 miles east of Booneville. Natural features include rocky hills and steep gorges that have created some cooler microclimates that support rare ferns, lichens, and rock plants, including the lobed spleenwort.

Willie F. Brown Nature Preserve
Hancock County
This 185-acre tract of longleaf pine forest, live oak–mixed hardwood forest, swamps, bogs, and creeks support the rare bog button, pitcher plant, and yellow pipewort.

MISSOURI

Missouri contains the following natural divisions: the Ozark, Ozark Border, Glaciated Plain, Big River, Mississippi River Lowland, and Osage Plain.

The Ozark occupies a large part of the southern half of Missouri. This area of igneous and sedimentary rocks has been uplifted and subsequently eroded to form hills, ridges, and hollows with occasional cliffs; soils are thin. Natural communities within the Ozark include shady glade, savanna, cliff habitat, pine and deciduous woodland, cave, and wetland.

Ozark Border terrains are wide transition zones between the Ozark and other regions to the north and east. Soils in the Ozark Border are deeper and more productive, and plant and animal ranges from different terrains overlap. Ozark Border terrain was originally covered by woodlands with lesser prairies, glades, and savannas; caves are common.

Glaciated terrain covers the northern half of the state. Glacial action leveled the topography and deposited thick layers of silt, sand, and gravel. Wind-blown silt and clay formed fine-grained loess deposits and dunes. Presettlement vegetation included deciduous forest, tallgrass prairie, and oak savanna. Agriculture and development have eliminated all but a few tiny patches of original prairie.

Big River terrain comprises parts of the Missouri, Mississippi, Grand, and Des Moines Rivers and includes their floodplains and terraces. This terrain originally consisted of sloughs, wetlands, sandbars, prairies, and forests. The rivers include threatened species of fish, and bald eagles nest along their banks during winter.

The Mississippi River Lowland was originally extensive swamps and lowland forests; soil cover is deep. The area has since been greatly affected by timbering, agriculture, and the drainage of wetlands. The terrain contains coastal plant and animal species.

Osage Plain comprises Missouri's prairie region. Before settlement, it was largely prairie with lesser forests, wetlands, and savannas. The terrain is gently rolling, with soils derived from the underlying bedrock. Most of the plant and animal species are the same as those in the glaciated plains.

Special Habitat Sites

Accola Woods
Lewis County
Located on 40 acres about 2 miles south of Benjamin, this Nature Conservancy property protects an excellent remnant of a once-sprawling white oak swamp savanna, a rare ecosystem today. The oak and hickory trees that dominate the hardwood forest are very old, approaching 200 years. Rare or unusual prairie grasses and forbs are part of the groundcover and include Culver's root, Nottoway brome, and Pennsylvania sedge. In fact, Accola Woods is one of only two places west of the Mississippi River where Nottoway brome can be found. The displays of wildflowers are very colorful during spring.

Allred Lake
Butler County
About 15 miles south of Poplar Bluff, this 76-acre tract contains Allred Lake, Missouri's best example of undisturbed cypress pond. Part of the Mississippi Lowland, the acreage is largely swamp and bottomland forest. Old cypress and tupelo trees surround the pond. Allred Lake is home to the endangered swamp darter and taillight shiner. Bottomland forest includes locusts, sweet gums, pecans, willows, chestnuts, and the rare Nuttall oak.

Altvater Corry Flatrocks
Dade County
Located on 40 acres about 3 miles west of Dadeville, this Nature Conservancy preserve encloses a rare sandstone glade community. The property protects one of the largest known populations of a vary rare species of wildflower that is only known from a few sites in southwestern Missouri and Arkansas. The rock surfaces are covered in delicate mosses and lichens as well as spikemoss. Rare or unusual plants include selenia, large-flowered coreopsis, sedges, and violet collinsia. Lichen grasshoppers can also be seen here.

Altvater Grasshopper Hollow
Reynolds County
This 120-acre property near Corridon protects one of the largest fen complexes within unglaciated terrain in the United States. Natural communities include fen, prairie fen, and deep muck fen. The fen complex is developed on a

substrate of dolomite, which creates calcium- and magnesium-rich groundwater. The rare prairie fen is covered by prairie cordgrass, big bluestem, rough-leaf goldenrod, New England aster, Michigan lily, and Indian grass. Unusual plants for Missouri are marsh fern, grass-of-Parnassus, and swamp thistle. Five rare plants and two rare animals have been found here: marsh bellflower, marsh blue violet, tussock sedge, wild sweet William, heart-leaf plantain, wood frog, and four-toed salamander. Beavers have built dams and lodges in the open water of the fen.

Altvater La Barque Hills
Jefferson County
Part of the Ozark Border, this 335-acre Nature Conservancy preserve is located about 6 miles south of Pacific. The topography is rocky and rugged. Hills are separated by deep canyons, streams, and waterfalls. Natural communities that exist here are sandstone cliff, sandstone glade, and dry-mesic forest. The rare prickly pear cactus, mosses, and various ferns thrive in the microenvironments among the rock faces. A rare population of heart-leaf plantain can be found in the lower wet areas. Prairie plants, such as bluestem, various mosses and lichens, and the threatened Fremont's leather flower, grow in the drier rocky conditions of the sandstone glades. Glade communities in general are among the rarest natural communities found in Missouri.

Altvater Pondberry Preserve
Ripley County
Located on 105 acres about 3 miles south of Naylor, this Nature Conservancy property protects a rare type of wet bottomland forest that has developed on ancient low sand dunes. The vegetation is unique to the changing water levels as the growing season progresses. Rare or unusual trees include bald cypress, swamp white oak, Nuttall oak, gum, and swamp cottonwood. The very rare pondberry shrub is found here, as are some rare sedges. A variety of vines and ferns also thrives here. The uncommon green tree frog breeds on the site.

Altvater Shut-In Mountain Fens
Shannon County
This 520-acre Nature Conservancy property about 8 miles east of Eminence preserves undisturbed representative habitats typical of the Saint Francois Mountains. Natural communities include three fens and xeric forest developed on igneous bedrock. The mineralized groundwater supports a high diversity of fen plants, including false loosestrife, tussock sedge, grass pink orchid, and other uncommon species. Black oak and hickory composes the xeric igneous forest; the rock outcroppings also support lichens, mosses, ferns, and wildflowers.

Ashe Juniper
Stone County
The Ashe juniper, a close relative to the eastern red cedar, is a rare tree in Missouri (it is usually found in west-central Texas, Oklahoma, and northern Arkansas). This 31-acre area of Ozark terrain southwest of Branson contains an Ashe juniper forest that overlies dolomite bedrock.

Bagley Woods
Mercer County
Located about 10 miles northwest of Mercer, this 20-acre part of the Glaciated Plain contains rare old-growth oak-history forest, including the state-endangered northern pin oak.

Bald Hill Glade
Ripley County
Located within Mark Twain National Forest 3 miles northwest of Gatewood, this 332-acre dolomite glade complex is part of the Ozark. A variety of rare or unusual species has been identified here, including the Missouri coneflower, purple prairie clover, heliotrope, finger coreopsis, and the endangered umbrella plant.

Bat Cave
Miller County
Part of the Ozark, this small cave system near Saint Elizabeth is a nursery site for a colony of endangered gray bats. The entrance is located in the middle of a dolomite cliff. The cave is closed to the public.

Bennett Spring Savanna
Laclede County
This 160-acre Nature Conservancy preserve located west of Lebanon protects a high-quality remnant of dry-mesic chert savanna. The tall-prairie grasses are broken by groups of old oak and hickory trees. The prairie is studded with a colorful array of wildflowers during spring and summer. Rare or important species on the property include the downy blue gentian, pink milkwort, stiff aster, dogface, Baltimore checkerspot butterfly, and various snakes. Prairie birds can also be found here.

Big Buffalo Creek Fen
Benton County
This 40-acre tract about 10 miles southwest of Stover is part of the Ozark. Features include a spring-fed fen, dolomite glade, and upland forest. Fen plants include blue flag, coneflower, Riddell's goldenrod, and the rare bushy aster.

Big Oak Tree Area
Mississippi County
This 940-acre tract about 5 miles northwest of Dorena is part of the Mississippi Lowland. It is a pristine remnant of bottomland forest and swamp, including 80 acres of virgin wet-mesic bottomland forest. The tree canopy averages about 120 feet in height. Some of the state's tallest trees grow here, including a 133-foot persimmon tree. The Big Oak area is home to a variety of bird species, including the rare Mississippi kite and Swainson's warbler.

Big Spring Pines
Carter County
This 345-acre area about 5 miles southeast of Chicopee is one of the few outstanding pine-oak forests remaining in the Ozark. Trees include shortleaf pine, oaks, sassafras, black cherry, dogwood, and black gum. Sedges, grasses, and wildflowers cover the forest floor. The rare heart-leaf plantain and petaltail stonefly can be found near springs on the site.

Blair Creek Raised Fen
Shannon County
The only known raised fen in Missouri, this 3-acre site is about 8 miles south of Bunker in the Ozark. It is within the Mark Twain National Forest. Located at the base of a wooded slope in the Blair Creek valley, it contains rare marsh blue violet, glossy-leaf aster, and sedges.

Bona Glade
Dade County
Located a mile west of Bona, this 20-acre site is a sandstone glade in the Ozark. Plant species include Nuttal sedum, widow's cross, fame flower, rushfoil, pineweed, and an extremely rare wildflower.

Brush Creek
Saint Clair County
Located in the Ozark, this half-mile stretch of pristine Brush Creek is critical habitat for the federally endangered Niangua darter. It and 40 other species of fish live in the runs, riffles, and pools of this well-defined channel. The creek is surrounded by upland forest and dolomite cliffs. The protected area is 26 acres in size.

Burr Oak Woods
Jackson County
A 33-acre part of the Burr Oak Woods Conservation Area about a mile north of Blue Springs contains an old-growth stand of chinquapin oak trees. The forests, part of the Glaciated Plain, include basswood, Kentucky coffee, and black walnut trees.

Carman Springs
Howell County
Three outstanding creeks are located within the 2,810-acre Carman Springs Natural Area about 7 miles west of Burnham. Part of the Ozark, it also includes high-quality cliff, bottomland, and upland forest communities. Rare plant species include the barren strawberry, queen-of-the-prairie, showy lady's slipper, and five others.

Cash Swamp
Dunklin County
This 310-acre wetland, part of the Mississippi Lowland, is a high-quality remnant of tupelo-cypress bottomland forest located about 5 miles northwest of Senath, along the Saint Francis River. Indigenous species include planer, tupelo, and cypress trees, river otters, and swamp rabbits.

Castor River Shut-Ins
Madison County
This 209-acre area around a pink granite shut-in (a narrow gorge-shaped part of a wider valley) of the Castor River in northeastern Madison County offers glade and woodland communities. Rare species include prickly pear, fame flower, and the eastern collared lizard.

Chloe Lowry Marsh
Mercer County
Located 2 miles northwest of Princeton, this area includes 40 acres of marsh and 15 acres of wet-mesic prairie. It is part of the floodplain of the Weldon Fork of the Grand River within the Glaciated Plain. One of the best natural marshes in the state, it contains a number of rare species, including sedges, marsh skullcap, star duckseed, tufted loosestrife, and the northern leopard frog.

Cook Memorial Meadow
Barton County
Located about 3 miles northwest of Golden City, this 280-acre preserve encloses a high-quality remnant of dry-mesic prairie. More than 300 native plants have been identified, including the endangered Harvey's beak-rush and fringed nut-rush. The prairie is an important habitat for prairie birds, such as dickcissels and certain sparrows. The wildflowers are especially colorful during May and June. The area is managed through controlled burning.

Coonville Creek
Saint Francois County
Located in Saint Francois State Park in Ozark Border terrain about 15 miles northwest of Farmington, this 2-mile stretch of Coonville Creek is home to 18 species of fish, including the fantail and orangethroat darter. Coonville Creek is a permanent headwaters stream that is fed by numerous seeps. Meadows and fens in the creek valley contain rare plants, such as queen-of-the-prairie and grass pink.

Cupola Pond Area
Ripley County
This site is an old sinkhole pond ringed with old tupelo gum trees. The 160-acre tract sits within Mark Twain National Forest (in the Ozark) about 8 miles north of Bardley. Several varieties of sedge and moss species and large pin oaks and red maples are also found here. The site is a registered National Natural Landmark.

Current River Natural Area
Shannon County
This 10-acre parcel 15 miles south of Bunker is part of the Ozark. Within it is a forest of old-growth white oak trees that are more than 300 years old.

Danville Glades
Montgomery County
Part of the Ozark Border, this 48-acre tract about 6 miles west of New Florence contains limestone glade and upland forest communities. Glade plants include little bluestem, sideoats grama, pale purple coneflower, and stickleaf. Many of the trees are oaks (including chinquapin oak), maples, and red cedars, many of which are more than 100 years old.

Dark Hollow
Sullivan County
This 308-acre tract about 5 miles north of Green City contains old-growth mesic and dry-mesic upland forest as well as remnants of original prairie in the Glaciated Plain. Ground flora includes yellow lady's slipper, bloodroot, and golden seal.

Diamond Grove Prairie
Newton County
Located southeast of Joplin, this 570-acre remnant of upland tallgrass prairie is part of the Ozark. Interesting species include greater prairie chickens and other birds, royal catchfly, Barbara's buttons, blue star, and fringed poppy mallow.

Diggs Natural Area
Audrain County
This conservation area about 2 miles south of Martinsburg contains a 3-acre tract of large white oak trees that are more than 100 years old. It represents dry-mesic upland forest in the Glaciated Plain.

Dobbins Woodland
Nodaway County
This 40-acre parcel owned by The Nature Conservancy is located about 5 miles north of Barnard. It is a rare type of woodland within agricultural northwestern Missouri. Natural communities include mesic walnut-hickory-elm forest bottomland, dry-mesic oak-hickory upland forest, and mesic oak-hickory-mulberry upland forest. Rare or unusual plants on the site are the yellow giant hyssop, Kentucky coffee tree, and Indian pipe. Several interior forest birds, especially warblers and buntings, nest in the woods.

East Drywood Creek
Barton County
Located about 3 miles southwest of Liberal, this 50-acre tract consists of remnant dry, dry-mesic, and mesic prairie along a headwater in the Osage Plain. Threatened species include the northern harrier, Mississippi kite, Henslow's sparrow, yellow-headed blackbird, greater prairie chicken, and upland sandpiper.

Eck Memorial Natural Area
Texas County
Part of Mark Twain National Forest near Roby, this 265-acre tract of Ozark terrain contains old-

growth dry and dry-mesic upland forest. Trees include shortleaf pine, black hickory, mockernut hickory, and three oak species. Because this area has had few fires over the past 200 years, it is an important benchmark by which to compare other natural communities that burn more frequently.

Gamagrass Meadows
Vernon County
Located on 80 acres about 13 miles north of Nevada, this Nature Conservancy property encloses rare remnants of mesic and dry-mesic sand-shale prairie. The prairie is covered by tall gamagrass and other common grasses; more unusual plants are the prairie larkspur, downy blue gentian, and several other wildflowers. Several varieties of rapidly disappearing prairie birds, such dickcissels and sandpipers, frequent the property.

George A. Hamilton Forest
Lincoln County
This 40-acre remnant mesic and dry-mesic forest a few miles northeast of Troy is one of the finest in the Glaciated Plain. It is dominated by a mature white oak forest.

Grasshopper Hollow
Reynolds County
This 593-acre site 2 miles northwest of Reynolds contains one of the largest known fen complexes in unglaciated North America. Sixteen different fens, including deep muck, forest, and prairie fen, occur in this complex within the Ozark. Nine rare plant and animal species also are found.

Greenfield Glade
Dade County
This 47-acre Nature Conservancy tract about a mile west of South Greenfield encloses a limestone glade, a rare community in Missouri. The glade hosts healthy populations of the Missouri bladderpod, one of the world's rarest plants—fewer than 50 sites are known in the world. This tiny plant can be identified only by its brilliant yellow flowers that peak in early May. The glades are mixed with stands of Osage orange, locust, and oak trees. Parts of the rocky ground are covered in prairie grasses, including Indian grass and bluestem; rarer plants are the prickly pear cactus and beardtongue.

Ha Ha Tonka
Camden County
Located about 3 miles southwest of Camdenton, this 1,033-acre Ozark parcel contains complex limestone karst topography and savanna. The collapse of limestone caves created the unique Ha Ha Canyon and a natural bridge formation. The savanna occurs on rocky chert-sandstone slopes and is one of the largest savanna tracts left in Missouri. Native prairie and woodland species include royal catchfly and ladies' tresses.

Haden Bald Area
Ozark County
Located within Mark Twain National Forest near Rome, this 40-acre Ozark parcel is characterized by dolomite glades and upland forest. Interesting species include the smoke tree and the rare eastern collared lizard and Bachman's sparrow.

Helton Prairie
Harrison County
Located about 7 miles south of Mount Moriah, this high-quality remnant of mesic and wet-mesic prairie covers about 30 acres. Rare species include Mead's milkweed and the western prairie fringed orchid.

Hinkson Valley
Boone County
This 43-acre Nature Conservancy property, located just south of Columbia, protects a stretch of Hinkson Creek. Other natural communities are mesic forest, dry-mesic forest, and bottomland forest. Although logged in the past, there are some very old, tall elm, cottonwood, maple, and sycamore trees in the bottomland forest. The upland forest communities have developed on limestone rubble and are dominated by maple, chinquapin oak, pawpaw, and hickory. The rare wild leek can be found on the upper slopes as well.

Holly Ridge Natural Area
Stoddard County
This 84-acre tract 3 miles southeast of Bloomfield is part of the Mississippi Lowland. It contains an old-growth stand of American holly, beech, tulip poplar, and alder forest. Springs and small streams are home to the state-endangered goldstripe darter; numerous unusual plants thrive in the moist conditions. The surrounding area is covered by oak-hickory forest.

Horton Bottoms
Vernon County
This 300-acre complex of wet prairie, wet savanna, bottomland forest, and marsh near Horton is a complete floodplain ecosystem in the Osage Plain. Bald eagle and heron nesting sites are located along this portion of the Little Osage River. Plants include cordgrass, buttonbush, smartweed, and various sedges.

Hunkah and Tzi-Sho Prairies
Barton County
These two remnant prairie tracts, about 3 miles southwest of Liberal, total 332 acres. They represent high-quality examples of dry upland prairie and dry-mesic upland prairie. The prairie is a critical habitat for a number of disappearing animal species, including the upland sandpiper, prairie chicken, short-eared owl, Henslow's sparrow, prairie mole cricket, and northern harrier. Both properties are owned and managed by The Nature Conservancy.

Hyer Woods
Dent County
Located on 30 acres about 16 miles southeast of Rolla, this Nature Conservancy preserve encloses an old-growth upland forest. The property represents a former black oak savanna that has converted to an upland oak-hickory forest. The larger oak trees that were once part of the savanna are more than 200 years old. Other trees are blackgum, flowering dogwood, and elm. The rocky slopes host several rare or unusual ferns that are adapted to rocky conditions, such as bulbet fern, hairy lip fern, and purple cliffbrake. Horse gentian and Culver's root can also be found.

Isley Park Woods
Clay County
This 15-acre tract near Excelsior Springs hosts an old-growth upland forest. Part of the Glaciated Plain, the forest includes a variety of oaks, black walnut, basswood, and maple. The site is also used by migrating warblers and other songbirds.

Jacks Fork Natural Area
Shannon County
This 966-acre area along the Jacks Fork River about 6 miles northwest of Birch Tree encloses a variety of special habitats. Part of the Ozark, it includes dolomite glades, savannas, cliffs, caves, and forests. The 3-mile stretch of Jacks Fork River offers aquatic habitat for about 13 species of fish and crayfish endemic to the Ozark. Jam Up Cave is a bat habitat. Rare plants, relicts from the last glaciation, survive on the steep north-facing slopes.

Jamerson McCormack Loess Mounds
Holt County
Located about 5 miles south of Mound City, this 158-acre Nature Conservancy property protects an excellent remnant of rare loess hill prairie. The fine sediments that make up the loess mounds were deposited by winds at the end of the last glaciation. The hills are covered by typical prairie grasses. Rare or special species are the downy painted cup, rock satin grass, dalea, thimbelweed, skeleton plant, and plains pocket mouse. The area is managed by prescribed burning. The wildflowers form a colorful background during spring and summer.

Johnson's Shut-Ins Fens and Glades
Reynolds County
A moist seep forest dominated by ash and locust trees, a dolomite glade, and a calcareous fen cover these 30 acres 5 miles north of Lesterville. Unusual plants include closed gentian, silky willow, Missouri evening primrose, Englemann's adder's tongue fern, sandwort, and southern blue flag. The groundwater is calcareous from nearby limestone and dolomite rocks. The area is part of the Ozark.

Lichen Glade
Saint Clair County
This 30-acre tract about 6 miles west of Osceola preserves sandstone glade and sandstone cliff communities along the rocky bed of Salt Creek. Small or stunted post oak, blackjack oak, and hickory are scattered across the grass-covered glade. Little bluestem is the dominant grass. Rare or unusual species found here include the prickly pear cactus, widow's cross, lichen grasshopper, rattlebox, and fame flower. Small hardy plants exist in the sterile dry areas underlain by rock, such as nits-and-lice, a tiny orange flower. Hidden among the patchy lichens on the sandstone rocks arelicen grasshoppers

Lily Pond
Reynolds County
This 8-acre property about 5 miles southeast of Centerville protects a sinkhole pond–shrub swamp. The property is unique because it shel-

ters relict plants that were part of a large swampy region that existed millions of years ago. Sinkholes, the surface expression of underground caves that have collapsed, have preserved wetland conditions in an otherwise dry terrain. The standing water is thick with swamp rose, Walter's Saint John's wort, sphagnum moss, and buttonbush. Rare species on the preserve are swamp loosestrife, water violet, epiphytic sedge, ribbon-leaved pondweed, and Engelmann's quillwort. The widespread magenta color of the swamp loosestrife is best viewed in July.

Little Bean Marsh
Platte County
Located along the Missouri River about 10 miles southwest of De Kalb, this natural marsh, slough, and bottomland forest is part of the Big River terrain. It is an important breeding habitat for a number of species and an important migration site for a number of birds and waterfowl, including hawks and bald eagles.

Little Osage Prairie
Vernon County
This 80-acre Nature Conservancy tract about 4 miles south of Nevada encloses a high-quality remnant of dry-mesic prairie. The site is relatively undisturbed and has retained it original high biodiversity. The groundcover is dominantly bluestem and Indian grass; the rapidly disappearing Mead's milkweed and prairie mole cricket can also be found here.

Locust Creek
Linn County
Part of Pershing State Park, this 330-acre tract in the Glaciated Plain is one of the few remaining examples in Missouri of a meandering river with oxbow slough, swamp, and floodplain forest communities. Located about 4 miles southeast of Laclede, it offers complete aquatic communities that are rare elsewhere in the state. Trees along the creek include old-growth shellbark hickory and a variety of oaks. The rare ostrich fern can also be found.

Marmaton River Bottoms Wet Prairie
Vernon County
Located about 3 miles northwest of Nevada, this 568-acre site encloses wet prairie, wet savanna, oxbow slough, and wet bottomland forest communities. The 6-foot-tall prairie cordgrass that covers most of the prairie has never been plowed. Other native plants are the prairie milkweed, sedges, and rushes. Pecan, hickory, and oak groves dot the stretches of wet savanna. The wet bottomland forest along Marmaton River is populated by silver maple, elder, elm, and hackberry. The unusual carnivorous humped bladderwort and great bladderwort can be found in the oxbow wetlands.

Mount Vernon Prairie
Lawrence County
This 40-acre property is located about 4 miles northeast of Mount Vernon. Owned and managed by The Nature Conservancy, it encloses a high-quality remnant of dry prairie. Typical grasses are porcupine grass, little bluestem, and Indian grass. Several varieties of wildflowers, including rattlesnake master, color the prairie from roughly May to October. The prairie is an important habitat for a number of threatened or special-concern species, such as the leopard frog, greater prairie chicken, and upland sandpiper.

Mud Creek
Butler County
About 10 miles northeast of Poplar Bluff, this 1,038-tract is representative of wet-mesic, mesic, and wet bottomland forest communities in the Ozark. Upland forests are dominated by oaks and hickories; floodplain forests are wet and swampy. Rare plants in the area include Nuttall oak and purple fringeless orchid.

Mudlick Mountain
Wayne County
Part of the Saint Francois Mountains in the Ozark, this 1,400-acre tract about 6 miles north of Patterson contains old-growth forest. Dominant trees are white and black oaks on ground that is strewn with igneous boulders. Deep canyons in the rocky ground have created glade communities that contain the rare yellowwood plant. The forest is more unique when the impacts of wind and lightning on Mudlick Mountain, one of the highest points in the Ozark, are considered.

Niawathe Prairie
Dade County
This 160-acre tract about 9 miles north of Lockwood is owned by The Nature Conservancy. It encloses an excellent remnant of dry-mesic upland prairie. The soil is rocky and covered with tall prairie grasses, such as big and little bluestem. Unusual landforms called mima

mounds occur across the property. More than 200 plant species have been identified, including the rare Mead's milkweed. Threatened animals are the prairie chicken, northern harrier, short-eared owl, Henslow's sparrow, and prairie mole cricket.

Osage Prairie/Little Osage Prairie
Vernon County
This 700-acre parcel in the Osage Plain 5 miles south of Nevada contains remnant tallgrass prairie. Its relatively large size allows the ecosystem to support a high diversity of plant and animal species. Animals include the greater prairie chicken, northern harrier, short-eared owl, upland sandpiper, Henslow's sparrow, coyote, and white-tailed deer.

Otter Slough
Stoddard County
Some 140 acres of Otter Slough Conservation Area 8 miles southwest of Dexter contain excellent examples of bottomland forest and swamp characteristic of the Mississippi Lowland. Unusual vegetation includes tupelo, bald cypress, and overcup oak trees, as well as copper iris and spider lily. The slough is rimmed with swamp tupeolo and buttonbush; the rare pugnose minnow and bantam sunfish live in the slough.

Paint Brush Prairie
Pettis County
This 74-acre tract of dry-mesic prairie is about 3 miles northeast of Ionia. Part of the Osage Plain, it is managed by haying, resting, and prescribed burns. Rare species include the endangered Mead's milkweed and the prairie mole cricket.

Pawhuska Prairie
Barton County
This 77-acre tract about 9 miles east of Lamar is a high-quality remnant of dry-mesic upland prairie. Amid the prairie violets and bluestem grow more uncommon plants, such as the rare Harvey's beak-rush. It is an important habitat for prairie birds. The wildflowers are especially colorful during late spring and early summer. The property is owned and maintained by The Nature Conservancy.

Pelican Island
Saint Louis County
Located north of Saint Louis, Pelican Island is one of the few remaining big islands in the Missouri River; most have disappeared as a result of channelization. Pelican Island is an excellent example of bottomland forest with riverine chutes and sloughs. Part of the Big River terrain, it is bordered by sandbars and mudflats that are ever changing.

Pickeralweed Pond
Lincoln County
Part of Cuivre River State Park 4 miles northeast of Troy, this quarter-acre upland sinkhole pond within the Glaciated Plain is an important breeding habitat for amphibians rarely found north of the Missouri River. Rare plant species include naiad and American lotus.

Pickle Creek/Hawn State Park
Sainte Genevieve County
About 600 acres in and around Hawn State Park 15 miles northeast of Farmington host a diverse range of communities and habitats, all part of the Ozark Border. Communities include mature upland forests with shortleaf pine, sandstone and igneous outcrops, sandstone glades and savannas, and shaded headwater streams. Several rare species have been documented, including ground cedar, white smooth violet, shining nut-rush, hay-scented fern, Whitlow grass, and numerous orchids and ferns.

Powder Mill Cave
Shannon County
Part of the Ozark, this 100-acre tract 12 miles east of Eminence protects Powder Mill Cave. The entrance is closed to preserve the fragile cave ecosystem and a population of federally endangered Indiana bats.

Prairie Slough
Lincoln County
This 406-acre wet-mesic bottomland forest, wet bottomland forest, and shrub swamp is part of the Big River terrain. It is located 5 miles northeast of Elsberry. It is noted as a winter habitat for the federally threatened bald eagle.

Quercus Flatwoods
Texas County
A flatwoods forest community covers this 48-acre tract about 4 miles northwest of Licking. The property is located in the Ozark, and most of the post oaks here are more than 200 years old.

Ripgut Prairie
Bates County
High-quality remnants of Osage Plain wet-mesic prairie and wet prairie occur in this 136-acre area along the Marais des Cygnes River, about 4 miles northeast of Rich Hill. The site has been affected by the manmade Bates County Drainage Ditch. Plant species include big bluestem, cordgrass, cottonwoods, and the threatened Sullivant's milkweed.

Roaring River Cove
Barry County
Part of the Ozark, this 120-acre site about 5 miles northwest of Eagle Rock contains old-growth dry upland forest habitat. Trees include various species of oak, shagbark hickory, and the Ozark chinquapin. The area is also noted for a diverse bird population.

Rock Spring Bluff
Barry County
Located about 2 miles southwest of Viola, this 20-acre tract of Ozark is one of the few sites in Missouri that contains all three upland conifer species: Ashe juniper, red cedar, and shortleaf pine. Some of the juniper trees exceed 500 years old. This area is representative of a dry open upland forest community.

Rockhill Prairie
Benton County
Located 5 miles south of Lincoln, this 68-acre Nature Conservancy preserve encloses an excellent remnant of original dry chert prairie. Dominant plants are big and little bluestem, porcupine grass, and broomsedge. More than 250 different plant species have been identified on the prairie, including the federally endangered Mead's milkweed, fringed poppy mallow, Missouri evening primrose, downy blue gentian, yellow coneflower, and New Jersey tea. The wildflowers are especially colorful during May and June. The property is maintained through controlled burning.

Saint Francis River Natural Area
Madison County
This 1.2-mile stretch of the Saint Francis River about 5 miles west of Fredericktown flows through Ozark terrain. The riverbanks and bottom are carved out of igneous bedrock. The rare eastern slim minnow and various crayfish species inhabit this section of the river.

Saint Francois Mountains Natural Area
Reynolds and Iron Counties
This 7,028-acre natural area about 10 miles west of Ironton was designated in 1996. Part of the Ozark, it offers a number of unique terrestrial and aquatic communities, as well as a variety of endangered plant and animal species, including the collared lizard and ribbed sedge. Community types include glade, savanna, and upland forest. The 132-foot Mina Saul Falls, the highest wet-weather waterfall in Missouri, is also found within this natural area.

Sand Ponds
Ripley County
This 202-acre conservation area 4 miles south of Naylor is located in the Mississippi Lowland. Dominated by mesic upland sand forest and wet and wet-mesic bottomland forest, it is a habitat for several endangered species, including pondberry, corkwood, and the western mud snake. Small sand dunes are also located on the property.

Sandy Island Eagle Sanctuary
Lincoln County
Located 3 miles east of Winfield, this Nature Conservancy property along the Mississippi River protects a stretch of riparian habitat dominated by large cottonwood and maple trees. More than 200 wintering bald eagles gather here to hunt along the Mississippi River. The eagles are most common from January to March.

Schell-Osage Prairie Relics
Saint Clair County
Located in the Osage Plain, these five small tracts (ranging from 4 to 15 acres) are remnants of the original prairie that existed in this area about 4 miles southwest of Taberville. State management of the buffer lands between these properties is attempting to restore the prairie.

Schnabel Woods
Boone County
Located about 2 miles southeast of McBaine, this 80-acre tract in the Ozark Border terrain is known for its old-growth forest. Tree species include maple, basswood, oak, and American elm.

Silas Dees Preserve
Saint Francois County
This 5-acre tract 10 miles southeast of Farmington protects a remnant of xeric chert forest and dry-mesic forest. Owned and man-

aged by The Nature Conservancy, the property is well known for its brilliant spring displays of flowering dogwoods and azaleas in the forest understory. Forest trees are large oaks, pawpaws, and maples. Groundcover includes black cohosh, wild ginger, and various ferns. It is an important habitat for reptiles, the three-toed turtle, and the uncommon pileated woodpecker.

Spring's End Forest
Phelps County
An old-growth bottomland forest and backwater slough occurs on this 15-acre tract 3 miles southeast of Saint James. Located along the Meramec River, this section of Ozark terrain contains large, old sycamore, hickory, persimmon, butternut, and oak trees.

Star School Prairie
Atchison County
This 70-acre tract 14 miles north of Rock Port is part of the Glaciated Plain. It consists of dry-prairie plants on steep loess mounds. Rare vegetation species include blue grama, hairy grama, large beardtongue, and scarlet gaura; rare animal species include the Great Plains skink and pocket mouse.

Taberville Prairie
Saint Clair County
This 1,360-acre site is Missouri's largest prairie natural area and a National Natural Landmark. Located about 3 miles north of Taberville, it contains a variety of communities, including upland prairie and a prairie headwaters stream. It is within the Osage Plain. An abundance of native grasses and wildflowers occurs here; one of the largest greater prairie chicken populations in the state lives here.

Tarkio Prairie
Atchison County
Located about 8 miles northeast of Tarkio, this 57-acre tract of dry-mesic and mesic prairie is in the Glaciated Plain. Rare plant species include the prairie dandelion and the western prairie fringed orchid.

Trice-Dedman Woods
Clinton County
This 60-acre parcel about 3 miles east of Plattsburg is owned and managed by The Nature Conservancy. Its most impressive feature is an old-growth oak forest. Many of the white oak trees approach 200 years old. The groundcover includes relict Culver's root, yellow pimpernel, coreopsis, and Michigan lily, now rare in Missouri. Other rare or unusual plants include ferns, orchids, and sedges. The woodland is an important nesting habitat for a variety of birds, including warblers and thrushes. The wildflowers are quite colorful during spring.

Victoria Glade
Jefferson County
This 101-acre preserve about 3 miles southeast of Hillsboro protects one of the rare glade habitats in the state. Part of the Ozark Border, this large, open, rocky, dry glade community hosts unique, specialized vegetation. Rare or unusual plants include Fremont's leather flower, Indian paintbrush, green milkweed, and glade heliotrope. The rocky terrain is an excellent home for scorpions, black widows, tarantulas, snakes, and several rare insects. These dry-climate plant and animal species are relicts that survived and reproduced 8,000 years ago, when the climate was much more arid.

Vilander Bluff
Crawford County
Containing a vertical dolomite bluff nearly 100 feet high, this 206-acre site along the Meramec River is in the Ozark. Cave communities, rare lichens, and 300-year-old cedars occur on the property.

Wah'Kon-Tah, Mo-Ko, and Monegaw Prairies
Saint Clair and Cedar Counties
Owned by The Nature Conservancy, these three remnant prairie tracts near El Dorado Springs total 1,316 acres. Representative of the vast prairie that covered much of Missouri during presettlement times, they have maintained their original diversity, with more than 200 plant species having been recorded. Rare or threatened species in the tall prairie grasses are Mead's milkweed, upland sandpipers, grass pink orchid, prairie chickens, Henslow's sparrows, and the prairie mole cricket.

Wah-Sha-She Prairie
Jasper County
This 160-acre parcel about 2 miles north of Asbury is a remnant of the very rare hardpan

prairie. It is underlain by a clay-rich, impermeable soil layer that keeps water from draining downward. Dominant plants on the prairie are little bluestem and switchgrass. Several varieties of wildflowers, including Indian paintbrush and coreopsis, add color to the tall grasses. It is an important habitat for the endangered prairie mole cricket.

Washington State Park
Washington County
Just north of Fertile, this park in the Ozark Border consists of upland hardwood forest. The larger overstory trees are 150–200 years old. Species include northern red oak, chinquapin oak, white oak, and the Kentucky coffee tree. A wide variety of wildflowers cover the ground during spring.

Wells Branch Fen
Ripley County
Part of Mark Twain National Forest, this 2-acre open prairie fen in the Ozark is about 3 miles northwest of Doniphan. Endangered plant species in the fen include yellow-eyed grass and false loosestrife.

Westport Island
Lincoln County
Located in the Mississippi River, this island contains a large and undisturbed old-growth bottomland forest that is part of the Big River terrain. The large trees include maple, cottonwood, sycamore, pin oak, and pecan. The sloughs that wind through the forest are habitats for a variety of birds, herons, turtles, and fish.

Wildcat Glen
Newton County
A chert glade on the outskirts of Joplin, it is one of the best examples of this rare community type in Missouri. The glade is formed on cherty rock. Glade species include lichens, prickly pear, wild onion, rock pink, Barbara's buttons, and lichen grasshoppers that have adapted to the color of the lichen-covered rocks.

Williams Memorial Woods
Douglas County
Located on 40 acres about a mile south of Sweden, this Nature Conservancy preserve encloses high-quality dry, dry-mesic, and mesic sandstone forests, an acidic seep, springs, and a running stream. The forests contain mature oak and shortleaf pine trees, azaleas, fire pink, and fly poison. The sheltered springs and seeps host a variety of ferns, including some uncommon species such as Christmas fern.

Zahorsky Woods
Crawford County
This 56-acre Nature Conservancy property is located along the Meramec River about 10 miles southwest of Steelville. Natural communities here include glade, savanna, wet bottomland sycamore-maple forest, and dry oak-hickory upland forest. The small openings of glade and savanna host a variety of specialized plants, such as Missouri black-eyed Susan, lead plant, dittany, and sideoats grama. The woodlands is also an important habitat for wild turkeys. The colorful spring wildflowers are another attraction.

MONTANA

The western third of Montana is dominated by the Rocky Mountains, Bitterroot Mountains, and some 50 smaller ranges. Natural communities change according to elevation, soil chemistry, and precipitation. Going from the valley floor to the top of the ranges, the communities are: sage-grassland valley floor, juniper-pinyon pine woodland (in the foothills), ponderosa pine forest, Douglas fir forest, subalpine Engelmann spruce–whitebark pine–subalpine fir forest, and alpine tundra with lake and sedge meadow. Quaking aspen groves occur at various levels in draws and canyons. The alpine zone grows at elevations greater than 10,500 feet and is vegetated with low plants that tolerate harsh climatic conditions.

The eastern two-thirds of Montana is covered by gently rolling prairie grassland, occasionally eroded into deep canyons, bluffs, or mesas. The dominant grasses on this shortgrass prairie are buffalo grass, blue grama, and western wheat grass. Eastern Montana has occasional prairie pothole wetlands that are important habitats for waterbirds.

Lowland river forest forms the green riparian corridors along the rivers and streams that wind through the prairies and western mountains. Tall cottonwood, sycamore, willow, and green ash trees shade the water and provide protective habitats for bird and mammal species.

Special Habitat Sites

Babcock Mountain
Granite County
Located near Clinton, this 8,000-acre area is marked by steep grassy hillsides, rocky cliffs, and lesser coniferous forest. The property is home to a herd of bighorn sheep that numbers about 150 animals. Moose inhabit the forest. Golden eagles and bald eagles hunt along the cliffs, and trout fishing is available in Rock Creek.

Beartooth Wildlife Management Area
Lewis and Clark County
This 32,000-acre area about 10 miles east of Wolf Creek is dominated by tallgrass prairies and sagebrush hills that merge with rocky, forested uplands. It is a critical habitat for elk, deer, bighorn sheep, and peregrine falcons. It is well known for its interesting prairie dog towns.

Benton Lake National Wildlife Area
Cascade County
Rich, shallow prairie marsh, surrounded by agriculture, covers this 12,400-acre property about 15 miles north of Great Falls. One of the most productive waterfowl refuges in the United States, it is an important stopover for waterfowl during March-April and September-October migrations. More than 200 species have been identified, including ducks, tundra swans, snow geese, and Canada geese. Burrowing and short-eared owls, deer, and pronghorn antelope inhabit the site.

Big Prairie
Glacier County
This 2,600-acre tract of sagebrush and buck-grass prairie is surrounded by spruce-fir forests. Located about 3 miles north of Polebridge, it is an important winter range for elk and white-tailed deer. Efforts are also being made to increase the threatened Rocky Mountain gray wolf population that resides here.

Big Sheep Creek Area
Beaverhead County
These isolated 12,000 acres about 10 miles southwest of Lima are located in a spectacular mountain valley. Wildlife includes bighorn sheep, raptors, mule deer, antelope, and elk. It also has a large population of pygmy rabbits, which can only be found in the southeastern corner of Montana; this area is also known as an important sage grouse strutting ground.

Bighorn Canyon National Recreation Area
Bighorn and Carbon Counties
This reservoir on the Bighorn River about 30 miles southwest of Saint Xavier is edged by limestone cliffs. The clear water supports rainbow and brown trout and is a winter habitat for waterbirds. Eagles, a variety of ducks, redheads, grebes, teals, and cormorants are commonly seen. Roosting sites for bald eagles and peregrine falcons are located in the cliffs.

Black Coulee National Wildlife Refuge
Blaine County
Located on about 1,500 acres 10 miles south of Turner, this substantial prairie marsh is surrounded by grass- and sagebrush-covered draws and coulees that drain into the marsh. It is an important habitat for a variety of plant and animal species, including waterfowl, mule deer, golden eagles, pronghorn, tundra swans, sharp-tailed grouse, and gray partridges.

Black Pond
Powder River County
Part of Custer National Forest, this 45,000-acre tract encloses a timbered plateau that rises up from grass-covered prairie. Located about 20 miles southeast of Ashland, the ponderosa pine forest is home to deer, turkeys, and songbirds. Grouse, coyotes, and antelope reside on the prairie. Six bat species, including the rare pallid bat, live in rocky openings in the plateau.

Blackleaf Wildlife Management Area
Teton County
This 19,500-acre area 17 miles west of Bynum along the Rocky Mountain Front encloses a variety of natural communities, including fir-limber pine forests, rocky cliffs, rolling grasslands, and extensive marshes. Wildflowers are spectacular during spring and summer. It is an important habitat for black bears, grizzly bears, elk, mountain goats, and a number of birds and raptors, including the sharp-tailed grouse.

Bob Marshall Wilderness
Flathead and Lewis and Clark Counties
Located in Flathead and Lewis and Clark National Forests, this million-acre wilderness is

about 50 miles southwest of Choteau, Montana. Located on the Continental Divide, its elevations range from 4,000 feet to 9,000 feet. Natural communities range from sweeping spruce-alpine forest to low-lying alpine vegetation and meadows on the high, rocky ridges. This expansive wilderness is a critical habitat for large mammals, such as grizzly and black bears, mountain goats, mountain sheep, moose, elk, and deer. Other important species are various raptors and interior forest birds.

Bowdoin National Wildlife Refuge
Phillips County
This 15,500-acre tract about 7 miles east of Malta contains short-grass prairie, wetland, scattered woodland, shrub field habitats. It is an important migration stop along the Central Flyway for many waterfowl during spring and fall, especially for white pelicans. The wetlands are well known for their large nesting populations of Franklin's gulls, night herons, ibis, white pelicans, and cormorants. The prairie uplands shelter grouse, pheasant, partridge, coyotes, rails, bitterns, grebes, and pronghorn.

Browns Lake
Powell County
This 1,540-acre site about 5 miles southeast of Ovando contains a 500-acre shallow lake surrounded by open sagebrush grasslands and scattered aspen groves. The area is critical habitat for bald eagles, ospreys, grebes, and herons. Sandhill cranes visit during spring. Elk and moose can also be seen.

Canyon Ferry Wildlife Management Area
Broadwater County
This 5,000-acre preserve near Townsend encloses exceptional wetlands and forested river bottom along the Canyon Ferry Reservoir. It is an important nesting habitat for a variety of bird life, including bald eagles that feed on spawning kokanee salmon.

Charles M. Russell National Wildlife Refuge
Fergus and Phillips Counties
At more than a million acres, this is one of the largest wildlife refuges in the continental United States. Located about 65 miles northeast of Lewistown, it is probably the most pristine remaining remnant of the northern Great Plains. It is therefore highly diverse, with rich populations of indigenous plant species and wildlife, including mule deer, white-tailed deer, grouse, prairie dogs, antelope, hawks, and eagles. The mating rituals of grouse are best observed from March through May. It also contains the largest remaining prairie elk herd in North America.

Clearwater Chain-of-Lakes
Missoula County
Located near the town of Seeley Lake, this area of lakes and wetlands hosts a variety of waterfowl, including red-necked grebes, ospreys, and the largest population of nesting common loons in the western United States. The loons nest on small islands or in marshy sections of the lakes.

Cliff and Wade Lakes
Madison County
Sheltered by high cliffs, these lakes near Ennis are spring-fed and cover about 900 acres. The wetlands and rock faces are an important habitat for nesting raptors, such as prairie falcons, bald eagles, and ospreys. Moose, deer, elk, and bears are also seen here.

Comertown Pothole Prairie Preserve
Sheridan County
Located northeast of Plentywood, this 1,130-acre Nature Conservancy preserve encloses an excellent prairie pothole complex. The wetlands are a critical nesting habitat and migratory stopover for thousands of birds, including the very rare whooping crane and piping plover. The prairie grasses protect many grassland birds, including the very rare Baird's sparrow. It may be the largest remnant of undisturbed pothole prairie remaining in Montana.

Crown Butte Preserve
Cascade County
This 376-acre Nature Conservancy site about 30 miles southwest of Great Falls encloses a steep butte that rises 900 feet above the foothill prairies along the Rocky Mountains. The difficult access up the steep rocky walls has kept the tallgrass prairie on top from being disturbed. Different plant communities in the prairie are bluebunch wheatgrass, blue grama, rough fescue, and Idaho fescue. Good displays of wildflowers can be seen during spring and summer. The butte is an important habitat for mule deer, rock wrens, and various raptors.

Dancing Prairie Preserve
Lincoln County
Located about 4 miles northwest of Eureka, this 680-acre Nature Conservancy property protects an important piece of undisturbed prairie. The preserve is a critical habitat for the rapidly diminishing Columbian sharp-tailed grouse, comprising one of the largest mating grounds in the intermountain West; the elaborate mating rituals can be witnessed during spring. Groves of trees on the prairie help protect the young from predators. The globally rare Spalding's catchfly also grows here.

Eastlick Pond/Big Lake Waterfowl Management Area
Stillwater County
This 700-acre area about 6 miles west of Molt contains Eastlick Pond, Big Lake, and related wetlands and prairie. The Eastlick prairie pothole lake is well known for its population of nesting great blue herons. Large numbers of Canada geese visit during spring and fall. Nesting burrowing owls are also present. Big Lake, a large alkaline lake, often goes dry during summer months. In spring, or during wet seasons, it is filled with white pelicans, tundra swans, cormorants, and gulls. Rare scoters and peregrine falcons can also be spotted.

Freezeout Lake
Teton County
This 11,400-acre wetland complex about 10 miles south of Choteau often has the largest waterfowl concentration in Montana. Up to a million birds traveling the Pacific Flyway visit during migration peaks (March-May, September-November). Rare birds include snow geese, tundra swans, sandhill cranes, herons, grouse, bald and golden eagles, gyrfalcons, and snowy owls.

Great Bear Wilderness
Flathead County
Located on 285,000 acres in northwestern Montana about 70 miles east of Flathead Lake, this wilderness area is critical grizzly bear and black bear habitat. With elevations ranging from 4,000 feet to 8,700 feet, natural communities include dense spruce-fir forest (at the lower elevation) and low-lying alpine vegetation on the windswept ridges above timberline. Other important species are mountain goats, mountain sheep, mountain lions, elk, moose, wolverine, and a variety of raptors. Several rare or unusual alpine-subalpine plants can be found at the upper elevations.

Hailstone National Wildlife Refuge
Stillwater County
Located 30 miles north of Columbus, this 1,913-acre refuge 5 miles northeast of Rapelje consists of a large alkaline lake surrounded by expansive short-grass prairie, rocky outcrops, and small grassy hills. Yucca and cacti grow in the flatlands. The lake is an important migratory stopover for thousands of birds during spring and fall. Antelope and grouse can be found in the upland areas, which are vegetated with sagebrush, greasewood, and native grasses. Other prominent species are black-tailed prairie dogs, golden eagles, and peregrine falcons.

Judith River Management Area
Judith Basin County
This 5,000-acre tract about 12 miles south of Utica protects a variety of prairie and rugged ponderosa pine forest natural communities. It is a critical wintering ground for elk and deer. The area is also well known as a nesting area for more than 100 bird species, including raptors, owls, and songbirds.

Kelly Island
Missoula County
Located about 10 miles southeast of Missoula, this 631-acre undeveloped island in the Clark Fork River is covered by a unique mix of mature cottonwood bottoms, large meadows, and ponderosa pine forests. It hosts a remarkable diversity of wildlife and is a historic nesting area for great blue herons, Canada geese, and red-tailed hawks. Bald eagles are also present.

Kootenai River
Lincoln County
About 15 miles from Libby, this 17-mile stretch of Kootenai River is noted for hosting populations of bald eagles and ospreys that nest and feed in the area. Feasting on kokanee salmon, up to 160 bald eagles can be sighted in a single day. Eagle migration starts around the first week of October, peaks during the first week of November, and tapers off around mid-December.

Lake Mason National Wildlife Refuge
Musselshell County
This extensive prairie marsh covers nearly 17,000 acres about 7 miles northwest of Roundup. It hosts nesting populations of

numerous waterfowl and shorebirds. Migrating geese, gulls, and white pelicans crowd the area during spring and fall; migrating bald eagles and peregrine falcons also stop. Antelope live on the drier uplands, which also have a prominent black-tailed prairie dog town. Burrowing owls can also be found.

Lee Metcalf National Wildlife Refuge
Granite County
This 2,800-acre tract just east of Stevensville protects a diverse combination of wetlands and forested riverbottom habitat along the Bitterroot River. Part of Bitterroot National Forest, it is critical nesting habitat for ospreys, Canada geese, and American bald eagles. More than 200 species of birds have been documented here.

Lewis and Clark Caverns
Jefferson County
Located on 2,800 acres about 15 miles west of Three Forks, these limestone caves are the residence of the only known maternity colony of western big-eared bats in Montana. The limestone rocks and crevasses near the caves host prairie rattlesnakes, bullsnakes, and racers. Raptors, such as golden eagles and red-tailed hawks, are often seen floating in the sky.

Lindbergh Lake Pines
Missoula County
Located in the Swan Valley, this 40-acre parcel protects a rare old-growth stand of towering ponderosa pine, Douglas fir, and western larch. The trees are at least 200 years old. Part of Flathead National Forest, it is also an important grizzly bear habitat.

Logan Pass Boardwalk
Glacier County
Located in Glacier National Park about 20 miles northeast of Lake McDonald, this site protects a very rare and fragile alpine habitat known for its spectacular bloom of unusual wildflowers. It is one of the few places in the United States where the ptarmigan grouse can be seen. Other special animals include hoary marmots, mountain goats, and grizzly bears. Access is restricted to a boardwalk that protects the delicate habitat.

Long Pines
Carter County
Part of Custer National Forest, this 19,200-acre segment near Capitol encloses ponderosa pine forest, aspen-lined ravine, sagebrush prairie, and rocky hill natural communities. An excellent habitat for raptors, Long Pines has the nation's highest reported nesting density of merlines. Hawks, kestrels, falcons, and bald eagles are usually seen gliding on the thermals.

Manning Corral/Prairie Dog Town
Fergus and Phillips Counties
Located within the Charles Russell Wildlife Refuge, this 1,000-acre tract about 50 miles northwest of Mosby contains a huge black-tailed prairie dog town in the native prairie grasslands. Pronghorn, mule deer, burrowing owls, and a variety of raptors also live here. The prairie dog town may be selected as a site to reintroduce the endangered black-footed ferret.

Medicine Lake National Wildlife Refuge
Sheridan County
This remote, 31,500-acre tract about 10 miles southeast of the town of Medicine Lake contains plains pothole wetlands, shrublands, prairie, and a large lake. It is an important migration route, and up to 100,000 ducks and 2,000 white pelicans nest here. Other nesting birds are cormorants, blue herons, and gulls. Sandhill cranes and rare whooping cranes can be seen during spring and fall. The prairies support large herds of white-tailed deer and pronghorn. Upland species include sharp-tailed grouse, lark buntings, and burrowing owls. The refuge hosts one of the largest white pelican rookeries in the country, as well as a nesting population of the endangered piping plover.

Medicine Rocks State Park
Fallon County
Located on 316 acres about 25 miles south of Baker, this park protects unique sandstone rock formation, prairie grassland, and ponderosa pine forest natural communities. Teddy Roosevelt said it was as "fantastically beautiful a place as I've ever seen." The sandstone cliffs provide excellent nesting habitat for a variety of raptors, including hawks, kestrels, harriers, eagles, merlines, prairie falcons, and cliff swallows. Grouse, mule deer, antelope, and prairie dogs frequent the prairie environment. Juniper-covered hills and cottonwood bottoms can be seen along the Powder River.

Mission Mountains Wilderness
Missoula County
Located about 25 miles southeast of Ronan in Flathead National Forest in northwestern

Montana, this 74-acre property contains 4,500–9,000 foot elevations. Natural communities include alpine habitat, lake, meadow, snow-fed stream and riparian corridor, rock face, talus, spruce-subalpine fir forest, alpine larch forest, and Engelmann spruce–Douglas fir forest. This wild, rugged terrain is an important habitat for large mammals, including mountain goats, grizzly bears, black bears, elk, and mule deer. Bald eagles and golden eagles frequent the cliff tops, and pileated woodpeckers and other old-growth–loving birds inhabit the forest.

Missouri Headwaters State Park
Gallatin County
This 527-acre property about 3 miles north of Three Forks is located where the Madison, Jefferson, and Gallatin Rivers join to form the Missouri River. The park protects extensive mature cottonwood riverbottoms and rocky, forested uplands. The limestone cliffs provide an important nesting habitat for raptors, such as ospreys, nighthawks, golden eagles, and prairie falcons. Great blue heron and cormorant rookeries have also been established on the site.

Mount Haggin Wildlife Management Area
Deer Lodge County
This 54,000-acre preserve is Montana's largest wildlife management area. Located about 20 miles southwest of Anaconda, it is dominated by aspen-covered hills, willow bottoms, and forested mountains. It is well known for its exceptional diversity of plants and animals, including moose, elk, bald eagles, and nesting sandhill cranes.

National Bison Range
Sanders and Lake Counties
Part of the National Wildlife Refuge system, this 18,500-acre preserve near Moiese encloses steep grassy hills and mature coniferous forest. About 500 bison inhabit the range, as do pronghorn, elk, bighorn sheep, and mountain goats. Interesting bird species, including bald eagles and golden eagles, can also be seen. A 19-mile loop drive enables visitors to view wildlife.

Ninepipe National Wildlife Refuge
Lake County
This 5,000-acre preserve about 4 miles south of Ronan is an exceptional wetland complex. It contains more than 800 glacial potholes and a 1,770-acre reservoir. More than 180 bird species inhabit the area. It is also an important stopover point for migrating birds during spring and fall. Cormorants, herons, Canada geese, swallows, and yellow-headed blackbirds are frequent visitors.

Northern Yellowstone Winter Range
Park County
This sprawling winter range about 5 miles northeast of Jardine hosts the largest and most diverse group of hoofed animals in the Lower 48 states. Natural communities include rugged, forested mountain, aspen-dotted foothill, and grassy prairie. From December to May this area is wintering ground for the nation's largest elk, mule deer, pronghorn, bison, and bighorn sheep populations. Bald eagles also frequent the area, which can be accessed via a 12-mile scenic drive.

Our Lake
Teton County
One of the few alpine lakes along the Rocky Mountain Front, this 700-acre area about 35 miles northwest of Choteau is an important raptor habitat. Other important species that live in the area are mountain goats, Yellowstone cutthroat trout, grizzlies, mule deer, coyotes, and bald eagles.

Pine Butte Swamp
Teton County
This Nature Conservancy preserve located about 27 miles west of Choteau encloses high-quality examples of marsh and fen wetlands, prairie grassland, bottomland and upland pine forest, and riparian habitat along Teton River and Willow Creek. Covering 18,000 acres, the property is an important habitat for grizzly bears, elk, and bald eagles. The wetlands host rare plants, including yellow lady's slipper and Macoun's gentian. Outcrops contain dinosaur fossils.

Pine Butte Swamp Preserve
Teton County
This 18,000-acre preserve about 15 miles west of Choteau is owned by The Nature Conservancy. It is the largest wetland complex along the Rocky Mountain Front; it is also the last stronghold on the Great Plains for the grizzly bear. Dominant natural communities are wetland, forested mountain, foothill, and

prairie grassland. It is an important habitat for elk and bear, and more than 150 species of birds have been sighted. Large sharp-tailed grouse strutting grounds are active during spring.

Red Rock Lakes Natuiobak Wildlife Refuge
Beaverhead County
Located around the town of Lakeview, this large refuge is 42,500 acres in size, about a third of which is wetland—shallow lakes, ponds, and marshes. The outlying areas are grassland prairie. Some 300–500 trumpeter swans live and breed here; this is the largest breeding population of this species in the Lower 48 states. Moose, raptors (including bald eagles and peregrine falcons), pronghorn, antelope, deer, and elk are all plentiful here.

Safe Harbor Marsh
Lake County
This 132-acre Nature Conservancy preserve encloses an important wetland habitat about 5 miles northeast of Polson. Natural communities include cliff face, pond, marsh, bog, fen, lowland conifer forest, prairie, and upland forest. Hundreds of bird species have been identified on the property. The wetlands are a critical nesting habitat and migratory stopover for thousands of birds, including waterfowl, interior forest birds, such as woodpeckers and songbirds, bald eagles, and other raptors.

Skalkaho Game Preserve
Ravalli County
This 23,000-acre tract about 15 miles east of Hamilton consists of heavily forested mountains with large, open meadows. Closed to hunting, it is critical habitat for large mammals, including elk, deer, bears, mountain goats, and moose.

Smith Lake Area
Flathead County
About 7 miles southwest of Kalispell, this 1,040-acre complex of wetland, marsh, and forest displays a high diversity of plants and animals. Several varieties of shorebirds nest in the area, including phalaropes, yellowlegs, killdeer, ducks, widgeons, pintails, gadwalls, Canada geese, grebes, gulls, tundra swans (late March and early April), and American bitterns. Sandhill cranes visit during spring and fall. It is also critical nesting habitat for bluebirds.

South Fork Madison Preserve
Gallatin County
This 81-acre Nature Conservancy tract about 5 miles west of West Yellowstone protects diverse riparian habitat along the South Fork of the Madison River. Rare species that inhabit this area include bald eagles and sandhill cranes. The river is an important spawning ground for rainbow and brown trout and may be a reintroduction site for the arctic grayling. The property's most important role is as a buffer for Yellowstone National Park, sheltering wildlife that migrates outside of the park, especially during winter.

Square Butte
Choteau County
This 3,000-acre tract protects a prominent volcanic feature about 15 miles southwest of Square Butte. A topographic high point rising 1,700 feet above surrounding plains, its steep slopes provide habitat for deer, antelope, and elk near the butte; a herd of mountain goats lives on top of the butte. The steep cliffs provide outstanding nesting habitat for raptors, including prairie falcons, golden eagles, and a variety of hawks. Permission to cross private land must first be obtained.

Sun River Canyon
Lewis and Clark County
This 4,000-acre property near Augusta is where the rolling prairies of the northern Great Plains abruptly meet the rugged peaks of the Rocky Mountains. Steep cliffs dominate the rugged terrain and are home to one of the largest herds of bighorn sheep in North America (up to 1,000 animals). It is also important habitat for elk, deer, and a variety of raptors. Just to the southeast is Montana's 20,000-acre Sun River Wildlife Management Area, winter home to one of the state's largest elk populations.

Swan River National Wildlife Refuge
Lake County
Located just south of the town of Swan Lake, this 1,600-acre undeveloped area contains extensive wetlands and coniferous forest. It is important habitat for elk, deer, moose, grizzlies, and black bears. More than 171 bird species have been identified, including bald eagles, blue herons, and black terns.

Swan River Oxbow Preserve
Lake County
Located about 3 miles south of Swan Lake, this 392-acre Nature Conservancy preserve encloses a complex mix of wetland and forest natural communities along Swan River. Marsh, birch carr (fen), sedge fen, and cottonwood lowland forest communities occupy the area around a sinuous, water-filled oxbow lake. Several rare or uncommon plants have been found here, including lichens, round-leaved pondweed, northern bastard toadflax, buckler fern, and water howellia. The preserve is a corridor between mountain ranges for grizzly bears, elk, moose, and deer. Bald eagles and interior forest birds, such as the western tanager, nest on the property.

Vinal Creek
Lincoln County
Located in Kootenai National Forest about 10 miles northeast of Yaak Valley, Vinal Creek is well known for its impressive old-growth timber. Examples of the western larch (tamarack) grow up to 4 feet in diameter and are more than 300 years old. Huge red cedars are more than 25 feet in circumference. This 6-mile stretch of Vinal Creek provides excellent habitat for pileated woodpeckers and barred owls.

NEBRASKA

The major vegetation types in Nebraska are tallgrass prairie, mixed-grass prairie, sandhill prairie, deciduous forest, and pine woodland.

Now mostly agricultural land, the eastern third of the state contains some tallgrass prairie. Pieces of remnant prairie consist of tall grasses, such as big bluestem, Indian grass, switchgrass, and Canada wildrye, with hundreds of different wildflowers. Less than 5 percent of this prairie type remains.

Much of the north-central and western parts of Nebraska are dominated by sandhill prairie. The sandy soil supports sand bluestem, prairie sandreed, dropseed, little bluestem, needle-and-thread, and hairy grama. This portion of the state is largely devoted to raising cattle.

Mixed-grass prairie is prominent in the south-central and northwestern parts of the state. A variety of short-, mid-, and tallgrass species grows on dry flatlands and rolling hills; major species include needle-and-thread, little bluestem, western wheatgrass, sideoats grama, and buffalo grass. About a third of this prairie region has avoided development.

Bottomland deciduous forest lines the drainage of the Missouri River and the lower parts of some of its tributaries, such as the Platte and Republican Rivers. Forest trees include oak–hickory–ash–American elm–basswood–black walnut.

Scattered areas of pine woodlands cover parts of extreme western Nebraska. Ponderosa pine, along with lesser limber pine and red cedar, cover the rocky hills and plateaus that begin to appear in the North Platte and Niobrara River drainage basins.

Special Habitat Sites

Arapahoe Prairie
Arthur County
This Nature Conservancy property covers 1,298 acres of high-quality upland prairie. More than 100 plant species grow here; colorful wildflower displays occur during spring and summer. The prairie is an important habitat for small grassland birds, mammals, and insects.

Big Bend Reach (Platte River)
Phelps, Buffalo, and Hall Counties
The Nature Conservancy owns about 8,500 acres of riparian habitat along Platte River between Overton and Chapman. During March migration, this site hosts the world's greatest concentration of sandhill cranes. It is also a critical roosting and feeding habitat for the federally endangered whooping crane and interior least tern, as well as the federally threatened piping plover.

Fliesbach Wildlife Area (Facus Springs)
Morrill County
Important alkaline wetlands are protected on this 422-acre Nature Conservancy preserve near Chimney Rock, about 5 miles south of Bayard. In addition to attracting a variety of nesting birds, the wetlands, a critical resting and feeding site, attract thousands of waterbirds during spring and fall migrations.

Graves Ranch
Garden County
This 840-acre Nature Conservancy property contains the world's largest population of fed-

erally endangered blowout penstemon, a rare prairie species.

Jumbo and Pullman Valley Fens
Cherry County
This 2,727-acre preserve is located in the Nebraska sandhills near the headwaters of Loup River, southeast of Steverson Lake. In addition to high-quality upland prairie, the site contains rare fen wetlands that are fed by mineral-rich springs and groundwater. The fens support a host of rare or unusual plants that thrive on the calcium- and magnesium-rich nutrients in the soil.

Kiowa Wildlife Management Area
Scotts Bluff County
This 531-acre wildlife management area about 10 miles south of Morrill encloses upland grasslands and fresh and alkaline wetlands. It is an important nesting site for a number of bird species; tens of thousands of birds visit during spring and fall migrations.

Lillian Annette Rowe Sanctuary
Buffalo County
Located along the Platte River just southeast of Kearny, this 2,200-acre sanctuary encloses riverine wetlands, tallgrass prairie, channels, and wet meadows. These important wetlands are breeding and feeding grounds for nesting and migratory birds, including the rare sandhill crane, piping plover, terns, and the federally endangered whooping crane. More than 500,000 sandhill cranes gather here during spring and fall.

Little Salt Fork Marsh
Lancaster County
Located a few miles north of Lincoln, this 179-acre Nature Conservancy preserve protects a continental saline wetland and marsh, a rare type of wetland. The wetland supports a group of plant, insect, and animal species, including the very rare Salt Creek tiger beetle, that are tolerant of the high salt content in the soil and water. The open water attracts thousands of waterbirds during spring and fall migrations.

Muskrat Run
Lincoln County
Owned by The Nature Conservancy, this 302-acre riparian and wetland habitat is located along the North Platte River. It is a very important sandhill crane habitat during spring.

Niobrara River Sanctuary
Rock County
Owned and managed by the Audubon Society, this 218-acre property about 20 miles north of Bassett is located along the Niobrara River. Natural communities include the river and riparian corridor, vertical rock cliff, native prairie, and upland mixed pine-hardwood forest. The cooler canyons and gullies create microclimates that support some unusual plant species. Numerous birds reside here, including sandhill cranes, prairie chickens, the rare peregrine falcon and long-billed curlew, and various sandpipers and terns. The prairie retains much of its presettlement vegetation and is colorful during summer.

Niobrara Valley Preserve
Cherry, Brown, and Keya Paha Counties
Located about 20 miles northwest of Ainsworth, this 55,000-acre Nature Conservancy property fronts a 25-mile stretch of the Niobrara River. Communities include springbranch canyon, ponderosa pine forest, eastern deciduous forest, birch woodland, riparian wetland, and upland prairie. The use of fire, bison, and cattle to maintain the quality of this prairie is a valuable model for grassland management nationwide.

Pawnee Prairie
Pawnee County
The Nature Conservancy owns this 320-acre piece of remnant tallgrass prairie, with unplowed bluestem and Indian grass. Biodiversity is high and includes several rare or unusual birds, mammals, and insects, including rare prairie butterflies.

Rainwater Basin
Hamilton, Fillmore, Franklin, and York Counties
Once one of the largest wetlands on the Great Plains, the Rainwater Basin wetlands covered more than 100,000 acres 200 years ago. Today, because of agriculture and development, less than 10,000 acres remain. Led by The Nature Conservancy, a cooperative project with the U.S. Fish and Wildlife Service and Nebraska Game and Parks Commission is restoring about 2,500 acres of wetland about 20 miles southeast of Grand Island; the goal is to restore

25,000 acres, one-quarter of the original wetland complex.

Rulo Bluffs
Richardson County
Located on 444 rugged acres overlooking the Missouri River, this preserve contains part of the ecotone between the hardwood forest ecosystem of the eastern United States and the open prairie (loess hill) environment of the Great Plains. There is a high biodiversity of animal and plant life, including some rare, unusual, or relict species in canyons and on rocky bluffs.

Signal Hill Research Natural Area
Thomas County
Located southwest of Halsey in Nebraska National Forest, this 707-acre tract protects a rare section of Nebraska sandhill prairie along Middle Loup River. The grassland is dominated by sand lovegrass, sandhill bluestem, prairie sandreed, little bluestem, blowout grass, Indian grass, and sandhill muhly. This pristine prairie includes native shrubs of wild plum, cloke cherry, and leadplant that have not been destroyed by grazing. Unique or uncommon species include the pronghorn, rattlesnakes, Ord's kangaroo rat, the plains pocket gopher, the greater prairie chicken, and several butterflies.

NEVADA

Nevada is located in the heart of the Basin and Range, a parallel series of rugged mountains and flat, arid valleys. The various natural regions are defined primarily by elevation, soil chemistry, and moisture content. Named for the dominant vegetation that thrives in the different elevation-temperature-moisture conditions, the dominant vegetation types are shadscale scrub, sagebrush flat, pinyon-juniper woodland, mid- to upper-elevation mixed pine forest, alpine, and narrow riparian corridors.

The dry, windswept valley floors in Nevada are dominated by shadscale shrub and sagebrush. The shadscale shrub zone is typically underlain by alkaline soils; it may also include greasewood, hopsage, and rabbitbrush. Sagebrush is abundant in slightly wetter areas; it can grow 8 feet tall. Other plants in the sagebrush zone include rabbitbrush and bitterbrush.

The lower slopes (2,000–5,000 feet elevation) of the mountain ranges that rise from the edges of the valley floors are often covered by patches of pinyon-juniper woodland. This abundant forest type is typically low and scrubby, rarely reaching heights of 20 feet or more. These areas receive a foot or more of annual precipitation.

Higher-elevation forests are often restricted to canyons and plateaus. Common trees include quaking aspen and conifers. Mixed forests below 8,000 feet elevation are usually composed of incense cedar, white fir, ponderosa pine, and Jeffrey pine; red fir dominates between 8,000 feet and 9,000 feet; higher elevations support scattered stands of whitebark pine and mountain hemlock. Bristlecone pine, whitebark pine, and limber pine can be found between 9,000 and 11,000 feet.

Above timberline, on Nevada's highest, most windswept mountains, patches of alpine zones can be found. The tiny, hardy, drought-resistant plants that grow here form mats on the rocky substrate. The plants are stunted because of the harsh environmental conditions: variable water, intense heat and wind, and dramatic temperature changes.

Special Habitat Sites

Arc Dome Wilderness Area
Lander County
Located in the Toiyabe Mountain Range, this 115,000-acre wilderness area southeast of Austin is the largest in the state. The landscape consists of rocky canyons, quaking aspen–lined riparian corridors, high-elevation grassy meadows, and mixed pine forests. It contains Nevada's largest stand of limber pine, a critical habitat for blue grouse. More than 40 different species of nesting birds inhabit the wilderness, including the red-naped sapsucker and prairie falcon. The rocky higher elevations are critical habitats for the rare Belding's ground squirrel, desert bighorn sheep, mountain lions, bobcats, and the occasional elk. The endemic Lahontan cutthroat trout can be found in several streams that traverse the area.

Ash Meadows National Wildlife Refuge
Nye County
This 23,447-acre oasis-type wetland 10 miles southeast of Amargosa Valley supports 24 very

rare endemic plants and animals, one of the highest concentrations of endemic species in the United States. Four endemic fish thrive in spring-fed wetlands, including three types of desert pupfish and speckled dace. Several varieties of birds are attracted to the open pool and riparian vegetation, including herons, grebes, egrets, sandpipers, Bewick's wrens, and blue grosbeaks.

Ash Springs
Lincoln County
About 100 miles north of Las Vegas, this BLM property encloses a rare warm spring-fed pool. The wetland rim of the pool is edged by saltgrass and tall ash and cottonwood trees. The greenery and water amid the desert attract a variety of wildlife and bird species. Within Ash Springs are two very rare fish species—the endemic White River springfish and the Paranagat roundtail chub, endemic only to Ash Springs.

Brown Knoll
White Pine County
Located about a mile south of Lund, this property is some of the most critical winter range for mule deer and elk. The high-quality natural communities that attract these large mammals are black sage scrubland, pinyon-juniper woodland, and mountain brushland (bitterbrush-cliffrose-serviceberry). Sometimes 3,000 animals can be seen here during winter.

Cathedral Gorge State Park
Lincoln County
Located about 4 miles northwest of Panaca, this 1,600-acre property is a narrow valley edged by high cliffs. Natural features include sand dunes with dune primrose and Indian ricegrass, yucca–juniper–greasewood–salt bush on valley floor, shadscale scrub, and juniper. The gorge is an important raptor and rock wren habitat and is a stopover for migrating songbirds and the rufous hummingbird.

East Humboldt Wilderness Area
Elko County
Located south of Wells, this 36,000-acre wilderness area is part of the East Humboldt Mountains. Natural communities include sagebrush flat, lake, stream, riparian corridor, alpine meadow with brilliant wildflowers, stands of aspen, and conifer forest. It is an important habitat for large mammals, such as mountain goats, bighorn sheep, mountain lions, and mule deer. The endemic and federally threatened Lahontan cutthroat trout can be found in the streams that drain the wilderness.

Grant Range Wilderness Area
Nye County
Located in Humboldt National Forest, this 50,000-acre wilderness area is located about 120 miles southwest of Ely. It includes some of the highest mountains in Nevada, including the 11,289-foot Troy Peak. The sharp, rocky ridgelines have sparse vegetative cover, with occasional stands of limber and bristlecone pine, some of which are nearly 1,000 years old. The windswept terrain is home to an indigenous herd of desert bighorn sheep, mountain lions, reptiles, and raptors. There are no trails that traverse this wilderness, making it truly undisturbed.

Great Basin National Park
White Pine County
Located in the Snake Range 10 miles southwest of Baker, this park encloses typical Basin and Range topography. Natural communities include sagebrush flat, desert grassland, pinyon-juniper forest with cliffrose, aspen groves along Lehman Creek, mountain mahogany–manzanita–prickly pear cactus shrubland, Engelmann spruce–Douglas fir forest, limber pine–spruce forest, ancient bristlecone pine forest, and alpine meadow. The bristlecone pines are 3,000 years old. Flowers in the meadows are buttercups, Fremont geraniums, and Indian paintbrush. The rugged country is prime habitat for raptors, turkey vultures, chukars, woodpeckers, pine siskins, mountain lions, and bighorn sheep.

Lake Tahoe–Nevada State Park
Douglas County
Located on 13,468 acres, this park encloses four pristine lakes, ponderosa pine–Jeffrey pine–limber pine forest, white fir–incense cedar–aspen–red fir forest, and high alpine meadows with abundant wildflowers, including Indian paintbrush. Interesting birds are raptors, Williamson's sapsuckers, woodpeckers, tanagers, and various warblers. The mountains are critical habitats for black bears, mountain lions, golden-mantled ground squirrels, and the rare and endemic Sierra Nevada red fox.

Mahogany Creek Natural Area
Humboldt County
Located about 30 miles southwest of Denio Junction, this 12,316-acre BLM property con-

tains a stretch of Mahogany Creek that supports a population of the endangered Lahontan cutthroat trout. Natural communities include stream, montane wet meadow and riparian aspen-willow corridor, and montane mountain sage–bunchgrass communities. Sage grouse nest in the upland sage areas, and the water attracts pronghorn antelope, mule deer, wild horses, and a variety of bird life, including eagles and falcons.

Mount Charleston Wilderness Area
Clark County
Part of Toiyable National Forest, this 43,000-acre parcel about 50 miles northwest of Las Vegas protects rugged montane- to subalpine-forested terrain in the Spring Mountains. Unique to this wilderness is an 18,000-acre forest of bristlecone pines, the largest such community in the western United States. Some of the oldest living organisms in the world, these trees are typically more than 1,000 years old. The forests are an important habitat for a variety of birds, reptiles, and mammals, including bighorn sheep and mountain lions.

Mount Moriah Wilderness Area
White Pine County
Part of Humboldt National Forest in the Schell Creek Mountains, this 82,000-acre wilderness area is about 40 miles east of Ely. It protects high-elevation mountain features, such as plateaus, caves, and interesting rock formations. At this extreme elevation, the dominant communities are stands of bristlecone, Douglas fir, and limber pines. Some of the bristlecone pines are more than 1,000 years old. Several sensitive subalpine-alpine plants occur here, as does the sensitive Bonneville cutthroat trout, a Great Basin endemic species.

Pyramid Lake
Washoe County
This body of water about 50 miles north of Reno is 28 miles long and 11 miles wide, the remnant of a larger ancient lake. Its shoreline is rimmed by interesting rocky spires. Long a spiritual Native American site, the lake supports two endemic fish: the cui-ui sucker and Lahotan cutthroat trout, a landlocked species. Golden eagles and waterfowl are abundant and include California gulls, grebes, and the country's largest colony of American white pelicans.

Quinn Canyon Wilderness Area
Nye County
Part of Humboldt National Forest, this 27,000-acre tract is about 120 miles southwest of Ely. Elevations range from about 6,000 feet to more than 10,000 feet and support a variety of natural communities, including riparian corridor, sagebrush shrubland, pinyon-juniper woodland, and higher-elevation white fir–aspen–mountain mahogany forest. Ancient bristlecone pine trees can be found at the highest subalpine elevations. Several rare endemic plant species also grow here.

Red Rock Canyon National Conservation Area
Clark County
Located 15 miles west of Las Vegas, this BLM property encloses a portion of Red Rock Canyon and its 3,000-foot sandstone cliffs. The area is watered by more than 40 springs and seeps, which fill lower-elevation shallow basins. The canyon floor is dominated by Joshua trees and several shrubs and patches of grass; the higher elevations are dominated by pinyon pine–juniper woodlands. A variety of birds can be found in the canyons of Pine Creek and First Creek, including sharp-eyed raptors. The property is well known as a critical habitat for desert bighorn sheep and the federally threatened desert tortoise. Wild horses and burros also frequent the canyon.

Santa Rosa–Paradise Peak Wilderness Area
Humboldt County
This rugged wilderness area protects a section of Nevada's high-desert mountain environment. Covering 31,000 acres about 30 miles north of Winnemucca, this rocky, steep, sparsely vegetated terrain is an important habitat for large mammals, such as California bighorn sheep, mule deer, bobcats, and mountain lions. Raptors, especially golden eagles and red-tailed hawks, nest in the cliff walls. Varied wildflowers grace grassy meadows on the flatter slopes. Stream canyons contain quaking aspen or cottonwood and support healthy populations of the endemic and federally threatened Lahontan cutthroat trout.

Sheldon National Wildlife Refuge
Washoe and Humboldt Counties
Located near the Oregon border 30 miles southwest of Denio, this expanse of sagebrush-covered flatland supports 2,000 prong-

horn antelope. It also provides critical habitats for mule deer, bighorn sheep, bobcats, sage grouse, and mountain lions. Scattered seeps, springs, and small wetlands attract a variety of birds, including the rare Wilson's phalarope; raptors are common.

Soldier Meadows
Humboldt County
Located on 307 acres south of Summit Lake Indian Reservation, this warm-spring desert aquatic ecosystem is the only place in the world where the federally threatened desert dace survives. The rare basalt cinquefoil grows near the warm springs. Nearby creeks are important spawning streams for Lahontan cutthroat trout. The surrounding desert is covered by sagebrush-greasewood-saltgrass-shadscale vegetation. Several rare hydrobiid snails have also been discovered on the site.

Spring Mountain Recreation Area
Clark County
Part of Toiyable National Forest, this property 20 miles south of Indian Springs in the rugged Spring Mountains contains communities that range from desert to alpine. Eighteen rare or endemic flowering plants are recorded here, including penstemon, milk vetch, chain ferns, angelica, and arching pussytoes, an alpine species. Other endemic species include eight butterflies and a chipmunk. Desert bighorn sheep and raptors are commonly seen. Subalpine bristlecone pine trees on the higher slopes are thousands of years old.

Stillwater Wildlife Management Area
Churchill County
The largest marsh in the state and a principal stopover along the Pacific Flyway, the 162,000-acre Stillwater Marsh near Fallon has lost much of its original size because of water diversion for agricultural use. Water levels are highest during spring. Nearly 300,000 ducks, teals, geese, tundra swans, egrets, ibis, herons, owls, Wilson's phalaropes, Virginia rails, and cliff swallows visit or nest here. The outlying flats are covered with greasewood-saltbush scrub. Like any desert wetland, the Stillwater Marsh is a critical habitat for a diverse array of mammals, birds, reptiles, amphibians, and insects that depend on it for food, water, and shelter.

NEW HAMPSHIRE

The major natural communities in New Hampshire are classified as palustrine, terrestrial, riparian, estuarine, and lacustrine.

Palustrine systems include open wetlands and forested wetlands. Open wetlands consist of marshes, ponds, lakeshore margins, coastal wetlands and marshes, graminoid marshes and sedge meadows, various bogs and peatlands, fens, and shrub swamps. These occur across the state and vary according to elevation, substrate, and location. Forested wetlands consist of floodplain forests, basin swamps, seepage swamps, and vernal pools. Northern white cedar seepage–basin swamp and Atlantic white cedar seepage–basin swamp are two of the more important communities along the Coastal Plain. The hardwood and bottomland swamps are dominated by red maple and black gum trees.

Terrestrial systems are classified according to the percentage of tree cover: Barrens communities have less than 25 percent tree cover and include rocky summits, talus barrens, shrub thickets, and sand-gravel barrens in river drainages. Woodlands contain 25–60 percent tree cover and include rocky summit woodlands, coastal rocky headlands, and talus woodlands. Forests have greater than 60 percent tree cover and are classified by the dominant trees, which are dependent on elevation, temperature, moisture content, and soil chemistry. Dominant types are mountain spruce-fir forest, lowland spruce-fir forest, northern mixed hardwood-conifer forest (for example, beech-hemlock), and pitch pine–red pine–transitional oak forest.

The vegetation in riparian corridors is again partially dependent on elevation. Typical communities include riverside seep, riverside outcrop, meadow, shrub thicket, dry riverbluff, and mesic riverbluff. Coastal communities are beach strand, inland beach, coastal dune, inland dune, and pitch pine–scrub oak barren.

At alpine elevations, forests rarely exist. The terrain is dominated by diaspensia, heath-meadow, sedge meadow, subalpine krummholz, fellfield, alpine cliff, herbaceous snow-bank meadow, and alpine bog.

The estuarine system comprises tidal and subtidal communities: various tidal marsh, brackish marsh, coastal salt pond, tidal creek, intertidal mudflat, eelgrass, channel bottom, and other subtidal communities.

The lacustrine system comprises lakes and ponds, from alpine elevation to sea level; streams and rivers are considered riverine elements.

Special Habitat Sites

Alpine Gardens Research Natural Area
Coos County
Located in White Mountain National Forest, this 100-acre tract is about 12 miles south of Berlin. It was established in 1989 to protect an outstanding example of alpine tundra habitat in the Presidential Range. Common species include mountain cranberry, Labrador tea, bearberry willow, sedges, mountain avens, Iceland lichen, reindeer moss, and sandwort. The list of rare and specialized species includes arctic-alpine plants, such as krummholz and Boott's rattlesnake-root. Nearly 50 (state) rare, threatened, or endangered species can be found here.

Beaver Brook Natural Area
Hillsborough County
Located about a mile southwest of Hollis, this 1,700-acre preserve is owned by the Beaver Brook Natural Association. Habitats include upland and bottomland forest, meadow, and various wetlands. The older trees are an important habitat for the endangered spotted owl. Biodiversity is high, with nearly 600 plant, mammal, and bird species recorded. Interesting species include dogwoods, orchids, gentians, and leatherwood.

Bowl Research Natural Area
Grafton County
This 510-acre RNA in White Mountain National Forest encloses a section of presettlement virgin forest that has not been logged or burned. Many of the trees are more than 150 feet tall and 500 years old. Natural communities include balsam fir–red spruce, paper birch, sugar maple–beech, and red spruce forests. Rare or unique species include squirrel corn, globally rare mosses, and old-growth forest bird species, such as the pileated woodpecker.

Cooper Cedar Woods
Strafford County
Located about halfway between New Durham and Rochester, this 30-acre tract property encloses high-quality examples of swamp, upland forest, heath bog, and cedar swamp natural communities. Some mature, and a few old-growth, cedar species remain. Upland forest trees are white pine, pitch pine, white oak, hemlock, and yew. The heath bog is dominated by rhodora, leatherleaf, blueberry, and sheep laurel. Several rare or unusual plant species grow here.

Durham Point Sedge Meadow
Strafford County
Located near Durham, this 18-acre reserve protects a wetland-bog-meadow complex, critical habitat to the globally rare banded bog skimmer. It is owned and managed by The Nature Conservancy.

Fourth Connecticut Lake
Coos County
Located near Pittsburg, this 78-acre Nature Conservancy preserve encloses a rare northern tarn–floating bog community at the headwaters of the Connecticut River. It shelters a number of rare plants, including the carnivorous pitcher plant and sundew.

Frank Bolles Preserve
Carroll County
This 247-acre preserve about 4 miles northeast of Chocorua preserves a highly diverse combination of forest, swamp, stream, and kettlehole natural communities. More than 160 plants and animals have been identified here, including some rare species. It is an important habitat for moose, black bears, and snowshoe hares. It is owned by The Nature Conservancy.

Green Hills
Carroll County
This 2,822-acre Nature Conservancy property encloses wetlands and forested uplands near North Conway. Natural communities include stands of mixed mature hardwood-pine forest, conifer forest, stream, pond, bog, and marsh. Unique characteristics include a high-elevation stand of red pine forest and a number of rare or endangered plant species. The wetlands are an important nesting site for blue herons, wood ducks, and swallows.

Hemenway State Forest
Carroll County
Located about 6 miles west of Chocorua, it includes Lovejoy Marsh and Big Pines Natural

Area. Conifers here are nearly 200 years old and 100 feet tall.

Hoyt Wildlife Sanctuary
Carroll County
Located near East Madison on the south shore of Purity Lake, this 168-acre property includes a kettlehole bog. Interesting or rare species include sphagnum moss, carnivorous sundews and pitcher plants, and the Virginia chain fern, a more southern species. It is also an important nesting habitat for several birds.

Hurlbert Swamp
Coos County
Located about 10 miles north of Colebrook, this 313-acre complex protects an excellent example of rapidly disappearing northern white cedar swamp. Other species include tamarack as well as white, black, and red spruces. This boreal swamp also contains shallow peat bogs. Old-growth trees are 60 feet tall, 2 feet in diameter, and hundreds of years old. The forest and wetlands are important moose habitat. Rare species include varieties of mosses and lichens, orchids, the black-backed three-toed woodpecker, and several warblers. The property is owned by The Nature Conservancy.

Mount Teneriffe
Strafford County
Located near Milton, this 170-acre Nature Conservancy tract protects a variety of rare or endangered species and communities. The rocky hills are covered by hardwood forests. It is one of the five most important sites in the world for the globally rare small whorled pogonia orchid. Other species are squaw root and American ginseng. It is also an important habitat for songbirds and other neotropical birds, moose, bears, and upland game birds.

Nancy Brook Research Natural Area
Grafton and Carroll Counties
Located on 1,385 acres about 15 miles south of Passaconaway, this RNA is located in White Mountain National Forest. Natural communities include old-growth red spruce–balsam fir forest, northern hardwood forest, fen, and rock face. Several rare species can be found here, including mountain avens, certain sedges, and the northern three-toed woodpecker.

Norton Pool
Coos County
Owned by The Nature Conservancy, this 427-acre preserve about 10 miles north of Pittsburg is the only virgin lowland spruce-fir forest remaining in New Hampshire. Specimens of virgin balsam fir, red spruce, and black spruce approach heights of 100 feet. It is a critical habitat for a variety of plant and animal species.

Ossipee Pine Barrens
Carroll County
This 573-acre tract about 2 miles north of Madison preserves an important part of the 3,000-acre Ossipee Pine Barrens. It is covered by a rare pitch pine–scrub oak–blueberry natural community. It is a critical breeding habitat for at least 12 species of rare butterflies and moths, rufous-sided towhees, and brown thrashers. The property is owned by The Nature Conservancy.

Packer Bog
Rockingham County
This 23-acre Nature Conservancy wetland near Portsmouth preserves a remnant of Atlantic white cedar swamp. It is also critical habitat for some rare butterfly species.

Pawtuckaway State Park
Rockingham County
Located about 8 miles northwest of Raymond, this 5,500-acre state park preserves a rare black gum swamp community, undisturbed for centuries. Several very old eastern hemlock trees that are more than 100 feet tall and 200–300 years old grow here. The swamp is an important habitat for a variety of birds, mammals, amphibians, and reptiles, some of them unusual or rare. The wildflowers are colorful during spring.

Pine River State Park
Carroll County
Located in the town of Effingham, about 15 miles northeast of Lake Winniposo, this 3,000-acre park protects a globally rare pitch pine–scrub oak barren. The understory consists of lowbush blueberry. Several varieties of unusual or rare species can be found here, including rare moths, brown thrashers, and the state-listed common nighthawk and rufous-sided towhee.

Sheldrick Forest
Hillsborough County
Located about 8 miles west of Milford, this 228-acre old-growth forest is an important nesting habitat for interior forest birds, such as the red-headed woodpecker. Some of the trees are very large and more than 150 years old. The forest consists of hemlock-pine, beech–black birch–red oak, and butternut stands. The understory also contains mountain laurel. The area is known for its display of wildflowers during spring and summer. It is owned and managed by The Nature Conservancy.

Stamp Act Island
Carroll County
This 100-acre Nature Conservancy parcel a few miles northeast of Wolfeboro protects an island in Lake Wentworth that is covered by old-growth forest, swamp, and beach communities. The largest trees are beech and red oak. The island is an important nesting site for various waterbirds, including great blue herons, loons, ducks, and bitterns.

Sunapee State Park
Merrimack County
Located a few miles south of Mount Sunapee, this 2,800-acre state park encloses a remnant stand of old-growth forest. This sugar maple–yellow birch–white ash hardwood forest reflects the "Big Woods" that once covered the state during presettlement times. The largest trees are centuries old, nearly 100 feet tall, and have breast-height diameters reaching 4 feet. The forest is an important habitat for birds and mammals that depend on old-growth forest. A few rare or special-concern plant species also grow here.

Wales Preserve
Hillsboro County
This 48-acre parcel 5 miles south of Peterborough protects important old-growth forest and stream habitats. Gridley River, a trout stream, runs through the property in a series of small waterfalls. Hemlock, maple, and white pine trees, some of which are nearly 100 years old, dominate. The area is best known for its diverse understory of flowers, shrubs, and herbs along the river, including royal and sensitive ferns, wood asters, Jack-in-the-pulpit, bloodroot, nightshade, bunchberry, and painted trillium. It is owned and managed by The Nature Conservancy.

Warwick Preserve
Cheshire County
This Nature Conservancy preserve is located about a mile northwest of Westmoreland. Its 36 acres protect a rocky, rugged terrain covered by hemlock, maple, oak, beech, hickory, birch, butternut, and white pine trees. Less common trees are round-leaved dogwood and mountain maple. The preserve's assemblage of wildflowers is one of the most diverse in the state and includes the showy orchid, spring beauty, maidenhair spleenwort, and Dutchman's breeches.

West Branch Pine Barrens
Carroll County
Located near Madison on 341 acres, this Nature Conservancy preserve encloses an ecosystem, critical habitat for a variety of rare insect species, including butterflies and moths. This is one of the best examples of pine barren ecosystem in the northern United States.

White Lake State Park
Carroll County
Located a mile west of West Ossipee, this 72-acre National Natural Landmark preserves a high-quality old-growth pitch pine forest on the western shore of White Lake. This is an extremely rare community. Several varieties of specialized plant and animal species inhabit the forest.

NEW JERSEY

New Jersey divides its major natural communities into the following systems: marine, estuarine, palustrine, terrestrial, and subterranean.

The marine system consists of subtidal and intertidal communities that exist along the Atlantic coast. This is characteristically a narrow strip of land that may extend a greater distance inland along creeks and rivers. Some of these communities include deepwater and marine subtidal aquatic bed, such as eelgrass, and marine intertidal gravel–sand beach and coastal sand dune. Sand dunes are most abundant from Sandy Hook to Cape May.

The estuarine system includes saline and brackish subtidal aquatic bed, saltwater marsh complex, and brackish tidal marsh complex; these can extend for miles up streams that drain into the ocean before freshwater plant assemblages become dominant. These are typically found in the Coastal Plain and sometimes the Piedmont physiographic provinces. Good examples are the Passaic, Hackensack, and Delaware Rivers.

The palustrine system is divided into open-canopy and forested wetlands. Open-canopy wetlands include freshwater tidal marsh complex, interdunal marsh, seep, fen, shrub swamp, Coastal Plain pond, various bogs and marshes, pine barren savanna, and pine barren shrub swamp. Forested wetlands include freshwater tidal swamp, seepage swamp, Atlantic white cedar swamp, floodplain forest, pine barren hardwood swamp, pitch pine lowland forest, black spruce swamp, and red maple swamp. Many of these are found in the relatively flat-lying Coastal Plain. As the terrain becomes more rugged toward the west, stream channels become narrower, and the wetlands become more restricted in size. Elevation also becomes a factor in wetland type (for example, the boreal bogs that occur in the Highlands Province of northwestern New Jersey).

Terrestrial systems are also classified as open-canopy or closed-canopy. Open-canopy communities include rock face, glade, coastal dune grassland-shrubland-woodland, pine plain, pitch pine–scrub oak forest on exposed ridges, talus slope, and riparian corridor. Closed-canopy communities comprise the major forest types in the state: chestnut oak, Virginia pine–oak forest, mesic hemlock-hardwood, inland mixed oak, and pine-oak. The distribution of these forest types depends on a number of factors, including elevation, climate, soil chemistry, and precipitation. The Coastal Plain is dominated by pine-oak barrens that have developed in the dry, sandy soil. Moving inward, the Inner Coastal Plain is dominated by mixed white oak–black oak, and beech-oak, forests. The northwestern, more rugged Piedmont, Highlands, and Ridge and Valley Provinces contain more fertile soil and more moisture; forests change to sugar maple–mixed hardwoods, mixed oak, chestnut oak, and hemlock–mixed hardwood forest in shaded ravines and valleys.

Special Habitat Sites

Absegami Natural Area

Burlington County

This 128-acre nature preserve is located in Bass River State Forest about 5 miles northeast of New Gretna. It is centered on a remnant Atlantic white cedar swamp and surrounding mature oak-pine upland forest.

Allamuchy Natural Area

Sussex County

Located about 10 miles northeast of Hackettstown, this 2,440-acre tract represents about a third of Allamuchy Mountain State Park. Natural communities include mixed oak, mixed hardwood, and swamp hardwood forests, marsh, calcareous fen, and open water. Rare fen species are eastern smoke grass, water avens, and meadow willow. In addition to the barred owl, other rare animal species include certain turtles, the timber rattlesnake, ospreys, and red-headed woodpeckers. The wetlands are an important waterfowl habitat.

Batsto Natural Area

Atlantic and Burlington Counties

Located near Pleasant Mills, this 9,449-acre tract is part of Wharton State Forest. It encloses pineland swamp, bog, and floodplain bottomland forest. The Batsto and Mullica Rivers flow through the area. The largest pitch pine lowland forest in the world occurs here. Other communities are hardwood swamp, Atlantic white cedar swamp, and mixed pine-hardwood forest. About 150 rare plant and animal species have been identified on the preserve, including the globally rare bog asphodel, endemic only to the New Jersey pine barrens.

Others are Pickering's morning glory, pine barren boneset, pine barren gentian, and several sedges.

Bear Swamp
Cumberland County
Located about 7 miles south of Millville, this 3,369-acre tract is an important woodland-wetland complex. Natural communities include mature hardwood swamp forest, beech-hardwood forest, pitch pine forest, oak-pine forest, white cedar swamp, and tidal marsh. Some of the forests are old-growth and contain three distinct levels of canopy, very rare in the eastern United States. Some of the larger red maple specimens are more than 150 years old, some of the oldest trees in the state. Rare or threatened species include the bald eagle, barred owl, cranefly orchid, purple bladderwort, fringe tree, least tern, and tiger salamander. The wetlands are critical nesting habitats for waterbirds and raptors.

Bearfort Mountain Natural Area
Passaic County
Located on 1,325 acres near Highland Lakes, this natural area encloses excellent examples of a number of natural communities: marsh, open water, oak-hardwood forest, bottomland forest, hemlock-hardwood forest, and upland chestnut oak forest. Biodiversity is high, and some rare or endangered plant and animal species occur on the property. It is an important nesting habitat for birds, especially forest interior species, such as woodpeckers and owls. It is also a migratory stopover for warblers and vireos.

Belleplain State Forest
Cape May County
This 13,000 acres of mature pine-oak forest is located just south of Belleplain. It includes rare white cedar swamp, mixed hardwood forest, pond, and marsh. It is an important bird breeding habitat, especially for the rapidly diminishing prothonotary warbler and ovenbird. Old-growth trees are favored by owls and red-headed woodpeckers.

Bennett Bogs Preserve
Cape May County
Located about 2 miles from Cape May, this 24-acre Nature Conservancy property protects a series of vernal ponds and swampwoods that flood during spring and dry up during summer. Most such ponds have been drained and filled for development purposes; consequently, surviving examples represent a rare ecosystem. Such ponds can also fail if groundwater levels drop due to domestic water use. Several rare plants can be found here, including pine barrens gentian, Muhly's dropseed, Small's yellow-eyed grass, snowy orchid, stinking fleabane, and southern greenbrier carpet. It is also an important bird habitat.

Black River Natural Area
Morris County
Located just northwest of Chester, this 1,820-acre preserve is part of Black River Wildlife Refuge. The marsh, swamp hardwood forest, and mixed hardwood upland forest are habitats for several rare plant and animal species.

Blair/Cape May Migratory Bird Refuge
Cape May County
Owned by The Nature Conservancy, this 187-acre preserve a few miles from Cape May protects critical bird habitat just inland from the Atlantic Ocean. Natural communities include ocean beach, dune, freshwater marsh, saltwater marsh, pond, and meadow. The wetlands are a critical feeding and resting site for thousands of birds that use the Atlantic Flyway. Nearly 100,000 birds congregate here during fall, representing nearly 400 different species of land birds, shorebirds, and waterfowl. The property is also a nesting habitat for the rare or endangered peregrine falcon, Cooper's hawk, upland sandpiper, least tern, and piping plover.

Bull's Island
Hunterdon County
This 20-acre area about 5 miles west of Rosemont is located along the northern bank of the Delaware River. It preserves good examples of mature floodplain forest, mixed oak-hardwood forest, and stream-riverbank communities. Dominant trees are sycamore, silver maple, and black walnut. Rare or endangered species include broad-leaved waterleaf, low sand cherry, few-fruited sedge, Carolina whitlow-grass, wood turtle, certain salamanders, and the cerulean warbler.

Cape May Bird Observatory
Cape May County
Owned by the Audubon Society, this property along the Atlantic Flyway includes beach–sand

dune, coastal pond, freshwater marsh, and woodland natural communities. It is thus a critical migratory stopover for herons, egrets, bitterns, herons, saw-whet and barn owls, wood storks, Bell's vireos, rock wrens, ospreys, and golden and bald eagles.

Cape May Point Natural Area
Cape May County
This 200-acre tract at Cape May Point contains a variety of unique coastal habitats critical to wildlife and shorebirds: freshwater, coastal, and brackish wetlands, sand dune, marsh, shrub thicket, beachfront, and hickory-oak-sassafras forest. Located along the Atlantic Flyway, it is also an important residential and migratory stopover for thousands of birds along the eastern U.S. coast. Several varieties of endangered or threatened bird, animal, and plant species are found here. These include the bald eagle, Cooper's hawk, and other raptors, upland sandpiper, red-headed woodpecker, great blue heron, salamanders, swamp pine, Long's rush, and rattlesnake master. The property is owned and managed by The Nature Conservancy.

Cape May Wetlands Natural Area
Cape May County
A relatively pristine tidal salt marsh is protected on this 2,600-acre preserve near Cape May Point. It is a critical breeding habitat for fish and shellfish.

Cedar Swamp Natural Area
Burlington County
Located 4 miles south of Pemberton, this preserve contains pine oak forest, pitch pine lowland, white cedar swamp, pine barren swamp, and open pond communities. Several rare or endangered plant and animal species inhabit the site, including swamp pink, bellwort, pine barren reed grass, pine barren gentian, the pine barrens tree frog, and a rare moth. The thick cedar forest is cool and dark and protects rare orchids, sundews, and pitcher plants. The turkey beard, a grasslike plant of the pitch pine lowland, displays brilliant white flowers during May and June.

Champagne Island
Cape May County
Located in Hereford Inlet between Stone Harbor and North Wildwood, this property is a critical nesting site for scores of endangered and threatened bird species, including black skimmers, least terns, and the piping plover. Beach, dune, scrub, and maritime forest communities cover the island.

Cheesquake Natural Area
Middlesex County
Located about 5 miles south of South Amboy, this 450-acre parcel protects an important wetland complex called Cheesquake Marsh. Natural communities include saltwater marsh, freshwater marsh, white cedar swamp, bog, upland pine-oak forest, and oak-hardwood forest. The complex relationships among these habitats create a high biodiversity. Bog and white cedar swamp are rare communities and host specialized flora. Rare species include the white milkweed, boneset, yellow fringed orchid, large marsh pink, pitcher plant, milkwort, Canada lily, marsh aster, pyxie moss, bald eagle, peregrine falcon, black skimmer, and vesper sparrow.

Delaware Bayshore Preserve
Cape May County
This tiny Nature Conservancy property protects a small portion of the Delaware Bay shoreline, a critical habitat for bird life. Natural communities include beach, dune, and grassland. Delaware Bay is the most important resting site in North America for the ruddy turnstone and red knot. Other species that feed here are migrating shorebirds, laughing gulls, sanderlings, dunlines, and certain sandpipers. The laughing gulls feed on the eggs of horseshoe crabs on the beaches during spring.

Dennis Creek Wildlife Management Area
Cape May County
Located just outside Dennisville, this 5,109-acre property protects the undisturbed tidal salt marshes of Dennis Creek. Salt marshes are one of the most nutrient-rich of any habitat and support an abundance of plants and organisms. Cordgrass is a common marsh plant. Dennis Creek is an important spawning ground for saltwater fish and shellfish. Several varieties of waterfowl and shorebirds either inhabit the marsh or visit during spring and fall migrations. More common species include the American wigeon, shovelers, common goldeneye, night herons, and black skimmers. Between 1 million and 2 million birds, includ-

ing the rare piping plover, pass through every year.

Dix Wildlife Management Area
Cumberland County
Located near Sea Breeze, this 2,395-acre preserve shelters pristine salt marsh. Important natural communities include grass salt marsh, creek, and pine-oak forest stand. As part of the Delaware Bay wetlands along the Atlantic Ocean, Dix is a critical nesting or migratory habitat for thousands of shorebirds, including laughing gulls, ruddy turnstones, red knots, and sandpipers. Rare bald eagles also nest in trees on the site and migrate here during winter.

Dryden-Kusler Natural Area
Sussex County
This 1,451-acre preserve about 10 miles northwest of Sussex contains oak forest, hemlock-hardwood forest, Atlantic white cedar swamp, pitch pine forest, and red maple swamp natural communities; rock outcrops also occur. Various threatened, endangered, or rare plant and animal species include cinquefoil, dwarf mistletoe, sedges, wheatgrass, slender panic grass, Cooper's hawk, timber rattlesnake, barred owl, wood turtle, and long-tailed shrew. Some old-growth stands of large hemlock, white pine, black spruce, and southern white cedar are northern relicts.

Dunnfield Creek Natural Area
Warren County
This 1,085-acre section of Worthington State Forest is located about 5 miles northwest of Columbia, along the Delaware River. It contains mature chestnut oak forest, wetland, and hemlock-mixed hardwood forest natural communities. Rare or endangered plant and animal species have been identified here.

Eldora Nature Preserve
Cape May County
Part of the famous New Jersey pine barrens, this 170-acre Nature Conservancy preserve is located near Dennisville. Natural communities are open pitch pine barren, wetland, salt marsh, and southern Coastal Plain hardwood forest. The Nature Conservancy established this preserve to protect rare moths and other insects, such as the plum sphinx moth and royal walnut moth; five species of moth are endemic only to New Jersey. Rare bald eagles and other raptors winter on the site. It is an important migratory stopover for a variety of other birds during spring and fall.

Farney Natural Area
Morris County
This 589-acre tract about 3 miles north of Denville is part of Farney State Park. It contains mixed oak-maple-chestnut-basswood forest and red maple–yellow birch swamp forest, streams, riparian corridors, and minor wetlands. Raptors nest in the taller trees of the upland forest. Rare or endangered plant and animal species occur here, such as the wood turtle, barred owl, red-shouldered hawk, bald eagle, goshawk, osprey, and three bird orchids.

Forsythe National Wildlife Refuge
Ocean County
Located along the Atlantic Ocean, this 36,000-acre property is critical bird habitat. Natural communities include tidal marsh, freshwater marsh, upland mixed forest, lowland red maple swamp, rare Atlantic white cedar swamp, and wet meadow. Rare birds that inhabit the site are the endangered peregrine falcon, northern harrier, and piping plover. Tens of thousands of geese, swans, cranes, and shorebirds rest and feed here during spring and fall migrations.

Gateway National Recreation Area (Sandy Hook)
Monmouth County
This 160-acre park is located on a 7-mile sand spit in Atlantic Ocean, about 20 miles south of New York City. Natural communities include surf zone, beach, dunes, salt marsh, and interior scrubland and forest. A rare American holly–black cherry forest, nearly 200 years old, grows in the sand. The area is an important site for nesting and migrating birds, including hawks and raptors, great horned owls, ospreys, and least terns. Songbirds congregate during spring and summer. Several rare turtles nest on the beaches.

Great Bay Boulevard Wildlife Management Area
Ocean County
Consisting of a peninsula, islands, and wetlands in Little Egg Harbor Bay, in the Atlantic Ocean, this 4,845-acre property south of Tuckerton contains diverse and very productive ecosystems. Natural communities include

tidal creek, salt marsh, swamp forest (marsh elder–bayberry), and upland pitch pine–scrub oak forest. The marshes are dominated by cordgrass and glasswort. The wetlands are an important spawning ground for saltwater fish and shellfish and provide critical habitats for nesting and migrating birds, such as ospreys, least terns, black skimmers, and rails.

Great Swamp National Wildlife Refuge
Morris County
Located on 7,200 acres about 5 miles east of Basking Ridge, this famous wetland complex consists of hardwood swamp, open marsh, shrub thicket, upland forest, and riparian corridor communities. The swamp is incredibly diverse, with more than 1,000 species of flora and fauna, including rare warblers, nesting wood ducks, raptors, woodpeckers, and woodcocks. Rare or uncommon mice, voles, turtles, snakes, and salamanders live along the edges. A few stands of virgin timber remain, including a 500-year-old beech tree with a circumference of 14 feet. Some of the oldest trees in New Jersey live deep in this swamp.

Greenwood Forest Wildlife Management Area
Ocean County
This lowland wetland complex near Webbs Mill includes bog, sphagnum moss hummock, bottomland forest, and other wetland communities. Rare species include the evergreen leatherleaf, pitcher plants, sundews, six kinds of bladderworts, swamp pink, bog aspodel, clammy azalea, and bog aster. Nearly all of these flowers are uncommon, threatened, or endangered.

Hacklebarney Natural Area
Morris County
Located on 465 acres, this portion of Hacklebarney State Park about 5 miles north of Pottersville contains swamp, wetland, hemlock-hardwood forest, and hardwood forest communities. Rare or endangered plant and animal species have been found here.

Hamilton-Trenton Marsh
Mercer County
Just a few miles south of Trenton, this 1,250-acre wetland complex exists amid industry, urban sprawl, and interstates. It contains marsh, pond, shrubland, riparian corridor, wet forest swamp, and mature second-growth upland forest. Considering its location and history of use, it remains remarkably diverse, with more than 700 plant species and nearly 250 bird species. It contains important habitat for reptiles, amphibians, and mammals and is an important migratory stopover.

Highbee Beach Wildlife Management Area
Cape May County
Located on Delaware Bay, this 634-acre property is an important bird habitat. Natural communities include beach, sand dune, forest, pond, and meadow. More than 200 species of songbirds rest here during fall. The maritime forest is composed of holly–red cedar–pitch pine–beach plum–persimmon. More than 100 species of butterflies, including monarchs migrating to Mexico, frequent the grasses and flowers.

Hirst Pond Preserve
Atlantic County
This 80-acre tract protects an undisturbed vernal pond, which is a rapidly vanishing ecosystem in New Jersey. Natural communities include the pond, swamp forest (Atlantic white cedar–red maple), and mixed upland forest. A variety of rushes, sedges, and grasses, some of them rare or unusual, can be found in and around the pond. Other rare plants are the carnivorous sundew, slender blue flag iris, tickseed, Canby's lobelia, linear-leaved ludwigia, floating heart, and the crested yellow orchid. The rare pine barrens tree frog also inhabits the site. The property is owned and managed by The Nature Conservancy.

Huber Woods
Monmouth County
Located near Middletown, this 256-acre tract protects an excellent remnant of Inner Coastal Plain forest. Older-growth oak trees that are more than 100 years old dominate the canopy; other trees are beech, birch, tulip, and cherry. Understory trees and shrubs are sassafras, chestnut, mountain laurel, and blueberry. The forest is dense and contains numerous vines. The colorful wildflowers on the forest floor peak during late April. The property is an important habitat for forest birds, including songbirds and woodpeckers; several rare warblers have been sighted.

Hutcheson Memorial Forest
Somerset County
Located about a mile east of East Millstone, this 181-acre forest preserves the only tract of virgin forest remaining in New Jersey. The tallest trees are 400-year-old oaks, followed by American beech and hickory. The woodland hosts a variety of birds, especially songbirds, woodpeckers, and owls. Because of its tremendous age, the woodland is an important research center for studying the natural succession of forest vegetation.

Island Beach Natural Area
Ocean County
Located east of Toms River in the Atlantic Ocean, this barrier island, 10 miles long, is one of the country's largest and least disturbed vegetated barrier islands. Natural communities include sand dune, forest, bayshore marsh, sand flat, saltwater marsh, bog, and intertidal flat. Rare or endangered plants include seabeach sandwort, sickle-leaved golden aster, and curly-grass fern. The endangered loggerhead turtle has also been discovered on the island. Its location on the Atlantic Flyway makes it an important stopover for hundreds of thousands of migrating birds; more than 250 bird species have been recorded here, including rare sparrows and warblers.

Lebanon State Forest
Burlington County
This 29,472-acre property about 20 miles southeast of Browns Mills is co-owned by New Jersey and The Nature Conservancy. It protects a large expanse of the New Jersey pine barrens. Natural communities include stunted pitch pine barren, full-height pitch pine barren, oak tree stand, hardwood (gum-maple) swamp, and rare Atlantic white cedar swamp. Rare plant species that grow here are curly-grass fern, pine barrens gentian, pyxie moss, sundew, and turkey beard. The woodlands are important nesting habitats for nighthawks, eastern bluebirds, and a variety of finches and grosbeaks. Mink inhabit the wetlands.

Liberty Park Natural Area
Hudson County
Located near Jersey City, this 36-acre tract preserves a remnant of relatively undisturbed tidal salt marsh. Dominant plant species include cordgrass, glasswort, and salt meadow hay. This represents one of the few remaining high-quality marshes in the Hudson River estuary. It is an important breeding ground for fish and shellfish and a nesting site and migratory stopover for thousands of waterbirds and shorebirds, as well as some other rare or endangered species: the diamondback terrapin, peregrine falcon, least tern, black skimmer, osprey, heron, savanna sparrow, American shad, and Atlantic tomcod.

Lockwood Gorge Natural Area
Hunterdon County
This 311-acre tract is part of Ken Lockwood Gorge Wildlife Management Area, near Bunnvale. It contains an excellent remnant of hemlock–mixed hardwood upland forest. The forest is an important habitat for interior forest birds, including the pileated woodpecker. Rare or endangered plant and animal species can be found here.

Lost Brook Preserve
Bergen County
This important, 379-acre tract is the largest woods within 10 miles of downtown New York City. Despite the pressure from surrounding development, it remains relatively undisturbed. Natural communities include mixed oak-pine upland forest, red maple–sweet gum swamp forest, and riparian habitats along small streams. The property is a critical habitat for flying squirrels, the rare pileated woodpecker, and large populations of orioles, tanagers, and thrushes, especially during spring and fall migrations.

Manahawkin Natural Area
Ocean County
Located near Robbinsville, this 64-acre tract protects a mature bottomland hardwood forest community and several rare plant and animal species.

Manumuskin River Preserve
Cumberland County
Located near Port Elizabeth, this 1,400-acre property protects a pristine stretch of Manumuskin River. Natural communities include tidal freshwater mudflat, wild rice meadow, maple–sweet gum swamp forest, Atlantic white cedar swamp, and mixed pine-oak upland forest. Because of its unaltered state, the riverine habitat shelters many rare or endangered flora and fauna. Rare plants include the federally endangered sensitive joint

vetch (only 11 sites worldwide), resinous boneset, Atlantic white cedar, Parker's pipewort, waterworts, stiff tick trefoil, and Virginia hedge hyssop. The least tern, corn snake, southern gray tree frog, pine barrens tree frog, marsh rice rat, and Allegheny mound ant are rare or endangered species found here. The wetland complex is an important nesting habitat and migratory stopover for more than 200 bird species, some of them rare. The property is owned and managed by The Nature Conservancy.

North Brigantine Natural Area
Atlantic County
This 677-acre tract is located on a barrier island about 5 miles northeast of Brigantine and preserves rare coastal dune and tidal salt marsh communities. This site is one of the few remaining undeveloped dune–salt marsh complexes remaining in New Jersey. Communities include beach, dune-and-swale, salt marsh, intertidal flat, and tidal creek. Several rare plant or animal species can be found here, including four nesting pairs of the rare piping plover.

Oswego River Natural Area
Burlington County
Located about 4 miles north of Egg Harbor City, this large tract of wetland contains pitch pine lowland forest, Atlantic white cedar swamp, dry pine–oak forest, pine barren hardwood swamp, and pine barren savanna. The habitats are home to a variety of animal species. Endangered, threatened, or rare plant and animal species include bog asphodel, pine barren boneset, false asphodel, bladderwort, Barratt's sedge, curly-grass fern, pine barren reed grass, northern pine snake, pine barrens tree frog, moths, and the pine barren bluet.

Parvin Natural Area
Salem County
This 465-acre area preserves a section of fringe forest that borders the famous New Jersey pine barrens. Located a few miles north of Deerfield, it protects pine barren hardwood swamp, pine-oak forest, Atlantic white cedar swamp, and pitch pine lowland natural communities. Rare species identified on the property include the barred owl and swamp pink, a globally rare plant species. The endangered eastern tiger salamander, as well as rare moths and butterflies, are thought to inhabit the site.

Poricy Park
Monmouth County
Located near Middletown, this 30-acre preserve protects a stretch of Shrewsbury River. Natural communities include mature oak-beech hardwood forest, sedge–black willow wetland, rose-blackberry-briar patch, and meadow. The woodland and briar patch are important nesting habitats for rapidly diminishing birds, such as cuckoos, brown thrashers, and gray catbirds; the wetlands are a migratory stopover for thousands of birds during spring and fall. The property is owned and managed by The Nature Conservancy.

Ramapo Lake Natural Area
Passaic and Bergen Counties
Located about 5 miles north of Oakland, this 1,417-acre tract encloses a mature mixed hardwood upland forest. The woodlands are an important habitat for interior forest birds, such as woodpeckers.

Rancocas Natural Area
Burlington County
Located about 5 miles southeast of Rancocas, this 58-acre property encloses part of a relatively undisturbed freshwater tidal marsh ecosystem. Wild rice grows here. Several rare plant and animal species have also been identified.

Readington Natural Area
Hunterdon County
Located on 36 acres, this state natural area is located about 2 miles east of White House Station. Natural communities include Rockaway Creek, meadow, and surrounding pine, floodplain, and mixed hardwood forests. Rare or uncommon species include various salamanders, the bobolink, songbirds, flycatchers, skipper and sulfur butterfly species, and the upland sandpiper.

Sedge Island
Monmouth County
This 7-acre island in the Manasquan River about a mile east of Brielle is owned by The Nature Conservancy. The mouth of the river is a tidal salt marsh, which empties into the Atlantic Ocean. The island is fringed with cordgrass and other wetland plants and is covered by grass and shrubs. It is an important wintering habitat for local waterfowl and is a resting place for migrating shorebirds.

Strathmere Natural Area
Cape May County
Located 6 miles southwest of Ocean City within Corson's Inlet State Park, this 98-acre property protects a rare coast dune community and its unique or endangered species. These include coastal dune grass (seaside goldenrod and sea rocket) and coastal dune shrubland (red cedar–beach plum). Rare or threatened species that nest in the dunes are the piping plover, least tern, and black skimmer. It is an important stopover for thousands of migrating waterfowl during spring and summer.

Sunfish Pond Natural Area
Warren County
Located in Worthington State Forest about 5 miles northwest of Columbia, this 258-acre parcel encloses a kettle lake, surrounding lowland forest, and upland chestnut oak forest. Rare species have been identified here.

Swimming River Natural Area
Monmouth County
Located about 5 miles east of Holmdel, this 133-acre tract encloses part of Swimming River and Pine Brook, hardwood swamp, tidal marsh, meadow, and mixed oak natural communities. Dominant trees are oak, hickory, poplar, beech, red maple, black tupelo, and white pine. The woodlands and wetlands are important habitats for a variety of bird and animal species. It is suspected that rare turtles, amphibians, and snakes may inhabit the property.

Tillman Ravine Natural Area
Sussex County
Located on 525 acres near Kittatinny Lake, this mixed hemlock-hardwood forest and wetlands in Stokes State Forest contains rare species. Natural communities include riparian corridor, cliff face, and mature to old-growth hemlock forest, in addition to the abundant wildflowers. Some of the hemlocks are more than 100 feet high and exceed 175 years old. An important habitat for a variety of songbirds, owls, and woodpeckers, the cooler microclimate also protects rare plant species, relicts from the last Ice Age. The site is also on a major migration route for ospreys, falcons, and eagles. Rare plants include blunt-lobed grape fern, bunch flowers, mountain holly rock spikemoss, two-leaved toothwort, white mandarin, and balsam ragwort. Timber rattlesnakes inhabit the site.

Troy Meadows Natural Area
Morris County
Bounded by development and interstates, this 407-acre parcel near Troy Hills protects swamp, bottomland forest, and the largest cattail freshwater marsh in New Jersey. It is an important nesting habitat for least and American bitterns, as well as various rails. Rare or unusual species are the bog turtle, blue-spotted salamander, and swamp sparrow.

Valhalla Hemlock Glen
Morris County
Located about a mile north of Montville, this 30-acre Nature Conservancy property protects a section of bedrock gorge along Glen Brook, north of Lake Valhalla. The cooler and damper microclimate of the rocky gorge has preserved an old-growth boreal forest remnant that contains towering eastern hemlock trees that are hundreds of years old. The groundcover consists of mountain laurel and a variety of relict ferns, including shield, Christmas, and maidenhair.

Wawayanda Hemlock Ravine
Sussex County
This 501-acre tract about 5 miles east of Vernon is within Wawayanda State Park. The natural area preserves hemlock–mixed hardwood forest, mixed oak forest, and red maple swamp. The hemlocks shade steep slopes, creating a cooler microclimate that harbors rare plant species; the forest is an important bird habitat. Rare species include the timber rattlesnake, salamanders, barred owl, great blue heron, Dewey's sedge, white-grained mountain ricegrass, and witch hobble.

Wawayanda Swamp Natural Area
Sussex County
This 2,167-acre forest-wetland complex is located near Highland Lakes. Communities include rare Atlantic white cedar swamp, bog, mixed hardwood forest, hemlock–mixed hardwood forest, and swamp forest. Several rare or unusual plant and animal species occur here, including boreal relicts from the Ice Age. Bobcats and black bears are occasionally seen here.

West Pine Plains Natural Area
Burlington County
This 3,830-acre preserve about 10 miles northeast of New Gretna is part of Bass River State Forest. It shelters a globally rare dwarf pine–oak forest community and several rare species. Pitch pines normally grow to 40 feet; these are 6–8 feet. Unusual species in the understory are sheep laurel, pyxie moss, sand myrtle, dangleberry, and the very rare broom crowberry. Rare moths also breed here.

Whittingham Natural Area
Sussex County
Located on 153 acres about 3 miles southwest of Springdale, this property is part of the Whittingham Wildlife Management Area. Natural communities include high-quality northern swamp and floodplain forest. Rare relict plants from the Ice Age grow in the understory, a reminder of the once colder climate.

NEW MEXICO

A dramatic range of natural communities exists in New Mexico. Their distribution is largely dependent on elevation, soil chemistry, and precipitation. Vegetation types range from low-level desert scrub to alpine tundra at elevations that exceed 10,000 feet.

The majority of the eastern half of the state, especially the northeastern corner, is covered by Plains Mesa Grassland. This dominant community is composed of bluestems, sideoats grama, wheat grass, threeawn, buffalo grass, saltgrass, and cheatgrass. Shrubby elements are sagebrush, saltbush, winterfat, and rabbitbrush. An isolated area of this grassland also occurs in the center of the state.

The southern quarter of the state is variably covered by desert grassland with lesser Chihuahuan Desert creosotebush scrub. Scattered patches also occur in northwestern New Mexico. Vegetation is composed mostly of western wheatgrass, Indian ricegrass, galleta, gramas, and various shrubs. The northwestern corner of the state is covered by sagebrush–saltbush–shadscale–sparse grass (Great Basin desert shrub).

Juniper savanna or woodland occurs with sagebrush at middle elevations to low elevations on the slopes of mesas and mountain ranges in the central, west-central, and northwest parts of New Mexico. Juniper-pinyon savanna is generally a transition zone into montane conifer forest.

Montane conifer forest can be found at middle elevations on the slopes of mountain ranges in the central and western portions of the state. Various forest associations include blue spruce, Douglas fir ponderosa pine, white fir–Douglas fir, Douglas fir–southwest white pine, ponderosa pine, Douglas fir–Gambel oak, and ponderosa pine–pinyon.

Above the montane forests are subalpine conifer forests, between 9,500 and 12,000 feet. This steep terrain supports forests of Engelmann fir, corkbark fir, bristlecone pine, various shrubs and herbs, and some white fir, limber pine, and Douglas fir.

Alpine tundra occurs in isolated zones on the tallest peaks in central and north-central New Mexico. This vegetation zone is marked by krummholz communities, dwarf Engelmann spruce and subalpine fir trees, shrubby cinquefoil, cushion plants, and wet alpine hairgrass meadows.

Special Habitat Sites

Aden Lava Flow
Doña Ana County
Located on 4,000 acres, this RNA is centered on a solidified, jagged lava flow that contains rare animal species. Located about 15 miles southwest of Las Cruces, the natural communities consist of rock face and shrub-grass plant assemblages. Shrubs include creosotebush, snakeweed, yucca, saltbush, juniper, black grama, and sideoats grama. Rare or unusual species found here include several bats, a darkly colored rock pocket mouse, a darkly colored black-tailed rattlesnake, and nesting raptors and turkey vultures. The darker pocket mouse and rattlesnake are endemic only to this site because of the extreme isolation of the property.

Bitter Lake Research Natural Area
Chaves County
This 300-acre property encloses Bitter Lake, a shallow basin that dries up to become a playa by late summer. About 14 miles northeast of Roswell, the site becomes an important water-

fowl habitat during winter, especially for nesting sandhill cranes. Natural communities include seepweed, sea lavender, glasswort, and iodinebush, which can tolerate the alkaline conditions. Bulrushes, cattails, reeds, stoneworts, and muskgrasses edge the shallow lake during wintertime. Three rare fish occur in nearby streams on the property: the Pecos gambusia, greenthroat darter, and Pecos pupfish.

Bosque del Apache National Wildlife Refuge
Socorro County
This 7,000-acre riparian marshland along the Rio Grande is 10 miles southeast of Socorro. Natural communities include grassy marsh, lake, lowland forest, meadow, cottonwood–black willow riparian corridor, and desert shrub–steppe. This is a fantastic bird habitat; species include snow geese, Mexican mallards, kestrels, songbirds, sandhill cranes, and rare whooping cranes. More than 300 bird species and 60 mammal species have been identified here.

Bottomless Lakes State Park
Chaves County
Sinkholes, ponds, lakes, and saltcedar-mesquite salt flats occur on this property 12 miles southeast of Roswell. Seven lakes are hydraulically connected via underground caves and sinkholes. Other plants that love the saline soils are bush muhly, fourwing saltbush, snakeweed, creosote, and saltgrasses. Mesquite, snakeweed, and saltbush are scattered across the slightly higher uplands. A variety of wildlife inhabits the park, including the rare barking frog.

Buck Mountain
Lincoln County
Located in Lincoln National Forest, this 727-acre property encloses spruce-fir forests at 10,000–11,000-foot elevations. This rugged, undisturbed terrain contains excellent examples of old-growth corkbark fir–cardamine groundsel forest. These are the largest corkbark fir trees in the United States and are 400–500 years old. Several varieties of mammals and birds, including mountain lions and raptors, depend on this forest.

Cibola National Forest
Catron and Socorro Counties
Elevations in this national forest range from the valley floors to more than 7,000 feet. The interior peaks are fairly pristine and include Apache Kid Wilderness. Vegetation types include juniper-pine woodland, ponderosa pine forest, old-growth Douglas fir–spruce forest, and high meadow. Unique species that inhabit this wilderness are spotted owls, tasseleared squirrels, Gila woodpeckers, elk, deer, mountain lions, bears, eagles, several rare warblers, and harlequin quails.

Colin Neblett Wildlife Area
Colfax County
Just east of Eagle's Nest, this 33,000-acre tract is the largest wildlife management area in New Mexico. Its gorges, riparian corridors, and rugged upland spruce-pine forest is prime habitat for big-game animals, such as elk, bears, mountain lions, and deer. Several rare bird and reptile species can be found here.

Dripping Springs
Doña Ana County
About 5 miles east of Las Cruces, this 2,852-acre Nature Conservancy property protects a number of natural communities in rocky canyons with free-flowing streams. Rare or unusual plant and animal species include velvet ash, Arizona white oak, netleaf hackberry, Organ Mountain evening primrose, red-naped sapsucker, black-chinned sparrow, canyon wrens, and prairie falcons.

Gila National Forest
Grant, Sierra, and Catron Counties
Gila National Forest includes a variety of topographic features and natural communities, such as desert, oak woodland, oak-juniper woodland, pinyon-juniper forest, ponderosa pine forest, spruce-fir forest, subalpine grassland, and riparian corridor. Ponderosa pine dominates between 6,000 and 7,000 feet; some old-growth stands remain. Some of the rare or endangered, threatened, or sensitive species found in this biologically diverse area include the endemic Gila trout, Rio Grande cutthroat trout, Chihuahua chub, roundtail chub, spikedace, Chiricuahua leopard frog, southwestern blackhead snake, Arizona coral snake, mountain lion, and various bats.

Gila Riparian Preserve
Grant County
Owned by The Nature Conservancy, this 7,308-acre preserve protects a stretch of riparian habitat along Gila River. Located about 7

miles east of Cliff, the specialized communities shelter a variety of rare plant and animal species, including the Gila woodpecker, southwest willow flycatcher, loach minnow, Chiricuahua leopard frog, black hawk, Gila trout, Gooding's willow, and spike dace. The property contains the most diverse broadleaf deciduous woodland in the state; it attracts hundreds of different bird species. An impressive one-third of all North American bird species have been sighted here.

Gila River Research Natural Area
Grant County
This 402-acre RNA 12 miles south of Cliff protects a section of a 600-foot-deep canyon in the Big Burros Mountains. It encloses good-quality desert shrub, pinyon-juniper woodland, and a lush riparian corridor that is dominated by walnut, box elder, cottonwood, ash, and sycamore. The RNA was established to protect the riparian corridor as critical habitat for several rare birds: the peregrine falcon and other raptors, Rivoli's hummingbird, yellow-billed cuckoo, Gila woodpecker, green heron, and Montezuma quail. Rare native fish that live in the river are the spikedace, loach minnow, longfin dace, gila coarse-scaled sucker, and Gila mountain sucker. The rare river otter may still survive here.

Gray Ranch
Hidalgo County
Owned by the Animas Foundation, this sprawling, 321,000-acre preserve about 25 miles southeast of Animas includes almost all of the Animas Mountains. Twenty different ecosystems are on the property, ranging from riparian to high-altitude subalpine. This is one of the most diverse natural areas in the United States. More than 25 rare animals and 70 rare or endangered plant species can be found in the mountains, canyons, and flatlands, including the white-sided jackrabbit, Mexican checkerspot, Emory oak, Animas Peak woodsnail, Mexican and Sanborn's long-nosed bats, lowland leopard frog, Baird's sparrow, and pronghorn.

Guadalupe Canyon
Hidalgo County
Located in Coronado National Forest near the Mexico border about 35 miles southwest of Animas, this steep canyon is lined with cottonwoods, sycamores, grassland, pinyon-juniper woodlands, and Arizona white oak forest. The oak trees are an important habitat for the Cous white-tailed deer. More than 160 bird species inhabit the canyon, including the rare and beautiful elegant trogon, fan-tailed warbler, even-eared trogon, flycatchers, hummingbirds, and Strickland's woodpecker. The southwest yellow bat roosts only in the fronds of the yucca. The rare Colorado toad also lives in the canyon.

Haynes Canyon Research Natural Area
Otero County
This pristine mixed conifer forest covers 610 acres in Lincoln National Forest east of Alamogordo. Located at elevations of roughly 8,000 to 9,500 feet, the rugged slopes are covered with open forests of white fir–Rocky Mountain maple, white fir–Gambel oak, or white fir–big-toothed maple. Old-growth patches of these trees are more than 100 feet tall and centuries old. It offers important habitats for mammals and birds, including mountain lions, bears, and raptors.

Inkpot Research Natural Area
Chaves County
Part of Salt Creek Wilderness, this 2-acre RNA is located about 15 miles northeast of Roswell. It was established to protect a sinkhole 90 feet deep with vertical walls. Natural communities include the surrounding snakeweed-creosote shrubland, prickly pear, saltcedar, the sinkhole rock face, and open water. The sinkhole pond supports the rare Pecos gambusia minnow and out-of-place marine algea. Raptors and owls nest on the interior walls of the sinkhole.

Jornada del Muerto Research Natural Area
Socorro County
Part of Bosque del Apache National Wildlife Refuge, this 11,000-acre tract is about 10 miles south of San Antonio. Its natural communities include sand sagebrush–broom pea–dropseed shrubland, mesquite hummock, creosotebush shrubland, and sand dunes. There are small patches of junipers and soapweeds. It is an important habitat for pronghorn, mule deer, golden eagles, and burrowing owls.

Lake Saint Francis Research Natural Area
Chaves County
Located about 12 miles northeast of Roswell, this 700-acre tract protects a series of sinkholes

in the underlying gypsum-rich bedrock. Several still contain water, such as Lake Saint Francis, which is the largest at 200 feet in diameter. The surrounding salt flats are dominated by saltgrass, saltcedar, sea lavender, sea purslane, and centaury. The flanking uplands are covered with grama, dropseed, snakeweed, saltbush, and gyp ringstem. The lake supports the rare and endangered Pecos gambusia minnow, Pecos pupfish, and several algea species. The lakes are an important water source for many desert birds and mammals.

Largo Mesa Research Natural Area
Catron County
Part of Gila National Forest near Quemado, this 300-acre tract preserves a high-quality pinyon–one-seed–juniper woodland that grows on an 8,000-foot-high mesa. Much of the woodland is old-growth with very large specimens (30 feet tall). The understory consists of occasional mountain mahogany and blue grama. The site shows little disturbance and is one of the best pinyon-juniper woodland remnants in the state.

Little Water Canyon
Cibola County
Located near Mount Taylor in Cibola National Forest, this 950-acre RNA about 15 miles northeast of Grants protects old-growth stands of blue spruce, a relatively rare community in the bottom of Little Water Canyon. These trees were never logged due to the inaccessibility of the canyon, and the spruce trees are very large and hundreds of years old. The higher reaches of the canyon support ponderosa pine, Gambel oak, and Douglas fir. The blue spruce forest is an important habitat for the various bird species that inhabit the canyon.

Lower Box
Grant County
This canyon, with its nearly vertical 600-foot-high walls, is 10 miles north of Lordsburg along the Gila River; it shelters 11 endemic species. The lush riparian corridor is lined with mature Arizona sycamores, walnuts, cottonwoods, willows, grasses, and flowers. Some of the trees are 100–200 years old. Surrounding vegetation includes desert upland grassland, grama grass, creosotebush, and mesquite. More than 100 bird species inhabit the canyon, including wintering bald eagles, the southwest willow flycatcher, the rare Bell's vireo, and black hawks. The rocks harbor gila monsters and the rare Mexican garter snake.

Mathers Research Natural Area
Chaves County
Rare relatively undisturbed shinnery oak shrubland is preserved on this 242-acre RNA about 40 miles east of Roswell. Grazing has seriously reduced the amount of shinnery oak vegetation, which once covered vast areas of New Mexico and Texas; only small portions remain, and much of those have been degraded by grazing. The high-quality community here includes 2-foot-high shinnery oak, sagebrush, and soapweed, with patches of grasses—mostly sand bluestem and giant dropseed. It is critical habitat for a variety of species, including the endemic sand dune lizard and the lesser prairie chicken, which nests on the site.

Maxwell Research Natural Area
Colfax County
Being a small wetland in an arid landscape, the Maxwell RNA, about 7 miles northwest of Maxwell, offers critical habitats to a variety of mammals and birds. Its 3,270 acres enclose upland western wheatgrass–galleta–ring muhly–wildrye grassland, hummocks of sacaton and saltgrass, and patches of saltgrass and seepweed. More than 150 nesting or migrating species have been identified here, including the rare sandhill crane, golden eagle, bald eagle, burrowing owl, and several songbirds.

Mimbres River
Grant County
This 160-acre tract along the Mimbres River protects a stretch of delicate riparian habitat. The shoreline is wooded with cottonwood, walnut, and alder. The aquatic community is home to the extremely rare Chihuahua chub and Chiricuahua leopard frog. Special plants include the Arizona walnut tree and New Mexico alder. The property is owned and managed by The Nature Conservancy.

Monument Canyon Research Natural Area
Sandoval County
This 640-acre tract of rugged, forested canyon in the Jemez Mountains is located about 20 miles southwest of Los Alamos. Elevations are about 8,200 feet. Natural communities include high-quality tracts of mature ponderosa

pine–Douglas fir forest and Douglas fir–white fir–white pine–Gambel oak–aspen forest. The canyon's walls are steep rock faces with narrow spires, or hoodoos. Among the several rare or endangered species that live here is the Jemez Mountain salamander, endemic only to this mountain range.

Pajarito Plateau
Los Alamos County
Located 15 miles northwest of Santa Fe, this wilderness is co-owned (private and state). Natural communities include cottonwood-box elder-birch-shrub riparian corridors, grasslands, pinyon-juniper woodland, oak forest, ponderosa pine forest, and mixed alpine conifer forest. This area is a critical habitat for several rare songbirds, the tassel-eared Albert's squirrel, elk, bears, and mountain lions. The high montane grasslands are flowered with abundant native species.

Pecos Wilderness Area
Santa Fe County
Part of Santa Fe National Forest, this 240,000-acre area is 20 miles northeast of Santa Fe. Its rugged canyons and steep mountains are heavily forested with spruce and pine and ancient stands of 1,000-year-old bristlecone pines. Its forests, lakes, and meadows are critical habitats for large mammals, such as bears and mountain lions, as well as various raptors.

Rattlesnake Springs
Eddy County
This 13-acre Nature Conservancy property about 26 miles south of Carlsbad protects an oasislike habitat in a desert environment. The springs and wetlands provide habitat for the rare Bell's vireo and green herons; buntings also frequent the site. Other rare species include the plain-bellied water snake, Pecos western ribbon snake, eastern barking frog, Blanchard's cricket frog, nesting turkey vultures, and the only known population of the very rare Texas emperor butterfly.

Rio Grande Marsh Research Natural Area
Socorro County
This 97-acre wetland complex along Rio Grande is part of the Bosque del Apache Refuge, about 10 miles south of San Antonio. Natural communities include tule marsh, oxbow lake, willow-saltcedar lake edge, and broom pea–four-winged saltbush–mesquite–juniper shrubland. The wetland is a critical habitat for nesting birds as well as thousands of migrating birds during spring and fall. More than 295 species have been identified here, including sandhill cranes and whooping cranes. Several varieties of amphibians, reptiles, and mammals, some of them rare or unusual, are attracted to the water.

Rio Nutria
McKinley County
This 1,200-acre tract protects a section of Rio Nutria that flows through a rugged sandstone canyon. It is rare as an undisturbed riparian ecosystem that shelters the endangered Zuni bluehead sucker. Mature narrowleaf cottonwood and aspen trees shade the river in the canyon.

Vegosa Research Natural Area
San Miguel County
Part of Las Vegas National Wildlife Refuge, this 537-acre site is about 3 miles northeast of Las Vegas, New Mexico. Its dominant natural communities consist of high-quality remnants of grassland and juniper-oak savanna. A few stands of ponderosa pine also exist. It is a critical habitat for a variety of bird and mammal species, including mule deer, bobcats, mountain lions, shrews, and numerous raptors, including golden and bald eagles.

NEW YORK

The major natural communities in New York are categorized into the following systems: marine, estuarine, riverine, lacustrine, palustrine, and terrestrial.

New York's marine system grades from the marine subtidal communities (such as eelgrass meadow and other submerged aquatics) through intertidal mudflat, rocky coast, and gravel-sand beach. This system is generally fairly narrow and parallels the Atlantic coast.

The estuarine system also has subtidal and intertidal components. Moving inland, the communities grade from tidal stream to salt shrub, salt marsh, coastal salt pond to brackish tidal marsh, mudflat, and marsh to freshwater swamp, mudflat, and marsh. These communities hug the coastline.

Riverine systems are flowing water or streams. Different parts of a riverine system include rocky or marsh headwaters, main channel, backwater slough, intermittent stream, and Coastal Plain stream. Riverine systems occur throughout the state and tend to have well-defined channels.

Standing bodies of water are representative of the lacustrine system. These vary tremendously in size, from the Great Lakes to small ponds on the Coastal Plain. Associated natural communities include deepwater and shallow shoal vegetation, oligotrophic or mesotrophic lake, inland salt pond, oxbow lake, and eutrophic pond.

Different kinds of wetlands can be found throughout New York and vary according to climate and substrate chemistry. Open wetlands consist of marshes, shrub swamp, wet meadows, lakeshore vegetation, sinkhole wetland, interdunal wetland, peatland, fens, and bog. Forested wetlands are typically floodplain forest, red maple swamp, white oak swamp, spruce-fir swamp, or mixed conifer-hardwood swamp.

The terrestrial system is the largest and is subdivided into open upland, barren and woodland, and upland forest. These different terrestrial communities cover a large section of the state. Examples of open upland include beach, dune, shrubland, heathland, grassland, alpine meadow, and various cliff and talus plant assemblages. Barrens and woodlands have some tree cover—typically less than 60 percent—creating large openings, called barrens, or a woodland with widely spaced trees. Examples of these include serpentinite and pitch pine–scrub oak barren, oak opening, pitch pine–oak heath woodland, glade, and rocky summit. Depending on the elevation, climate, and precipitation, forested uplands range from maritime oak–holly or red cedar to mixed pine-hardwood to mesic beech-maple, northern hemlock–hardwood, mountain spruce–fir, mountain fir, and subalpine forests.

Special Habitat Sites

Accabonac Harbor Preserve and Merrill Lake Sanctuary
Suffolk County
This 230-acre Nature Conservancy preserve in Springs Township protects a well-developed salt marsh along Accabonac Harbor. Part of an estuary, the preserve is an important habitat for a variety of wildlife and aquatic life. Affected daily by the tides, the vegetation progresses from cordgrass to salt marsh hay to beach plum and bayberry. Rarer species are silver weed and bushy rockrose. Thousands of birds nest in the preserve or visit during annual migrations; rare or endangered species include the piping plover, least tern, and black-crowned night heron.

Albany Pine Bush
Albany County
This 2,200-acre Nature Conservancy preserve encloses a globally rare scrub oak–pitch pine barren ecosystem. It has developed on a sand plain; other communities are scrub oak forest and heathlands (blueberry–huckleberry–wild blue lupine). Rare species include the Karner blue butterfly, Jefferson salamander, spadefoot toad, hognose snake, and inland barrens buckmoth. The Karner blue butterfly caterpillar feeds only on lupine, which is abundant on the preserve.

Atlantic Double Dunes
Suffolk County
Located in Amagansett Township, this 200-acre Nature Conservancy property protects double dunes and the freshwater swales between them. Natural communities include swampy pool, cranberry bog, ocean beach, and vegetated sand dune. Specialized plants, such as American beach grass and beach pea, grow on the windswept dunes. Vegetation in the swales is relatively diverse and includes sedges, sphagnum moss, reed grass, and rare orchids. The extremely rare Fowler's toad can be found in the swales.

Barberville Falls
Rensselaer County
This 119-acre Nature Conservancy preserve about 8 miles east of Troy protects a section of riparian habitat along pristine Poesten Kill. Natural communities include rock face, Poesten Kill, waterfall, and mixed upland hemlock-maple-birch-ash forest. Rare lichens, mosses, and ferns can be found among the rocks. The forest is an important nesting habitat for numerous interior forest birds. The wildflowers are impressive during springtime.

Barcelona Neck
Suffolk County
Bounded by Little Northwest Creek, Sag Harbor Bay, and Northwest Creek, this 341-acre peninsula is owned by The Nature Conservancy. Natural communities include pine-oak upland forest, red maple–hardwood lowland forest, wetlands, and salt marsh. Although the upland forest is less than 100 years old, the preserve includes several plant and animal species of special concern: piping plover, least tern, osprey, southern leopard frog, diamondback terrapin, wild pink, beach knotweed, and bushy rockrose. Rocky bluffs along the shore are host to mosses, lichens, and ferns. The salt marsh is an important spawning ground for shellfish.

Bear Swamp
Albany County
Located about 20 miles southwest of Albany, this 310-acre preserve is owned and managed by The Nature Conservancy. The wetland-peatland complex has been developing for at least 13,000 years and has formed a peat island. Among the rare or unusual plants found here are large relict flowering rhododendron trees. The wetland is an important habitat for various birds, amphibians, and reptiles. The rhododendrons usually bloom in early July. The surrounding hemlock-oak forests have some impressive old-growth specimens that are hundreds of years old.

Black Pond Wildlife Management Area
Jefferson County
This property on the eastern shore of Lake Ontario protects a series of rare communities: beach, sand dune, interdunal swale, backdune, and inland wetland or forest. These dunes are the second highest in the northeastern United States and protect wetlands that have developed behind them. Certain plants that grow in the windy and sandy conditions anchor the dunes and prevent their erosion; these include American beach grass, the rare dune willow, dune grape, poison ivy, and cottonwood saplings. The higher sand ridges support red oak and red maple; the backside of the dunes are anchored by dense alder thickets. Numerous birds reside here or visit in large numbers during spring and fall migrations. Several rare or endangered plant, animal, and bird species occur here.

Butler Memorial Sanctuary
Westshester County
These former grazing lands and woodlots were heavily used during the late 1800s. Communities include upland deciduous forest and red maple–hardwood swamp, which are critical habitats for the endemic Broad Brook salamander and the spotted salamander. The Nature Conservancy owns and manages the property.

Caswell Cliffs
Suffolk County
Owned by the Nature Conservancy, this 36-acre property encloses a quarter-mile stretch of ocean beach with steep, sandy bluffs. Other natural communities are peat bog, maritime grassland, and maple-hardwood swamp. Rare animals, such as the blue-spotted salamander, spotted turtle, and eastern hognose snake, inhabit the wetlands. Rare species of sedge can be found within a remnant of maritime grassland.

Chaumont Barrens
Jefferson County
This unique Nature Conservancy property northwest of Watertown near Chaumont covers an alvar barrens landscape. Alvar grasslands are characterized by shallow soil and alternating floods and drought, creating a harsh growing climate. Natural communities, some of them rare, include coniferous limestone woodland, alvar grassland, rubble slope, rock crack and cliff, calcareous pavement barren, shrub savanna, and deciduous limestone forest. Rare or unique plant and animal species include the prairie warbler, prairie smoke, reindeer lichen, certain mosses, and lichens. The woodlands are an important nesting habitat for interior forest birds.

Clintonville Pine Barrens Preserve
Clinton County
Owned by The Nature Conservancy, this 900-acre property northeast of Ausable Forks encloses an excellent remnant of relatively undisturbed pitch pine–heath barrens (blueberry–huckleberry–sheep laurel) forest. A community that depends on fire for propagation, the pitch pine forest is a dwindling habitat due to modern-day fire suppression. The barrens is flanked by oak-maple-beech hardwood forest. Other groundcover plants are the sweet fern, bearberry, pipsissewa, and wintergreen. Four

rare or endangered species, including the prairie redroot and two species of noctuid moths, can be found here.

Coon Mountain Preserve
Essex County
Located a few miles northeast of Westport, this 246-acre Nature Conservancy property encloses a rugged portion of Champlain Valley. Natural communities include rocky cliff face, rocky upland oak-hickory forest, hemlock forest, shady deep ravine, and vernal pool. Rare or unusual species here include black bears, wood frogs, certain salamanders, and raptors. The woodlands are an important nesting habitat for interior forest birds such as the pileated woodpecker. The trees in this old-growth climax forest are hundreds of years old.

Cranberry Bog County Preserve
Suffolk County
This 165-acre tract southwest of Riverhead is part of the Peconic River corridor. Natural communities include pine-oak barren, pond, Atlantic white cedar swamp, freshwater marsh, bog, and bog heath. A host of rare plants and insects are found here, including leatherleaf, swamp honeysuckle, a variety of orchids, several caddis flies, and the local bog copper moth, which thrives on cranberries. Arctic relict plants in the barrens are reindeer lichen, beach heather, and bearberry. Rare bog plants include water hornrush, spike rush, shining whip grass, pitcher plant, round-leaved sundew, and bladderwort. More than 100 bird species have been recorded on the property.

Deer Creek Wildlife Management Area
Oswego County
This property on the eastern shore of Lake Ontario protects delicate and rare beach habitats. Natural communities include beach zone, sand dune, interdunal swale, backdune, and inland wetland or forest. The sand dunes here are some of the highest in the eastern United States and protect the nearby wetlands from the harsh effects of the lake climate. Up to 300 different plant species help anchor the sandy substrate, including the rare dune willow, American beach grass, poison ivy, dune grape, and cottonwoods. The more landward parts of the dunes are covered by scrubby red oak–red maple trees and alder thickets. Numerous birds nest here; many others rest and feed here during spring and summer migrations.

Eastern Lake Ontario Dunes and Wetlands
Jefferson and Oswego Counties
This ongoing Nature Conservancy project, near Richland and Ellisburg, is protecting a section of New York's only freshwater barrier beach ecosystem. Natural communities include freshwater beach, dune-and-swale complex, marsh, fen, and swamp. Rare or unique species found here include the black tern, northern harrier, bog buckmoth, dragon's mouth orchids, sand cherry, and several rare migrating birds.

El Dorado Nature Preserve
Jefferson County
Located about 20 miles west of Adams along Lake Ontario northwest of Woodville, this 360-acre property is a critical bird habitat. Shoreline, beach, sand dune anchored by beach grass, rare dune willow, as well as sand cherry, wetland, and red cedar forest communities offer sanctuary to thousands of migrating birds every year. Some of the more rare species include terns, yellowlegs, warblers, bitterns, northern harriers, herons, and the occasional bald eagle. Hordes of monarch butterflies gather here during their southward migration during fall. The property is owned and maintained by The Nature Conservancy.

Emmons Pond Bog
Delaware County
This 150-acre Nature Conservancy property about 4 miles southeast of Oneonta completely encloses an undisturbed 7-acre open water acidic bog and its surrounding swamp and shrub forest. The bog is at least 11,000 years old and has a floating sphagnum moss mat. Healthy populations of typical northern boreal plants can be found here, including pitcher plant, buckbean, cottongrass, sheep laurel, and leatherleaf. The surrounding forest is mature to old-growth, with large specimens of oak and pine. The bog environment supports a variety of animal, amphibian, and bird species, some of them rare or endangered.

Everton Falls Preserve
Franklin County
Located about 7 miles east of Saint Regis Falls, this Nature Conservancy property encloses red spruce–balsam fir–white pine coniferous forest, mixed hardwood forest, northern white cedar swamp, tamarack bog, mountain meadow, rock exposure, the East Branch of Saint Regis

River and its riparian corridor, and waterfall communities. The mature trees are nearly 100 years old; some specimens may be older. Many interior forest birds, such as the rare pileated woodpecker, inhabit the site. During spring an abundance of colorful wildflowers, including Canada mayflower and wood-sorrel, adorn the forest floor. Several ferns, some of them rare or unique, can also be found.

Five Ponds Wilderness Area
Saint Lawrence County
Located on 95,525 acres about a mile south of Wanakena, some of this rugged tract consists of virgin white pine–hemlock–red spruce forest. These trees are 300 to 400 years old and nearly 150 feet tall. This deep wilderness is a prime habitat for large mammals, such as bears, deer, and lynx, and resembles presettlement forest.

Four Brothers Islands
Essex County
Located in Lake Champlain, this 17-acre Nature Conservancy property consists of four small islands. They are important because more than 15,000 pairs of herons, gulls, and cormorants nest here every year.

French Creek Preserve
Chautauqua County
This 90-acre Nature Conservancy property protects part of French Creek watershed in southwestern New York. Natural communities include oak-maple-hemlock upland forest, bottomland forest, and marsh in addition to French Creek and its riparian corridor. This highly diverse watershed supports a variety of plants, birds, mammals, reptiles, insects, and aquatic life. Many of these are rare or endangered, such as the bald eagle, northern riffleshell mussel, and clubshell mussel. The mature upland forest is an important habitat for nesting interior forest birds such as woodpeckers.

Fresh Pond Preserve
Suffolk County
Located in Amagansett Township, this 96-acre parcel protects a wetland complex. Its natural communities include stream and pond, bottomland red maple forest, and upland oak-hickory forest. The moist conditions support a variety of different ferns and the state-endangered yellow fringed orchid. This is also one of the few areas in New York where the eastern spadefoot toad can be found. The preserve is owned and managed by The Nature Conservancy.

Freund Wildlife Sanctuary
Columbia County
This property is located about 30 miles southeast of Albany. Although heavily used in the past, the 57-acre tract contains a maturing oak-pine-hickory forest as well as old fields. Some white pines and oaks that escaped lumbering are hundreds of years old. Plants in the understory include dogwood, spicebush, witch hazel, various ferns, and the unusual clubmoss. Several small streams cross the property. It is an important habitat for numerous bird species, including the rare blue-winged and golden-winged warblers. The property is owned and managed by The Nature Conservancy.

Gadway Sandstone Pavement Barrens
Clinton County
Located near Mooers Forks, close to the Canadian border, this 335-acre preserve encloses a globally rare sandstone pavement barrens community; less than 20 examples occur worldwide. The flat-lying rock beds and thin, rocky soil support a stunted jack pine forest with an understory of blueberry-huckleberry. Several lichen and moss species cling to the exposed rock outcrops. Trees are more than 100 years old, but typically they are less than 5 feet tall. Several rare groundcover plants also grow here.

Greenfield Center Wild Area
Saratoga County
This 27-acre property is about 5 miles north of Saratoga Springs. Owned by The Nature Conservancy, it encloses a rare spring-fed swamp. Some of the plants found here are ferns, watercress, wood lily, foamflower, red trillium, pink lady's slipper, and starry Solomon's seal. It is also an important habitat for amphibians, reptiles, forest birds, and wetland birds.

Hands Creek
Suffolk County
This 32-acre Nature Conservancy site protects intertidal salt marsh and shoreline adjacent to Three Mile Harbor. Dense tree communities include alder thickets and patches of oak. A variety of sedges grows in the marsh, including

a rare spike rush. The property also includes important Native American habitation sites.

Hannacroix Ravine Preserve

Albany County

This 440-acre tract is about 2 miles from Clarksville. It contains Hannacroix Creek, a deep gorge, rocky cliffs and ledges, mature beech–birch–oak–maple–red pine–hemlock upland forest, and seeps. Many old-growth trees in the steep ravine along the creek escaped logging activities and are quite large. The forest floor and creek bed boast colorful displays of wildflowers during spring; species include columbine, bloodroot, and Dutchman's breeches. Ferns, mosses, and lichens can be found along the shady, moist rock edges. Several rare or endangered species are sheltered here. The property is owned and managed by The Nature Conservancy.

Harriman–Bear Mountain

Orange and Rockland Counties

Located about 15 miles southwest of West Point, this 50,000-acre tract contains high-quality mature second-growth forest. Wetlands include more than 20 lakes, brooks, swamps, bogs, and marshes. More than 300 bird species have been identified here, including 100 species that nest here regularly. Rare or unusual birds include the pileated woodpecker and a variety of warblers.

Hempstead Plains

Nassau County

This 19-acre preserve on eastern Long Island north of Uniondale is the only surviving remnant of what was once a 60,000-acre grassland. Located amid urban development, this piece of prairie has more than 250 different plant species, including the very rare sandplain gerardia (11 known sites worldwide). A variety of rare insects also inhabits the site. Two uncommon prairie birds—the upland sandpiper and grasshopper sparrow—have also been observed on the site. The property is owned and managed by The Nature Conservancy; prescribed burning is being used to restore the prairie to its original condition.

Hither Hills State Park

Suffolk County

The 1,755-acre state park preserves Goff Point Peninsula, between Napeague Bay and Napeague Harbor, and its associated wetlands and dunes. Natural communities are oak–pitch pine scrub forest, beach, dune, cranberry bog, and beach plume–bearberry–pitch pine shrub forest. Sandy meadows are anchored by beardgrass and switchgrass. Rare beach flowers include the beach knotweed and lyster-catcher. Wildflowers bloom in the grassy areas during summer. A variety of orchids, some rare, grows in the bog.

Hither Woods/Koppleman Nature Preserve

Suffolk County

Located on 1,340 acres in Montauk, this nature preserve encloses bluff, pond, pine barren, dune heath, dune, and maritime grassland natural communities. Parts of the pine barrens remain undisturbed and reflect the oak-holly understory from presettlement times. The rare pine barrens sandwort can be found here, as can various sedges and ferns. A remnant of a once-vast hair grass–bluestem–Indian grass maritime grassland still exists. Rare or state-threatened birds include the red-shouldered hawk and northern harrier. The preserve is carefully managed using prescribed burns.

Hudson River Ice Meadows

Warren County

Located along the Hudson River northwest of Warrensburg, this ice meadow is a globally rare community created by the buildup of ice during the winter and its scouring action on the floodplain. Eight rare plants grow here.

Iroquois National Wildlife Refuge

Orleans and Genessee Counties

Located about 15 miles south of Medina, this 20,000-acre property supports an abundance of wildlife. Natural communities include wetland, marsh, upland forest, and wet and dry meadows. Hosting an important blue heron rookery, the property also offers excellent habitat for songbirds, such as nesting prothonotary warblers, scarlet tanagers, and vireos. Ospreys and waterfowl frequent the marshes.

Kenrose Preserve

Albany County

Located 22 miles west of Albany, this Nature Conservancy property preserves high-quality mixed pine-hemlock-maple-beech-oak upland forest, streams, and pine forest. Some of the oak and beech trees are old-growth, probably more than 200 years old. The property is an important habitat for a variety of northern flora and fauna, including some rare species.

Labrador Hollow State Nature Preserve
Cortland County
Located about 20 miles north of Cortland, this property encloses a steep-sided rocky valley forested with northern hardwoods. Shaded areas support a subalpine plant community that is more typical of high mountain bogs. Unique plants include Labrador tea, sundews, pitcher plants, and other bog species. Some of the hardwoods are quite old. The forest and wetlands are an important nesting and migratory habitat for a variety of birds, especially raptors.

Lake Maratanza
Ulster County
Located north of Ellenville, this lake and surrounding vegetation is on a flat-lying, windswept ridge of sandstone and conglomerate. The lake is enclosed by an unusual pygmy forest called dwarf pine plains, globally a very rare community. Pitch pines trees are no taller than 5 feet (they typically reach 35 feet) because of the strong winds and thin topsoil. Several other rare plant species can be found here.

Lakeview Marsh Wildlife Management Area
Jefferson County
This property is located along the eastern shoreline of Lake Ontario and protects a number of rare communities: beach, sand dune, interdunal swale, backdune, and wetland or lowland forest. This dune complex is one of the highest in the eastern United States and protects the inland vegetation and wetlands from the harsh effects of the lake. The dunes are anchored by plants that tolerate the windy and sandy conditions of the dunes; these include American beach grass, cottonwood, the rare dune willow, poison ivy, and dune grape. The higher parts of the dunes support scrubby red oak–red maple forests; the backdune is anchored by impenetrable alder thickets. More than 300 plant species, some of them rare or endangered, are found here. Numerous birds nest in this environment or visit to rest and feed during spring and fall migrations.

Letchworth State Park
Wyoming and Livingston Counties
Located about 5 miles southeast of Perry along the Genesee River, this 14,300-acre property contains waterfalls, canyons, gorges, cliff faces, and other shaded microclimates. The steep topography has saved some forest stands from the lumberman's axe; these include 400-year-old hemlocks and 600-year-old eastern red cedars with breast-height diameters reaching 4 feet. Nearly 30 varieties of nesting warblers can be found here.

Limestone Rise Preserve
Albany County
This 61-acre site protects a variety of habitats about 20 miles west of Albany. Natural communities include limestone cliff, rocky ledges, deep fissure, pond, wetland, young red maple–aspen–white pine lowland forest, and older hemlock–white pine–maple upland forest. The rock exposures host a variety of clinging vegetation, such as ferns, mosses, and lichens, a few of them rare. The woodlands are an important nesting habitat for numerous birds. Talus at the bottom of the rock ledges often contains marine fossils from about 500 million years ago. The preserve is owned and managed by The Nature Conservancy.

Lishka Kill
Schenectady County
Located on 108 acres about 16 miles northwest of Albany, this property encloses forested ravines, hills, and rocky exposures. Natural communities include flowing stream, intermittent wetland, mature hemlock–white pine–oak upland northern forest, and young birch-aspen-sumac forest. The understory includes a variety of majestic ferns, including the ostrich fern and Christmas fern. Other colorful plants and wildflowers are the selfheal, fly honeysuckle, white wood aster, meadow rue, and skunk cabbage. The woodlands are an important nesting habitat for a variety of birds, including woodpeckers. The property is owned and managed by The Nature Conservancy.

Livingston Marsh Sanctuary
Greene County
Located about 20 miles south of Albany on Hudson River, this Audubon Society property encloses 140 acres of tidal wetland. Natural communities include open water, marsh, bottomland swamp, and wet meadow. It is an important bird habitat; nesting species include the rare wood duck, black ducks, and other waterbirds. Ospreys hunt along the river.

Loines Preserve
Warren County
This 31-acre Nature Conservancy property along Lake George is about 15 miles north of Lake George Village. Natural communities include mature beech-oak upland forest, white-pine-ash lowland forest, rocky ledges, wetlands, and a half-mile stretch of undisturbed lakefront. Some of the trees are quite large. Rare plant or animal species seen here include various mosses and lichens, Missouri rock cress, timber rattlesnakes, and raptors, such as ospreys and eagles.

Long Pond Greenbelt
Suffolk County
Located on 459 acres in the town of Bridgehampton, this Nature Conservancy project is a unique wild area on Long Island. The preserve encloses a series of Coastal Plain ponds and their associated wetlands. The incredible plant diversity includes more than 30 globally rare species, such as the drowned horned rush, rosy tickseed, and short-beaked baldrush. Carniverous sundew and bladderwort plants grow in the ponds. The rapidly disappearing tiger salamander has also been seen here. It is a nesting habitat for a variety of birds. Located on the Atlantic flyway, it is also an important resting place for thousands of migrating birds during spring and fall.

Lordsland Conservancy
Otsego County
Located about 55 miles west of Albany, this Nature Conservancy property contains forested ridges, meadows, wetlands, streams, and beaver ponds. Older-growth maple-beech-hemlock forest occurs on the steeper ridges. Several rare plants can be found here, including the globally rare Jacob's ladder. The woodlands are also a prime habitat for a variety of birds, especially during spring and fall migrations.

Malloryville Fen
Tompkins County
Located near Malloryville, this 310-acre property encloses a diverse fen-bog complex. Its lush habitats include rich sloping fen, shrub fen, silver maple–ash swamp, and tamarack bog. Rare or unusual plants that grow here include a variety of orchids and the spreading globeflower. Several varieties of birds, some of them rare or diminishing, nest in the area. The property is owned and managed by The Nature Conservancy.

Mashomack Preserve
Suffolk County
This 2,039-acre Nature Conservancy preserve covers the southeastern part of Shelter Island. Natural communities on the site include tidal creek, ocean coastline, bottomland red maple–pine forest and swamp, upland oak-beech forest, and salt marsh. The vegetation in the pine swamp is rooted in a layer of sphagnum moss that covers the water; this is a rare habitat in New York. The swamp is about 4,000 years old and relatively undisturbed by humans. Hummingbirds use a special lichen that grows in the swamp to construct their nests. A variety of birds inhabits the preserve, including interior forest birds and ospreys. It is also an important resting and feeding site for thousands of waterbirds that migrate during spring and fall.

Mecox Bay and Dunes Preserve
Suffolk County
This 15-acre parcel protects important shoreline, beach, sand dune, and salt marsh communities at Mecox Bay in Water Mill. The dunes are stabilized by such vegetation as American beach grass, dusty miller, beach pea, seaside spurge, and seaside goldenrod. On the landward side of the dunes beach plum and lichen species begin to take hold. The salt marsh is dominated by salt hay and cordgrass. This area is an important feeding and nesting site for thousands of waterbirds. Rare or threatened species include the least tern, roseat tern, and osprey. The property is owned and maintained by The Nature Conservancy.

Mianus River Gorge Preserve
Westchester County
This Nature Conservancy property is located east of Bedford Village. Natural communities include deep shaded gorge, upland forest, lowland forest, wetland, running stream, and glade. The cliff faces are adorned by mosses and lichens, some of them rare or unusual. A variety of ferns grows in the glades. Rare species include Fowler's toads, green frogs, and eastern painted turtles. Stands of old-growth oak, beech, and hemlock are 300–400 years old.

Moccasin Kill Sanctuary
Schenectady County
This 81-acre site about 25 miles northwest of Albany protects an excellent example of second-growth chestnut oak–red oak–white oak–hickory–maple–beech forest. Other communities are small stream, waterfall, mossy rock ledge, and intermittent wetland. Many different plants, animals, birds, and reptiles thrive here, including some rare or unusual species. A host of wildflowers, as well as flowering dogwood and azalea in the understory, provide dazzling color during spring. The property is owned and managed by The Nature Conservancy.

Montauk Mountain
Suffolk County
This 4-acre property protects one of the most outstanding remnants of maritime grassland and maritime heathland in the state. It has rapidly vanished because of its attractiveness to developers. In addition to protecting these ecosystems, the property contains the rare New England blazing star, bushy rockrose, and Nantucket shadbush. Other important plants that anchor these sandy areas are bearberry, wavy hairgrass, and little bluestem. The property is owned and managed by The Nature Conservancy.

Montezuma National Wildlife Refuge
Seneca County
This 7,000-acre tract of wetlands is located northeast of Seneca Falls along the Atlantic Flyway. It is a critical stopover and feeding ground for more than 300 species of birds during spring and fall migrations. Birds include Canada geese, mallards, snow geese, ospreys, hawks, and bald eagles.

Montezuma Wetlands Complex
Wayne, Cayuga, and Seneca Counties
This 36,000-acre wetland complex is the focus of a larger effort to restore the wetland to its original size. Located about 10 miles southeast of Clyde, this property is critical habitat for many nesting birds; it is also an important stopover during spring and fall migrations (especially Canada geese, snow geese, ducks, tundra swans, and wood ducks). More than 300 different species of birds have been identified here. Other important communities include open pool, marsh, dry marsh, mudflat, and bottomland hardwood forest (red maple–swamp white oak). The rare bald eagle and the great blue heron nest in the wetland.

Moonbeams Sanctuary
Orange County
Located on 150 acres near the town of Wallkill, this property contains red maple–black gum swamps, hemlock–chestnut oak forest, red oak–white pine upland forest, and upland meadows. The wetlands are an important habitat for a variety of birds and dragonflies. Black bears and rare bald eagles have also been sighted. The property is owned and maintained by The Nature Conservancy.

Morgenthau Preserve
Westchester County
Located near Pound Ridge, this 30-acre tract along Blue Heron Lake includes undeveloped lake shoreline, an island, vernal ponds, wetlands, lowland forest, and streams. Some of the white pine trees are hundreds of years old and up to 3 feet in diameter at breast height. The property is an important habitat for waterbirds and interior forest birds; the pileated woodpecker nests on the site. The preserve is owned by The Nature Conservancy.

Napeague Cranberry Bog/Promised Land
Suffolk County
Located on 45 acres, this Nature Conservancy tract protects the following natural communities, which are rapidly disappearing because of development: pine-oak scrub forest, bog, wetland, dune, and beach. The acidic bog contains a number of rare or unusual plants, such as the carnivorous sundew, cranberry, orchids, and sedges. One of the rarest plants in the northeastern United States, the crested fringed orchid thrives here.

Nellie Hill Preserve
Dutchess County
Located near Dover Plains, this 144-acre tract encloses a rock cliff capped by a rare red cedar community. Other communities are bottomland and upland forest, spring, wetland, and meadow. At least ten rare plants occur on the property, including a globally rare mint. Raptors can be seen gliding on the thermals along the cliff face. The property is owned by The Nature Conservancy.

Neversink Preserve
Orange County
This 205-acre Nature Conservancy tract near Godeffroy encloses a pristine stretch of

Neversink River. Communities include mature floodplain forest, upland forest, wetland, and riparian habitat along the river. A tremendous variety of rare, threatened, or endangered species has been found here, including the globally rare dwarf wedgemussel, bald eagles, osprey, certain dragonflies, and bat species, including the globally rare small-footed bat. A variety of wildflowers and ferns adorns the riparian communities along the river.

Otter Creek Preserve
Westchester County
Located near Mamaroneck, this 27-acre Nature Conservancy property encloses a tidal creek, tidal marsh, and lowland forest in the Lower Hudson region. Otter Creek empties into Long Island Sound. In addition to being an important breeding habitat for fish and other aquatic organisms, the wetlands are critical nesting and migratory habitats for thousands of waterbirds. The rare purple gallinule and osprey have been sighted here.

Poultney River
Washington County
This 281-acre Nature Conservancy tract protects a pristine stretch of Poultney River a few miles northwest of Whitehall. Natural communities include the river and its riparian corridor, extensive wetlands, cliff face, and mixed hardwood upland forest. Several rare plant and animal species occur here, including the endangered eastern snail darter, logperch, channel darter, silver plamprey, blackchin darter, black pondshell mollusk, pink heelsplitter mollusk, and rock-clinging orchids. The wetland is an important habitat for nesting birds and thousands of migrating birds during spring and summer.

Ram Isthmus
Suffolk County
Surrounded by encroaching development, this 7-acre section of Ram Island Peninsula in Coecles Harbor protects a wetland-dune complex. Delicate natural communities here are harbor beach, sand dune, and salt marsh. Some uncommon plants grow in the wetlands and swales. It is an important bird habitat (including nesting ospreys). The property is owned and maintained by The Nature Conservancy.

Rome Sand Plains
Oneida County
This rare inland pine barren ecosystem is preserved on this Nature Conservancy property west of Rome. Community diversity is high and includes northern bog, pitch pine barren forest, mixed northern hardwood (pine-hemlock-oak) forest, and Wood Creek and its riparian corridor. Rare or unusual plants include the pitch pine, blue lupine, certain ferns, grass pink orchid, and the pitcher plant; rare or endangered fauna include the pine warbler and Nashville warbler, red-shouldered hawk, buckmoth, and spotted turtle.

Rush Oak Openings
Monroe County
This ongoing preservation effort among The Nature Conservancy and several other organizations is protecting oak openings, or savanna, a very rare natural community in New York. Mature chinquapin oak trees form groves amid a lush prairie of various grasses, sedges, and wildflowers. Special species than can be found here include Indian grass, walking fern, Nebraska sedge, eastern bluebirds, and Jefferson's salamander.

Sagg Swamp Preserve
Suffolk County
Owned by The Nature Conservancy, this 90-acre tract encloses a major wetland complex on Long Island. Natural communities include upland oak-pepperbush forest, pond, stream, bog, and swamp. The swamp has been virtually undisturbed by development and contains stands of Atlantic white cedar, a rare occurrence along the Eastern Seaboard. More than 200 plant species, some rare or unusual, have been identified here. The wetlands are an important nesting site for a number of birds. The wildflowers are colorful during spring and summer.

Sandy Pond Beach Natural Area
Oswego County
Located along Lake Ontario, this property about 10 miles west of Lacona protects sand flats, wetlands, sand beach, and sand dunes. These rare communities host a variety of rare or unusual plants and animals, including the dune willow. Owned and managed by The Nature Conservancy, the preserve is a critical nesting habitat and migratory stopover for waterbirds.

Scallop Pond Preserve
Suffolk County
This 35-acre Nature Conservancy tract protects part of the Sebonac Creek estuary, one of the last undisturbed wetlands on Long Island. The salt marsh estuary supports a large population of fish and shellfish. It is an important breeding habitat and migratory stopover for thousands of birds. Rare birds that breed here are the piping plover and least tern. Nearly 150 species have been identified, including the black-crowned night heron and snowy egret. Marsh vegetation consists of salt marsh cordgrass, saltgrass, and salt hay. The pond is edged by oak forest. The greatest threats to this ecosystem are toxic runoff and sedimentation from construction and development.

Scotia Island Preserve
Schenectady County
This tiny Nature Conservancy preserve is a small island in Collins Lake near Scotia. Natural communities include woodland forest and marsh. The island is an important nesting site for the imperiled great blue heron.

Shawangunk Mountains
Ulster County
This impressive ridge and its vertical cliff faces are popular for hang gliding and rock climbing. The ridge itself runs roughly from Port Jervis to New Paltz. Deep cracks in the sandstone and conglomerate outcrops preserve ice well into the summer months. Natural communities include dwarf pitch pine–huckleberry–gray birch–wild raisin ridge top forest, pitch pine–blueberry peat swamp, and a pitch pine–oak–heath rocky summit assemblage. All of these are relatively rare communities. One of the world's oldest stands of pitch pine is estimated to be 300 years old. Bats hibernate in the many cracks, fissures, and old mine sites along the ridge. Located along the flyway, the steep mountain scarps attract many different birds, including raptors. Rare or unusual northern relict plants can be found in the colder microclimates near the ice-filled fissures. The Nature Conservancy has been involved in purchasing tracts or negotiating easements with property owners along the ridge. These lands are now part of the Virginia Viney Smiley Preserve and Minnewaska State Park west of New Paltz.

Silver Lake Bog
Clinton County
This 61-acre property west of Ausable Forks is owned and managed by The Nature Conservancy. Natural communities include prisine shoreline along Silver Lake, pine upland forest, hemlock-hardwood forest, white cedar swamp, and black spruce–tamarack bog. Typical bog vegetation consists of ferns, cedar, sphagnum moss, pitcher plants, and several rare species. Clubmoss grows on the rocky slopes. Some of the larger hardwood trees compose stands of old-growth forest. The woodland is critical habitat for a number of birds, including pileated woodpeckers, crossbills, and various warblers.

Spring Bog Preserve
Franklin County
Located about 10 miles west of Tupper Lake, this 4,200-acre Nature Conservancy site protects a large peatland-wetland complex. Natural communities include upland northern hardwood maple-beech-cherry forest, black spruce–tamarack swamp, lowland pine–spruce–fir forest, acidic sphagnum bog, fen, and sedge marsh. The peatlands show a well-defined pattern of ridges and depressions (strings and flarks, respectively). More than 130 species of wetland and forest birds inhabit the preserve. Rare or endangered species found here include pitcher plant, several other rare plants, the black-backed woodpecker, owls, and raptors. The property is owned and managed by The Nature Conservancy.

Stewart Preserve
Rensselaer County
Located about 15 miles east of Albany, this 123-acre Nature Conservancy property preserves wooded rolling hills with minor wetlands. The forest is dominantly mature upland oak–sugar maple–beech. Individual beech and oak specimens have 100-foot canopy spreads and breast-height diameters greater than 4 feet; some are probably more than 500 years old. Mature hemlock growths can also be found. Groundcover consists of abundant ferns (22 species), mitewort, foamflower, baneberry, hepatica, and starflower. The forest is an important nesting site for rapidly disappearing interior forest birds, including woodpeckers.

Summit Lake Swamp
Otsego County
Located about 60 miles west of Albany, this 36-acre property protects a pristine lake-wetland complex. Natural communities consist of Summit Lake and its shoreline, graminoid fen, shrub fen, and northern white cedar swamp. All of these habitats are globally rare. Some old-growth cedar remains. Rushes, cattails, sedges, and various grasses are supported in the fen; rare sedge species, spreading globe-flower, orchids, and ram's head lady's slipper are found here. The wetland is an important habitat for birds, amphibians, and mammals, some of them rare or unusual. The property is owned and managed by The Nature Conservancy.

Swyer (Lewis A.) Preserve
Columbia County
This 95-acre Nature Conservancy site is located about 10 miles from the village of Castleton. Natural communities include a rare freshwater tidal swamp along the Hudson River, freshwater tidal marsh, and related intertidal mudflat. The trees that grow in this wet environment are ash, maple, elm, and white oak; shrubs include buttonbush and spicebush. In addition to hosting several rare or unusual plant species, the wetlands are an important habitat for the rare green-backed heron, bald eagles, and various other raptors.

Target Rock National Wildlife Refuge
Nassau County
Located on Long Island, this 80-acre property protects subtidal, tidal flat, beachfront, sand dune, brackish pond, and mixed oak-pine forest communities. Located on the Atlantic Flyway, it is an important resting and feeding stop for hundreds of thousands of migrating birds. The unusual olive hairstreak butterfly depends on a dense stand of eastern red cedar trees as a food source for its caterpillars. Prickly pear cactus and salt spray rose can be found on the upper beaches. High bluffs overlooking the ocean are critical nesting sites for the bank swallow; the federally threatened piping plover nests on the beach.

Thompson Pond Preserve
Dutchess County
Located near Pine Plains, this 491-acre Nature Conservancy property encloses a rare, undeveloped kettle pond, surrounding wetlands, and upland forest. Biodiversity is high, with more than 400 plant species being identified. The wetland complex is an important habitat for fish, amphibians, mammals, and birds, including the rare bald eagle, golden eagle, and certain warblers and waterbirds.

Turtle Cove
Suffolk County
Although less than 2 acres in size, this Nature Conservancy property on Shelter Island protects a salt marsh cove that is an important nesting habitat for the diamondback terrapin. Hundreds of these rapidly vanishing turtles mate here during spring. The salt marsh is surrounded by thick brush, catbrier, honeysuckle, and dense thickets of cedar and bayberry.

Waitecliff Preserve
Albany County
This 75-acre tract about 2 miles south of Altamont protects a critical habitat for red-tailed hawks. Natural communities include pasture, upland forest developed on rock talus, and steep cliff faces along the Helderberg escarpment that are covered with scrubby vegetation. During spring and early summer, red-tailed hawks nest in the trees at the base of the cliffs. Other raptors, as well as the federally endangered Indiana bat, can be seen in the area. The property is owned by The Nature Conservancy.

Wellborn Woods Conservancy
Saratoga County
Located about 15 miles west of Saratoga Springs, this 105-acre property is owned by The Nature Conservancy. It protects an excellent example of mixed northern hardwood forest. Although some logging occurred during the 20th century, many impressive old-growth trees remain. The larger beech and white pine trees are probably at least 200 years old. As a mature hardwood forest, it is excellent habitat for many interior forest birds.

West Branch Nature Preserve
Delaware County
Located about 10 miles southwest of Delhi, this 446-acre Nature Conservancy site protects some old-growth forest that is at least 200 years old. Riparian habitat and wetlands occur along the West Branch of the Delaware River, which borders one side of the preserve. The larger trees are maple, oak, and beech. Riverweed, a rare plant species in New York, can be found

along the riverbank. Numerous birds nest here or visit during migration.

Wolf Swamp Preserve
Suffolk County
This 20-acre parcel flanks Big Fresh Pond in the community of North Sea. The property encloses a mature xeric oak-beech forest and red maple swamp. Berry bushes and mountain laurel form the understory. A variety of wildflowers bloom during spring and summer. The wetland is an important nesting habitat and migratory stopover for a variety of birds, including bitterns, wood ducks, and the state-protected black-crowned night heron. The property is owned and maintained by The Nature Conservancy.

NORTH CAROLINA

North Carolina can be divided into three major physiographic provinces that trend roughly from northeast to southwest: the Coastal Plain Province, the Piedmont Province, and the Blue Ridge (Appalachian) Province.

The Coastal Plain occupies the eastern third of North Carolina and includes subtidal, beach, dune-and-swale, and other coastal communities. Forests in the Coastal Plain tend to be mesic mixed hardwood (beech-tulip-poplar-oak-hickory) with a dense, junglelike understory of shrubs and vines. Dry to mesic oak-hickory forest is also abundant. The unique Coastal Plain marl community consists of black walnut–maple–pawpaw forests that are developed on flat-lying outcrops of limestone. The calcium-rich soil also supports a wide variety of rare plant species such as Carolina spleenwort. Coastal communities include maritime grassland, maritime upland forest (beech–loblolly pine–hickory–American holly), dry longleaf pine, scrub oak, maritime wetland forest, and blackwater and brownwater Coastal Plain wetlands. The unique peatland pocosin and Carolina bay communities occur only in the Coastal Plain and harbor several rare species.

The Piedmont Province occupies the central third of the state and consists of a series of rolling hills, valleys, and occasional steep, rock-walled monadnocks. Its forest cover is typically mesic mixed hardwood or basic mesic forest, including beech–tulip poplar–red oak. Dry white pine or Virginia pine, pitch pine, and chestnut oak heath forest can be found on rocky ridgelines, where soil is generally dry. Dry to mesic oak-hickory forests are very common on upland slopes and ridge tops. Wetlands in the Piedmont tend to be restricted to bottomland floodplains along rivers and streams (sycamore–river birch–aspen).

The Appalachian, or Blue Ridge, Province forms the western third of the state and is dominated by the Blue Ridge, Great Smoky, and Appalachian Mountains. Red spruce–Fraser fir or purely Fraser fir forests cloak the rugged slopes at elevations of 5,500 feet and higher. At these high elevations grassy or heath balds form open, windswept meadows of mountain oatgrass, sedges, and low shrubs like rhododendron or mountain laurel. Northern hardwood forests (beech-birch-buckeye-maple) can be found at intermediate elevations, often on north-facing slopes. Forests of white oak, chestnut oak, and montane oak–hickory occur on lower slopes at montane elevations. Specialized communities, such as rock face plant assemblages, bogs, and fens, can be found at various elevations.

Special Habitat Sites

Bluff Island and East Beach
Brunswick County
Located east of Southport in the Atlantic Ocean, Bluff Island is a relatively undisturbed barrier island. Its natural communities include tidal flat, beach, dune, maritime live oak–laurel oak–loblolly pine forest, saltwater marsh, and freshwater marsh. Several rare or threatened species inhabit the island, including the bald-head blue curl, eastern woodrat, nesting loggerhead sea turtle, brown pelican, and peregrine falcon. The island attracts large numbers of birds during spring and fall migrations.

Boone's Cave State Park
Davidson County
This 100-acre property about 2 miles west of Churchland is in the Piedmont along the Yadkin River. Forest types include successional pine forest, successional pine–mixed hardwood forest, mesic mixed hardwood forest, and the unusual Piedmont–mountain levee forest community. Located in the river floodplain, the Piedmont–mountain levee community contains some giant old-growth cottonwood trees,

one of which may be the largest in the state. Several rare plant species grow in the area.

Bushy Bay State Natural Area
Cumberland County
Located near Autryville, this site encloses Bushy Lake, a peat-filled Carolina bay wetland. Pocosin evergreen shrub, bay tree, stunted pond pine, open water, and swamp forest are the dominant natural communities. Longleaf pine forest grows on the sand rim around the lake. A noted black bear habitat, this state natural area also supports several rare plants, including white wicky, rough-leaf loosestrife, and Venus flytrap.

Buxton Woods
Dare County
The largest maritime forest in the state occurs on Roanoke Island, a barrier island off the Atlantic coast. Sand dunes up to 55 feet tall are crested with live oak, loblolly pine, and laurel oak. Interdunal swales support swamp dogwood, southern bayberry, Carolina willow, grasses, and cattails. The thick forest canopy attracts numerous birds as shelter, including raptors, brown pelicans, and the rare worm-eating, prothonotary, and prairie warblers. The trees are short and twisted from the wind and salt spray. Dwarf palmettos also grow here.

Carolina Beach State Park
New Hanover County
Located along Cape Fear River about a mile north of Carolina Beach, coastal communities here include longleaf pine, live oak, and turkey oak forest on sand dune, interdunal bog and wetland, longleaf pine–wiregrass savanna, pond pine–covered pocosin (raised bog), gum–mixed hardwood swampland forest, brackish tidal marsh, and dwarf cypress forest. Calcium-loving plants rim the limestone sink ponds. Rare plants include orchids, Venus flytrap, pitcher plants, sundews, bladderworts, and butterworts. Illegal poaching has reduced these populations. Rare or special animals are the red-cockaded woodpecker, alligator, and Carolina gopher frog.

Chattooga River Gorge/Ellicott Rock
Jackson and Macon Counties
Located along the Chattooga River near Highlands, this site is centered on part of the Chattooga River gorge. High-quality natural communities include montane oak–hickory forest, acidic cove forest, rock face, river, and riparian corridor. Several rare species depend on these habitats; these include the Biltmore sedge, divided-leaf groundsel, endemic Blue Ridge bindweed, rock gnome lichen, several salamanders, and three rare fish species. Some of the largest trees in the forest are several hundred years old. The property is an important habitat for a variety of nesting birds, including some rare warblers and woodpeckers.

Craggy Gardens
Buncomb County
Located in the Great Craggy Mountains on the Blue Ridge Parkway 19 miles northeast of Asheville, this 700-acre property encloses rhododendron forest, rocky heath and grass balds, spruce-fir forest, and old forests of stunted beech and yellow birch. A few stands of virgin eastern hemlock remain in steep draws and ravines. Rare plants, including the one-flowered rush, an endemic plant, grow on the rock cliffs and lower riparian corridors.

Croatan National Forest
Craven and Jones Counties
Bordering Neuse River near Riverdale, this 150,000-acre property includes shrub bog, lake, pine forest, bottomland mixed hardwood forest, and various wetland communities. Some of the state's best examples of pocosins, or raised bogs (dense masses of evergreen shrubs growing over peat beds), occur here. A remnant stand of Atlantic white cedar grows along Catfish Lake, one of the largest natural lakes in North Carolina. Pine flatwoods surround the pocosins. Rare or special species include sphagnum moss, pitcher plant, red-leaved parrot plant, green-leaved trumpet, cottongrass, sundew, alligators, and black bears.

Crowders Mountain State Park
Gaston County
Located in the southwestern Piedmont about 5 miles south of Gastonia, this park enclosed Crowders Mountain, a rugged monadnock with rocky slopes. The unique geology and geomorphology have created a series of microclimates and associated plant communities. Plant diversity is therefore high and includes rare species such as bear oak, Bradley's spleenwort, and ground juniper. Plant communities include chestnut oak forest and Virginia pine–mixed xeric shrub. Among nearly 200

mammals, birds, amphibians, and reptiles are several rare species, including the Carolina pigmy rattlesnake and several rare millipedes.

Dismal Swamp State Natural Area
Camden County
Located near South Mills, this state natural area is part of the Great Dismal Swamp. Natural communities include a variety of wetland types, white cedar swamp, and evergreen pocosin forest. Several rare plants (including ferns) grow here. The swamp is also a critical habitat for deer and black bear.

Eno River
Orange and Durham Counties
Located a few miles from Efland, this aquatic preserve encloses part of the Eno River, one of the headwaters of the Neuse River basin. Among the riffles and pools are a number of rare or endangered species, including the green floater, yellow lampmussel, Atlantic pigtoes, squawfoot, eastern lampmussel, Neuse River waterdog, Carolina madtom, notched rainbow, Roanoke bass, panhandle pebblesnail, clubtail dragonfly, and several other insects.

Fort Macon State Park
Carteret County
Located about 4 miles east of Atlantic Beach, this park encloses a number of high-quality communities associated with a sea-to-sound barrier island system. They include intertidal mudflat, salt marsh (cordgrass) beach, dune and dune grassland (sea oats and seaside little bluestem), maritime shrub, coastal red cedar stand, and stunted maritime forest (live oak–loblolly pine–wax myrtle). These relatively undisturbed communities are important habitats for several rare species, including nesting loggerhead turtles, the least tern, seabeach amaranth, beach morning glory, and seabeach buckwheat.

Goose Creek State Park
Beaufort County
Part of the Lower Coastal Plain, this 375-acre state natural area about 10 miles southeast of Washington is located along Pamlico River. Natural communities include freshwater and brackish marshes, riparian woodland, and upland longleaf pine forest and mixed pine-hardwood forest. Some stands of longleaf pine are considered old-growth and are quite rare in the southeastern United States. The park supports three colonies of the federally endangered red-cockaded woodpecker, as well as osprey nests. The marshes are an important nursery for fish, shellfish, and other marine species.

Grandfather Mountain
Caldwell County
This Nature Conservancy preserve in the Blue Ridge Mountains about 6 miles northeast of Linville protects very rugged terrain. Natural communities include upland spruce-fir forest, rhododendron-blueberry heath, and lower-elevation mixed hardwood forest. Changing elevations and rock chemistry have resulted in very high biodiversity. Twenty rare plants grow here, including pink shelled azalea, Heller's gayfeather, Gray's lily, and balsam groundsel.

Green Swamp Preserve
Brunswick County
This 15,722-acre Nature Conservancy property is located about 6 miles north of Supply. Natural communities include wetland, pond pine wood, flatwood, pocosin, and longleaf pine savanna. Rare species include 14 carnivorous plants, the federally endangered red-cockaded woodpecker, and ten orchids. Green Swamp is the remnant of a great wetland complex that once covered 200,000 acres.

Hammocks Beach State Park
Onslow County
Located about 20 miles southeast of Jacksonville and centered on Bear Island, a pristine barrier island, this state park encloses high-quality barrier island, salt marsh (cordgrass–black needlerush), tidal estuarine creek, and lowland forest (loblolly pine–red cedar–live oak) communities. The sparsely vegetated island has tall sand dunes and is an important nesting habitat for some shorebirds. Loggerhead turtles and the rare least tern nest on the site; the rare seabeach amaranth can be found growing here.

Hanging Rock State Park
Stokes County
Located about 2 miles southwest of Danbury, this park includes steep rock scarps and montane vegetation. Natural communities include Virginia pine–pitch pine–Table Mountain pine forests with a heath understory on the dry

ridge tops, mixed oak-hickory forest on the slopes, cliff face plant assemblages, and Carolina hemlock–rhododendron stands in the shaded ravines. Some of these trees escaped logging and are quite old. Among the rare or endangered species found here are several spleenworts and sandworts, dwarf bear oaks, and mosses and lichens.

Jones Lake State Park
Bladen County
Located about 6 miles northwest of Elizabethtown, this property encloses two large Carolina bays with raised sand rims. Natural communities include bald cypress swamp, xeric sandhill scrub, wet pine flatwood, pond pine woodland, high and low pocosins, and Atlantic white cedar forest. These diverse and unusual habitats shelter several rare species, including the red-cockaded woodpecker and white wicky.

Joyce Kilmer Memorial Forest
Graham County
Located about 20 miles west of Robbinsville in the northwestern section of Nantahala National Forest, this property shelters old-growth forest. Some oak and hemlock trees are greater than 6 feet in diameter, more than 100 feet tall, and hundreds of years old. The complete watersheds of two streams are within the park's boundaries and are totally protected. Other communities include acidic cove forest, montane oak–hickory forest, chestnut oak forest, white pine forest, northern hardwood forest, and heath bald. Several rare plants grow here. The old-growth forest contains critical habitat for important mammal and bird species, including red-cockaded and pileated woodpeckers and great horned owls.

Kitty Hawk Woods
Dare County
Located near Kitty Hawk, this property encloses relatively pristine coastal ecosystems near a developed area. Natural communities include a mixed hardwood–loblolly pine maritime forest and oak-sweet gum-blackgum maritime swamp forest, both considered to be quite rare. This property reflects the presettlement vegetation that once covered this part of coastal North Carolina. Several rare or threatened salt-tolerant plant species that thrive in dry, sandy soil can be found here.

Lake Phelps State Lake
Tyrrell and Washington Counties
Located about 5 miles southwest of Creswell, this 16,000-acre lake is the second largest natural lake in North Carolina reportedly formed by two joined Carolina bay lakes. Spring-fed and shallow, its water levels shift dramatically during the year. In addition to being a rare and unique habitat, this large Carolina bay harbors some rare plants, including bladderworts, watermilfoils, spotted pipewort, and the slender arrowhead.

Lake Waccamaw State Park
Columbus County
Lake Waccamaw is a rare Carolina bay lake about 6 miles southwest of Bolton. Communities include the sandy rim–beach, second-growth longleaf pine forest on sand ridge, pocosin, cypress-gum swamp, and the headwaters of Waccamaw River. This is some of the cleanest water in the state. Some of the special endemic fish and mollusk species here are the Waccamaw killifish, Waccamaw silverside, Waccamaw darter, spike mussel, and lance. The water's unique neutral pH is a result of the surrounding limestone substrate.

Medoc Mountain State Park
Halifax County
Located about 2 miles northwest of Ringwood, this state park straddles the border between the Piedmont and the Coastal Plain. High-quality examples of mature chestnut–shortleaf pine, riparian beech-sweet gum-birch, dry-mesic oak-hickory, and bottomland water oak–green ash–red maple–willow oak forest predominate. Large beech trees may be 200 years old or older. Several varieties of wildflowers are visible during spring and summer. The rare Neuse River waterdog inhabits Fishing Creek, which flows across the property.

Merchants Millpond State Park
Gates County
Located near Sunbury, this property encloses high-quality natural communities, including swamp forest, stream, virgin cypress–gum–water tupelo, and mesic mixed hardwood forest. Some of the cypress trees are nearly 1,000 years old; the largest water tupelo tree in the state grows here. Rare or unusual species include the red-cockaded woodpecker, American featherfoil, pale mannagrass, Virginia least trillium, and southern Twayblade. The site is an important

habitat for nesting waterbirds, raptors, wild turkeys, and bobcats.

Mitchells Millpond State Natural Area
Wake County
Located a few miles southeast of Rolesville, this property is on the transition between the Piedmont and the Coastal Plain. This site is unique, as a stream that flows across a flat, exposed rock surface supports the rare endemic Neuse River waterdog. Rare plants that cling to the outcrops are several mosses, the endemic Small's portulaca, and the Piedmont quillwort. A dry mixed pine-hardwood forest surrounds the streambed and offers shelter to various birds and mammals.

Morrow Mountain State Park
Stanly County
Located in the south-central Piedmont about 10 miles east of Albemarle, this park is located near Lake Tillery, a reservoir on the Pee Dee River. Mount Morrow is a volcanic monadnock with steep rock sides. Natural communities include excellent examples of white oak–hickory forest, chestnut oak–hickory forest, laurel health, beech forest, and red cedar–Virginia pine–shortleaf pine successional forest. The riparian edge along the Pee Dee River is dominated by large specimens of sycamore, sweet gum, and green ash.

Mount Jefferson State Park
Ashe County
Located on the Blue Ridge Highlands about 3 miles southeast of Jefferson, this park encloses high-quality examples of mature northern hardwood forest. The larger trees are red oak, mountain ash, and yellow birch; some are 100–200 years old. Large-toothed aspen is a rare species that also occurs here. The lower slopes are dominated by mature oak and oak-hickory forest. These forest communities, and the rocky cliffs and slopes, support about a dozen rare or endangered plant species.

Mount Mitchell State Park
Yancey County
Located high in the Blue Ridge Mountains about 20 miles north of Montreat, this state park encloses Mount Mitchell, the highest point in North Carolina. Natural communities include very rare virgin Fraser fir forest and lower-elevation Fraser fir–red spruce–dwarf yellow birch. A host of rare plants and animals lives here, including mountain paper birch, Arctic bent grass, spreading avens, purple fringeless orchid, and New England cottontail rabbit.

Nags Head Woods Preserve
Dare County
Located 5 miles south of Nags Head along the Outer Banks, this Nature Conservancy property covers 1,300 acres of coastal communities, including rare maritime forest, virgin hardwood forest, freshwater pond, marsh, dune, sandhill, and pine hammock. Ancient southern red oaks, loblolly pines, beeches, hickories, and hollies are nearly 400 years old. The property is, of course, an important habitat for waterbirds, as well as several rare salamanders and turtles.

New River State Park
Ashe County
Located along the South Fork of New River about 3 miles east of Jefferson, this park follows a meandering course to New River, is one of the oldest in North America. The park encloses high-quality examples of many of the natural communities in the county, harboring 40 rare or special bird, mammal, amphibian, and reptile species. A variety of rare plants also occurs here, including Virginia spiraea. Biological inventories are continuing on this unique property.

Paint Hill
Moore County
Located about 2 miles south of Southern Pines, this site is part of Weymouth Woods State Natural Area. Part of the Sandhills of the Upper Coastal Plain, it is covered by a dense forest of scrub oak–longleaf pine forest and xeric sandhill scrub with wiregrass. The unusual creeping blueberry shrub forms low mats in the sand, as does the rare Sandhills pyxie-moss. The pine trees may be current nesting habitat for the red-cockaded woodpecker.

Pettiford Creek Savannas
Carteret County
Located near Peletier in Croatan National Forest, this natural area encloses pocosin, lowland swamp, mesic pine flatwood, and pine savanna natural communities. Several rare species depend on this habitat, including the endangered red-cockaded woodpecker, Bachman's sparrow, branched gerardia,

Hooker's milkwort, scale-leaf gerardia, Carolina goldenrod, and gopher frog.

Pettigrew State Park
Washington County
This 16,000-acre park is situated on the northern shore of Lake Phelps, North Carolina's second largest natural lake, and a Carolina bay about 5 miles southwest of Creswell. Lake Phelps is unusual as it is located on a topographic high, and its only inflow is from precipitation. Its waters support the rare endemic Waccamaw killifish. Other communities include mixed wetland hardwoods as well as old-growth bald cypress, sweet gum, poplar, and black gum. Some of the state's largest cypress trees (10 feet in diameter) and oldest trees (more than 800 years) grow here. Nearly 125 bird species depend on this habitat for nesting, food, or rest during spring and fall migrations.

Pine Island Sanctuary
Currituck County
Part of Currituck Sound, this 5,221-acre Audubon Society preserve protects a pristine area along North Carolina's coast. Currituck Sound is located at the end of a series of barrier beaches. Natural communities include high-quality brackish marsh, freshwater marsh, maritime forest, shrub thicket, and sand dune. The salt marshes, along the Atlantic Flyway, are important habitats and feeding grounds for nesting and migrating waterfowl.

Raven Rock State Park
Harnett County
Located at the ecotone between the Piedmont and Coastal Plain about 10 miles northwest of Lillington, this property contains a mix of rugged and lower-elevation communities. These include river, bordering bluff, wetland, and lowland and upland mixed pine-hardwood forest communities. A small remnant of old-growth longleaf pine forest also occurs here. Rare or unusual species include the buttercup phacelia, Atlantic isopyrum, red-cockaded woodpecker, and bald eagle.

Richland Balsam/Beartrail Ridge
Jackson County
Located along the Blue Ridge Parkway where it crosses the Haywood County–Jackson County boundary, this site encloses the rocky ridgelines of the Great Balsam Mountains. Dominant natural communities include high-quality mature red spruce–Fraser fir forest, northern hardwood (oak-hickory-maple-beech) forest, and small seep wetland. Many rare or endangered species can be found here, including the northern flying squirrel, saw-whet owl, the endemic ground beetle, Mary Alice's small-headed fly, large-leaved grass-of-Parnassus, and various moonworts, clubmosses, and liverworts. The unusual pink azalea graces some of the open meadows on the property.

Singletary Lake State Park
Bladen County
Located about 10 miles southeast of Elizabethtown in Bladen Lakes State Forest, this park consists of a rare Carolina bay lake and its associated sand rim, outlying sand ridges, and swales. Pond cypress trees are old-growth and exceed 400 years. Turkey oak trees dominate the sandhills. Rare or unique species include Venus flytrap, white wicky, Edisto crayfish, red-cockaded woodpecker, and several other rare plants. The woodpeckers inhabit a stand of closed-canopy longleaf pine old-growth forest.

South Mountains Jacob Fork Watershed
Burke County
Located about 20 miles south of Morganton, this park, on the southern border of Burke County, encloses a rugged section of the South Mountains. Natural communities include chestnut oak forest, acidic cove forest, montane oak–hickory forest, pine-oak heath, cliff face, and a spray cliff assemblage at the bottom of High Shoal Falls. Some of the larger trees are 200 years old. Among the rare or endangered species that thrive here are bear oak, Blue Ridge bindweed, large witch-alder, shale-barren blazing star, Bradley's spleenwort, and Diana fritillary. It is also an important black bear habitat.

Stone Mountain State Park
Wilkes and Alleghany Counties
Located just north of Traphill, this park is centered on a granitic monadnock; its steep walls rise 700 feet above the surrounding countryside. It is considered to be one of the best examples of a monadnock in the entire U.S. Piedmont. The ranges in elevation and soil types have led to a diverse array of natural communities: rock face, xeric pine forest,

chestnut oak–mockernut hickory forest, white oak–chestnut oak heath, covered hardwood forest, stream, waterfall, and riparian corridor. Several open exposures of outcrop have well-developed biotic communities, many of them rare, growing on the rock face.

Theodore Roosevelt State Natural Area
Carteret County
Located about 8 miles west of Atlantic Beach, this state natural area encloses a remnant of rare maritime forest and shrubland that reflects presettlement vegetation. Communities include stunted maritime evergreen forest, maritime swamp forest, interdunal swamp, freshwater marsh, brackish marsh, and saltwater marsh. Two endangered plants are found in the salt marsh. Alligators and raptors (including the osprey) can be seen near the marshes.

Weymouth Woods/Sandhills Nature Preserve
Moore County
Located about a mile southwest of Southern Pines, this property protects old-growth longleaf pine–wiregrass savanna and turkey oak–wiregrass savanna communities. There is also a small grass-sedge bog on the property. Old-growth longleaf pine forest represents critical habitat for the rare red-cockaded woodpecker. Other rare or unique species include the state-endangered bog spicebush and the pine barrens tree frog.

Whiteside Mountain
Jackson County
Part of Pisgah National Forest 2 miles south of Cashiers, this vertically walled mountain at the headwaters of the Chattooga River is known as an important raptor habitat. The rock cliffs contain vertical drops of nearly 800 feet. The rock faces are home to nesting ravens, rare peregrine falcons, eagles, and vultures. Several unique moss-lichen assemblages also grow here.

William B. Umstead State Park
Wake County
Located just west of Raleigh, this state park preserves several high-quality natural communities. These include old-growth beech-mixed hardwood forest, stream, riparian corridor, floodplain forest, and mixed hardwood–purple rhododendron–nestronia forest. The largest hardwoods are nearly 100 feet tall and 200–300 years old. Rare or endangered species that thrive in this habitat are the four-toed salamander, Lewis's heartleaf, and ginseng.

NORTH DAKOTA

North Dakota is covered by four major natural regions: Lake Agassiz Plain, Northern Glaciated Plain, Northwestern Glaciated Plain, and Northwestern Great Plains.

The Lake Agassiz Plain was deposited underneath a giant glacial lake about 10,000 years ago. It occupies roughly the eastern fifth of the state. Once covered by tallgrass prairie, it has now been converted to agriculture. Grasses include big and little bluestem, switchgrass, and Indian grass. A few small lakes and pothole wetlands also occur here.

The Northern Glaciated Plain form a northwest-southeast band that trends through central North Dakota. This gently rolling glacial topography is covered by a mixed-grass transition prairie consisting of green needlegrass, porcupine grass, big and little bluestem, western wheatgrass, and needle-and-thread. Stands of trees (aspen-oak-birch) edge wetlands and pothole prairies.

West of there is the northwest-southeast trending Northwest Glaciated Plain, which occupies roughly a quarter of the state. This hummocky terrain is the result of moraines deposited by retreating glaciers, creating significant wetlands. The prairie is dominantly mixed-grass with green needlegrass, needle-and-thread, western wheatgrass, little bluestem, and junegrass.

The southwestern corner of the state is occupied by the Northwestern Great Plains. Elevation differences are a little more extreme, resulting in some deep canyons, badlands, and buttes. Dominant grasses are little bluestem, prairie sandreed, green needlegrass, western wheatgrass, bluegrama, and needle-and-thread.

Special Habitat Sites

Bluestem Meadow
Ransom County
Part of Sheyenne National Grassland near McLeod, this 80-acre tract has been proposed as an RNA. The section here is an excellent

remnant of tallgrass prairie. Rare or sensitive species on the site are the small white lady's slipper, regal fritillary, greater prairie chicken, and Dakota skipper butterfly.

Bullion Butte
Billings County
This 19,000-acre prairie-capped plateau rises above surrounding flatlands. The area is undeveloped and roadless, and 60 bighorn sheep live here. The site is part of Little Missouri River National Grassland, about 20 miles southwest of Medora. The native prairie grasses are a mix of tallgrass and short-grass species.

Cross Ranch Nature Preserve
Oliver County
This 6,000-acre Nature Conservancy property is about 6 miles southeast of Hensler. It protects one of the most diverse areas in the state, including upland prairie and floodplain forest. Typical species are cottonwood, elm, and bur oak trees; more than 100 plant species have been identified, including wheatgrass, grama grass, and needlegrass. Tallgrass prairie exists in moister areas. A herd of bison also inhabits the site. Rare or endangered birds are the bald eagle, Baird's sparrow, and Sprague's pipit. The area is also well known for its colorful displays of wildflowers during spring and summer.

Davis Ranch Preserve
Sheridan County
This Nature Conservancy property about 20 miles north of Wing is a cattle ranch and wildlife sanctuary. It preserves a 7,137-acre prairie pothole complex. The shallow lakes and ponds are surrounded by mixed-grass prairie. The wetlands are important nesting habitats and migratory stopovers for thousands of shorebirds and waterfowl. The rare Baird's sparrow can also be found here. Rare wetland plants include the yellow lady's slipper, wood lily, and golden Alexanders.

Edward M. Brigham Alkali Lake Sanctuary
Stutsman County
Located near Spiritwood, this 1,750-acre Audubon Society property encloses a remnant native prairie, prairie pothole wetlands, and cropland that is being restored to prairie. The prairie is highly diverse, with numerous wildflowers during spring and summer. The wetlands support every species of waterfowl that is known to nest in the state, including some rare or unusual species.

Fritillary Meadow
Ransom County
Part of Sheyenne National Grassland, this 240-acre tract preserves a remnant of original tallgrass prairie. Rare or unusual species include the western prairie fringed orchid, various butterflies, silky aster, and the greater prairie chicken. This area has been proposed for RNA status by the North Dakota Natural Heritage Program.

Kinley Plateau
Billings County
Part of Little Missouri River National Grassland north of Medora, this 21,000-acre topographic feature is part of the rocky, desolate Badlands. It is known for vibrant hues in red, brown, and orange rocks. The narrow stream corridors are lined with cottonwood–ash–wild plum–chokecherry vegetation. This is one of the last wild areas in Billings County that has not been affected by oil and gas exploration. It is also home to a healthy, reproducing herd of bighorn sheep.

Kulm Wetlands
Kidder County
Located southwest of Medina, this site comprises three wetland complexes. Natural communities include prairie, marsh, seep, calcareous fen, and wet meadow; some of the marsh vegetation is unusual or rare. The wetlands are an important feeding site for millions of migrating birds because of its abundance of amphipods, algae, snails, shrimp, and insects.

Little Missouri State Park
Dunn County
About 10 miles north of Kildeer, this 5,900-acre tract encloses rocky canyons, buttes, prairie, and forested riparian corridors. Prairies are covered in wheatgrass, little bluestem, little rice grass, and silver sagebrush. Juniper stands are scattered across the north-facing slopes, with cacti and yucca on the warmer south-facing slopes. This dry landscape is critical habitat for bald and golden eagles, falcons, several other birds species, bobcats, and rattlesnakes.

Lostwood National Wildlife Refuge
Burke County
Located about 10 miles north of Lostwood, this rolling short-grass and mixed-grass prairie with

pothole lakes covers about 27,000 acres. Most of the prairie is rare in that it has never been plowed and retains much of its original composition. The prairie wetlands are an important habitat for prairie birds and waterfowl, including sharp-tailed grouse, Baird's sparrow, Sprague's pipit, upland sandpiper, the rare piping plover, Wilson's phalarope, various raptors, and the short-eared owl.

Oak Hills
Richland County
This 385-acre proposed RNA preserves oak savanna habitat. Natural communities include bur oak savanna, bur oak woodland, shrub thickets, sand prairie, and mixed-grass prairie. Bur oak savanna is a rapidly disappearing ecosystem in the Midwest.

Pembina Gorge
Cavalier County
This rocky shale canyon along Pembina River is one of the most biologically diverse locations in the state. Natural communities include woodland, prairie, wetland, shrubland, and riparian corridor. Nearly 100 breeding bird species (including rare warblers) and 500 different plants have been recorded here. Because its forest is one of the largest in the state, it provides vital habitats for moose, Manitoba elk, bears, and lynx.

Pigeon Point Preserve
Ransom County
Located in northeastern Ransom County, this 560-acre tract protects a number of natural communities, including wetland, prairie expanse, prairie thicket, and forested upland. It is well known for its biodiversity and contains numerous rare plant and butterfly species. The property is owned and managed by The Nature Conservancy.

Platanthera Prairie
Ransom County
This 400-acre tract is a proposed RNA that preserves excellent examples of tallgrass prairie, sedge meadow, and mixed-grass prairie. The grassy hills shelter the federally threatened western prairie fringed orchid, the greater prairie chicken, and the skipper and fritillary butterflies.

Sheyenne Sandhills
Ransom and Richland Counties
Located about 30 miles southwest of Fargo, this 135,000-acre tract encloses one of the largest examples of pristine tallgrass prairie in the country. The rolling grassy hills, sand prairie, and wooded savanna stretch along Sheyenne River. Parts of the prairie have never been plowed. Biodiversity is high and reflects a large number of native species, colorful wildflowers, and grassland mammals, birds, insects, and reptiles, some of them rare or threatened (greater prairie chickens, several orchids, and sandhill bluestem).

Sheyenne Springs Research Natural Area
Ransom County
Located in Sheyenne National Grassland, this 55-acre RNA is about 50 miles southwest of Fargo. The property encloses a spring-fed wetland complex, including a calcareous fen. Natural communities include a variety of wetland plants, such as bog birch, water sedge, willow, slough sedge, wild sarsaparilla, bur oak, chokecherry, and green ash. This site has one of the highest concentrations of rare or endangered plants in North Dakota, including a number of rare ferns and sedges, orchids, marsh horsetail, and liverwort. The rare western wood-pewee, pileated woodpecker, and eastern bluebird are among more than 60 bird species that nest in the wetland. Other threatened species are the gray tree frog and several fish and mollusks.

Smith Grove Wildlife Management Area
McLean County
This little-known 23-acre property near Washburn protects extremely rare old-growth cottonwood trees. Some specimens are 300 years old, 15 feet in circumference, and more than 100 feet tall. These trees were already a century old when Lewis and Clark first explored the region during the early 1800s.

Two Top and Big Top Mesas
Billings County
This RNA shelters two flat-topped mesas in the Badlands of North Dakota. Located about 25 miles north of Belfield in Little Missouri River National Grassland, the steep-sided mesas pre-

serve remnant virgin prairie grassland that has not been grazed, plowed, or burned. These rare remnants consist of western wheatgrass, blue grama, thickspike wheatgrass, junegrass, needle-and-thread, threadleaf sedge, and green needlegrass. This ungrazed site is especially important in studies that investigate how grazing has impacted prairie grasslands.

Williams (John E.) Memorial Preserve
McLean County
Important wetlands are protected by this 1,440-acre site about 4 miles north of Mercer. Owned by The Nature Conservancy, it protects a series of alkaline lakes that are surrounded by mixed-grass prairie. It contains the largest U.S. breeding population of the piping plover, a rare bird species. The lakes are an important stopover for migrating shorebirds during spring and fall. The endangered Baird's sparrow also resides here.

OHIO

Ohio contains the following major natural communities: Teays River channel, boreal forest, prairie, oak opening, old-growth forest, Lake Erie shoreline, and Northern Allegheny woodland.

Roughly 10 million years ago a large waterway, Teays River, flowed from the Appalachian Mountains in North Carolina across Ohio. It emptied into the ancient Mississippi River. It existed for millions of years and introduced many species northward into what is now the state of Ohio. The river's channel is best represented by such plants, which exist today: rhododendrons, flame azaleas, and magnolias.

As a result of the last glaciation, about 18,000 years ago, outwash sedimentation formed hills and depressions, or kettles. Boreal forest communities inhabited the lakes, bogs, fens, and peatlands. Various species include orchids, sundews, and fringed gentians. The boreal communities contain the highest numbers of rare or threatened species in the state. Fen communities in west-central Ohio contain prairie species, such as big bluestem, spiked blazing star, and queen-of-the-prairie.

During presettlement times, rolling expanses of head-high prairie grasses were common. Prairie grasses also began to invade the boreal communities. Native grasses were big bluestem, Indian grass, and prairie cordgrass. The prairie would be interrupted by openings of oak or hickory trees. Due to farming and development, only a few remnants of original prairie survive. Prairie species in southwestern Ohio are more related to Missouri ecosystems and escaped the glaciation in the northern part of Ohio.

Oak openings are found in northwestern Ohio. The terrain consists of sandhills and ridges that support stands of stunted oak trees. The fine, yellowish sand was deposited as beach and sandbar by a series of glacial lakes. Today it alternates between dry sandhill and depressions filled with wet meadows, ponds, and swamps. Several rare plants survive here.

Presettlement old-growth deciduous forests covered much of Ohio. Only a few remnants remain. Typical younger forest types today are mixed beech (beech-maple-oak-ash-tulip) and mixed oak (oak-hickory).

Important marsh and wetland communities occur along the western shoreline of Lake Erie. They host a variety of important aquatic species. Farther east along the lake are sand beaches that support Atlantic shoreline flora that, thousands of years ago, existed along a seashore. These species, such as American beach grass, beach pea, and inland sea rocket, survive on protected stretches of this ancient beach.

Allegheny-type woodland occurs in northeastern Ohio, representing the southwestward expansion of this forest type from New York. Rolling woodlands are poorly drained and support hemlcok, yellow birch, gray birch, and white pine. Rarer ground species, such as hobble-bush, blue bead lily, painted trillium, starflower, Carolina spring beauty, and robin-run-away grow in the wetter parts of this ecosystem.

Special Habitat Sites

Acadia Cliffs
Washington and Athens Counties
This 277-acre tract about 3 miles southwest of Cutler contains unusual sandstone cliff communities. Several ferns and other rare plant

species, including the endangered Bradley's spleenwort, cling to the rocks. This is the only protected site for that species in Ohio.

Adams Lake Prairie
Adams County
Located about a mile north of Union, this 22-acre tract protects a remnant of cedar barren prairie. Prairie grasses are thin but support a large population of prairie dock. Unusual plants include spider milkweed, shooting star, and Carolina buckthorn. The rare Edwards' hairstreak butterfly is also found on the site.

Baker Woods
Mercer County
This 47-acre tract encloses a rare remnant of old-growth mixed forest. Located a few miles northwest of Coldwater, it contains very tall specimens of shagbark and shellbark hickory, bur oak, white oak, and the rare chinquapin oak. Rare ground species include Muskingum sedge and grove sandwort.

Beck Fen
Portage County
This 10-acre tract of boreal fen is located in western Portage County. Alkaline in nature, the fen supports a number of rare or threatened species, including cinquefoil, sedges, bayberry, showy lady's slipper, and round-leaved sundew.

Beck Preserve
Hocking County
Geologic formations and a number of interrelated forest and aquatic natural communities are preserved on this 2,234-acre tract near Rockbridge. More than 800 vascular plants have been identified, representing Allegheny, mixed-forest, boreal forest, and Teays River influences. Several warbler species nest here.

Betsch Fen
Ross County
Located about 10 miles north of Chillicothe, this 90-acre tract preserves a rare cinquefoil-sedge fen that formed more than 10,000 years ago. The groundwater is alkaline. A prairie community is also present. Plants include Ohio goldenrod, queen-of-the-prairie, small fringed gentian, shrubby cinquefoil, big bluestem, and Indian grass. Rare animals are the spotted turtle and Baltimore checkerspot butterfly.

Bigelow Cemetery Prairie
Madison County
This tiny tract 8 miles west of Plain City encloses a pioneer cemetery and virgin remnants of presettlement prairie. Tallgrass prairie species include big bluestem, Indian grass, purple coneflower, gray-headed coneflower, bergamot, whorled rosinweed, prairie dock, stiff goldenrod, and the very rare royal catchfly.

Boord
Washington County
This 89-acre tract 5 miles east of Bartlett preserves a deep gorge in sandstone that contains eastern hemlock. Rare species that live here are rock skullcap, narrow-leaved toothwort, and golden-knee.

Brown's Lake Bog
Wayne County
Part of a boreal forest community, this 100-acre tract 2 miles west of Shreve encloses a relict sphagnum bog around a natural kettle lake. Interesting species include alder, highbush blueberry, poison sumac, pitcher plant, tawny cottongrass, sundews, cranberry, orchids, and buckbean. Fifteen state-threatened plants grown here. The property is owned and managed by The Nature Conservancy.

Buzzardroost Rock Preserve
Adams County
East of West Union, this Nature Conservancy property includes a high dolomite bluff covered by prairie grasses and mixed mature hardwood forest. The woodland is a vital habitat for neotropical migrating birds.

Campbell Preserve
Lucas County
This 170-acre tract just east of Toledo contains oak opening, sand barren, swamp forest, and wet sedge meadow communities. Among the 34 rare species reported here are grass-pink orchid, sundew, soapwort gentian, Kalm's Saint John's-wort, plains puccoon, wild lupine, cow-wheat, and Missouri ironweed. Rare animals are the spotted turtle, Blanding's turtle, eastern hognose snake, and various butterfly species.

Cedar Bog
Champaign County
This 427-acre tract 4 miles south of Urbana is the largest and most pristine calcareous bog in

Ohio. It contains boreal, fen, and prairie communities. It is noted for stands of white cedar. Rarer plant species are grass-pink orchid, sundew, smaller fringed gentian, swamp birch, shrubby cinquefoil, valerian, prairie dock, and big bluestem. Rare animals that inhabit the bog are the spotted turtle, massasauga rattlesnake, Milbert's tortoise shell butterfly, and swamp metalmark butterfly.

Chaparral Prairie
Adams County
A blackjack–post oak prairie opening and cedar barren prairie occupy this 67-acre site about 3 miles northwest of West Union. Dominant plants are little bluestem, prairie dock, and spiked blazing-star. The rare Edwards' hairstreak butterfly has been reported on the prairie. Rare species include spider milkweed, prairie false indigo, pink milkwort, and Carolina buckthorn.

Clear Fork Gorge
Ashland County
This 29-acre tract about 3 miles southwest of Loudenville is known for its old-growth white pine and hemlock forest community. It also shelters rare nesting species of warblers, including the chestnut-sided warbler.

Clifton Gorge
Greene County
The limestone gorge on this 269-acre tract just southwest of Clifton is a National Natural Landmark, known for impressive cliff walls and waterfalls. The cool temperatures and moist conditions of the north-facing walls shelter rare species, such as baneberry, Canada yew, arbor-vitae, mountain maple, and snow trillium.

Compass Plant Prairie
Lawrence County
This remnant piece of prairie occupies 16 acres about 2 miles north of Aid. In addition to the more common bluestem species, milkweed, and gentians, it hosts the only known location of the rare compass plant in the state.

Cooperrider/Kent Bog
Portage County
This 43-acre tract just south of Kent contains a pristine late successional sphagnum kettlehole bog. The bog is covered by unique plant species, such as sphagnum moss, Virginia chain fern, gray birch, leatherleaf, sedges, tawny cottongrass, and the largest stand of tamarack in the state. The tamarack stand is probably the southernmost occurrence of tamarack in the continental United States. The bog is also home to the rare spotted turtle.

Crabill Fen
Clark County
This 1-acre fen near Springfield is one of the few remnants of the wetland-prairie complex that existed along Buck Creek. Plants include queen-of-the-prairie, gentians, Indian plantain, Ohio goldenrod, Riddell's goldenrod, and big bluestem. There are at least eight other endangered plant species in the fen, including shrubby cinquefoil.

Cranberry Bog
Licking County
Possibly the world's only floating cranberry bog, this 12-acre National Natural Landmark is about a mile south of Hebron. It is a remnant from a larger bog system that was destroyed by development in the 1800s. The bog is primarily a sphagnum moss meadow that contains cranberry, pitcher plant, rose pogonia orchid, grass-pink orchid, tawny cottongrass, mud sedge, and buckbean.

Crane Hollow
Hocking County
Located in Laurel Township, this privately owned 1,112-acre preserve encloses geologic formations, cliffs, and waterfalls in a hemlock-filled gorge. There are 14 rare or threatened plant species that grow in these natural communities, including northern monkshood.

Culberson Property
Clinton County
A remnant of the original lowland swamp forest, this 238-acre property about 3 miles southeast of Clarksville contains numerous woodland and swamp communities. Trees include oak, maple, hickory, swamp white oak, elm, and sweet gum. Rarer plants are the purple fringeless orchid, cardinal flower, and northern fox grape.

Davis Memorial
Adams County
This 87-acre tract a few miles east of Peebles protects a variety of natural communities,

including dolomite cliffs, caves, talus slopes, sinkholes, wetlands, and forest. Several unusual plants occur here, such as American aloe, dwarf hawthorn, hairy wing-stem, sideoats grama, purple coneflower, larkspur, Adder's tongue fern, and Walter's violet.

Drew Woods
Darke County
This 15-acre tract about 2 miles northwest of Ansonia is known for old-growth stands of white, swamp white, bur, and Shumard's oak. Tall white ash and shagbark hickory trees are also present. Rare species are pumpkin ash, grove sandwort, and the tiger salamander. The spring wildflowers are spectacular.

Dupont Marsh
Erie County
Located along Huron River about 2 miles south of Huron, this 114-acre marsh is one of the best examples of riverine marsh in Ohio. Several aquatic, marsh, and lowland communities are present, changing character according to water depth. In addition to hosting more traditional vegetation, such as cattails and spatter-dock, the marsh is home to the state-endangered leafy blue flag.

Eagle Creek
Portage County
Diverse wetland, floodplain, bog, and upland forest communities coexist on this 441-acre tract about 2 miles northeast of Garrettsville. The fern plants are especially spectacular, including the 5-foot-tall ostrich fern. The uplands are dominated by beech forest with lesser maples and oaks. Less common trees are the cucumber magnolia and yellow birch. Rare wetland plants are large cranberry, winterberry holly, cotton sedge, round-leaved sundew, and northern pitcher plant. Spotted turtles also inhabit the site.

Erie Sand Barrens
Erie County
This 32-acre parcel 9 miles south of Sandusky is covered by glacial sandy hills and beaches. The sand is well drained and supports prairie plants, such as milkweed, partridge pea, and sand panic grass. Wet depressions contain sedge meadows with rarer plants, including lance-leaved violet, Virginia meadow-beauty, twisted yellow-eyed grass, and the only Ohio occurrence of Saint John's-wort.

Flatiron Lake Bog
Portage County
This well-preserved sphagnum kettlehole bog is located on 97 acres about 15 miles east of Akron. It displays classic acidic pool–sphagnum moss–shrub–tamarack–swamp forest zonation. Special bog plants include cranberry, leatherleaf, northern pitcher plant, sundew; rare species are the white fringed orchid, Carolina yellow-eyed grass, and two-scaped bladderwort.

Fowler Woods
Richland County
Old-growth forest, wetland, and swamp forest communities are found on this 133-acre tract 6 miles northwest of Olivesburg. Tall beech and maple trees dominate the old-growth stands. The swamp forest contains buttonbush, ferns, ginseng, and saxifrage. It is one of the best sites in Ohio for viewing lush displays of spring wildflowers, especially marsh marigolds.

Frame Lake Fen
Portage County
Numerous communities, including marsh, fen, wetland, and upland hardwood forest, are on this 125-acre tract about 2 miles southwest of Streetsboro. Frame Lake is an artificial lake created in 1953; the fen communities have developed along the sides of the lake. Rare species within the fens are reproducing tamarack trees, cinquefoil, sedges, sundews, bunchflower, bayberry, and wand-lily. It is also an important habitat for herons and marsh wrens.

Gahanna Woods
Franklin County
Located on 54 acres about 3 miles northeast of Columbus International Airport, this mature beech forest and oak-maple swamp forest is well known for its bloom of wildflowers, including hyacinth, saxifrage, skunk cabbage, and trillium. The swamp forest contains abundant buttonbush and the only Ohio occurrence of cypress-knee sedge. There is also a rare thicket of prickly ash, an important food source for the larva of the giant swallowtail butterfly.

Gallagher/Springfield Fen
Clark County
This 213-acre property just east of Springfield contains one of the best prairie fen complexes in Ohio. The fen is surrounded by mixed oak-

hickory woodlands. Plant species in the fen include the rare cut-leaf prairie dock, shrubby cinquefoil, round-leaved sundew, grass-of-Parnassus, Canada burnet, and flat-leaved bladderwort. Spotted turtles also live in the fen. The drier gravel ridges around the fen are covered by dry prairie grasses, such as little and big bluestem, sideoats grama, and hoary puccoon.

Goll Woods
Fulton County
Located about 2 miles north of Archbold, an old-growth wood covers this 321-acre site. The oldest trees in the state occur here. Bur, white, and chinquapin oak trees are up to 400 years old, with diameters that exceed 4 feet. Rare ground species include the spotted coral-root and three-birds orchid.

Gott Fen
Portage County
One of the best examples in Ohio of a boreal fen occurs on this 45-acre site just west of Streetsboro. Its rare species include the showy lady's slipper, white wand-lily, round-leaved sundew, grass-of-Parnassus, fen cottongrass, cinquefoil, tamarack, and northern bayberry. The rare spotted turtle also inhabits the fen.

Greenville Falls
Miami County
This 79-acre tract about 2 miles southwest of Covington protects a steep limestone gorge, rock cliffs and arches, waterfalls on Greenville Creek, and rare cliff plants. The cliff microcommunities include rocksandwort, harebell, tufted hairgrass, limestone savory, and wand-lily. Prairie plants, such as coneflower and butterfly weed, also occur.

Gross Memorial Woods
Shelby County
A mix of swamp forest and old-growth mixed woods occurs on this 49-acre tract about 3 miles northwest of Jackson Center. The property is known for large specimens of bur oak, basswood, shagbark hickory, walnut, beech, and sugar maple. The rare pumpkin ash tree can also be found on the property.

Headland Dunes
Lake County
Located just north of Painesville on Lake Erie, this 16-acre area of natural beach hosts a variety of unusual Atlantic Coastal Plain plants. These coastal plants arrived when an arm of the Atlantic Ocean covered this area during the last glacial period. The list of rare species includes sea rocket, beach pea, seaside spurge, beach grass, purple sand grass, winged pigweed, clammy weed, and sand dropseed. This is also an excellent location for viewing migrating monarch butterflies during fall.

Hueston Woods
Preble and Butler Counties
This 200-acre tract 5 miles north of Oxford beside Acton Lake is known as one of the best old-growth beech forests in the state. The largest trees are beech and sugar maple; the understory is dominantly spicebush and pawpaw. Spring wildflowers include Dutchman's breeches, bloodroot, and trillium.

Indigo Barrens Preserve
Adams County
This 200-acre tract about 4 miles north of West Union contains oak-hickory and prairie communities. Rare species include green and spider milkweed, prairie falso indigo, large summer bluets, false scurf-pea, Carolina buckthorn, and few-flowered nut-rush. This is one of the largest populations of the rare prairie false indigo in Ohio.

Irwin Prairie
Lucas County
Wet prairie and sedge meadow occupy these 223 acres about 10 miles west of Toledo. Several natural communities overlap because of changing water levels, including tallgrass wet prairie, sedge meadow, shrub swamp, and swamp forest. Several rare species are found on the site, including red baneberry, sedges, gentians, Kalm's Saint John's-wort, Riddell's goldenrod, grass-leaf arrowhead, sedge wren, Bell's vireo, least bittern, golden-winged warbler, purplish copper butterfly, spotted turtle, and Blanding's turtle.

Jackson Bog
Stark County
Although small, this 6-acre fen complex about 5 miles north of Massillon contains 22 rare plants. A remnant of a boreal fen, it is only one of two places in the state that hosts the rare northern pitcher plant. Other rare species include shrubby cinquefoil, smaller fringed gentian, tufted hairgrass, false asphodel, arrow-

grass, round-leaved sundew, grass-of-Parnassus, and Kalm's lobelia.

Johnson Woods
Wayne County
This piece of virgin old-growth oak-hickory-beech forest occupies 206 acres about 4 miles north of Orrville. The oldest trees are more than 400 years old, more than 120 feet tall, and about 5 feet in diameter. It is also well known for good displays of spring wildflowers.

Karlo Fen
Summit County
Boreal and prairie fen communities are found on this 15-acre tract about 12 miles southeast of Akron. Rare species include the spotted turtle, rose pogonia, calopogon orchid, pitcher plant, round-leaved sundew, tamarack, and Canada burnet.

Kendrick Woods
Allen County
This mixed swamp forest occupies about 160 acres 10 miles west of Lima. It also contains wetlands and a flowing artesian sulfur spring. Rare species include saxifrage, Virginia bluebell, greed dragon, fire pink, and blue cohosh. The woods and water are an attractive stopover for migrating birds during spring and fall.

Kiser Lake Wetlands
Champaign County
This alkaline fen and associated marsh cover 51 acres about 4 miles north of Paris. Two fens are the last remnants of Mosquito Lake Bog, mostly destroyed by the construction of Kiser Lake. Prairie and fen species include shrubby cinquefoil, Kalm's lobelia, grass-of-Parnassus, smaller fringed gentian, big bluestem, queen-of-the-prairie, Ohio goldenrod, and poison sumac.

Kitty Todd
Lucas County
Reflecting the oak opening vegetation type, this 400-acre site in Harding Township contains a number of communities, including sand dune, oak savanna, sedge meadow, and sand barren prairie. Biodiversity is high as a result. There are more than 70 threatened or rare species here, including wild lupine, Karner blue butterfly, prickly pear cactus, yellow fringed orchid, wood lily, grass-pink orchid, and Skinner's foxglove.

Kyle Woods
Mahoning County
Located about 2 miles west of Canfield, this 82-acre tract of scattered old-growth trees includes giant white oaks that are more than 300 years old.

Lake Katharine
Jackson County
Located along the trace of the former Teays River, this 1,850-acre preserve 2 miles west of Jackson surrounds a clearwater lake. The preserve contains the state's largest populations of the threatened bigleaf and umbrella magnolias. Other Appalachian species found in the sandstone hills around the lake are rhododendron, mountain laurel, Virginia pine, and oak. Shaded, moist springs and ravines protect a variety of orchids.

Lakeside Daisy
Ottawa County
Located about a mile southeast of Marblehead along Lake Erie, this 19-acre tract protects the extremely rare lakeside daisy. It is the rarest plant in Ohio, and this is the only naturally occurring population of the flower in the United States. The site is actually located within the perimeter of an active quarry. Other rare prairie species include spiked blazing-star, rock sandwort, balsam squaw-weed, and tall cinquefoil.

Little Rocky Hollow
Hocking County
An impressive, steep-walled gorge is preserved within this 259-acre tract about 12 miles south of Logan. The shaded, moist conditions support eastern hemlock, Canada yew, yellow birch, partridge berry, creeping phlox, Sullivantia, round-leaved catchfly, and a variety of ferns. Appalachian species found in the oak-hickory upland forests include Virginia pine and mountain laurel.

Lynx Prairie Preserve
Adams County
Just south of Lynx, this Nature Conservancy property encloses alkaline glades. It contains typical prairie vegetation, such as big and little bluestem, Indian grass, shooting star, and coneflower.

Mantua Bog
Portage County
A National Natural Landmark, this excellent example of boreal alkaline fen is located on 64

acres 2 miles east of Mantua. Rare plant species are bunchflower, northern rose azalea, cranberry, bog bedstraw, marsh spear-grass, Crawe's sedge, and dragon's mouth orchid. Shrubby cinquefoil and Canada burnet can also be found here.

Milford Center Prairie
Union County
This 7-acre strip of former railroad right-of-way harbors a rare remnant of original prairie. It contains nearly 60 different prairie plants, including big bluestem, royal catchfly, smooth rose, prairie dock, stiff goldenrod, and coneflower. The most spectacular blooms occur during July and August.

Mud Lake Bog
Williams County
One of the few remaining kettle lakes with associated bogs in the state, this 74-acre site a few miles north of Edon contains a variety of rare bog and fen species. It contains an alkaline fen and an acidic sphagnum bog. Nearly 30 threatened species are present on the site. Rare plants include tamarack, round-leaved sundew, small purple fringed orchid, hoary willow, shrubby cinquefoil, lesser bladderwort, and large cranberry; rare animals are the northern copperbelly watersnake and the lake chubsucker.

Newberry Wildlife Sanctuary
Hamilton County
Located in Colerain Township, this 50-acre preserve encloses a wooded valley with waterfalls and cliff communities. More than 50 different wildflowers have been identified here, including the federally endangered running buffalo clover.

Old Woman Creek
Erie County
This 571-acre site about 3 miles east of Huron encloses one of the few remaining freshwater estuaries along Lake Erie. Natural communities include open water estuary, marsh, barrier beach, and lowland and upland forests. Relict prairie plants are little and big bluestem and whorled rosinweed. The preserve is best known for its diverse wildlife, including bald eagles, tundra swans, spotted turtles, Blanding's turtles, and the eastern fox snake.

Owens Fen/Liberty Fen
Logan County
Located near West Liberty, this delicate prairie fen community is one of the best in Ohio. Nineteen acres in size, it contains a high diversity of prairie species, including some Atlantic Coastal Plain species. Rare plants include spiked blazing-star, queen-of-the-prairie, smaller fringed gentian, false dragonhead, goldenrod, shrubby cinquefoil, whorled rosinweed, and the very rare prairie valerian.

Portage Lakes Wetland
Summit County
Located a few miles south of Akron in a heavily developed area, this 6-acre wetland is a remnant of a much larger tamarack swamp. Now a sphagnum bog, it contains rare species, such as arrow-wood, swamp birch, tamarack, alder-leaved buckthorn, and smooth gooseberry.

Prairie Road Fen
Clark County
Located about a mile north of New Moorefield, this alkaline fen covers 95 acres. It contains a variety of fen and prairie plants, including false gromwell, smaller fringed gentian, prairie dropseed, spiked blazing-star, prairie rattlesnake root, queen-of-the-prairie, Canada burnet, Baltic rush, grass-of-Parnassus, round-leaved sundew, and cinquefoil. It is also host to the rare spotted turtle and massasauga rattlesnake.

Pymatuning Creek Wetlands
Ashtabula County
This wetland complex and its numerous communities cover 108 acres in northern Wayne Township. It includes a high-quality alkaline fen that hosts rare species, such as round-leaved sundew, fen cottongrass, closed gentian, and yellow sedge. The rare bearded wheat grass, small purple fringed orchid, and Hill's pondweed also occur on the preserve. It also has Ohio's largest population of the rare spreading globe flower, as well as several rare mussels.

Rhododendron Cove
Fairfield County
This rugged, 70-acre tract 6 miles south of Lancaster contains Ohio's largest population of the state-listed great rhododendron. Other interesting plants are chestnut oak, mountain laurel, heaths, hemlock, and yellow birch. The

site is distinctive for its overlap of Appalachian and Alleghenian flora.

Rothenbuhler Woods
Monroe County
Located in Adams Township, this 44-acre tract of mixed forest contains old-growth red and white oak trees that are hundreds of years old and up to 4 feet in diameter. Impressive hemlock stands occur within the moist, shaded ravines on the property.

Sears Woods
Crawford County
This 99-acre tract a mile southwest of Bucyrus contains old-growth beech-maple forest along Sandusky River. It is well known for its display of spring wildflowers.

Sheick Hollow
Hocking County
Located about 12 miles south of Logan, this 151-acre tract protects unique geologic features, waterfalls, cliff communities, and moist and dry forests. The less common hemlock, Canada yew, and yellow birch are found along the steep, cool, shaded rock slopes. Several types of ferns and wildflowers also inhabit the damp cliffs and ravines, including trailing arbutus, round-leaf catchfly, rattlesnake plantain, pink lady's slipper, and polypody. It is also an important nesting site for a number of thrushes, warblers, and vireos.

Sheldon Marsh
Erie County
Sheldon Marsh is located along the shore of Lake Erie. About 2 miles west of Huron, the 463-acre tract protects barrier beach, marsh, and swamp forest communities. Rare beach plants include wormwood, inland sea rocket, seaside spurge, silverweed, bushy cinquefoil, and sand dropseed. The eastern fox snake and Blanding's turtle also inhabit the site. The marsh is an important migratory stopover and feeding ground for songbirds, waterfowl, and shorebirds crossing Lake Erie.

Smith Cemetery
Madison County
This pioneer cemetery from the 1800s west of Plain City contains a tiny and rare remnant of the original prairie known as the Darby Plains. It is well known for beautiful purple coneflower, big and little bluestem, prairie cordgrass, whorled rosinweed, prairie false indigo, flowering spurge, wild bergamot, and New Jersey tea.

Springville Marsh
Seneca County
Located about a mile west of Springville, this 201-acre marsh is the only remnant of a much larger wetland–wet prairie complex called Big Spring Prairie. Natural communities include open water wetland, marsh, calcareous sedge meadow, and swamp forest. Among the rare plant species are twig-rush, spiked blazing-star, shrubby cinquefoil, small fringed gentian, swamp birch, and Ohio goldenrod. It is also an important habitat for songbirds, ducks, and turtles (including the rare spotted turtle).

Stage's Pond
Pickaway County
This 178-acre kettle-hole lake about 5 miles north of Circleville is an important migratory stop for many birds, mostly waterfowl. Marsh species include buttonbush, water smartweed, and swamp mallow.

Strait Creek Prairie Bluff
Pike County
This 220-acre tract contains cedar glade, prairie, cliff, floodplain woodland, and mixed woodland communities. It is located southeast of Sinking Springs. Rarer species include American blue-hearts, Darlington's spurge, Michaux's leavenworthia, narrow-leaved summer bluets, bluestem, Indian grass, sideoats grama, Indian paintbrush, and purple coneflower.

Swamp Cottonwood
Medina County
A mixed swamp woodland consisting of 21 acres 3 miles southwest of Homerville supports one of the state's best populations of swamp cottonwood, a rapidly disappearing tree species. The swampy center is usually surrounded with excellent wildflower displays during spring.

Thomas Preserve
Franklin County
This 319-acre tract of mixed oak-hickory woods about a mile west of Westerville contains a mixture of natural communities: ravine bottom, upland forest, and swamp forest. Interesting trees include the Ohio buckeye,

black walnut, pawpaw, and red elm. The upland forest contains a variety of old-growth oak trees, some of which are more than 250 years old.

Travertine Fen
Green County
An alkaline prairie fen is located on this 21-acre tract about a mile north of Spring Valley. A clear-running spring has deposited calcium carbonate deposits, called travertine, that form a series of steps. Unusual fen and fen meadow species include Indian grass, Riddell's goldenrod, false gromwell, marsh arrowgrass, shrubby cinquefoil, fen Indian plantain, and midwest spikemoss.

Triangle Lake Bog
Portage County
This 61-acre tract a few miles west of Kent contains an excellent example of a kettle bog, complete with a floating sphagnum mat. Virtually pristine, it shows the classic pattern of vegetation zonation for boreal bogs. The water is dark and acidic. Rare plant species include northern pitcher plant, round-leaved sundew, leatherleaf, highbush blueberry, cranberry, and catberry.

Warder-Perkins
Hamilton County
A wooded stream valley and open meadows are protected within this 26-acre preserve a few miles east of North Bend. It is known for spectacular displays of wildflowers. Rare plants include synandra, Virginia dayflower, and fern-leaf scorpion-weed. A large population of the federally endangered running buffalo clover also grows here.

Zimmerman Prairie
Greene County
This 3-acre remnant of a once-extensive wet prairie fen complex is located about 10 miles northwest of Xenia. It is one of the southernmost fens in Ohio. Rare species include the spotted turtle, queen-of-the-prairie, Riddell's goldenrod, prairie dock, big bluestem, Indian grass, and Ohio goldenrod.

OKLAHOMA

Major vegetation types in Oklahoma include tallgrass prairie, short-grass prairie, mesquite grassland, sand-sage grassland, post oak–blackjack forest, cypress bottom forest, shinnery oak–grassland, bottomland forest, oak-pine forest, oak-hickory forest, loblolly pine forest, and pinyon-juniper-mesa forest.

The eastern fifth of Oklahoma is dominated by various oak-hickory and oak-pine communities trending north to south. Oak-pine forest is concentrated in the southern section, over the Kiamichi, Jackfork, and Sans Bois Mountains. Stands of post oak–blackjack savanna can also be found here.

The central section of the state is dominated by a mix of post oak–blackjack forest or savanna and tallgrass prairie (big and little bluestem, Indian grass, switchgrass). A few zones of short-grass Great Plains prairie appear sporadically toward the east. Major river drainages that cross this central part are lined by corridors of bottomland forest (cottonwood-sycamore-elder-willow) and various grasses.

Toward the west, a north-south strip along the northern border with Texas is best described as mixed-grass eroded plains. Several different communities occur here, including tallgrass prairie, Great Plains short-grass prairie, sand-sage grassland, shinnery oak–grassland–savanna, and mesquite grassland. This zone marks the transition between the tallgrass and short-grass prairies.

The Panhandle is dominated by Great Plains tallgrass prairie trending east to west. The river corridors are lined with cottonwood-sycamore-willow forest and sand-sage grassland. In the extreme northwestern corner, pinyon-juniper woodlands appear as elevations shift at Black Mesa.

Special Habitat Sites

Arkansas River Bald Eagle Preserve
Tulsa County
The Nature Conservancy organized the registration of private land along the Arkansas River south of Tulsa as a bald eagle habitat. The site is located between Keystone Dam and Fort Gibson Dam. Migrating bald eagles roost and nest during winter.

Arkansas River Least Tern Preserve
Tulsa County
This 1,400-acre Nature Conservancy–managed property south of Tulsa protects a 9-mile stretch of the Arkansas River, providing critical nesting habitat for the endangered least tern. Nesting terns can be seen from about May through August.

Beaver River Wildlife Management Area
Beaver County
Located about 10 miles west of Beaver, this state-operated area encloses a stretch of Beaver River. Natural communities include wooded riparian corridor and grassland. Interesting or unusual species that are protected here include burrowing owls, prairie dogs, and the rapidly vanishing lesser prairie chicken. Several rare insects are also found here. Bountiful wildflowers bloom during spring and summer.

Beavers Bend State Park
Beaver County
This state park a few miles northwest of Beaver shelters a variety of natural communities and rare species. Community types include mature pine-oak wood, stream, pine forest, oak-hickory-pine forest, wet meadow–marsh, and rocky slope. Bald cypress trees (rare in Oklahoma) grow at the water's edge. Rare plants include waterfall's sedge, blue jasmine leather-flower, jewelflowers, spiderwort, and the Ozark chinquapin oak. Beaver River supports the rare peppered shinner and leopard darter; the pileated woodpecker can also be found here.

Beech Creek Natural Area
Le Flore County
Located south of Talihina, this 7,500-acre preserve encloses the entire watershed of Beech Creek. The lack of development has helped maintained the pristine quality of the water. The upland slopes from the riparian corridor consist of mature beech–black walnut–Ozark chinquapin forests; some larger specimens are hundreds of years old. The Rich Mountain salamander, an endemic species, lives in this area. The dense forests are an important nesting habitat for the rare cerulean warbler as well as other forest birds.

Black Kettle National Grassland
Ellis County
Located along the Texas–Oklahoma Panhandle border, this 30,710-acre grassland is located in and around Cheyenne. The major community is mixed-grass prairie consisting of big bluestem, Indian grass, little bluestem, sideoats grama, and buffalo grass. The prairie is an important habitat for a variety of insects, birds, mammals, and reptiles, including the dickcissel and grasshopper sparrow. A mature shinnery oak woodland occurs along Washita River, which flows through the property. The rare Bell's vireo can be found nesting in the trees. The endangered least tern also nests on the site.

Black Mesa
Cimarron County
This 1,600-acre Nature Conservancy preserve in western Oklahoma is located a few miles north of Kenton. The site protects bluestem-grama shortgrass atop a steep volcanic mesa, juniper–scrub oak communities on the talus slopes, and surrounding short-grass (bluestem, grama grass) prairie. Straddling a transition zone between the Rocky Mountains and the Plains, this preserve contains 23 rare plant species and eight rare animal species. Raptors are also a common sight.

Black Mesa at Lake Etling
Cimarron County
This preserve protects rocky, arid grassland, short-grass prairie, mixed-grass prairie, and lake communities. These communities are in excellent condition and very diverse. Rare species found on the site include yucca, various buckwheats, rabbitbrush, and the brush mouse.

Boehler Seeps and Sandhills
Atoka County
Located near McGee Creek State Park about 4 miles northwest of Farris, this tract protects a rare intact Gulf Coastal Plain ecosystem. Natural communities include bluejack oak forest, seep, and marsh. Twenty-six rare species grow here, including spikemoss, clammyweed, beardtongue, nut-rush, pipewort, dwarf pipewort, and beak-rush. The rare green tree frog, as well as pygmy rattlesnakes and several rare wetland plants, can be found in the marshes. This property is owned and managed by The Nature Conservancy.

Canadian River Least Tern Preserve
Cleveland County
Located between Oklahoma City and Purcell, this preserve comprises lands owned by a

number of individuals dedicated to the protection of the least tern, a federally endangered shorebird. The reserve protects several nesting sites along a 16-mile-stretch of Canadian River. The migrating terns arrive in May and depart in September. The area is managed by The Nature Conservancy.

Cucumber Creek Nature Preserve
Le Flore County
Located in rugged country near the Arkansas border, this property is owned by The Nature Conservancy. It protects a high-quality portion of Cucumber Creek. The forested areas contain magnolias, black gum, sycamore, hickory, and shortleaf pine. The forest is a critical nesting habitat for neotropical birds, including the rare cerulean warbler. Other rare animals include a variety of salamanders and the Ouachita Mountain shiner. The Ouachita Mountain leadplant, bluet, and a variety of sedges are among the rare or endangered plant species on the property.

Dripping Springs State Park
Okfuskee County
Rocky canyons forested with oak and hickory trees are preserved on this tract. The shaded, moist conditions in the canyon support a variety of fern, moss, and lichen species; the forest contains a healthy population of the rare Ozark chinquapin oak.

E. C. Springer Prairie Preserve
Garfield County
This tract of Nature Conservancy land is located about 10 miles north of Garber. Well-preserved tallgrass prairie is dominated by head-high big bluestem and switchgrass. The area is well known for its brilliant displays of wildflowers during summer.

Hochatown State Park
McCurtain County
This site about 20 miles south of Sherwood protects mature oak-hickory-pine forest on rocky hills dissected by ravines and gullies. A running stream through the property supports the rare waterfall's sedge along the banks.

McCurtain County Wilderness Area
McCurtain County
Located about 15 miles east of Sherwood, this 14,000-acre preserve is part of the Ouachita Mountains. Within its borders are one of the last mature shortleaf pine–oak forests in the country; other natural communities include grassland, glade, and maple-oak-hickory forest. More than 110 bird species have been sighted here; it is one of the best breeding habitats in the country for the endangered red-cockaded woodpecker and Bachman's sparrow.

McGee Creek State Natural Area
Atoka County
This 8,900-acre site near McGee Creek Lake encloses well-developed oak-hickory-pine forest, oak-pine forest, tallgrass prairie, and stream communities. The forests contain a rare species of rapidly disappearing sandgrass. Shortleaf pine–oak woodland forms the riparian corridor. A biodiverse relict northern bog supports a number of rare or unusual species from a cooler period.

Pontotoc Ridge Preserve
Pontotoc County
Located south of Ada, this 2,900-acre preserve protects high-quality forested bottomland, scrub, tallgrass prairie, short-grass prairie, and stream communities. The woodland is an important habitat for a variety of songbirds. Raptors can also be seen here. The property is noted for its abundance of rare or endangered butterflies, such as fritillaries and skippers.

Quartz Mountain State Park
Greer County
Located about 6 miles southeast of Granite, this park protects a variety of important natural communities along Red River and Altus Lake. Community types include post oak–blackjack pine forest, short-grass prairie, meadow, oak forest, rocky cliffs, and vegetated granite talus slopes. Riparian corridors are lined with cottonwoods and willows. In addition to being a critical habitat for a variety of animal species, the site contains the rare Jeff Davis cholla cactus and a strong population of long-hair phlox, a rare forb.

Redbud Valley
Tulsa County
Located in northeastern Oklahoma, this 82-acre Nature Conservancy property protects a scenic stretch of Bird Creek and its tributaries. Natural communities include free-flowing stream, limestone cliff, cave, and upland sugar maple forest. Rare or unusual plants found here include ferns and other cliff species, cactus, yucca, and

smoke tree. The woodlands are an important bird habitat and are colored by wildflowers during spring and summer.

Sequoyah National Wildlife Refuge
Haskell and Muskogee Counties
Just south of Webbers Falls, this 20,800-acre property is at the western end of Robert S. Kerr Lake, near the confluence of the Canadian and Arkansas Rivers. Natural communities include lake, steep shoreline, bottomland slough, wetland, and upland oak-hickory-elm-pecan forest. The refuge is a critical wintering habitat for bald eagles. Other birds that live here are white pelicans, yellowthroats, Harris's sparrows, and several raptors.

Springer Prairie Preserve
Garfield County
The Nature Conservancy acquired this 40-acre tract about 14 miles east of Enid during the 1980s. About two-thirds of the site is virgin prairie that has never been plowed. Dominant plant species are big bluestem, sideoats grama, Indian grass, and switchgrass. The property supports diverse plant and animal populations.

Tallgrass Prairie Preserve
Osage County
Located about 10 miles east of Shidler, this 37,000-acre Nature Conservancy property protects an excellent remnant of tallgrass prairie. Its undisturbed condition has resulted in high biodiversity of plant and animal species; more than 250 plants have been identified. It is also an important nesting site for the greater prairie chicken, dickcissel, and grasshopper sparrow. Bison herds and prescribed burns are used to regenerate the prairie ecosystem. Vivid displays of wildflowers can be seen during spring and late summer. The natural grasses, such as Indian grass and big and little bluestem, reach heights of 8 feet.

Twin Cave
Delaware County
This property near Twin Oaks protects delicate natural communities within Twin Cave. Owned by The Nature Conservancy, it contains the largest cave chamber in northeastern Oklahoma. The cave shelters the rare white Ozark cavefish and cave crayfish. The endangered gray bat also lives in the cave system.

OREGON

Oregon can be divided into nine major natural communities: coastal rain forest, hemlock forest, the Willamette Valley community, subalpine fir forest, ponderosa pine forest, juniper-pinyon woodland, desert shrub, shrub steppe, and steppe.

Towering Sitka spruce trees dominate the strip of coastal rain forest along the Pacific coast. Only 2–5 miles wide, this zone is a result of mild temperatures and high precipitation. Other trees include the western hemlock, western redcedar, and Douglas fir. Some of North America's most impressive old-growth timber is found here. The lush understory is graced by a variety of tall ferns and delicate cryptogams.

Interior to the coastal rain forest lies a western hemlock–mountain hemlock forest, which covers much of the Cascade Mountains. Occupying roughly the western third of the state, this zone is more temperate and has more variable precipitation. Other trees include Douglas fir and western hemlock. Grass balds and talus slope communities are also part of this mountain landscape. Forming a central north-south corridor in the hemlock forest is the Willamette Valley, which is dominated by Douglas fir–grand fir–bigleaf maple–western hazel forest, with scattered grasslands and meadows.

Subalpine forests cover a north-south strip in central Oregon, as well as roughly one-half of the northeastern quarter. These forests occur at montane-subalpine elevations, within wetter, cooler conditions. Dominant trees are grand fir, Douglas fir, Pacific silver fir, subalpine fire, Shasta red fir, and the occasional mountain hemlock. Lush herbs and shrubs compose the understory.

The shrub steppe and steppe zones occupy the southeastern quadrant of Oregon, as well as sections within the north-central interior. Shrub steppe grows in the rain shadows east of the Cascade Mountains. Vegetation includes various bunchgrass-sagebrush associations, with some perennial grasses. The shrub steppe zone is ringed by steppe, which is dominated by wheatgrass-bluegrass and wheatgrass-fescue associations, with some sagebrush and bitterbrush. Patches of desert shrub or salt desert shrub (shadscale–salt sagebrush–greasewood) can be found along ancient lake beds in the southeastern corner.

Open ponderosa pine forest and pinyon-juniper woodland grow in the south-central section. Ponderosa pines occur on the edges of steppe country and may be interspersed with aspen, cottonwood, and lodgepole pine. Various shrubs and grasses (mostly fescue) form the groundcover in both types of woodlands.

Special Habitat Sites

Ashland Research Natural Area
Jackson County
Located on 1,408 acres in Rogue River National Forest southwest of Ashland, this U.S. Forest Service RNA was established to protect old-growth Pacific ponderosa pine and ponderosa pine–Douglas fir forests. These rare communities, which are more than 500 years old, occur in a steep, rocky valley on the southwestern slope of the Siskiyou Mountains. Lesser trees include California black oak, Pacific madrone, Pacific yew, and bigleaf maples. The largest trees are about 150 feet tall, with 5-foot breast-height diameters. Many birds, mammals, reptiles, and amphibians inhabit the forest, some of them rare, unusual, or of special concern.

Boardman Research Natural Area
Morrow County
This 2,560-acre property is located about 5 miles southwest of Boardman, within the Columbia River floodplain. Important natural communities include high-quality native Columbia Basin bunchgrass grassland and dry prairie. A fragile moss-lichen cryptogam groundcover may be present in the bunchgrass community. Plant species include various wheatgrasses, cheat grass, several fescues, buckwheat, Russian thistle, prairiestar, various milk vetches, pale evening primrose, long-leaf phlox, and sand-dune penstemon. The site is an important bird habitat for short-eared owls, burrowing owls, and various raptors. A large population of long-billed curlews nest here. Rare animals include the northern grasshopper mouse, grasshopper sparrow, and Washington ground squirrel.

Brewer Spruce Research Natural Area
Josephine County
Located on 210 acres in the rugged Siskiyou Mountains, this RNA, about 10 miles northeast of Cave Junction, was established to protect high-quality high-elevation mixed conifer forest. Natural communities include Brewer spruce–Douglas fir forest, ponderosa pine–sugar pine–fir forest, Port-Orford-cedar forest, montane chaparral, brushfield, and rock face assemblages. The largest trees are typically 300–500 years old and about 150 feet tall. Several rare or uncommon species can be found here, including red tree voles, mountain lions, and wolverine.

Bushnell-Irwin Rocks
Douglas County
Located in Oregon's western interior, this 958-acre ACEC shelters the upper drainages that enter Umpqua Valley, as well as pristine oak-madrone-conifer woodlands. The streams are unpolluted and contain several uncommon species of aquatic life; two special plant species have been identified in the woodland understory. The oak-madrone-conifer woodland is an important habitat for a variety of birds and mammals.

Butte
Yamhill County
This 40-acre RNA is located about 30 miles northwest of Salem. Its natural communities include mixed habitats of young and old-growth Douglas fir, dogwood, bigleaf maple, and Oregon white oak forest, as well as an oak-grass ecotone. Some of the old-growth trees are nearly 200 feet tall and 500 years old. Several rare or special-concern plant species have been discovered in the understory. Old-growth forests are critical habitats for forest birds such as the spotted owl.

Carolyn's Crown
Linn County
Part of the western Cascades, this 260-acre property about 50 miles southeast of Salem protects mature to old-growth forest and talus and cliff face communities. This climax forest has no fire history, and its huge old-growth Douglas fir, western redcedar, and western hemlock trees are nearly 1,000 years old and more than 200 feet tall. Several interesting plant species can be found in the fragile talus and cliff face environments.

Cascade Head Preserve
Lincoln County
Located a few miles north of Lincoln City along the Pacific coast, this Nature Conservancy

property shelters a coastal headland. Natural communities include beach, cliff face, upland Sitka spruce forest, and virgin grassland (prairie). Rare, endangered, or threatened plants grow in this special environment. The rare Oregon silverspot butterfly can also be found here.

Cherry Creek Research Natural Area
Coos County
Located about 35 miles southwest of Roseburg, this 570-acre RNA is part of the Oregon Coast Range. It was designated to protect an extensive stand of old-growth Douglas fir, western hemlock, and western redcedar forest that is more than 500 years old. Some of these trees approach 200 feet in height and diameters of 5 feet or greater at breast height. Species that depend on old-growth forests as a critical habitat, such as the spotted owl and several species of woodpeckers, can be found here.

Coquille River Falls Research Natural Area
Coos County
This 500-acre RNA on the southwestern slope of the Oregon Coast Range was established to protect virgin old-growth Port-Orford-cedar and Douglas fir forests in a rugged canyon. Located about 18 miles south of Powers, the natural communities include Port-Orford-cedar–Douglas fir forest, western hemlock–grand fir forest, stands of tanoak–golden chinkapin–Pacific madrone. The tallest trees approach 200 feet in height and are 700 years old. A variety of fauna can be found in the forest, including several rare salamanders as well as the red tree vole, Pacific jumping mouse, ringtail cat, mountain lion, and elusive marten. Roosevelt elk also browse the understory.

Crane Prairie
Deschutes County
About 30 miles southwest of Bend, this site protects a piece of pristine remnant short-grass prairie and wetlands. It is biologically diverse, with a variety of different bird, mammal, reptile, and insect species. Teals, goldeneyes, wood ducks, bald eagles, and ospreys are some of the birds that inhabit the site. The rare Cascade spotted frog can also be seen. The grassland is a foraging habitat for elk and deer. Old-growth ponderosa pines are nearby.

Cummins Creek Wilderness Area
Lincoln and Lane Counties
South of Yachats, this montane-alpine–level wildlife management area is known for its impressive old-growth timber. Sitka spruce, western redcedar, and Douglas fir trees are nearly 9 feet in diameter and more than 200 feet tall. The old-growth timber is a critical nesting habitat for the marbled murrelet, a threatened seabird, and the endangered spotted owl.

Dry Mountain
Harney County
This 2,084-acre RNA about 10 miles northwest of Riley protects a rare ponderosa pine community that has never been logged and only lightly grazed a century ago. Thus, the native plant assemblages are relatively intact. Some of the ponderosa pines are nearly 200 feet tall, more than 300 years old, and up to 4 feet in diameter at breast height. The site is an important forest and grassland habitat for a variety of animal and bird species, some of which are rare or threatened.

Eagle Cap Wilderness
Wallowa County
Located 20 miles east of La Grande, this 361,000-acre tract is in Wallowa-Whitman National Forest. Natural communities include montane-alpine ponderosa pine forest, spruce-fir forest, lake, meadow, and stream with narrow riparian corridor. Some of the lakes carry naturally endemic fish populations. The old-growth ponderosa pines are more than 200 feet tall. Indian paintbrush and lupine grace the high-elevation meadows. The rugged wilderness is prime habitat for bears, mule deer, elk, mountain lions, and eagles.

Fox Hollow Research Natural Area
Lane County
Located on 150 acres in the Willamette Valley foothills about 8 miles southwest of Eugene, this property straddles Willamette Valley and the Coast Range. The site was designated an RNA to protect an old-growth remnant of ponderosa pine and Pacific Douglas fir forest. The oldest trees are nearly 200 feet tall with a diameter at breast height of nearly 5 feet. Important bird species that use this habitat are the spotted owl, pileated woodpecker, northern goshawk, and raptors.

Gold Lake Bog Research Natural Area
Lane County
A rare subalpine bog community is protected on this 463-acre tract in Willamette National Forest a few miles west of Odel. Natural communities include pristine subalpine bog, old-growth Engelmann spruce-subalpine fir forest, and upland mountain hemlock-subalpine fir forest. Large trees are nearly 200 feet tall and 5 feet in diameter at breast height. Several rare bog plants grow here, including carnivorous plants. The Cascade frog and western spotted frog can both be found here.

Grass Mountain
Benton County
Located about 50 miles southwest of Salem, this 730-acre BLM-administered ACEC protects a high-quality Coast Range grassy bald complex. The grassy balds are surrounded by mature noble fir and Douglas fir forests. Several of the taller specimens exceed 200 years in age. Rare species inventories are currently under way.

Grassy Knob Wilderness Area
Curry County
Part of Siskiyou National Forest, this wilderness area 5 miles east of Port Orford is known for its old-growth Douglas fir and Port-Orford-cedar forests. Some of the trees are more than 500 years old. It is a critical habitat for large mammals and old-growth–dependent species, such as the spotted owl. Creeks and streams that traverse the wilderness carry salmon and steelhead.

Harney Lake
Harney County
This 30,000-acre RNA 25 miles south of Burns totally encloses a large, internally drained Oregon alkaline lake (or playa). Natural communities include the lake itself as well as alkali desert, sagebrush steppe, cold and hot springs, dune, and marshland. It is a critical habitat for avocets, terns, golden and bald eagles, snowy plovers, and thousands of other migrating birds. The RNA protects more than 40 birds and mammals that are listed as rare, threatened, or endangered, including the Ord's kangaroo rat, northern grasshopper mouse, and Malheur shrew.

High Peak–Moon Creek Research Natural Area
Tillamook County
Located southeast of Tillamook, this 1,526-acre RNA protects high-quality mature western hemlock, Douglas fir, riparian hardwood, and swordfern and salal natural communities. Clear-cutting is occurring around the tract, which includes some old-growth timber stands of coastal hemlocks and Douglas firs that are more than 500 years old. Several other rare plant species occur in the understory.

Holton Creek
Jackson County
Located on 423 acres in the Siskiyou Mountains about 10 miles south of Medford, this property encloses excellent Douglas fir–white fir, dwarf Oregon grape–madrone, and ponderosa pine–Douglas fir–madrone communities. It encloses the entire watershed of Holton Creek, which is therefore clear and unpolluted. Several stands of timber are old-growth and nearly 200 feet tall, offering important habitats for old-growth–associated flora and fauna such as the spotted owl.

Honeycombs Natural Area
Malheur County
This 11,930-acre tract occurs on a series of isolated volcanic cinder cones and ash flows in a remote section of Malheur County, about 50 miles south of Vale. Within the dominant plant community of sagebrush-bunchgrass are four rare endemic plants specific to the rocky volcanic surface. This pristine area is an important habitat for a variety of birds, reptiles, insects, and mammals, some of them rare or unusual.

Island Research Natural Area
Jefferson County
Part of the High Lava Plains, this 195-acre parcel in southwestern Jefferson County protects western juniper, big sagebrush–bitterbrush, bluebunch wheatgrass, and Idaho fescue communities. The site is remarkably pristine, with only limited grazing about 100 years ago. These communities are in high seral condition. Several rare lichens can also be found on the rock outcrops. Many birds, insects, and mammals, some of them uncommon, inhabit the site.

Jordan Crater Research Natural Area
Malheur County
Located about 30 miles northwest of Jordan Valley, this 32,000-acre tract protects lithified lava flows and spatter cones in arid southeastern Oregon. Natural communities include pond with riparian edge, wet meadow, and pristine bluebunch wheatgrass–sagebrush. Two rare ferns can be found growing on the volcanic outcrops. The small wetlands are an important nesting site and migratory stopover for birds and support some rare or unusual amphibians and reptiles. This property is one of the most biologically diverse in southeastern Oregon.

Kalmiopsis Wilderness Area
Curry County
Part of Siskiyou National Forest, this site is located along the headwaters of Chetco River, an important salmon habitat. Much of the old-growth forest that covers the steep, rocky slopes consists of Douglas fir or sugar pine that is up to 8 feet in diameter, more than 200 feet tall, and probably 400–700 years old. The endemic kalmiopsis shrub is found only here.

Ladd Marsh Wildlife Area
Union County
This is the only original wetland left in Grande Ronde Valley and the largest bulrush marsh in eastern Oregon. South of La Grande, it is a very important waterfowl habitat, especially during spring and fall migrations.

Leslie Gulch–Honeycombs Area
Malheur County
Located about 25 miles southwest of Nyssa, this property is part of the barren, volcanic Mahogany Mountains. The landscape is dry and windy and marked by tall, rocky spires, volcanic rocks, and canyons. Some 300-year-old mountain mahogany groves occur among the rocks; there is also a stand of ponderosa pine. Rare or interesting species include blazing star, bighorn sheep, golden eagles, pronghorn, and occasional herds of wild horses.

Little Blitzen Research Natural Area
Harney County
This 2,539-acre research natural area is located at middle to high elevations on Steens Mountain, about 90 miles southeast of Burns. Vernal pond, subalpine stream and riparian corridor, and alpine fescue grassland are the dominant natural communities on these rocky heights. Although the area was heavily grazed until 1960, the fragile alpine communities are intact and contain some extremely rare plants.

Little Sink Research Natural Area
Polk County
A mature Douglas fir forest with wetlands and beaver ponds is protected on this 80-acre RNA about 22 miles southwest of Salem. Within the forest are several old-growth stands of towering trees that are nearly 500 years old and 6 feet in diameter at breast height. Several uncommon understory species occur here, in addition to rare or threatened birds and reptiles.

Lower Table Rock Preserve
Jackson County
Located northeast of Medford, this Nature Conservancy property shelters a variety of habitats and communities: upland oak forest, upland pine forest, rocky basalt mesa, grassland, and vernal pool. The area is well known for its high biodiversity, including the globally rare and endemic dwarf wooly meadowfoam.

Mohawk
Lane County
Located in Oregon's western interior about 10 miles northeast of Eugene, this 292-acre property encloses old-growth Douglas fir forest, western hemlock forest, and wet sedge-alder meadow. The old-growth trees are nearly 200 feet tall and more than 500 years old. The wet meadow is a critical habitat for the rare red-legged frog, as well as several plant species. Old-growth forests are critical habitats for some rapidly diminishing species such as the spotted owl.

Neskowin Crest Research Natural Area
Tillamook County
Located along the Pacific coast south of Neskowin, this 1,190-acre tract protects pristine Sitka spruce–western hemlock forest. Other communities include beach and dune, grassy headland, and two complete stream drainages. Several old-growth stands remain, reaching heights of 240 feet and breast-height diameters of nearly 7 feet. Several rare or sensitive species may inhabit the area, including several species of bats.

Oregon Caves National Monument
Josephine County
Located about 25 miles southwest of Grants Pass, this property is an important habitat for

eight species of bats, including the endangered Townsend's big-eared bat. The cave system is being restored after years of unregulated public visitation. Above the ground, a lush forest of old-growth Douglas fir, with specimens nearly 12 feet in diameter and more than 1,000 years old, reaches to the sky.

Oregon Dunes National Recreation Area
Coos County
About 12 miles north of North Bend along the Pacific coast, this 32,150-acre property contains coastal strand and duna-and-swale natural communities. Anchored by patches of beach grass and spruce, some of the dunes are up to 500 feet high. Four hundred species of wildlife inhabit the area, including eagles, deer, herons, northern harrier hawks, western sandpipers, and the federally threatened western snowy plover. Elk can be spotted occasionally.

Oswald West State Park
Tillamook County
Located 10 miles south of Cannon, this old-growth coastal rain forest contains huge specimens of Sitka spruce, hemlock, and cedar that are more than 12 feet in diameter at breast height, 200 feet tall, and 600 years old. The rare purple checkermallow can be found in the rocky headland.

Pigeon Butte Research Natural Area
Benton County
Located on 70 acres in Willamette Valley, this RNA is about 10 miles south of Corvallis. It encloses a rare old-growth stand of Oregon white oak woodland-savanna. Natural communities include Oregon white oak closed forest, Oregon white oak savanna, and interspersed grassland. The oldest oaks are nearly 300 years old and are surrounded by younger stands. The site is an important habitat for more than 40 bird species, including several woodpeckers and the Oregon junco. Other inhabitants include the Townsend mole, several bats, foxes, and black bears.

Pipe Fork Research Natural Area
Josephine County
This 529-acre property about 10 miles from Applegate protects old-growth stands of Port-Orford-cedar forest. Other natural communities include Oregon grape–Port-Orford-cedar forest and salal. Although selective logging has occurred in the past, many of the trees are more than 500 years old and support plant and animal species that are specially adapted to old-growth timber. Pacific yew is common in the understory.

Poker Jim Ridge
Lake County
Located east of Plush in Hart Mountain National Antelope Refuge, this RNA preserves an excellent old-growth tract of juniper–big sagebrush savanna that is at least 200 years old. Bluebunch wheatgrass is the dominant grass species. The site is an important habitat for California bighorn sheep, mule deer, mountain lions, ground squirrels, bats, and pronghorn antelope.

Port Orford Cedar Research Natural Area
Coos County
Established in 1937, this 1,122-acre RNA in Siskiyou National Forest protects a virgin tract of Port-Orford-cedar. Located about 20 miles south of Powers, forest types include Douglas fir–Port-Orford-cedar, tan oak–golden chinkapin–bigleaf maples, and Oregon ash–California laurel. The largest trees are nearly 200 feet tall, 7 feet in diameter, and 500–800 years old. Unique species found here include several rare plants, Roosevelt elk, wolves, and various salamanders and newts, some of them rare or of special concern.

Round Top Butte
Jackson County
Located about 12 miles northeast of Medford, this 604-acre high-quality natural area contains oak-grass savanna, oak–ponderosa pine woodland, and grassland natural communities. Biodiversity is high, with nearly 300 species of flora and fauna being identified. This ACEC is an important winter range habitat for deer and elk and other large mammals. Three rare or unusual plant species have been identified here.

Saddleback Mountain
Lincoln County
This 29-acre RNA protects a very rare relict population of old-growth Pacific silver fir and western hemlock forest. Reminiscent of presettlement times, these giant trees are more than 400 years old. Nearby clear-cutting has exposed this tract to damaging winds.

Sandy River Gorge
Multnomah County
This Nature Conservancy property is located just southeast of Portland and protects a stretch of the Sandy River. Natural communities include rocky gorge wall and riparian old-growth forest. The woodlands are an important habitat for large mammals, such as bears, elk, and cougars. The river is an important spawning ground for various species of salmon.

Sauvie Island Wildlife Area
Multnomah and Columbia Counties
This 24,000-acre system of islands, about 10 miles from Portland, is bounded by the Columbia and Willamette Rivers. The predominant natural community is cottonwood-ash-willow floodplain forest; a few higher areas are covered by oak trees. The preserve is an important stopover for migratory waterfowl, including herons, loons, egrets, ibis, swans, goldeneyes, and the rare osprey and wood duck. Bald eagles, peregrine falcons, and kestrels also use the area. Rare vertebrates that depend on this habitat are the western pond turtle, fringed myotis bat, western big-eared bat, mountain lion, and black-tailed jackrabbit.

Scotch Creek Research Natural Area
Jackson County
Located in the eastern Siskiyou Mountains, this 1,797-acre RNA protects a high-quality example of chaparral. In addition to the typical chaparral vegetation, six rare plant species can be found here.

Shafer Creek
Linn County
Located about 55 miles southeast of Salem, this 961-acre site is part of the western slopes of the Cascade Range. It protects old-growth Douglas fir forest, western redcedar forest, western hemlock forest, lake, and wetland communities. With no fire history in the last 900 years, some of the tallest specimens are at least that age or older. Several rare relict plants from the most recent glacial period survive here.

Three Sisters Wilderness Preserve
Deschutes and Lane Counties
This 287,000-acre wilderness is about 30 miles west of Bend. Natural features include three tall volcanic peaks, alpine glaciers, waterfalls, streams, beargrass prairie, and various mixed conifer forest, some of it old-growth. The acreage is an important habitat for large animals that need extensive range, such as elk, deer, bears, bobcats, and mountain lions. Volcanic soils on the mountain slopes host several rare endemic Oregon plants.

Tom McCall Preserve
Wasco County
Located in Rowena, this Nature Conservancy property protects a portion of the transition zone between the Cascade Range and the Columbia River plateau. The natural communities include prairie grasslands and scablands. Ecotones (transition zones) tend to be biologically diverse, and the McCall Preserve is known for its variety of wildflowers and other plants. The globally rare Columbia lomatium, endemic only to the Columbia River gorge, occurs on the property.

Upper Elk Meadows
Douglas County
Located east of Sutherlin, this 203-acre property protects old-growth forest. Natural communities include ancient Pacific silver fir–grand fir–Douglas fir forest, wetland, and wet shrubland. Biodiversity is high; more than 200 plant species have been identified here. Wildflowers, some of them rare, are vivid during spring and summer. The tallest trees are nearly 200 feet tall with breast-height diameters approaching 5 feet. Several rare ferns, mosses, and lichens grow on some of the basaltic outcrop and talus. It is also an important habitat for species that depend on old-growth forests such as the spotted owl.

Wenaha Wildlife Management Area
Wallowa County
Located near Troy, this state wildlife area covers a diverse number of communities, including steep canyon, meadow, grassland, and heavy mixed-conifer forest, some of it very tall old-growth spruce and fir. Narrow riparian corridors are edged by aspen, cottonwood, and various grasses, depending on elevation. The property is a critical winter range for large herds of mule deer, Rocky Mountain elk, and bighorn sheep. Other species that depend on this habitat are black bears, bald eagles, turkeys, and raptors. Several rare plants have been identified at higher elevations.

Wheeler Creek Research Natural Area
Curry County
This 334-acre RNA in Siskiyou National Forest is about 2 miles south of Brookings. It was established to protect old-growth redwood and Douglas fir trees. Other communities include rock face, bigleaf maple–California laurel, and red alder stands. The largest trees on the site are more than 150 tall and 300–500 years old. Many mammals, birds, reptiles, and amphibians inhabit this area; these include the rare red tree vole, California slender and Del Norte salamanders, and ringtail cat.

Wildcat Mountain Research Natural Area
Linn County
Located on 1,000 acres along high-elevation slopes in the western Cascade Range northeast of Eugene, this RNA protects old-growth noble fir–Pacific silver fir forest, noble fir–Douglas fir forest, Pacific silver fir–mountain hemlock forest, rock outcrop, wet meadow, and brushland. Some of the oldest trees are more than 500 years old and nearly 200 feet tall with 5-foot breast-height diameters. Rare or special animals found here are the wolf, mountain lion, fisher, elk, Pacific jumping mouse, red tree vole, and several bats and shrews. Specialized plant communities, including several rare species, cling to the rock outcrops.

Willamette Flood Plain Research Natural Area
Benton County
This special property protects 400 acres of original Willamette Valley prairie a few miles south of Corvallis; most of the prairie has been converted to agricultural use. It contains a high number of original prairie plant species. The wetter portions of the prairie are some of the few remaining high-quality wetlands in the valley and are important habitats for a variety of birds, including migrating waterfowl.

PENNSYLVANIA

Pennsylvania is divided into the following major natural regions: Glaciated Northwest (northwestern corner of the state), Unglaciated Allegheny Plateau (north-central), Pittsburgh Plateau (southwest quadrant), Glaciated Northeast (northeast corner), Ridge and Valley (east-central), and Piedmont (southeast). The Appalachian Mountains swing through the state in a northeasterly direction. With the exception of the Piedmont, the entire state is covered by various forms of conifer-broadleaf terrestrial forest (typically hemlock–northern hardwood or Virginia pine–oak). Individual forest types are dependent on soil chemistry, moisture content, and slope aspect.

Broadleaf terrestrial forests across the state include dry oak–mixed hardwood. Northern hardwood forest (typically maple-beech-cherry, red maple, or aspen-birch) dominates across the regions in the northern half of the state. Dry oak–heath forests can be found on xeric acidic sites, often on steep rocky slopes or at the tops of ridges. Mixed mesophytic forest is found in the southwestern quarter of the state, mostly on the Pittsburg Plateau. These trees grow in rich, deep soils on gentle slopes. The distinctive black locust forest is common in the southern half of the state (Pittsburg Platuea, Ridge and Valley, and Piedmont).

Many types of wetland occur across the state: wetland forest, or woodland, wetland shrubland (typically buttonbush, alder, alder-sphagnum, or highbush blueberry), or herbaceous wetland. Black spruce–tamarack and hemlock peatland forests are common in the northern half of the state, as are leatherleaf bogs and leatherleaf-cranberry peatlands. Bottomland mixed hardwood wetland forests (especially sycamore–silver maple–elm–willow) form riparian corridors along streams and rivers.

Prairie grassland openings, calcareous glades, and meadows are common across the state. Plant associations include little bluestem–Pennsylvania sedge–common hairgrass, big bluestem–Indian grass–sideoats grama, and other grasses and herbs.

The Piedmont, in the southeastern corner of the state, has several distinct communities. Most of its forest associations lack hemlock. Unique forest types include serpentinite pitch pine–oak, Virginia pine–mixed hardwood, and the unusual tulip tree–beech–maple association. Other unique communities include red cedar–pine serpentinite shrubland, serpentinite grassland (with an abundance of rare or unusual plant species), sweet gum–oak, and red maple–magnolia bottomland forest.

Unique communities can also be found along the shoreline of Lake Erie: sand and cobble beach, beach shrubland, sand dune, lakeshore bluff seep, and woodland. Several rare plant species occur in this area.

Special Habitat Sites

Adams Swamp
Monroe County
Located in of the most pristine boreal swamps in the state, this 1,200-acre parcel protects conifer and acidic shrub swamp communities. Typical assemblages include peat–tamarack–black spruce and rhododendron–leatherleaf–Labrador tea. Wildflowers are colorful and abundant during spring. Rare plants that grow here include several orchids, pitcher plants, and sundews.

Alan Seeger Preserve
Huntingdon County
Located 6 miles north of McAlevy's Fort, this 390-acre tract contains stands of virgin hemlock and white pine, with an understory of rhododendron.

Alexander-King Nature Preserve
Lancaster County
Owned and managed by the Lancaster County Conservancy, this 24-acre property is located along Octoraro Creek about 7 miles southeast of Quarryville. Within the oak-hickory-maple forest and its understory of vines, fox grape, and wild yamroot, a variety of plants and wildflowers specific to the underlying serpentinite-derived soils can be found. These include the fame flower, star chickweed, and pink moccasin flower.

Algerine Swamp
Lycoming County
An exceptionally well-developed spruce-fir swamp community exists on this 84-acre parcel in Tiadaghton State Forest. Rare or unusual plants include sphagnum moss floating mats, sundews, and pitcher plants.

Anders Run
Warren County
Located 6 miles west of Warren, this 96-acre parcel contains old-growth white pine, eastern hemlock, and some hardwoods. The area is well known for its brilliant display of wildflowers during spring and summer.

Bear Meadow Natural Area
Huntingdon County
Located in Rothrock State Forest about 5 miles southeast of State College, this topographic depression is surrounded by hills and harbors a cool microclimate; nighttime temperatures at or below freezing typically occur for 9 months out of the year. A relict boreal bog is edged by oak–white pine–hemlock forest. Rare sedges, cottongrass, orchids, sundew, bog laurel, leatherleaf, and highbush blueberry grow here, as does a rare black spruce–balsam fir wetland forest. Rare butterfly species emerge from the grasses during spring.

Bear Run
Centre County
Located in Bald Eagle State Forest near Woodward, this 15-acre tract contains a small stand of old-growth eastern hemlock trees that are hundreds of years old.

Bristol Marsh
Bucks County
Located just south of Bristol, this wetland is one of the few relatively undisturbed freshwater tidal marshes remaining in the state. Rare or unusual plant species found here include Walter's barnyard grass, waterhemp ragweed, and Indian wild rice. Linked hydraulically to the Delaware River, this habitat is an important migratory stopover for tens of thousands of birds. The property is owned and managed by The Nature Conservancy.

Bruce Lake
Pike County
This 2,845-tract in Delaware State Forest north of Bushkill surrounds a rare glacial lake. Tamarack and balsam fir, relicts from the last Ice Age and rare species in Pennsylvania, occur near the lake.

Chrome Serpentinite Barrens
Chester County
This 340-acre Nature Conservancy property is located just east of Oxford and protects globally rare communities that grow only on serpentinite (iron- and magnesium-rich rock). Natural communities here are serpentinite prairie, savanna, and forest. Rare or unusual plant species include the serpentinite aster and round-leaved fame flower; six rare species of moths are also found here. The property is also

an important nesting site for forest birds, including woodpeckers and songbirds.

Cook Forest State Park
Jefferson and Clarion Counties
Located on 6,668 acres just north of Cooksburg, this state forest tract contains virgin white pine, hemlock, and beech trees up to 400 years old. Some pine trees are nearly 200 feet tall and 4 feet in diameter at breast height.

Cranberry Swamp
Clinton County
A rare mountain bog with native cranberry vegetation occurs on this 144-acre tract about 3 miles south of Renovo.

David Johnson Tract
Bucks County
Located in the Valley Forge State Forest near the New Jersey–Pennsylvania border, this 56-acre tract preserves a topographic plateau that has an unusually diverse assemblage of trees.

Delaware Water Gap Recreational Area
Pike County
Old-growth forest occurs here along Dingmans Creek, a high-quality trout habitat. Stands of tall hemlock trees are more than 300 years old. Interesting fern-moss-lichen communities cling to rock exposures along the creek.

Delhaas Woods
Bucks County
Just north of Bristol, this 181-acre Nature Conservancy property contains excellent remnants of Coastal Plain forest and bog. This property represents one of the best examples of Coastal Plain communities and contains one of the most diverse populations of relict Coastal Plain plants, including umbrella and sweet bay magnolias. More than 12 rare or endangered plants grow here, including Maryland meadow beauty, eastern blue-eyed grass, long-leafed aster, netted chain fern, sundews, bladderworts, and pitcher plants. A 200-year-old willow oak is one of the old-growth trees on the property.

Detweiler Run
Huntingdon County
This 463-acre tract in Rothrock State Forest contains virgin stands of hemlock and white pine, with a rhododendron understory.

Devil's Elbow
Lycoming County
Located northeast of Ralston, this 404-acre tract encloses a series of wetlands in addition to the headwaters of North Branch Rock Run. Several rare plants have been identified among the wetland natural communities.

East Branch Swamp
Clinton County
A dense mountain swamp, wetlands, and old-growth hemlock forest are protected on this 186-acre parcel in Sproul State Forest.

Edward Woolman Nature Preserve
Chester County
Located 5 miles south of Montrose, this wetland complex is one of the most significant wetlands in eastern Pennsylvania. Its 2,000 acres include freshwater marsh, submerged meadow, shrub swamp, swamp forest, and floodplain forest along Marsh Creek. Its undisturbed state has resulted in a high biodiversity, with more than 200 species of flowering plants, several of which are rare or endangered, such as Saint-John's wort. More than 150 bird species live in the woods and wetlands. Rare butterflies include the black dash and mulberry-winged skipper. The preserve is managed by The Nature Conservancy.

Enlow Fork Natural Area
Greene County
Several natural communities are found on this preserve a few miles south of West Finley, including old-growth stands of sycamore, buckeye, and elm along creeks, and the uncommon chinquapin oak. Unusual plants that thrive here are scarlet fire pink, blue phlox, Riddell's hedge nettle, and blue-eyed Mary. Rock ledges support ferns, lichens, and various mosses. Rare butterflies can be found here as well.

Ferncliff Wildlife Preserve
Lancaster County
Located on 65 acres about 5 miles south of Buck, this preserve became a National Natural Landmark in 1972. It is most known for a remnant old-growth forest that survives in a ravine along Barnes Run. Among the colorful wildflowers are several rare species, including putty root and the cranefly orchis. The old-growth forest consists of magnificent specimens of tulip, beech, hemlock, and oak trees that are

hundreds of years old. Pawpaw is common in the understory. Interesting fern communities cling to the wet rock cliffs in the ravine.

Glens Natural Area
Luzerne County
Located along Kitchen Creek, natural features here include steep gorge, waterfall, lush riparian corridor, and mixed conifer-hardwood forest with virgin hemlock, white pine, beech, and oak trees (some up to 500–600 years old, 100 feet tall, and 4 feet in diameter).

Goat Hill Serpentinite Barrens
Chester County
Located south of Lees Mill, this 602-acre tract in Valley Forge State Forest preserves a rare pitch pine–scrub oak barrens community developed on serpentinite rock outcrop, a globally rare community. Several rare plants have been discovered on the site, including serpentinite aster, round-leaved fame flower, and long-hairy barrens. Other uncommon, threatened, or endangered species include several moths and butterflies, the barred owl, and several warblers. The property is managed by The Nature Conservancy.

Hearts Content Scenic Area
Warren County
Part of Allegheny National Forest about 5 miles southeast of Cobham, this property, at the headwaters of Tionesta Creek, consists of 121 acres of remnant virgin forest. Majestic hemlock, white pine, beech, and maple trees are nearly 200 feet tall, 400–500 years old, and 4 feet in diameter at breast height. This parcel is a reminder of what the presettlement "Big Woods" looked like 500 years ago.

Heinz Wildlife Refuge
Bucks County
Located near Tinicum, this important wetland complex is the largest freshwater tidal marsh in the state. It is a critical nesting habitat for more than 50 birds and a feeding and resting stopover for nearly 300 species of waterfowl during spring and fall migrations. The property is owned and managed by The Nature Conservancy.

Hemlocks
Perry County
Virgin hemlock and mixed hardwood trees are found on this 131-acre parcel in Tuscarora State Forest about 2 miles south of Big Spring State Park. These trees are 400–500 years old, have breast-height diameters of 4–5 feet, and reach skyward as much as 150 feet. This strip of forest borders a 2-mile ravine along Patterson Run.

Holly Gap Preserve
Cumberland County
Located just west of Mount Holly Springs, this 913-acre Nature Conservancy property protects an important wetland complex. Natural communities include marsh, bog, and upland forest. Glade spurge is one of the rare or unusual plants that thrive here.

Hoverter and Scholl Box Huckleberry Natural Area
Perry County
Located 2 miles southeast of New Bloomfield, this 9-acre tract in Tuscarora State Forest protects an extremely rare box huckleberry colony. Located within an oak–hickory–eastern hemlock forest, the colony covers the entire 9-acre tract. It is actually a single plant with multiple runners and is thought to be 1,300 years old. This shrub occurs in limited numbers in only seven states. Other plants in the understory are Christmas fern and the fragile ebony spleenwort.

Jakey Hollow
Columbia County
Located on 58 acres near Bloomsburg, this old-growth remnant contains virgin white pine, hemlock, and hardwood trees that are more than 200 years old.

Johnson Run
Cameron County
This 216-acre tract of land a few miles north of Driftwood protects a stand of old-growth hemlock and hardwood trees.

Joyce Kilmer Woods
Union County
This 77-acre parcel about 6 miles west of Harleton contains stands of virgin white pine and hemlock trees that are hundreds of years old. The site is also an important nesting habitat for bird species.

Lacawac Sanctuary
Wayne County
Located north of Ledgedale, this 500-acre property includes part of Lake Wallenpaupack. A

glacial lake, Wallenpaupack is one of the cleanest in North America. Natural communities include swamp marsh, meadow, pond, quaking bog, and second-growth hardwood-pine forest. Several rare plant and animal species depend on these habitats, such as orange and green newts, pitcher plants, sundews, and centuries-old rock tripe that cling to steep cliffs.

Lebo Red Pine
Lycoming County
This 124-acre tract near Cammal includes a rare stand of old-growth red pine trees, with some birch and oak.

Lehigh Pond and Marshes
Wayne, Lackawanna, and Monroe Counties
Owned and managed by the Nature Conservancy, this 3,912-acre preserve is about 2 miles from Gouldsboro. Natural communities include glacial kettlehole pond, boreal sphagnum bog, tamarack–black spruce swamp and old-growth upland eastern hemlock–mixed hardwood forest. Many of these centuries-old trees are nearly 100 feet tall. Rare or endangered species found here include Labrador tea, many-fruited sedge, tamarack, pitcher plant, sundew, osprey, and several vireos and warblers. The property is located along a major flyway and is an important migratory stopover for bird species.

Little Mud Pond Swamp
Pike County
Located southeast of Germantown, this 182-acre tract contains pristine boreal conifer swamp and glacial bog. The bog contains a variety of rare or endangered plants endemic to that natural community.

Little Tinicum Island
Delaware County
Little Tinicum Island is located in the Delaware River near Essington. Two hundred acres in size, it hosts special-concern estuary and bottomland forest communities, as well as rare plant species.

Long Eddy River Edges
Wayne County
This 680-acre property along the Upper Delaware River protects old-growth forest, Delaware River riparian forest, cliff face, and waterfall communities. The cliffs support a variety of mosses, ferns, and lichens, some of them rare, such as the hairy lip fern and rock spikemoss. The rare miner's lettuce plant can be found along the waterfall. Wildlfowers are abundant during spring. The tall old-growth trees are dominated by hemlock, sugar maple, and beech, reaching heights of nearly 100 feet.

Long Pond Reserve
Monroe County
This 2,000-acre Nature Conservancy property near the town of Effort includes scrub oak–pitch pine barrens, heaths, wetland forests, bogs, and vernal ponds. Its biodiversity is extremely high and includes 40 threatened plant and animal species and 15 rare insects and butterflies. Some of these are native American azalea, pink-flowered rhodora, the fly poison bulb-borer moth, and two species of dragonflies. This property contains one of the highest densities of rare or unusual species in Pennsylvania.

Lost Lakes
Monroe County
Located on 858 acres, this landscape of thick bog, swamp, rhododendron thicket, bottomland, and upland conifer-hardwood forest is one of the wildest sites in Pennsylvania. Dominant communities are glacial bog, acidic shrub swamp, northern hardwood forest, and boreal conifer swamp. The acreage is an important wildlife corridor that connects other wildlife areas; it is thus often traversed by large mammals, such as black bears and deer. Rare plants that grow in the wetlands are various sedges and rushes, as well as the yellow cowlily.

Marian Brooks
Elk County
This 917-acre tract about 40 miles southeast of Saint Marys in Moshannon State Forest protects a high mountain bog, surrounding birch, oak, and laurel, and other natural features.

Mount Cydonia Ponds
Franklin County
Located in Michaux State Forest, this 183-acre tract shelters a series of ephemeral (vernal) ponds and numerous plant communities.

Mount Holly Preserve
Cumberland County
This 913-acre Nature Conservancy property is a few miles from Mount Holly Springs. Natural communities include upland mixed pine-hard-

wood forest, sedge meadow, marsh, and seepage swamp. The wetlands host a number of unusual plants, including a globally rare species. Several species of turtles and reptiles frequent the marsh, which also attracts many bird types, especially during spring and fall migrations.

Nottingham Serpentinite Barrens
Chester County
Located 25 miles southeast of Lancaster in Nottingham County Park, this 630-acre site is managed by The Nature Conservancy. A globally rare serpentinite barren ecosystem is preserved here. The toxic levels of metals and high temperatures on the bare rock defeat most species. Some, however, can tolerate the harsh environment; the rare serpentinite aster, moss pink, and fame flower are found on the serpentinite prairie and savanna. Serpentinite asters occur only here and in a few sites in Maryland.

Ohio River Islands
Beaver County
These wooded islands in the Ohio River were donated to the U.S. Fish and Wildlife Service by The Nature Conservancy. Thickly vegetated and often swept by floodwaters, the islands contain a diverse number of plant species. The safety of the wooded habitats attracts a variety of nesting species, including the bald eagle, peregrine falcon, and countless songbirds. Several rare or endangered mussels occur in the shallow waters around the islands.

Pen Run
Bucks County
Located on 104 acres along the Delaware River near Bensalem, this property protects an important riparian corridor of upland forest, riparian vegetation, freshwater tidal wetland, and forested wetland. Much of the forest is Coastal Plain forest, a vanishing ecosystem in Pennsylvania. Several special-concern species grow here, including the umbrella magnolia and subulate arrowhead. The rare barn owl, peregrine falcon, and osprey also nest along the river. A rare fish has been identified in the tidal marsh. Some of the upland hickory-oak-ash forest trees are old-growth, nearly 100 feet tall, and centuries old.

Pine Lake
Pike County
A high mountain bog, with complete vegetation zonation patterns, occurs on this 67-acre tract 2 miles north of Greentown.

Presque Isle State Park
Erie County
Located about 5 miles from Erie in Lake Erie, this property is actually a peninsula. Natural communities include sandbar beach, plains, dune, sand ridge with beach grass, cottonwood willow, rare climax oak-maple-hemlock forest, and wetland. More than 300 bird species frequent the park, including raptors, shorebirds, warblers, the endangered piping plover, vireos, tanagers, and wrens. Monarch butterflies stop here in September during their migration to Mexico. More than 500 plants have been identified, many of them rare, including the endangered Kalm lobelia.

Reynolds Spring
Tioga County
A high mountain bog, with related rare or unusual plant communities, is located on this 1,302-acre property. The bog is surrounded by tall white pine and pitch pine trees.

Richard O. Rowlands Preserve
Mifflin County
This 43-acre property encloses a cave system and the surrounding mixed-hardwood forest. The bat hibernaculum is a winter roosting site for the rare small-footed bat, Keens bat, and Indiana bat. A rare shrimplike isopod lives in an underground stream in the cave.

Salt Springs State Park
Susquehanna County
This 411-acre state park about 7 miles north of Montrose encloses magnificent stands of virgin hemlock and northern hardwood forest, very rare natural communities. The largest hemlock trees are estimated to be 700 years old; they are more than 100 feet tall and nearly 5 feet in diameter at breast height. Several unusual plants can be found growing beside the narrow streams and waterfalls that wind through the forest, an important habitat for interior forest birds.

Sheets Island Archipelago
Dauphin County
This preserve consists of a series of islands in the Susquehanna River near Harrisburg. The natural communities on the islands provide critical habitats for waterfowl and songbirds.

Sideling Hill Creek Natural Area
Fulton County
This 136-acre Nature Conservancy property is located near Emmaville, in the central Appalachians. It protects a rare Appalachian shale barren ecosystem (only 200 of which occur in the world). Few plants can tolerate the extreme temperature conditions and, consequently, are vary rare: Virginia spiderwork, hairy beardtongue, columbine, alumroot, and others. Rare wetland plants along the stream are golden club, Tennessee pondweed, and several sedges. The rare Potomac sculpine and several endangered mussels live in the waters of the stream.

Snyder-Middleswarth
Snyder County
Towering virgin white pine, hemlock, and pitch pine trees are found on this 500-acre tract within Bald Eagle State Forest about 5 miles west of Troxelville.

Spruce Swamp
Lackawanna County
Located a few miles from Thornhurst, this 97-acre tract in Lackawanna State Forest protects wetland, scrub, shrub, and forest natural communities. Rare plants have been identified on the site.

Sweet Root
Bedford County
This 1,403-acre tract in Buchanan State Forest about 2 miles west of Chaneysville contains stands of virgin hemlock and second-growth oak forest.

Tamarack Run
Sullivan County
Located near Laporte, this 234-acre property shelters swamp, bog, marsh, bottomland forest, and upland forest natural communities. Dominant tree types are conifer, hemlock, and hardwoods (oak and maple). Several rare plants have been identified in the bogs and wetlands.

Tannersville Cranberry Bog
Monroe County
Located on 775 acres about 12 miles northwest of Stroudsburg, this property encloses wetlands and a relict northern boreal bog. The bog is surrounded by black spruce–tamarack lowland forest and upland oak-maple forest. Shrubby thickets of heath plants, such as highbush blueberry and leatherleaf, and mats of sphagnum moss edge the bog. Rare plants are cottongrass sedge, pitcher plant, loosestrife, cranberry, orchids, bog rosemary, Labrador tea, goldthread, and sundew. The rare bog copper butterfly (its larvae feed on cranberry) hatches here. The bog turtle also lives in the wetlands. The property is managed by The Nature Conservancy.

Thomas Darling Preserve
Monroe County
Located on 2,200 acres along Two Mile Run, this pristine Nature Conservancy property protects virgin peatland, spruce, acidic shrub swamp, acidid fen, marsh, boreal spruce swamp, and wet meadow communities. Biodiversity is high, with a spectacular array of wildflowers during spring and summer. Interesting or rare plants that grow here include bog laurel, clubmoss, and tamarack. It is an important habitat for black bears, coyotes, barred owls, ospreys, and other birds.

Tionesta Research Natural Area
Warren and McKean Counties
Located in Allegheny National Forest southwest of Wetmore, this 2,113-acre RNA was established in 1940. It protects one of the last remaining stands of virgin hemlock-beech climax forest on the Allegheny Plateau. There is no record of logging or fire. Rare birds that inhabit the forest are Swainson's thrush and the yellow-bellied flycatcher.

Tucquan Glen
Lancaster County
Located 4 miles south of Martic Forge along Tucquan Creek, this dark, shaded forest includes birch–tulip–eastern hemlock–basswood forest that is estimated to be 200–300 years old; some trees are nearly 100 feet tall. Rhododendron forms a dense understory, and numerous plants grow along the streams and wetlands, including 40 different wildflowers, rare ferns, Bradley's spleenwort, mountain spleenwort, and rare entangled filmy fern.

Unusual moss-lichen-fern communities cling to dark, wet rocks along the creek.

Woodbourne Forest and Wildlife Sanctuary
Susequehanna County
Located a mile north of Dimock, half of this 648-acre property is virgin forest. Hemlock, beech, maple, and oak trees are up to 500 years old and more than 100 feet tall. The biologically diverse wetlands host more than 85 plant species, including rare or unusual mosses, sundew, pitcher plant, and leatherleaf. Rare salamanders and other amphibians can be seen along the marshy shore. Pileated woodpeckers and uncommon warblers live in the old-growth forest, as do other interior forest birds, such as owls.

RHODE ISLAND

Vegetation in Rhode Island is divided into the following groups: marine, estuarine, palustrine, riverine, lacustrine, terrestrial, and forested upland.

The marine group includes open ocean, coastline, and saline bay. Various communities are marine deepwater, tidal (eelgrass), intertidal, mudflat (algae), sand beach, and rocky shoreline, including associated flora and fauna.

The estuarine environments comprise areas where ocean water is diluted with inflowing fresh water, resulting in brackish conditions. Communities in this category include tidal river and creek, salt marsh (elder, groundsel, cordgrass), coastal salt pond that forms behind barrier dune, mudflat (low-growing aquatic plants), and tidal marsh with taller emergent vegetation.

Palustrine environments are freshwater wetlands that contain varying amounts and types of vegetation. Different communities include gravel bar within river, deep flooded marsh, shallow marsh, scrub wetland, floodplain forest, red maple swamp, and hemlock-hardwood swamp with nearly closed canopy. Peatland, fen, and bog communities are included.

Riverine communities occur in flowing freshwater streams that lack persistent emergent vegetation. Subtypes are determined by characteristics of the gradient and substrate.

The lacustrine group comprises lake and pond communities. The diversity of aquatic communities is determined by the site of the lake, the kind of vegetation, and the availability of nutrients.

Terrestrial systems occur on well-drained land and include open upland, beach, dune, maritime shrubland and grassland along the coastline, sand barren, barren woodland, pitch pine–scrub oak barren, and red cedar populations along rocky ridges. Forested upland occurs at different elevations and can be pitch pine–oak, white pine–oak, hemlock-hardwood, oak-hickory, chestnut oak, and beech-maple.

Special Habitat Sites

Black Rock
Washington County
This 200-acre site is part of Block Island in the town of New Shoreham. It preserves part of a long section of glacial moraine that formed about 15,000 years ago. Sections of the moraine have survived development and contain unique plant species, such as bayberry, arrowwood, shadbush, and bushy rockrose. The rare northern blazing star can also be found. Several rare insects have also been identified.

Black Spruce Bog
Kent County
Located near Penny Hill, this 10-acre bog occupies a glacial kettle depression. It is surrounded by plant species that are relicts from the last glaciation (10,000 years ago). This site contains a large population of black spruce, a rare tree in Rhode Island. Other species include leatherleaf, cranberry, pitch pine, and scrub oak. The area is surrounded by a rare pitch pine barren community and some of its species, such as golden heather and northern ricegrass. Rare moths and butterflies, as well as a variety of warblers, inhabit the bog.

Block Island Greenway
Washington County
Several natural communities are protected on this island about 12 miles off the coast of Rhode Island. More than 600 acres of state and private land are linked via trails and easements. The major natural communities are upland forest, maritime scrub, and rare maritime grassland. Rare or unique plant, animal, and insect species include the American burying beetle, northern blazing star, various raptors, and bushy rockrose.

Cedar Swamp Fen
Washington County
Located just north of Worden Pond near Tuckertown Four Corners, this 500-acre wetland complex is the state's largest freshwater wetland. It contains an unusual array of relict Coastal Plain species, such as white cedar. The wooded swamp contains red maple, black gum, swamp white oak, and Atlantic white cedar. Many of the larger cedars are nearly 100 years old. Rare species include Long's bulrush, bog sedge, Walter's sedge, redroot, and Plymouth gentian. Rare vertebrates include the spotted turtle, smooth green snake, various salamanders, and the possibly extinct southern bog lemming.

Diamond Bog
Kent County
About 50 acres in size, this wetland contains a variety of communities, including marsh, bog, open water, cedar swamp, graminoid fen, and sedge meadow. The bog is partially covered with a well-developed sphagnum moss mat. Unusual species include white cedar, twig-rush, tussock sedge, and woolfruit sedge. Rare species are grass pink, rose pogonia, pitcher plant, sundew, bladderwort, and bog aster. It is also one of the few sites in the state where buckbean grows. Endangered amphibians and salamanders frequent the marsh. The fen is a habitat for the very specialized banded bog skimmer dragonfly.

Ell and Long Pond Preserve
Washington County
This Nature Conservancy preserve is located in Hopkinton. It shelters a variety of intermixed natural communities: rocky cliff face, bottomland forest, marsh, pond, and sphagnum moss quaking bogs. Plant diversity is high as a result and includes rare mosses, lichens, ferns, sundews, and pitcher plants.

Fogland Marsh Preserve
Newport County
This 49-acre Nature Conservancy property in Tiverton encloses an undeveloped stretch of Atlantic Ocean beach and its flanking wetlands. Natural communities include beach, sand dune, mudflat, pond, and salt marsh as part of a tidal estuary connected to the Sakonnet River. Several rare plants grow here. Various fishes and shellfish use the estuary as a nursery. The wetlands are an important habitat for a variety of wading birds, including herons, sandpipers, and egrets.

Goosewing Beach
Newport County
This Nature Conservancy preserve is located west of Little Compton. Natural communities include open beach, mudflat, coastal pond, and salt marsh. The protected habitats host nesting sites for the rare piping plover and least tern. The wetlands and beach are important stopovers for thousands of migrating waterbirds.

Great Swamp
Washington County
This property encloses a beach-dune barrier island ecosystem. Communities include the ocean beach–dune complex, freshwater wetland, coastal grassland (little bluestem–Indian grass–switchgrass). Coastal salt ponds vary in salinity and are usually brackish. The rare upland sandpiper nests in the coastal grasses; the rare piping plover nests along the beach.

Lime Rock
Providence County
Located only about 5 miles north of downtown Providence, this acreage protects a rare forest community that developed on rocky, dolomitic soil. The groundwater is rich in calcium and magnesium. The unique groundwater-soil chemistry supports 30 rare or unusual plant species.

Stepstone Falls
Washington County
This 1,000-acre site in Exeter protects the state's best example of northern hardwood forest along a steep ridgeline. Trees include beech, oak, maple, yellow birch, and various hickories. Some of these trees are found in stands of old-growth timber. The area is well known for colorful displays of wildflowers during spring and summer. The rare pileated woodpecker and cerulean warbler are some of the bird species that inhabit this interior woodland. Rare dragonflies and butterflies, including the northern pearly-eye, also breed on the property.

SOUTH CAROLINA

South Carolina is divided into four major physiographic provinces: From east to west, they are: Coastal Plain, Sandhills, Piedmont, and Blue Ridge.

The Coastal Plain comprises the eastern third of South Carolina. This flat-lying area is highly diverse biologically. Its mosaic of communities include turkey oak and scrub oak barrens, xeric and mesic pine-mixed hardwood forests, magnolia forest, various wetlands including bald cypress–tupelo swamp, bottomland hardwood forest, hardwood-cane bottom, pine-wiregrass savanna, fresh to saline marshes, maritime forest and heath, dune shrubland, dune-and-swale, ocean beach, intertidal, and barrier islands.

The Sandhills comprise a narrow, northeast-to-southwest–trending zone across central South Carolina. Major vegetation types include turkey oak barren, scrub oak barren, xeric pine-mixed hardwood forest, pocosin, bay forest, and hillside bog. Barrens communities often contain a thick, shrubby understory. The dominant trees in the xeric pine-hardwood forest are loblolly pine and post oak. The pocosins and bay forests are densely vegetated lowlands where drainage is poor.

The Piedmont covers most of the western third of the state in a series of rolling hills, monadnocks, bluffs, and valleys. Vegetation systems include stream and pond riparian edge, red gum–tulip poplar floodplain forest, beech-oak-hickory-maple mesophytic hardwood forest, midslope oak-hickory forest, post oak–blackjack oak–ridge top forest, flat rock face, and chestnut oak–heath forest. The south-facing slopes—higher in elevation, rockier, and drier—are dominated by oak and pitch pine vegetation.

The rugged Blue Ridge Mountains are the dominant feature in the Blue Ridge, located in the extreme northwestern part of the state. Larger communities include alder-willow riparian corridor, floodplain forest (shortleaf pine, scrub pine, Virginia pine, sweet gum, and red maple), mixed mesophytic forest, xeric upland oak forest, pitch pine–Virginia pine forest, and rock face plant assemblages. Mixed mesophytic forests are dominated by hemlock, tulip poplar, basswood, and beech.

Special Habitat Sites

Aiken Gopher Tortoise Heritage Preserve
Aiken County
This 366-acre property about 5 miles southeast of New Ellenton protects an important section of the Coastal Plain that supports the rare gopher tortoise. Natural communities include mature longleaf pine forest, turkey oak forest, and wiregrass understory. Several rare plants also grow here.

Alexander Sprunt Sanctuary
Charleston County
Located near Charleston, this Audubon Society property includes Edisto Island and Deveaux Bank, a large sandbar in the North Edisto River. The property was purchased to protect two highly productive salt marshes that are important food sources for nesting and migrating waterbirds. Some of the species that nest on the Deveaux Bank are black skimmers, brown pelicans, various terns, and the American oystercatcher.

Ashmore Heritage Preserve
Greenville County
Located about 5 miles west of Marietta, this 496-acre property is part of the Mountain Bridge Wilderness and Recreation Area. Natural communities include rugged upland forest, rocky cliff, and the rare cataract bog (moist, vegetated rock along a sheetlike waterfall). The steep cataract bog environment hosts a variety of rare or unusual plants and animals, including mountain sweet pitcher plant, Indian paintbrush, grass-of-Parnassus, the green salamander, and the state-endangered Rafinesque's big-eared bat.

Bamberg County Cowbane Preserve
Bamberg County
Located about 9 miles south of Bamberg, this 58-acre Nature Conservancy property encloses an undisturbed Carolina bay wetland complex. Natural communities include oak-hickory forest, sandhill scrub, pond cypress savanna, pond cypress pond, swamp tupelo pond, open pond, sedge bog, and cypress–tupelo gum

swamp. Old-growth cypress trees can be found in the wetlands. Rare or endangered plant species include Canby's cowbane, Boykin's lobelia, slender blue flag iris, and panic grass. It is an important habitat for water-birds such as the little blue heron.

Bennett's Bay Heritage Preserve
Clarendon County
Located west of Foreston, this 620-acre property preserves a large Carolina bay wetland, a rare Coastal Plain pond for South Carolina. Natural communities include open water, wetland, shrubby swamp, lowland pond pine–loblolly pine forest, and upland pine forest. These unique habitats shelter rare or endangered flora and fauna, such as pocosin, fetterbush, red bay, and sweet gallberry. Prickly pear cactus can be found in the scrub oak that grows on the dry, sandy rim around the pond. Numerous bird species inhabit or visit the site.

Bird Key–Stono Heritage Preserve
Charleston County
This 20-acre island in the mouth of Stono River southeast of Sol Legare is an important brown pelican rookery. Natural communities include ocean beach, sand dune, high tidal marsh, and patches of maritime grassland. In addition to hosting thousands of federally endangered brown pelicans, it is an important breeding site for Louisiana herons, laughing gulls, egrets, and various terns. Thousands of birds visit the island during spring and fall migrations.

Black River Swamp
Georgetown County
Located northeast of Andrews, this 1,276-acre Nature Conservancy property protects old-growth forest vegetation along a blackwater river. Natural communities include bald cypress–tupelo gum swamp, mixed bottomland hardwood forest, and nine rare old-growth loblolly pine hammocks. Some of the pines are more than 100 feet tall and nearly 3 feet in diameter at breast height. Rare plants include sarvis holly, false dragonhead, and river quillwort; the threatened American alligator and endangered shortnose sturgeon also inhabit the site.

Brasstown Creek
Oconee County
Located on 461 acres near Sumter National Forest and the Tugaloo River, this property encloses highly diverse natural communities, including the rare cove hardwood forest and pine-oak heath. Seven rare or endangered plants can be found on the property. The cove wetlands and surrounding forest and heath are important habitats for a number of birds, mammals, amphibians, and reptiles.

Bunched Arrowhead Heritage Preserve
Greenville County
Located on 133 acres northeast of Travelers Rest, this property protects bottomland hardwood forest, upland pine-hardwood forest, flowing streams, and riparian habitats. The property is unique because it supports one of the best populations of the federally endangered bunched arrowhead plant in the United States. Several varieties of birds, including some diminishing species, nest in the forest.

Buzzard Roost Heritage Preserve
Oconee County
Located about 10 miles northwest of Walhalla, this 501-acre preserve encloses a rugged portion of Buzzard Roost Mountain. Natural communities include upland oak–hickory–Virginia pine forest, rock outcrop and cliff face, and scattered grassland-prairie. The rare Table Mountain pine tree can be found at the highest elevations. Although heavily logged some 60–70 years ago, the property still supports a number of rare, threatened, or endangered species. These include, in addition to the Table Mountain pine, pileated woodpeckers, roosting turkey vultures, raptors, wooly lipfern, blackstem spleenwort, purple cliffbrake, smooth coneflower, bindweed, and stoneroot. The woodlands are an important nesting habitat for interior forest birds.

Capers Island
Charleston County
This 2,000-acre barrier island in Charleston County is relatively undisturbed and protects a variety of coastal natural communities, including beach, dune, tidal flat, salt marsh, and interior forest. It is an important habitat for a number of species, including nesting waterfowl, some of which are rare. It is a migratory feeding and resting site for thousands of birds during spring and fall.

Cartwheel Bay Heritage Preserve
Horry County
Located on 680 acres west of Loris, this property encloses a rare Carolina bay wetland com-

plex. Natural communities include swamp, pocosin, streamhead pocosin, pine savanna, pine-wiregrass savanna, pine–scrub oak sandhill, and longleaf pine forest. Several rare or endangered plant species occur in these unique habitats, including Pixie moss, toothache grass, Venus flytrap, numerous orchids, and pitcher plants. The rare red-cockaded woodpecker nests in the 100-year-old longleaf pine forest.

Cathedral Bay Heritage Preserve
Bamberg County
A rare Carolina bay wetland complex is protected on this 58-acre parcel about 2 miles south of Olar. Natural communities include the dominant pond cypress pond with an edge of myrtle-leafholly-buttonbush shrubs, grasses, and various sedges. This unique wetland supports several rare and unusual plant species that are specially adapted to this kind of habitat.

Chandler Preserve
Greenville County
Located on 251 acres in Greenville County, this property protects a rare cataract bog. Within its lush vegetation are several rare plants, including the federally endangered mountain sweet pitcher plant, other carnivorous plants, the rare grass-of-Parnassus, Indian paintbrush, and the unusual divided-leaved groundsel.

Cliffs at Glassy Preserve
Greenville County
Located on 600 acres along Green Creek about 20 miles west of Gowensville, this Nature Conservancy property protects a variety of natural communities: rocky cliff and bluff, upland bog, stream, oak-maple-beech upland forest, hemlock–white pine grove in ravine, chestnut oak forest, and lowland oak-hickory forest. The bog is a rare habitat in this mountainous terrain and contains rare or unusual species, such as orchids, certain sedges, and a sphagnum moss mat. Some of the trees, especially the hemlocks, are mature to old-growth. Colorful wildflowers during spring are dominated by American lily of the valley, Carolina tassel-rue, and yellow lady's slipper.

Congalee Swamp National Monument
Richland County
Located about 18 miles southeast of Columbia, this extensive 22,000-acre wetland complex is an unspoiled remnant of the bottomland forests that once covered much of the southeastern United States. Natural communities include bottomland forest, creek, slough, canbrake, and marsh. Some of the tallest and oldest trees in the United States can be found here, including loblolly pines that are nearly 200 feet tall as well bald cypress, sycamore, and tupelo. The mix of wetlands and bottomland forest has resulted in an incredible biodiversity that includes more than 300 plant species and a variety of bird and animal life, including the endangered red-cockaded woodpecker, canebrake rattlesnake, and threatened bald eagle.

Deveaux Bank
Charleston County
This 15-acre property is located on a sandspit peninsula that has been an important shorebird rookery for decades. Its coastal communities offer shelter to the federally endangered wood stork, which nests on the island. It is also one of the largest royal tern rookeries in the state and attracts huge numbers of birds migrating along the Atlantic Flyway during spring and fall.

Dungannon Heritage Preserve
Charleston County
Located on 643 acres north of Rantowles, this preserve shelters mature longleaf pine forest and wetlands. These natural communities are a critical nesting habitat for the federally endangered wood stork. Other species that feed and roost in the area are great blue herons, storks, terns, and great egrets.

Eastatoe Creek Natural Area
Pickens County
This 373-acre preserve west of Rocky Bottom protects a 5-mile stretch of pristine Eastatoe Creek and three of its tributaries. Natural communities include rocky cliff face, ravine, remnants of old-growth hardwood forest, old-growth hemlock stand, and riparian creek habitat. The huge hemlock trees may be virginal remnants; several old-growth beech trees are more than 100 feet tall and 9 feet in diameter. The groundcover contains rare plants along the moist stream banks, including the rare bristle fern and dwarf filmy fern; this is the only site in North America where the tropical tunbridge fern can be found. Eastatoe Creek is also a self-sustaining rainbow trout habitat. A moss species in Eastatoe Gorge occurs nowhere else in the world.

Eva Russell Chandler Heritage Preserve
Greenville County
This 113-acre preserve about a mile north of the community of Little Bald Rock protects mature forest-wetland habitat that shelters a number of rare plant species. Natural communities include mixed Virginia pine–hardwood–mountain laurel upland forest, maple-oak-hardwood forest, flatwood forest, rock outcropping, Slickum Creek, and a sheet waterfall and cataract bog. Rare plants that grow in the fragile rock and bog communities are Mountain sweet pitcher plant, grass-of-Parnassus, divided-leaved groundsel, Indian paintbrush, numerous ferns, prickly pear, reindeer moss, and stonecrop.

Flat Creek Heritage Preserve
Lancaster County
Located on 598 acres a few miles south of Taxahaw, this state preserve protects part of the ecotone between the Piedmont and the Inner Coastal Plain. Ecotones generally have a very high biodiversity because of the two overlapping ecoregions. Natural communities on the property include rock outcropping, cave, waterfall, Piedmont cove forest, piedmont flatwood, chestnut oak forest, upland pine-hardwood forest, and wetland. Some of these communities harbor very rare plant species. The extremely rare pool sprite occurs in the shallow puddles that briefly occupy depressions in the bedrock surface. At least 10 rare or endangered wildflowers grow in abundance on the property. Animals and birds include beavers, deer, prairie warblers, songbirds, and raptors such as red-tailed hawks.

Francis Beidler Forest
Berkeley and Dorchester Counties
This 5,820-acre Audubon Society property is located about 40 miles northwest of Charleston. Part of Four Holes Swamp, it contains the largest remaining stand of virgin cypress-tupelo swamp in the world. These giant trees are about 120 feet tall and more than 1,000 years old. Rare or endangered fauna that inhabit the blackwater swamp are alligators, certain snakes and lizards, and the eastern big-eared bat.

Glassy Mountain Heritage Preserve
Pickens County
Located northeast of Pickens, this 65-acre tract protects a geologic monadnock, a rare jutting landform with steep rock faces. The exposed rocky top and sides are home to rare plant species, such as yucca, crotonopsis, thousand-leaf groundsel. Other natural communities are Virginia pine forest and upland oak-hickory forest. Some of the older trees form probable stands of old-growth timber that escaped logging. Wildflowers, some of them rare, are well-displayed during spring and summer on the dome of the monadnock.

Henderson Heritage Preserve
Aiken County
This 198-acre property north of Aiken protects a section of the Coastal Plain Sandhills. Natural communities include a mature longleaf pine forest with a wiregrass-dominant understory.

Hitchcock Woods
Aiken County
Located within Aiken, Hitchcock Woods encloses about 1,200 acres of nearly virgin old-growth longleaf pine–wiregrass community. Other trees include sweet bay and red bay, with diverse examples of shrubs, herbs, ferns, and mountain laurel. More than 350 plant and 100 bird species have been counted here. Some of the more rare, unusual, or endangered species that depend on these habitats include pileated and red-cockaded woodpeckers, several vireos, and numerous herbs and ferns.

Hunting Island State Park
Beaufort County
Located about 18 miles east of Beaufort, this 5,000-acre property encloses coastal dune, beach, estuary, maritime scrub forest, saltwater marsh, freshwater marsh, slash pine forest, palmetto forest, and tidal pool communities. The slash pine–palmetto forest is one of the few remaining undisturbed old-growth remnants in the state; the rest has been destroyed by development along the coast. The complex estuarine ecosystem protects rare or unusual species, such as the least tern, beach pigweed, loggerhead turtle, rare orchids, and the occasional bald eagle.

Janet Harrison High Pond Heritage Preserve
Aiken County
Part of the Coastal Plain of South Carolina, this 30-acre property a few miles northwest of Steedman protects a wetland meadow. Several

rare plant species grow in this unique environment, including the federally endangered harperella, regionally threatened pink tickseed, and state-concern Florida false loosestrife.

Johns Island Maritime Forest Preserve
Charleston County
Located on 160 acres about 15 miles southwest of Charleston, this property protects one of the few undeveloped islands left in the Charleston area. The most important natural community is the globally rare maritime forest, a transition community between saltwater-effected vegetation and freshwater-continental vegetation. Dry sand dune ridges are covered by loblolly pine, and the interdunal sways are occupied by oak trees. Ponds and wetlands also occur. The island is an important nesting and migratory stopover for songbirds and wading birds, including the snowy egret and great blue heron. The property is owned and managed by The Nature Conservancy.

Lewis Ocean Bay Heritage Preserve
Horry County
Located east of Conway, this 9,343-acre preserve encloses an important wetland complex of more than 20 Carolina bays, a rare community in South Carolina, as well as sand ridges. Rare plant species found in the bays include pitcher plant, sundew, spoonflower, Venus flytrap, and a variety of endemic orchids. Other natural communities are peatland, shrub swamp, lowland pond pine–loblolly pine forest (pocosin), and longleaf pine forest on the higher sand rims. The rare prickly pear cactus can be found in this sandy soil. The surrounding woodlands are an important black bear habitat.

Little Pee Dee River
Marion and Horry Counties
Located on four different tracts totaling some 8,000 acres, these preserves protect about 17 miles of Little Pee Dee River frontage, including high-quality examples of mature cypress-gum swamp, fluvial sand ridge, oxbow lake, and bottomland hardwood forest communities. These are important habitats for a variety of mammals and birds, including the pileated and red-cockaded woodpeckers.

Longleaf Pine Heritage Preserve
Lee County
Located on 843 acres, this property encloses seasonally wet longleaf pine forest and wetlands. Rare species that inhabit the preserve are the rare red-cockaded woodpecker, American chaffseed, and Canby's dropwort. The area is maintained with prescribed burns.

Lynchburg Savanna Heritage Preserve
Lee County
Located on 291 acres, this property about 5 miles south of Elliot protects an excellent remnant of seasonally wet longleaf pine savanna and pond cypress savanna. The preserve has one of the highest biodiversity rankings in the state. Among the many rare plant species are American chaffseed, coreopsis, mountain mint, primrose, hooded pitcher plant, yellow pitcher plant, certain orchids, milkwort, sundew, and fly-poison. The area is maintained through prescribed burns.

Matthews Creek Heritage Preserve
Greenville County
This 1,660-acre property is part of the Mountain Bridge Wilderness Area. Important natural communities include a rare mountain bog, mature oak-hickory upland forest, Virginia pine stands, and hemlock groves along creeks and streams. The larger specimens are old trees that escaped past logging. Rare plant species in the bog environment include swamp pink, rose orchid, climbing fern, and long sedge. The pristine headwaters of Matthews Creek and Julian Creek, two self-sustaining brook trout habitats, originate here. The forest is an important habitat for black bears and a variety of bird species.

Old Island
Beaufort County
Located on 400 acres, this property consists of a relict sand dune ridge-and-swale complex that is interspersed with freshwater ponds and salt marsh. Four rare plants grow on the island. It is an important nesting and feeding site for the federally endangered wood stork and the federally threatened bald eagle; numerous birds stop here during spring and fall migrations.

Pacolet River Heritage Preserve
Spartanburg County
Located near Spartanburg, this 257-acre property protects a stretch of riparian habitat along the Pacolet River. The dominant natural communities are mixed hardwood forest, pine forest, and bottomland forest. Colorful wildflowers can be viewed during spring and summer.

The woodlands are important nesting habitats for interior forest birds. In addition to these natural resource attributes, the property also protects two Native American soapstone quarries that are 3,000–5,000 years old.

Peachtree Rock Preserve
Lexington County
This 305-acre Nature Conservancy property west of Columbia is underlain by a rocky substrate. Natural communities are quite varied and include bog, sandhill pine-oak-sparkleberry scrub, sandstone outcrop, the only waterfall in the state's Coastal Plain, and longleaf pine forest. Rare or unusual plants that grow here are the federally endangered Rayner's blueberry bush, tit, crane-fly orchid, several ferns, galax, and sand myrtle. The property provides important nesting habitat for bird species.

Red Bluff Marsh Preserve
Jasper County
One of South Carolina's most productive estuaries is protected on this 1,740-acre preserve about 8 miles southeast of Hardeeville. Natural communities include an extensive unspoiled salt marsh, marsh island, mudflat, and patches of pine-oak-cedar maritime forest. The marsh is dominated by smooth cordgrass. The abundant fish and shellfish species in the estuary attract thousands of shorebirds and wading birds, such as egrets, herons, sandpipers, and the rare piping plover. Rare fishes that live in the marsh are the endangered shortnose sturgeon and Florida gar. The preserve is owned and managed by The Nature Conservancy.

Sandhill Bay Preserve
Clarendon County
Located on 132 acres, this Nature Conservancy property near Manning protects two Carolina bay habitats. Natural communities include the open water of the bays, maple-loblolly pine on the surrounding sand rims, and old-growth pond cypress-tupelo gum swamp. Several rare or threatened plants occur in these wetlands, including royal fern and lizard's tail. The spotted turtle also thrives here.

Sandy Island
Georgetown County
This 9,164-acre preserve southeast of Yauhannah Landing covers an island parcel between the Waccamaw and Great Pee Dee Rivers. This land is bordered by the Great Pee Dee River, Little Bull Creek, Bull Creek, and the Waccamaw River. Natural communities include bottomland hardwood forest, upland hardwood forest, longleaf pine forest, tidal bald cypress–tupelo gum swamp, pocosin, and freshwater swamp. The cypress and tupelo trees are large. The forest-wetland complex supports a number of rare or unusual plant species; the longleaf pine forest is a critical habitat for the federally endangered red-cockaded woodpecker. Many wading birds nest in the wetlands. The property is owned and managed by The Nature Conservancy.

Savage Bay
Kershaw County
This 93-acre property northeast of Camden shelters a Carolina bay community and a sand rim forested with longleaf pine–scrub oak. Prickly pear and Georgia bear grass grow in the sandy soil of the rim; rare vegetation in the bays includes pond cypress, fetterbush, carnivorous plants, and pondspice. Grasses, sedges, and redroot grow on the mud bottom when the ponds are dry.

Savannah Coastal Refuges
Jasper County
Part of Savannah National Wildlife Refuge south of Hardeeville, these properties enclose excellent examples of marshland, freshwater marsh, saltwater marsh, tidal creek, and bottomland forest. It is an important breeding habitat and migratory stopover for millions of birds. Often ablaze with water flowers, the wetland is an important habitat for the Mississippi and swallow-tailed kites, great blue herons, ibis, the rare wood stork, and a variety of uncommon warblers. It also has a robust alligator population.

Savannah River Bluffs Heritage Preserve
Aiken County
This 83-acre tract just outside the community of North Augusta protects about a quarter-mile of river shoals along the Savannah River. The wetlands, bottomland forest, and upland forest along the river shelter a diverse group of rare, threatened, or endangered plants, including the rocky shoals spider lily, bald cypress, yellowwood, false rue anemone, relict trillium, bottle-brush buckeye, and upland swamp privet. A foot trail allows easy access through the preserve.

Shealy's Pond Heritage Preserve
Lexington County
This 62-acre property a few miles east of Pelion protects a portion of the Coastal Plain Sandhills. Natural communities include mature Atlantic white cedar swamp, longleaf pine forest, turkey oak forest, and grassland. The cedar swamp supports several rare species of insectivorous plants, such as sundews and pitcher plants. The cedar trees are quite large. The property is an important habitat for a number of bird species.

South Williman Island
Beaufort County
Located on 2,765 acres, this island preserve encloses several high-quality coastal habitats, including intertidal mudflat, beach, dune-and-swale, heath, and upland maritime forest. Several rare plants grow here, and it is an important nesting and feeding site for thousands of waterbirds, some of them rare or endangered, such as the wood stork and piping plover.

Stevens Creek Heritage Preserve
McCormick County
Located on 138 acres a few miles northeast of Clarks Hill, this preserve contains excellent examples of a variety of habitats along the Savannah River. Natural communities include mature bald cypress–saw palmetto floodplain forest, rocky outcrop, hardwood (oak-hickory) midslope forest, and upland shortleaf pine–loblolly pine forest. The high biodiversity here includes a large number of rare, endangered, or threatened plant and animal species, including bald cypress, Florida gooseberry (wildflower), Webster's salamander, reindeer moss, the pileated woodpecker, ferns, and spiderwort. One of the state's best displays of wildflowers can be seen at the base of the uplands during spring and summer.

Tillman Heritage Preserve
Jasper County
Located on 953 acres 5 miles west of Tillman, this property encloses part of a sand ridge along the Savannah River that is covered with longleaf pine–turkey oak forest and wiregrass. Other communities are cypress–tupelo gum swamp and mixed hardwood bottomland forest. Rare gopher tortoises, slender glass lizards, and diamondback rattlesnakes live on the property. Rose dicerandra, narrow-leaved rushfoil, and pondspice are some of the rare plants found among the wetlands.

Tom Yawkee Wildlife Center
Georgetown County
Located near Georgetown, this 18,086-acre property contains a barrier island covered by maritime forest and mature longleaf pine flatwoods. The forest is maintained through prescribed burns. Biodiversity is high, with several rare plant and animal species, including the red-cockaded woodpecker.

Victoria Bluff Heritage Preserve
Beaufort County
Located on 1,111 acres about a mile from Hilton Head Island in South Carolina's Coastal Plain, this property shelters a complex wetland system. Dominant natural communities are longleaf pine–slash pine–pond pine–saw palmetto bottomland forest-swamp and upland live oak forest. Although it sits amid resorts and a four-lane highway, it is a relatively pristine ecosystem that harbors a variety of rare or unusual plants. It supports the largest U.S. population of the rare pondspice.

Waccamaw River Preserve
Horry County
Owned and managed by The Nature Conservancy, this 650-acre property southeast of Conway protects habitats along Waccamaw River. This blackwater river is unique in that it displays several brownwater river features, such as levees and active sloughs. Natural communities include tidal cypress-gum swamp and loblolly pine forest. Some of the cypress trees are quite old. Almost 30 rare or endangered plants have been identified, including dwarf fimbry. Rare vertebrates include the pygmy sunfish, shortnose sturgeon, and several rare bats. The preserve is an important corridor for large woodland mammals, such as the black bear.

Washo Preserve
Georgetown County
This 1,040-acre Nature Conservancy property southwest of Georgetown protects extensive wetlands along the Santee River. Natural communities are freshwater cypress lake, cypress–tupelo gum swamp, and lowland pine forest. Some of the cypress and tupelo trees are large and old. The oldest wading bird rookery in continuous use in North America is located on the preserve. One of the largest nesting

populations of osprey in the eastern United States also lives here. Several rare or unusual plants have also been found.

Watson Heritage Preserve
Greenville County
Part of the Mountain Bridge Wilderness Area, this 1,660-acre tract contains the headwaters of Matthews Creek. A pristine mountain bog occurs at montane elevations. Clear-running trout streams are bordered by lush riparian corridors. Unusual plants that thrive here are chokeberry, alder, sphagnum moss, ferns, swamp pink (nationally threatened), twayblade, rose orchid, climbing fern, and painted trillium. The surrounding forest is oak-hickory on the slopes, pine on the ridge tops, and hemlock in the cooler, wetter ravines. Some of the taller trees are old-growth, untouched by past logging due to the steep terrain.

Woods Bay Heritage Preserve
Clarendon and Sumter Counties
Located just north of Turbeville, this 368-property encloses a rare elliptical Carolina bay pond. Features include the associated sand rim, longleaf pine forest, marsh, and cypress-gum swamp. It is ranked as the fifth most significant Carolina bay in the state. It is home to several rare plants as well as alligators, red-cockaded woodpeckers, ospreys, and native South Carolina lizards, amphibians, and turtles.

SOUTH DAKOTA

South Dakota is divided into three major vegetation regions. These are, from east to west, eastern deciduous forest, the prairie grasslands of the Great Plains, and the Black Hills.

The eastern deciduous forest can be found along a north-south zone in the easternmost part of the state. Dominant trees include sugar maple, basswood, beech, and bur oak with a herbaceous understory. Much of this forest has been cleared for agriculture and development; today it can be found in state parks and along river corridors.

Most of South Dakota is covered by grass. The short-grass grasslands of the Great Plains cover roughly the western two-thirds of the state; the eastern third is covered by tallgrass prairie (big and little bluestem, cordgrass, and Indian grass). Streams and rivers are lined with cottonwood, sycamore, box elder, and willow trees, along with an abundance of grasses.

The billion-year-old rocks of the Black Hills protrude through the Great Plains grasslands in southwestern South Dakota. Similar to the Rocky Mountains farther west, the rugged Black Hills (3,000–7,500 feet in elevation) are covered by a series of coniferous forests and woodlands that are dependent on elevation, soil chemistry, and precipitation; these include pinyon-pine-juniper woodland, ponderosa pine forest, and limber pine communities. Stands of quaking aspen can be found along higher-elevation rocky draws and canyons.

Several prairie pothole lakes exist in the northeastern corner of the state. These shallow wetlands are rimmed by a variety of grasses, bulrushes, and sedges and are important staging areas for migrating birds along the Central Flyway.

Special Habitat Sites

Altamont Prairie
Deuel County
This 62-acre prairie near Altamont is owned and operated by The Nature Conservancy. It is known for its wetlands, which provide important habitats for nesting and migrating bird species.

Aurora Prairie
Brookings County
Located about 3 miles southeast of Brookings, this 30-acre high-quality remnant wet prairie is owned by The Nature Conservancy. Dominant plant species include big bluestem, Indian grass, and cordgrass. The wetlands are an important bird habitat. The area is well known for its summer displays of wildflowers.

Black Fox Canyon
Lawrence County
Located in the heart of Black Hills National Forest near Savoy, this property preserves high-quality natural communities. Savanna, ponderosa pine forest, white spruce forest (some of it old-growth), and spruce-pine forest can be found here. Old-growth trees are very rare in the Black Hills. The rugged canyons and uplands are an important habitat for large mammals such as mountain lions.

Clovis Prairie
Brown County
This pristine remnant of tallgrass prairie occupies 157 acres about 8 miles northwest of Columbia. It is an important habitat for a number of prairie birds and animals. The grasses are head-height during summer, with showy flowers.

Crystal Springs
Deuel County
This 2,080-acre expanse of tallgrass prairie is managed by The Nature Conservancy. Located about 10 miles northeast of Clear Lake, it also shelters wetlands, springs, and fens that are important habitats for nesting and migratory birds, especially waterfowl. Rare plant and animal species are also found on the property, such as skipper butterflies, waxy bogstar, and Riddell's goldenrod.

Custer State Park
Custer County
Located on 73,000 acres about 10 miles southwest of Hermosa, this area is part of the Black Hills. Within its borders are shrub-covered foothills, a limestone plateau, upland pine-spruce forest, mixed- and short-grass prairie on the valley floors, and mixed hardwood forest on the lower slopes. It has a high biodiversity, including several rare plant, bird, and animal species. Raptors nest in the rocky canyons.

French Creek Natural Area
Custer County
This 2,200-acre site about 5 miles east of Custer shelters some of the canyonland that surrounds French Creek. Natural communities include short-grass prairie and mixed pine-hardwood forest in the canyons. French Creek's waters are very clean and support a strong population of trout. Bighorn sheep, elk, mountain lions, and bison frequent the canyon. A prairie dog town is located in the flats. Raptors are abundant and are often seen gliding over the canyon or perched amid its rock walls.

Hansen Nature Preserve
Brown County
The Nature Conservancy owns this 800-acre wetland and slough in southeastern Brown County. The area is mostly riparian habitat centered along the current flow in Stratford Slough. The aquatic habitat is a spawning ground for more than 20 species of fish. The slough is a critical nesting environment for many waterbirds, including herons and cormorants. Rare plant and animal species have also been identified here.

La Framboise Island Natural Area
Stanley County
This property contains a wooded sandbar in the Missouri River just south of Pierre and is one of the few remaining undisturbed river islands. It supports a remnant stand of old-growth cottonwood as well as green ash, box elder, and a small prairie. Numerous birds, amphibians, and turtles, some of them rare, have been identified here.

Makoce Washte
Minnehaha County
Located about 10 miles west of Sioux Falls, this 40-acre tract is a remnant of tallgrass prairie. The dominant plant species is big bluestem. Prairie birds are also found here. The wildflowers are colorful during summer. The property is owned by The Nature Conservancy.

Ordway (Samuel H.) Preserve
McPherson County
This 7,800-acre parcel is a large remnant of the prairie that once covered the Great Plains. Located about 8 miles west of Leola, it is dominated by head-high big bluestem and cordgrass, little bluestem, porcupine grass, and buffalo grass. Species diversity is high, with more than 300 different plant species. Its more than 400 pothole lakes are important wetlands for thousands of nesting or migrating ducks, geese, godwits, sandpipers, Baird's sparrows, terns, herons, and cranes. Herds of bison and numerous raptors can also be seen. The property is managed by The Nature Conservancy.

Sage Creek Wilderness Area
Pennington County
Located just outside Scenic, this 64,250-acre property includes rolling grasslands and Badlands with scattered juniper-sagebrush scrub on the hillsides. Cottonwood-ash-elm line the edges of Sage Creek. More than 200 species of flowers grow here, including the rare milk vetch, yellow prairie coneflower, and evening primrose. Rock and canyon wrens as well as warblers, swallows, and eagles frequent the canyon. The federally endangered black-footed ferret lives along the creek.

Sica Hollow State Park
Marshall County
Sica Hollow State Park is located about 5 miles south of Claire City. It encloses remnant prairie, seep, calcareous fen, marsh, a rare red dogwood forest, and bog communities. The rusty-red water derives from the iron in the surrounding ground. Wildflowers are abundant on the little bluestem prairie. The marshes and creeks are lined with silver maple, box elder, and various shrubs. More than 150 birds have been sighted here, including the threatened bald eagle and golden eagle.

Sioux Prairie
Moody County
Located in southwestern Moody County, this 200-acre tract encloses prairie-covered hills and pothole wetlands. Plant and animal diversities are high. The pothole lakes are important habitats for nesting and migrating birds. The property is owned by The Nature Conservancy.

Upper Pine Creek Research Natural Area
Pennington County
Part of the Black Elk Wilderness Area, this 1,205-acre RNA is one of the few roadless and ungrazed sections in the Black Hills. Located a few miles south of Hill City on the northern slope of Harney Peak, its topography is extremely rugged. Natural communities include ponderosa pine forest, white spruce forest, aspen-lined riparian corridor, and patches of grassland and meadow. The understory consists of various low shrubs, kinnikinnick, timber oatgrass, and golden pea. It is an important habitat for elk, deer, mountain goats, and mountain lions.

Vermillion Prairie
Clay County
Located about 15 miles northwest of Vermillion, this 22-acre Nature Conservancy property preserves a remnant of hillside prairie. It displays vivid wildflowers during spring and summer and is an important habitat for grassland birds, mammals, insects, and reptiles. Several rare species occur here.

Wilson Savanna
Lincoln County
This 160-acre property protects a rare combination of bur oak savanna and prairie. Other trees include basswood and walnut. Several varieties of birds have been identified here, including some rare or unusual species that nest on the site. It is owned and maintained by The Nature Conservancy.

TENNESSEE

Tennessee can be subdivided into three general natural regions: the Western Plains, the Central Basin and Highland Rim, and the Mountains and Valleys.

The Western Plains comprise roughly western third of Tennessee. This includes the valleys of the Mississippi and Tennessee Rivers, as well as the Coastal Plain between. The flat alluvial plains and rolling loess hills are covered by wetlands, broad swampy bottoms, stands of cypress-tupelo swamp, bottomland hardwood (poplar–gum–sycamore–elm–hickory–red maple) forest, and upland mixed hardwood forest. Much of this area has been converted to agriculture, and only remnants of the once-extensive wetlands remain. It is now a mosaic of agricultural fields and communities, segmented wetlands, dry upland forest, and mesophytic forest.

The middle third of the state is a flat-lying basin (the Central Basin) underlain by limestone and edged by bluffs and ridges rising up to 1,000 feet (the Highland Rim). The Central Basin is largely agricultural today and contains only remnant patches of grassland and stands of eastern redcedar. The rim is forested with upland tulip poplar, maple, pine, oak, and hickory. Unique glades and fens thrive on the limestone substrate. Dry grassy ridge tops can be found along the Rim.

The Mountains and Valleys, the easternmost province, comprise the Cumberland Plateau, Unaka Mountains, and Ridge and Valley—all part of the Appalachian Mountain chain. The terrain is steep with deep gorges, high cliffs, waterfalls, steep descents, and narrow valleys converted to agriculture. Slopes are largely covered by second-growth mesophytic forest and mixed pine-hardwood forest, with conifer trees on the ridge tops. The highest mountains along the eastern border range from montane to subalpine. Upper-elevation forests are red spruce and Fraser fir forest, with grassy and shrubby balds and heaths. Occasionally, cool, wet cove forests with tall, dense stands of hemlock–tulip poplar–umbrella magnolia can be found in pockets on the lower slopes. Rugged cliffs and rock faces shelter a variety of rare plant communities.

Special Habitat Sites

Big Cypress Tree State Natural Area
Weakley County
Located along the Middle Fork of the Ohio River about 20 miles southeast of Reelfoot, this state natural area contains high-quality examples of bottomland hardwood forest and old-growth cypress swamp. The flanking hills are covered by upland forest and grassland. This is a very diverse area; more than 200 songbird species gather here, some of them rare or endangered. The cypress trees are hundreds of years old and more than 100 feet tall.

Big South Fork National River and Recreation Area
Scott, Pickett, and Fentress Counties
Located on 110,000 acres about 10 miles west of Oneida along the Big South Fork of the Cumberland River, this rugged terrain consists of rock face, second-growth pine and mixed hardwood upland forest, and occasional rare stands of towering old-growth forest communities. The narrow valleys and gorges have created safe havens for several rare plants, including Lucy Braun snakeroot, Cumberland rosewood, and Virginia spiraea. The rare or endangered Indiana myotis bat, bald eagle, spotfin chub, red-cockaded woodpecker, and peregrine falcon inhabit the park. The forest is also an important black bear habitat.

Carter State Natural Area
Franklin County
Located 2 miles southeast of Sewanee, Lost Cove Cave harbors the Tennessee cave salamander and the big-eared bat, two rare species. Cave systems, special habitats for a variety of bats, amphibians, and isopods, are easily degraded by public overuse and changing water chemistry.

Catoosa Wildlife Management Area
Cumberland and Morgan Counties
Located on 80,000 acres about 15 miles northeast of Crossville, this rugged canyon along Obed River includes high-quality examples of mature pine–mixed hardwood forest. The river waters support rare mussels, the federally threatened spotfin chub, the endemic Cumberland Plateau muskie, and the river otter; ospreys and peregrine falcons nest along its edges. Other important species are golden and bald eagles and the red-cockaded woodpecker, which prefers to nest in the old-growth stands of shortleaf pine forest on this property.

Cedars of Lebanon State Forest
Wilson County
This well-managed tract about 5 miles west of Norene is the largest red cedar forest in the United States, although it is second-growth. Occasional glades and openings have developed on limestone substrate, supporting about 20 unique species that are specially adapted to calcium-rich soils. Rare plants include the Tennessee purple coneflower, glade mint, breadroot, and Gettinger's purple prairie clover. The forest is an important bird habitat for common and rare species, such as pileated woodpeckers and various warblers.

Citico Creek Wilderness
Monroe County
Located on 15,891 acres about 5 miles east of Tellico Plains, this wilderness area is part of Cherokee National Forest. It is roadless, and its mature pine forests and mixed lowland-upland forests are a critical habitat for large forest mammals, such as the black bear.

Clark Conservation Area
Fayette County
An excellent remnant of bottomland hardwood forest is protected on this 330-acre conservation area along the Mississippi River. It contains high-quality examples of riparian and floodplain forest habitats. It is an important habitat for songbirds, woodpeckers, and several rare plants and amphibians. The property is owned and managed by The Nature Conservancy.

Coker Creek Scenic Area
Monroe County
Located in the Unicoi Mountains of eastern Tennessee near the town of Cokercreek, this state-managed natural area was established to protect old-growth hemlocks along Coker Creek. The trees are centuries old and escaped logging due to the steep terrain. Several rare plants and amphibians occur in this moist, shaded environment.

Cross Creek National Wildlife Refuge
Stewart County
Located just east of Dover along the Cumberland River, this refuge includes a vast wetland-forest complex. Natural communities include swamp, marsh, lake, bottomland forest, mixed hardwood upland forest, and cliff exposure. It is considered to be a critical habitat for nesting and wintering bald eagles, Mississippi kites, snow buntings, peregrine falcons, golden eagles, ospreys, and the federally endangered wood stork. Nearly a million ducks, geese, and other waterbirds rest and feed here during spring and fall migrations. Some of the oak-beech-hickory forests are old-growth and quite tall. Wildflower displays are prominent during spring and summer, reflecting the high plant biodiversity.

Frozen Head State Park and Natural Area
Morgan County
Located on 12,000 acres about 6 miles east of Wartburg, this natural area protects a high-elevation mix of cove forest and upland hardwood forest on the divide between the Cumberland and Tennessee Watersheds. It is a critical breeding site for the rare cerulean warbler, other songbirds, and veeries. The rare Black Mountain dusky salamander, an endemic species, and timber rattlesnakes can be found here. Specialized plant communities cling to rock outcrops and boulders.

Great Smokies National Park
Blount and Sevier Counties
This park, about 20 miles south of Sevierville, is one of the most biologically diverse areas in the United States. Nearly 2,000 different plant species from a range of natural communities have been recorded here; the variety is partly due to the topographic changes, rainfall variations, and changing rock and soil chemistries. More than 100 tree species grow on the forested slopes, and about 25 percent of the 520,000 acres is virgin timber (oak-pine, oak–hickory–red maple, spruce-fir) that existed during presettlement times. The vast stretches of unbroken forest are critical habitats for large mammals such as black bears. Nearly 30 different salamanders can be found along the riparian corridors here—the highest number anywhere in the world—including the very rare endemic red-cheeked salamander. Other special species include timber rattlesnakes, more than 200 types of birds, red wolves, and Appalachian salamander. Great Smokies National Park is one of the most heavily visited parks in the country.

Grundy Forest State Natural Area
Grundy County
Located about 5 miles south of Tracy City, this rugged natural area preserves clear stream, riparian corridor, waterfall, and mature mixed upland forest communities. Steep slopes deterred lumbering, and some sections contain rare towering old-growth forests that are nearly 500 years old. Several rare or unusual ferns, mosses, and lichens can be found on the rocks and in the canyons. The old-growth forest is an important habitat for species that depend on old-growth trees, such as woodpeckers, owls, and other interior forest birds.

Long Hunter State Park
Davidson County
Located about 10 miles east of Nashville, this 14,000-acre park includes lakeshore, limestone bluff, dry upland hardwood forest, red cedar forest, cedar glade, and prairie communities. The cedar glades shelter a variety of unusual and rare plants and wildflowers. The lake, wetlands, and forest are important bird habitats, especially for nesting ospreys. Rare glade-adapted land snails and several rare butterflies can be found here.

Meeman-Shelby Forest State Park
Shelby County
Located about 8 miles west of Millington along the Mississippi River, this property encloses mature bottomland hardwood forest, upland pine-hardwood forest, and various wetlands and sloughs. The forest is an important habitat for a great number of songbirds, owls, Mississippi kites, and flycatchers. The federally endangered least tern nests on nearby river islands. The rare banded whitelip snail can be found among the leaves and branches on the forest floor.

Mount View Cedar Glade
Davidson County
Owned and managed by The Nature Conservancy, this property protects a rare cedar glade community. This rocky, grassy opening supports a rare population of the federally endangered Tennessee coneflower.

Natchez Trace State Forest Resort
Carroll and Henderson Counties
This 48,000-acre tract just east of Yuma protects old-growth stands of oak-hickory forest and a well-developed dogwood-sassafras understory. Within its borders is the third largest pecan tree in the world; it is hundreds of years old and more than 100 feet tall. The forest is an important nesting habitat for more than 50 bird species, including the rare cerulean warbler, worm-eating warbler, pileated woodpecker, and several owls.

Nickajack Cave
Marion County
Near South Pittsburg, this small tract protects a cave system and its surrounding forested slopes. The cave is a nursery for 50,000 federally endangered female gray bats. Other rare species that depend on this specialized habitat are the blind Tennessee cave salamander, the Indiana bat, and several isopods.

Prentice Cooper State Forest
Marion County
Located on 26,000 acres about 10 miles northwest of Chattanooga, this rugged area of limestone cliffs is along the Tennessee River gorge. The dominant natural community is mature mixed upland hardwood forest. Rare plants, such as mountain mint, rose gentian, and mountain skullcap, thrive in cool, moist microclimates in the deep, rocky ravines and cliffsides.

Rankin Bottoms
Jefferson County
Located west of Newport, this 740-acre wetland complex beside Douglas Reservoir is an important bird habitat. Natural communities include the lake, mudflat, extensive marsh, slough, swamp forest, and mixed pine-hardwood forest. Shorebirds in great numbers visit during spring and fall migrations. Rare or unusual species include various sandpipers, bald eagles, northern harriers, ospreys, short-eared owls, and swamp and savanna sparrows.

Roan Mountain State Park
Carter County
This 21,156-acre tract is located about 20 miles southeast of Elizebethton. Its dominant natural communities include junglelike forests of rhododendron, low-elevation oak-beech-maple forest, higher-elevation spruce-fir forest, and grassy bald and shrub heath on the upper ridges. These bare, windswept openings are covered with grasses, herbs, heather, and other plants that thrive among rocky exposures. The site is an important habitat to more than 150 bird species and nearly 200 wildflowers, several of them rare or of restricted range, reflecting the high biodiversity here. Birds include flycatchers, veeries, rare warblers, and saw-whet owls. The endangered northern flying squirrel and New England cottontail also inhabit the forested slopes.

Savage Gulf State Natural Area
Grundy County
This 11,500-acre tract is about 5 miles northwest of Barkertown. Natural features include steep rocky ridges, karst topography (caves and sinkholes), underground streams, grassy meadows, seeps, and virgin stands of maple-hickory-yellow poplar-beech-hemlock forest that are more than 500 years old and more than 100 feet tall. Grassy balds occur on the ridge tops. The area is very diverse, with nearly 800 different plants being recorded, including some rare northern relict species. The rare false hellebore is endemic only to this part of Tennessee. Stands of old-growth trees are critical habitats for several bird and animal species, including the red-cockaded woodpecker.

Sneed Road Cedar Glade
Williamson County
Just southwest of Nashville amid heavy development, this small cedar glade supports a healthy population of the globally rare leafy prairie clover. The site is owned and managed by The Nature Conservancy.

South Cumberland State Recreation Area
Grundy County
Located a few miles north of Monteagle, this 841-acre property includes upland hardwood forest, cave, rare rock face, sink, freshwater marsh, pond, and glade communities. Many songbirds inhabit the woodlands during summer, including the rare worm-eating warbler. Other unique species include the timber rattlesnake and blind Tennessee cave salamander, an endemic species.

Washmorgan Hollow
Putnam County
Located near Cookville, this property encloses high-quality examples of mature mixed-hardwood bottomland and upland forest, wetlands,

and riparian habitats. Wildflowers are abundant during spring and summer, and the forest and wetlands are important nesting sites and migratory stopovers for a variety of bird species. The property is owned and managed by The Nature Conservancy.

TEXAS

Texas can be subdivided into nine major natural divisions: Pineywoods, Gulf Prairie and Marsh, Post Oak Savanna, Blackland Prairie, South Texas Plains, Edwards Plateau, Rolling Plains, High Plains, and the Trans-Pecos.

Pineywoods country is flat to slightly rolling. Located along the eastern edge of the state, its uplands are covered by mixed pine-hardwood forest. Pine species include loblolly, shortleaf, and longleaf. The lowland mixed hardwood forest is dominated by oak, hickory, maple, and beech.

Curving along the Gulf of Mexico is the Gulf Prairie and Marsh. This narrow strip of coastal vegetation includes extensive beach, dune, saltwater marsh, estuary, brackish marsh and pond, and freshwater marsh communities. Inland from these is the Gulf Prairie, an expanse of tallgrass prairie and post oak–live oak savanna dominated by little bluestem, Indian grass, and gulf cordgrass.

Just west of the Pineywoods is a mixture of post oak savanna and blackland prairie. The savanna is characterized by post oak, blackjack oak, occasional juniper, and tallgrass prairie. The prairie grasses are dominated by bluestem, Indian grass, dropseed, gramas, Texas wintergrass, and many blooming forbs.

The South Texas Plains can be found along the Rio Grande in south-central Texas. Originally a grassland and savanna, it is now rangeland and irrigated farmland. Vegetation types include seacoast bluestem, bristlegrass, buffalo grass, Indian grass, gulf cordgrass, and prickly pear. Brushy chaparral occasionally grows in the uplands.

Most of central Texas is occupied by Edwards Plateau. Flat and rocky, this terrain has many streams, drainages, ravines, caves, and sinkholes. The surface is covered by live oak–shin oak–juniper savanna, with tallgrass prairie (bluestem, Indian grass, and switchgrass). Some short-grass prairie is present in the western portion.

The Rolling Plains and High Plains occur in north-central Texas. Tall- and mid-grass prairies dominate to the east, with bluestem, gramas, wheatgrass, and some shin oak savanna. The western section (the High Plains) is more mixed-grass and short-grass prairie, including blue grama, buffalo grass, galleta, wheatgrass, and sideoats. Most of this area has been converted to agricultural use.

The Trans-Pecos is located in western Texas, a rugged area with mountain elevations up to 8,700 feet. From low elevation to high elevation, the dominant communities are: creosote-sage shrubland, desert grassland, pinyon–juniper woodland, Mexican pinyon woodland, ponderosa pine forest, and mixed spruce-fir forest.

Special Habitat Sites

Armand Bayou Nature Center
Harris County
Located on 2,500 acres on the southeastern edge of Houston, this property protects excellent remnants of woods, wetland, and one of the last remaining gulf coastal tallgrass prairies in Texas. Water levels, affected by groundwater pumping, are being restored. It is important habitat for a variety of birds, mammals, and reptiles, including bobcats and marsh rabbits. Four rare plants also grow here.

Attwater Prairie Chicken National Wildlife Refuge
Colorado County
Located about 5 miles north of Eagle Lake, this 8,000-acre property protects the endangered Attwater greater prairie chicken. Other birds, mammals, reptiles, and insects, some of them rare or unusual, live in the grasslands. The relatively unusual crested caracara can also be seen here.

Balcones Canyonlands National Wildlife Refuge
Travis, Burnet, and Williamson Counties
Located in Texas Hill country on 41,000 acres north of Lago Vista, this refuge is on Edwards Plateau. It is being developed to protect two of America's rarest birds: the black-capped vireo and golden-cheeked warbler. Natural communities include canyonland and karst caves and

sinkholes. The warbler uses the bark of the Ashe juniper for nests. More than half of the rare post oak–Ashe juniper woodland, which is critical habitat for these birds, has been destroyed by development and industry. Other rare species found here are Hellier's false-gromwell, Texabama croton, and Glass Mountain coral-root.

Barton Creek
Travis County
This 4,000-acre nature preserve protects an important part of Barton Creek, associated riparian habitat, and flanking communities, including upland forest. In addition to being a pristine stream with rare amphibians, it is a critical recharger of the underlying aquifer, a highly used water source. The rare Barton Springs salamander lives in the area, along with the black-capped vereo and golden-cheeked warbler. The property is owned and managed by The Nature Conservancy.

Bastrop State Park
Bastrop County
Located just outside Bastrop, this 3,500-acre tract contains loblolly pine forest, an unusual natural community for Texas. The rugged hills, streams, gullies, and riparian corridors are critical habitats for the endangered Houston toad, which is endemic to only seven counties in southeastern Texas. Pine warblers and pileated woodpeckers frequent the pine forest.

Bentsen–Rio Grande Valley State Park
Hidalgo County
This 588-acre tract along the Rio Grande just southwest of Mission preserves a riparian strip of cedars, elms, willows, and dry thorn scrub. It is important habitat for the 8-foot purple indigo snake, various kites, raptors, and rare chachalacas, green jays, and Texas tortoises.

Big Bend National Park
Brewster County
Part of the Chihuahuan Desert along the Rio Grande east of Study Butte, this 800,000-acre tract includes valleys, streambeds, and mountains. The flats are vegetated with sparse grasslands. Riparian corridors along streams consist of tall cottonwoods, willows, mesquite, buttonbush, and reeds. Sparse oak-juniper woodland (including the rare endemic Chisos oak), open ponderosa pine forest, and Douglas fir–quaking aspen forest grow on the upper ridges. The endemic Sierra del Carmen white-tailed deer inhabits the forest, as do mountain lions. More than 450 birds have been identified, including the very rare nesting Colima warbler, golden eagles, peregrine falcons, black-capped vireos, and Scott's orioles.

Big Bend Ranch State Park
Brewster County
Located on 277,000 acres in the Chihuahuan Desert along the Rio Grande, this park includes ten steep canyons that drain into the river. The dominant community is desert grassland (grama grasses, silver bluestem, and tobosa). Shrub zones of creosotebush, tarbush, mariola, and mesquite with some cacti, ocotillo, and acadias grow on the lower slopes. The endangered Hinckley's oak is part of the scrub oak–evergreen woodland that grows in the cooler, shaded canyons. Ash-willow-cottonwood stands form lush, shaded riparian corridors in the canyon bottoms, where cactus wrens nest. Mountain lions prowl the streambeds and rocky slopes. Other unique species include western mastiff bats (the largest bats in North America) that roost on cliff faces, kangaroo rats, and rare spadefoot toads. More than 400 bird species, including eagles and peregrine falcons, nest in the canyons.

Big Thicket Nature Preserve
Hardin, Newton, Polk, and Tyler Counties
Thirteen individual tracts totaling 86,000 acres around the town of Kountze contain very diverse natural communities that include swamp, bald cypress–water tupelo slough, a section of riparian corridor along the Neches River, and upland beech–southern magnolia–loblolly pine forest. The swamps contain sweet bay magnolia and gallberry holly. The wildflowers are spectacular during spring. The bogs and sloughs also support healthy populations of carnivorous sundews and pitcher plants.

Bolivar Flats Shorebird Sanctuary
Galveston County
Natural communities on this property on the outskirts of Port Bolivar include ocean beach, tidal marsh, and mudflat. Hundreds of thousands of shorebirds and waterfowl come here to feed and rest during spring and fall migrations. The more common species include gulls, terns, shorebirds, spoonbills, pelicans, and

avocets. Several rare or unusual birds can also be spotted here.

Buck Wildlife Management Area
Kimble County
This 2,155-acre property in the Hill Country along the South Llano River southwest of Junction preserves important oak-juniper woodland, riparian habitats, and cedar groves. The woodland is a critical nesting habitat for the federally endangered black-capped vireo. Rare Tobusch fishook cacti also grow on the site.

Caprock Canyons State Park
Briscoe County
This 14,000-acre tract about 15 miles southeast of Silverton includes a steep-edged escarpment that separates short-grass High Plains prairie from mixed-grass Rolling Plains prairie. The elevation difference along this ecotone is about 200 feet. The park is an important habitat for nesting raptors, such as golden eagles, red-tailed hawks, Swainson's hawks, and ferruginous hawks. Small caves in the escarpment are hibernating roosts for free-tail bats.

Chandler Independence Creek
Terrell County
A portion of Chandler Independence Creek is protected on this Nature Conservancy preserve about 30 miles south of Sheffield. The property encloses the creek, flanking riparian habitats, and upland forests. Chander Independence Creek is a pristine recoverable freshwater tributary to the Lower Pecos River and an important water source. Numerous rare or unusual species can be found here, including the prosperine shiner in the creek and the black-capped vireo in the surrounding trees.

Chihuahua Woods
Hidalgo County
This 243-acre property in southwestern Hidalgo County is located just south of Chihuahua. It encloses a relict habitat that shelters many species indigenous to the Rio Grande Valley and boasts a very diverse cacti population. More than 30 bird species nest here, including some rare and endangered species. It is also documented habitat for ocelots and jaguarundi. The property is managed by The Nature Conservancy.

Clive Runnells Family Mad Island Marsh
Matagorda County
This 7,000-acre Nature Conservancy preserve is located along West Matagorda Bay southwest of Collegeport. Natural communities that thrive on the property are coastal wetland and coastal prairie. Rare or unusual birds, such as sandhill cranes, nest on Mad Island. The wetlands attract thousands of waterbirds during spring and fall migrations.

Clymer Meadow Preserve
Hunt County
This Nature Conservancy property consists of about 500 acres 10 miles northwest of Greenville. Many regard it as the best remnant of pristine blackland prairie in Texas. The dominant vegetation is little bluestem–Indian grass, which covers gently rolling hills. Other species include big bluestem, switchgrass, and eastern gamma grass. A few oak and elm trees grow in the draws. Wildflowers are colorful and abundant and include Indian paintbrush and coneflower. The prairie is an important habitat for grassland birds, such as dickcissels and flycatchers, as well as small mammals and various insects.

Cross Timbers Research Natural Area
Wise County
Located on 370 acres near Pecan Creek, this site encloses remnants of the original vegetation of the western Cross Timbers and Grand Prairie regions of Texas. Although it has been disturbed by grazing, many of the original plant communities remain. The prairie is dominated by little bluestem and Indian grass, the woodland by post oak and blackjack oak. Some of the larger trees are more than 200 years old. The site is being restored through prescribed burns.

Dallas County Escarpment Preserve
Dallas County
Located on 630 acres, this undeveloped nature preserve near Dallas encloses mixed evergreen-deciduous woodland and rock cliff. The pinyon pine–juniper woodland is considered to be an excellent habitat for the endangered black-capped vireo and golden-cheeked warbler. Rare plant species include the Glass Mountain coral-root and Warnock's coral-root.

Davis Mountains State Park
Jeff Davis County
Located a few miles southwest of Fort Davis, this park encloses desert grasslands and higher-elevation oak-juniper forests. It is known as a critical habitat for the Montezuma quail, a shy and uncommon species. The higher ponderosa forests support mule deer and mountain lion populations; pronghorn graze on the lower grasslands.

Diamond Y Spring Preserve
Pecos County
Located near Fort Stockton, this 1,502-acre property is a rare saline *cienega* (spring). This site is the largest and most pristine *cienega* that remains in West Texas. The spring is rimmed by marsh, grass, and shrub species. This unique system is a critical habitat for two globally rare fish, a rare salt-tolerant plant (the puzzle sunflower), and three rare snails. It is the only site in the world where the federally endangered Leon Springs pupfish can be found.

Dolans Falls Ranch
Val Verde County
This Nature Conservancy property is located along Devil's River. Natural communities include Dolan Creek, flanking riparian habitat, rocky bluff, prairie, and forested upland. As part of the edge of Edwards Plateau, it has both upland and lowland species of flora and fauna, including the Texas snowbell, Tobusch fishhook cactus, Devil's River minnow, Conchos pupfish, Big Bend blackheaded snake, and black-capped vireo. Raptors can also be seen in the area. Dolan Falls is also the largest volume perennial waterfall in the state.

Eckert James River Bat Cave Preserve
Mason County
Owned and operated by the Nature Conservancy, this 10-acre property protects a rare cave system. It is one of the largest bat colonies in the world, with up to 6 million individuals from various species, including the rare Mexican free-tail bat, which winters in Texas. Nearly 4 million of the inhabitants are pregnant females, which give birth and nurse their young during spring before returning to Mexico. Bats are important for cross-pollinating plant species and controlling insects. Witnessing the "bat tornado" that emerges from the cave at dusk is a breathtaking experience in this unique natural site.

Enchanted Rock State Natural Area
Gillespie County
Located 15 miles north of Fredericksburg, this property includes a steep-sided dome of sparsely vegetated pink granite. Vernal pools collect in depressions on its surface and support a remarkable array of hardy plants, such as algae, quillworts, sedges, grass, wildflowers, and some rare species. Tiny fairy shrimp, which survive dry periods as eggs, hatch in abundance when the pools refill with rainfall. Rare endemic plants to this part of Texas are basin bellflower and rock quillwort.

Ezell's Cave Preserve
Hays County
Owned and managed by The Nature Conservancy, this 3-acre tract north of San Marcos protects a small fissure cave in central Texas. Although a small site, it encloses one of the most diverse cave biotic communities in the United States. Forty-three rare or unusual cave species inhabit the cave system and its stream. These include the globally rare Texas blind salamander, which occurs nowhere in the world except Hays County. The property is a National Natural Landmark.

Gene Howe Wildlife Management Area
Hemphill County
Located outside Canadian, this 5,821-acre tract near Lake Marvin protects high-quality upland sand sagebrush–mixed grass prairie, tallgrass prairie, and cottonwood-sycamore-willow riparian woodlands along Canadian River. Special species found here include the nesting interior least tern; the cottonwood–tallgrass series is an unusual assemblage in the state.

Godwin Ranch Preserve
Williamson County
This 106-acre tract on Edwards Plateau near Georgetown protects a cave-sinkhole system, including Red Crevice Cave and Twisted Root Cave. The surface is covered by a grass-forb savanna with stands of live oak–Ashe juniper. Several rare cave species can be found here, including the coffin cave mold beetle and bone cave harvestman. Bats hibernate in the caves during winter months.

Goose Island State Park
Refugio County
Part of the Gulf Islands, this 314-acre park just northeast of Lamar is at the southern tip of the

Arnasas National Wildlife Refuge. Natural communities include beach, saltwater marsh, brackish pond, and coastal live oak forest. Some of the live oaks are old-growth and nearly 1,000 years old, a very rare occurrence. The wetlands support a variety of marine life, alligators, and hundreds of bird species during spring and fall migrations.

Government Canyon State Park
Bexar County
This 4,717-acre park near Helotes protects critical habitat for two rare bird species. Its mature oak-juniper woodlands on rolling hills is considered an excellent habitat for the rare golden-cheeked warbler and black-capped vireo.

Gus Engleling Wildlife Mangement Area
Anderson County
Located about 21 miles northwest of Palestine, this 11,000-acre property encloses stream, riparian corridor, and high-quality post oak–hickory woodland. The dominant understory is greenbrier, dogwood, hawthorn, yaupon, and elm. Rare or endangered species that depend on this habitat are the alligator snapping turtle, sphagnum-beakrush series, three species of caddisfly, several sedges, pipewort, and the sandhill woolywhite.

Hamilton Pool County Park
Travis County
Located near Shingle Hills and Hammetts Crossing, this 232-acre park encloses several important habitats. Natural communities include a stream and riparian corridor, bald cypress–sycamore bottomland forest, oak savanna, prairie, seep, and spring. There is an extremely high diversity of native species, including the federally endangered golden-cheeked warbler and canyon mock-orange. The wetlands and forests attract a variety of other birds, mammals, and amphibians.

Hill Country State Natural Area
Bandera County
Consisting of 5,400 acres a few miles southwest of Bandera, this natural area protects pristine grassland and oak-juniper woodland communities. It is critical habitat for the shy, nocturnal ringtail cat as well as the federally endangered golden-cheeked warbler and black-capped vireo, both of which nest on the site.

Honey Creek State Natural Area
Comal County
Located near Anhalt, this 2,293-acre state natural area preserves an important upland ecosystem. Natural communities include plateau live oak woodland, stream, riparian corridor, and bald cypress–sycamore floodplain forest. The property straddles an ecotone between the upland ecosystem and a prairie grassland of Indian grass, little bluestem, and switchgrass. Several rare species can be found here, including the Comal blind salamander, Edward Plateau spring salamander, golden-cheeked warbler, Guadalupe bass, and Cagle's map turtle. Caves on the property harbor bats and unusual invertebrates.

Kerr Wildlife Management Area
Kerr County
This 6,493-acre property is located about 14 miles northwest of Hunt. Part of Edwards Plateau, the rolling terrain is covered by grassland, live oak–shin oak thickets, cedar brakes, springs, and mature Ashe juniper–live oak woodlands. The site is a critical nesting habitat for the federally endangered golden-cheeked warbler and black-capped vireo. The rare Tobusch fishhook cactus grows on the hillsides.

Kickapoo Cavern State Park
Kinney and Edwards Counties
Located 22 miles north of Brackettville, this 6,368-acre park on Edwards Plateau straddles the Kinney County–Edwards County boundary. Fifteen caves provide habitat for millions of Brazilian free-tail bats, which migrate here during winter. The surrounding Hill Country is covered by mature live oak, juniper, Texas persimmon, and pinyon pine forests, critical habitats for the rare black-capped vireo, barking frog, mottled rock rattlesnake, and Texas alligator lizard.

Lake Meredith National Recreational Area
Pooter, Moore, and Hutchinson Counties
Administered by National Park Service, this 17,000-acre parcel along Canadian River near Fritch borders Lake Meredith. Communities include open water, wetland, rock ledge, flatland of yucca–prickly pear–grass, and riparian cottonwood-willow-soapberry corridor. Located on a flyway, the lake is an important migratory stopover for a variety of birds during

spring and fall. Wintering bald eagles can be seen here, as can loons, grebes, ducks, sandhill cranes, and several unusual reptiles and amphibians.

Las Palomas Wildlife Management Area
Hidalgo County
This 7,700-acre area near Edinburg encloses excellent examples of mature Texas ebony–anacua subtropical woodland. This dense vegetation is critical habitat for the white-winged dove and provides nesting, feeding, and roosting sites. Rare species that are monitored here include Texas ebony, Runyon's water-willow, several cacti, Vasey's adelia, Montezuma bald cypress, and Bailey's ball moss.

Lennox Woods Preserve
Red River County
Located about 10 miles from Clarksville, this 375-acre tract is managed by The Nature Conservancy. Natural communities include old-growth bottomland hardwood forest, pine-oak upland forest, and a small pecan bayou along an intermittent stream. This is some of the oldest remaining forest in Texas. Some of the post oak trees are 300 years old; the loblolly pine trees approach 200 years old. This is an important habitat for species that prefer old-growth forests, such as pileated and red-cockaded woodpeckers. The rare Arkansas meadow-rue can also be found here.

Lime Creek Preserve
Travis County
This 100-acre site near Jollyville is part of the Edwards Plateau karst area. A series of caves and sinkholes occur on the property, including Rolling Rock Cave and Broken Arrow Cave. The surface is covered with stands of oak-juniper woodland and native grasses that may be suitable habitat for the federally endangered golden-cheeked warbler and black-capped vireo. Several endangered invertebrates inhabit the cave systems.

Lost Maples State Natural Area
Bandera County
A few miles north of Vanderpool, this state natural area protects a rare stand of mature bigtooth maple trees in a canyon of the Sabinal River. These trees are thought to be relicts from the last period of glaciation and survive because of the shaded, cooler microclimate in the canyon. Many rare flora and fauna inhabit the canyon, including Texas madrona, witch-hazel, sycamore-leaf snowbell, American smoke tree, Canada moonseed, golden-cheeked warblers, black-capped vireos, green kingfishers, Texas salamanders, Texas cliff frogs, Guadalupe bass, and barking frogs.

Matagorda Island
Calhoun and Refugio Counties
Natural communities on this undisturbed barrier island include beach, dune, and tidal flat communities dominated by saltgrass, glasswort, saltwort, and cordgrass. This highly productive marsh is critical habitat for various crabs, oysters, and other shellfish, which are important food sources for the more than 300 bird species that nest or visit here. It is wintering ground for the rare whooping crane. Other special species are the nesting brown pelican, threatened Texas horned lizard, peregrine falcon, Ridley turtle, short-eared owl, and piping plover.

McKittrick Canyon
Culberson County
Part of Guadalupe Mountains National Park about 100 miles east of El Paso, the cool, shaded interior of McKittrick Canyon supports rare relict species, such as Texas walnut, bigtooth maple, velvet ash, and the Texas madrona tree. Many species depend on this lush riparian corridor for food and shelter; raptors nest among the jagged rocks of the canyon walls.

Palmetto State Park
Gonzales County
Just southwest of Ottine, this 265-acre site along San Marcos River encloses wetland and bottomland forest on the river's floodplain. Natural communities include Rutledge Creek, seep, spring, bog, marsh, oxbow lake, and some peat deposit—all relicts from a cooler glacial climate. Oak and elm trees have an understory of dwarf palmetto; swamp flora include buttonbush and black willow. Like all wetlands, especially in arid climates, those in Palmetto State Park are important habitats for a variety of animals and birds.

Palo Duro Canyon State Park
Armstrong and Randall Counties
Located about 15 miles east of Canyon, this 16,400-acre property encloses a steep rock canyon. Natural communities include pinyon-

juniper woodland, shin oak woodland, and underlying grasses. Dominant groundcover consists of the cottonwood tallgrass–sideoats grama series, blue grama–buffalo grass series, and Harvard shin oak–tallgrass series. The rare and endemic Palo Duro mouse can be found here.

Parkhill Prairie Preserve
Collin County
This 436-acre remnant of virgin blackland tallgrass prairie is dominated by big and little bluestem, Indian grass, switchgrass, and eastern grama grass. Wildflowers are beautiful and abundant and include Indian paintbrush, prairie parsley, coneflower, and blazing star. The tall grasses are important habitats for prairie-loving birds, such as dickcissels and flycatchers, as well as small mammals and reptiles.

Pedernales Falls State Park
Blanco County
Located 10 miles northeast of Johnson City, this high-quality Ashe juniper–oak woodland along Pedernales River is critical breeding habitat for the federally endangered golden-cheeked warbler. The rare black-capped vireo and golden-fronted woodpecker also live in the riparian corridor.

Ruth P. Lehmann Preserve
Travis County
Located near Jollyville, this 160-acre property is centered on Long Hollow Creek. Natural communities include the creek, riparian corridor, spring and seep, cave, upland oak-juniper woodland, and heavily wooded bottomland forest. Biodiversity is high. Rare or endangered species that depend on these habitats include the golden-cheeked warbler, black-capped vireo, the endemic Jollyville Plateau salamander, and several cave amphipods.

Sabal Palm Grove Sanctuary
Cameron County
Owned by the National Audubon Society, this 172-acre sanctuary borders the Rio Grande near Brownsville. It represents one of last woodland stands of virgin to second-growth Texas sabal palm riverine forest. This habitat is almost extinct from development. Spiny hackberry, lime prickly ash, manzanita, and the uncommon David's milkberry grow in the understory. Rare ocelots, hummingbirds, and jaguarundi inhabit the sanctuary.

Sam Houston National Forest
Polk, Walker, and Montgomery Counties
This national forest is best known for its 130 breeding colonies of the red-cockaded woodpecker, a threatened species. Mixed hardwood forest, longleaf pine forest, slash pine forest, sweet gum–tupelo–red maple wetland, and occasional cypress swamp are the dominant communities. Several varieties of shrubs, including the uncommon pawpaw, form the understory. Other rare or unusual species that depend on this habitat are the western lesser siren, southern dusky salamander, Strecker's chorus frog, American alligator, Mississippi map turtle, six-lined racerunner, and a variety of snails.

Stephen F. Austin Experimental Forest
Nacogdoches County
Located on 2,760 acres near Clawson between Angelina River and Alazan Bayou, this property encloses rare old-growth bottomland hardwood forest and upland pine and pine-hardwood forest. Some of the oak trees are centuries old and nearly 100 feet tall. Rare plants that are monitored here include the water oak–willow oak series and the Texas screwstem.

Testudo Tube Cave Preserve
Williamson and Travis Counties
Located near Jollyville on the Edwards Plateau, this karst area includes a series of small caves and sinkholes. The surface is covered by stands of live oak–Ashe juniper, with patches of native grass. The cave system protects several rare cave invertebrates and many paleontological fossil sites.

Thick-Spiked Tridens Prairie
Lamar County
This 97-acre tract of remnant blackland prairie is located about 6 miles west of Paris. It is rare in that it is dominated by the unique tallgrass community of silveanus dropseed, a remnant of the presettlement prairie. It is an important habitat for a number of prairie-dependent birds, mammals, insects, and reptiles.

Wild Basin Wilderness Preserve
Travis County
Located in the city of Austin, this 227-acre preserve encloses woods, grasslands, and the riparian corridor along Bee Creek. High quality oak-juniper woodland grows along the creek, where rare orchids and shooting star grow. The woodland is an occasional habitat

for black-capped vireos and golden-cheeked warblers, two endangered species. Other trees include the Texas mountain laurel, cedar, elm, and the rare Texas madrona. The uncommon Heller's false-gromwell and the Glass Mountain coral-root can be found in the understory. More than 100 bird species have been identified here, including the painted bunting and golden-cheeked warbler.

UTAH

Utah is divided into three major physiographic regions: the Plateau Region of southern and central Utah, the north-south Wasatch Range in the center, and the Great Basin to the west. The Plateau stretches across Uintah, Emery, Grand, Wayne, Garfield, Kane, and San Juan Counties, with a few isolated ranges. The western front of the Wasatch Mountains borders Interstate 15, which runs from Ogden to Saint George. The Great Basin comprises a series of parallel, north-south–trending mountain ranges and flanking valleys that extend westward across the arid country west of Highway 15. Salt flats and playa lakes often develop in the centers of the valleys.

The vegetation zones are dependent on altitude, soil chemistry, and precipitation. Low-elevation communities include warm desert shrub, chaparral, salt desert shrub, sand desert shrub, cool desert shrub, and riparian corridors. Warm desert shrub (creosotebush–Joshua tree–blackbrush–cholla–sagebrush) are found in the southeastern corner of the state. Turbinella live oak chaparral can be found adjacent to the Wasatch Range in Washington County. Shadscale-saltbush-greasewood-sage (salt desert shrub) and sagebrush–rabbitbrush–blue grama (cool desert shrub) occur in the Great Basin. Sagebrush-yucca sand desert shrub can be found in the Great Basin and in southeastern Plateau country. Riparian corridors are commonly lined with cottonwood, birch, alder, willow, and box elder trees, as well as various grasses and sedges.

Middle-elevation vegetation zones include pinyon-juniper, mountain brush, and ponderosa pine communities. Open pinyon-juniper woodlands are common across the state, typically in the foothills. Gambel oak–bigtooth maple–mountain mahogany–big sagebrush (mountain brush) also occurs in the foothills, or in slightly wetter canyons. Open ponderosa pine forest has typically developed on drier sites.

Various spruce forests and alpine tundra dominate the higher elevations. White fir–Douglas fir, Engelmann spruce–alpine fir, and bristlecone pine forests represent the typical progression as altitude increase. Above the spruce-fir forests is timberline—and the windswept alpine tundra communities. Usually above 10,500 feet, these rocky summits and ridgelines are vegetated with hardy grasses and shrubs, which form mats and meadows.

Special Habitat Sites

Bryce Canyon National Park
Garfield County
Located about 15 miles east of Hatch, this property is known for its brilliant rock formations and high-elevation alpine habitats. Lower-elevation communities include riparian vegetation in the canyon bottom, big sagebrush–prickly pear flats, cliff face, pine-juniper forest, and Gambel oak–pinyon pine woodland. Upper elevations are covered by ponderosa pine (7,500–8,500 feet), greenleaf manzanita, Douglas fir–blue spruce, and subalpine bristlecone pines that are more than 2,000 years old. Several rare or unusual endemic plant species can be found here that are specially adapted to the lime-rich soils. Nearly 200 bird species have been recorded. The park is an important habitat for deer, mountain lions, Rocky Mountain elk, blue grouse, goshawks, peregrine falcons, black bears, Ord's kangaroo rats, and the rare ringtailed cat.

Cascade Springs
Wasatch County
Located near Orem, this wetland opening in the forested Gambel oak slopes consists of a series of shallow travertine pools, underlain by pearly calcium carbonate deposits. The alkaline groundwater and soil support unique plant communities, including wild roses, yellow monkeyflowers, Scouler's willows, watercress, and others. The pools are an important water source for the birds, mammals, and reptiles in the area.

Cedar Breaks
Iron County
This national monument about 15 miles southeast of Cedar City reaches an elevation of 10,000 feet and includes deep canyons, rugged forested hills, and high plateaus cloaked in native grasses and wildflowers. Some of the conifer forests are old-growth, including ancient bristlecone pine trees that are thousands of years old. Several varieties of lichens, mosses, and other specialized plants cling to these cold, windswept ridges.

Deep Creek Mountains
Juab and Tooele Counties
This mountainous wilderness area includes everything from flatlands to mountaintops. Natural communities include sagebrush-grass scrubland, juniper-aspen woodland, various mixed conifer forest, subalpine meadow, and alpine tundra. This is prime habitat for mountain lions, hawks, eagles, bighorn sheep, and other large mammals. Rare plants include several mosses, mints, Parry's primrose, certain wildflowers, and monkshood. Brooks contain the rare Bonneville cutthroat trout. Tundra plants at high elevations are cushion buckwheat, yellow sedum, and saxifrage. The highest elevations are graced with skypilot and moss campion.

Desert Peak Wilderness Area
Tooele County
Part of Wasatch-Cache National Forest, this area 10 miles south of Grantsville in the central Stansbury Mountains covers 25,508 acres. Typical Rocky Mountain and Great Basin ecosystems exist here, ranging from low-elevation grassland and shrubland to montane-level woodland to subalpine mixed conifer forests with meadows of coneflowers and lupines. Majestic old-growth Douglas fir trees that are 300–500 years old guard the upper slopes. Riparian corridors are lined with alder and box elder, with maidenhair ferns and snowberries. Specialized plant communities occur at timberline and above and include delicate white bladder campion and stunted trees. Bighorn sheep and herds of wild horses can often be seen.

Grand Gulch Primitive Area
San Juan County
This tributary to the San Juan River is about 35 miles northwest of Bluff. Its steep canyon walls are an important habitat for raptors. Different microclimates created by shade and water support highly diverse plant communities with several rare species. In fact, the rich biodiversity here originally attracted Pueblan farmers and villagers nearly 1,000 years ago. Several rare birds, reptiles, and amphibians live along the wooded riparian corridor.

High Uintas Wilderness Area
Summit and Duchesne Counties
High in the wilderness of the Uinta Mountains, this site is about 30 miles northeast of Hanna. Its rapid elevation changes create a series of different communities, ranging from scattered woodlands on the lower slopes to subalpine mixed conifer forests on the upper slopes. Three-quarters of all Utah's bird species, attracted to the extensive stands of rare old-growth timber that is more than 150 feet tall and centuries old, can be found here. These special trees are homes for the rare saw-whet owl, pileated woodpecker, northern goshawks, and other species that depend on old-growth forest. This is the only location in Utah where large, mature stands of lodgepole pine grow. This vast tract of wilderness is a key habitat for elk, moose, black bears, mountain lions, and bighorn sheep. The 13,440-foot Kings Peak and its barren stretches of tundra are within the borders of the wilderness area. Streams fed by snowmelt support several rare endemic fish, river otters, and bog vegetation.

Joshua Tree Forest
Washington County
Centered on the town of Shivwits, this property protects widely spaced Joshua trees, representing a rare desert natural community. These trees are important nesting sites for desert birds, such as the cactus wren, and its fruits and seeds nourish many desert animals.

La Sal Mountain State Forest
San Juan County
Located about 25 miles southeast of Moab, this property's mountains, plateaus, foothills, and canyons are covered by sagebrush, pinyon-juniper forest, Gambel oak forest, ponderosa pine–Douglas fir forest, aspen stand, subalpine fir-spruce, meadow, and bald communities. Some of the conifer stands are old-growth. The area is a critical habitat for big mammals such as mountain lions, as well as a variety of birds and reptiles, some of them rare or unusual.

Layton Wetlands Preserve
Davis County
Located on the edge of Great Salt Lake north of Salt Lake City and just west of Layton, this 1,472-acre Nature Conservancy property protects an important wetland complex. Natural communities include freshwater stream mouth, saline mudflat, saltwater, brackish, and freshwater marshes, and sloughs. Different zones of vegetation live and die according to the seasonal changes in lake level and salt content. For example, when fresh water dominates the marsh during spring, the salt-loving alkali bulrush is crowded out by hardstem bulrush, cattail, and other freshwater plants. The Great Salt Lake is one of the largest staging areas in the world for bird species; rare species, such as the snowy plover and Wilson's phalarope, can be found in the wetlands.

Matheson Wetland Preserve
Grand County
Just west of downtown Moab, this 875-acre property protects the only significant wetlands along the Colorado River in Utah. The site is well known as a habitat for nearly 200 different species of birds. Natural communities include wetland, riparian corridor, and desert floodplain. Endangered or threatened raptors, such as the peregrine falcon or bald eagle, can frequently be seen along the river. The wetlands are also an important stopover for migrating birds during spring and fall. The property is owned and maintained by The Nature Conservancy.

Mount Naomia Wilderness
Cache and Rich Counties
This 44,964-acre wilderness area is located about 15 miles northeast of Logan. The 4,800–9,980 foot elevation range supports both subalpine and alpine habitats. High-elevation meadows are soft, with lush grasses and abundant wildflowers, some of them rare species. Mature stands of maple-aspen, Douglas fir, subalpine fir, and old-growth limber pine forest grow on the upper slopes. Tundra communities can be found at the highest elevations. These wild areas are ideal habitats for moose, elk, bears, and mountain lions. The pristine headwaters of streams, which supply seven communities and many farmers, originate on this property.

Mount Nebo Wilderness Area
Utah County
Part of the Wasatch Mountain Range, this wilderness area about 3 miles east of Mona includes low-elevation oak-covered forest, aspen-fir-spruce forest, mature Douglas fir and subalpine fir–lodgepole pine forest, grassland, wet meadow, juniper-sagebrush woodland, and riparian corridor communities. More than 300 species of wildlife and fish can be found here, including big mammals like bears, mule deer, moose, mountain goats, and mountain lions. This property hosts 19 sensitive plant and animal species, including Bonneville cutthroat trout, Colorado cutthroat trout, eagles, and peregrine falcons.

Painter Spring
Millard County
This oasislike system of streams, springs, and small shallow wetlands is lined by tall cottonwoods, wild grapes, wild roses, and large willows. Several rare plants can be found here, including unusual mosses. Like all wetlands in arid areas, Painter Spring is an important habitat for desert wildlife, including rattlesnakes, hawks, eagles, and amphibians.

Pine Valley Mountain
Washington County
This highly diverse property straddles the Mojave–Great Basin ecotone. Located about 10 miles east of Central, elevation begins at 6,000 feet with scattered mixed-conifer woodland, then grades upward into high-elevation old-growth forests of bristlecone pine, limber pine, subalpine spruce, Douglas fir, and aspen forest with scattered, colorful meadows. The larger conifer stands are centuries old. The mountain is an important habitat for bears, cougars, raptors, and a variety of reptiles.

Red Cliffs Desert Reserve
Washington County
This 60,000-acre property about 10 miles northwest of Saint George includes desert landscapes, a 5-mile stretch of the Virgin River, and Snow Canyon State Park. This ecotone represents the merging of the Mojave Desert, Great Basin Desert, and Colorado Plateau. Biodiversity is high as a result. Natural communities include desert cliff and various shrub communities. Rare or unusual species include the desert tortoise, gila monsters, and several raptors, including bald eagles, red-tailed

hawks, and peregrine falcons. Mountain lions roam the hills. The dwarf bearclaw poppy also grows here.

Wild Strawberry River
Wasatch and Duchesne Counties
This 5,652-acre Nature Conservancy tract protects nearly 75 percent of Strawberry River in north-central Utah west of Duchesne. Communities include steep rock canyon face, river, riparian corridor, as well as upland scrub and forest. Tall narrowleaf cottonwoods and dogwoods shade the river. The waters support the Yellowstone cutthroat trout, and the riparian edge offers shelter to large mammals, such as bears, elk, deer, and mountain lions. Raptors nest on the canyon walls and hunt along the corridor.

VERMONT

Vermont can be broken into three general natural regions: spruce–fir–northern hardwood forest, northern hardwood–hemlock–white pine forest, and transitional hardwood–white pine–hemlock forest.

The spruce–fir–northern hardwood forest forms a north-south–trending zone that occupies roughly the middle third of Vermont (generally along the axis of the Green Mountains). This mesic-upland forest is dominated by red spruce and balsam fir, with occasional white spruce, beech, basswood, and yellow birch. Generally found at the highest elevations in Vermont, this zone includes compressed wetlands and bogs.

Northern hardwood–hemlock–white pine forest forms a narrow edge on both sides of the spruce-fir forest. It also crosses the spruce-fir forest along major river corridors, such as that of the Winooski River. This forest type is dominated by more northern hardwoods, primarily sugar maple, beech, yellow birch, basswood, and black cherry, with lesser amounts of hemlock and white pine. The conifers tend to occur more in restricted sites, such as narrow ravines, that contain cooler and moister microclimates.

The third type of forest—transitional hardwood–white pine–hemlock—encloses the northern hardwood–hemlock–white pine forest to form the eastern and western edges of the state. This transitional forest has more trees characteristic of the central hardwood zone, including oak and hickory and, occasionally, pitch pine. These species tend to grow at lower elevations in drier climates.

Wetlands include bog, fen, red maple–black ash swamp, hardwood-cedar swamp, spruce-fir-tamarack swamp, marsh, sedge meadow, and woodland seep communities. Alpine and subalpine bogs can be found at higher elevations; fens and other wetland types are scattered across the state, often along floodplains.

Rare specialized communities that occur sporadically are outcrop, lake sand dune and lake cobble-sand beach, glade, serpentinite barren, subalpine heath, talus slope, and black gum swamp.

Special Habitat Sites

Barr Hill Preserve
Orleans County
Just northwest of Greensboro, this Nature Conservancy property shelters mixed conifer and deciduous woods, meadow, fen, and wetland communities. Rare or unusual plants can be found within the fen. The site is also a nesting habitat for a number of bird species.

Battell Biological Preserve
Addison County
This research area for Middlebury College contains 400 acres of old-growth northern hardwood forest. The large trees are 200–400 years old and nearly 100 feet tall. Colorful wildflowers can be seen in the understory during spring and fall. This sizeable tract of old-growth forest is an important habitat for species that depend on large, old trees, such as woodpeckers and owls.

Big Falls of the Missisquoi
Orleans County
This 16-acre tract protects a spectacular gorge and waterfall on Missisquoi River. Several old-growth hemlock and white pine trees on the property are more than 100 years old. Rare plant species in the understory have also been identified.

Black Mountain Natural Area
Windham County
Located on a rocky summit about 6 miles northwest of Brattleboro, this Nature Conservancy preserve borders the West River. Natural communities found here are riparian habitat

along the river, rocky glade, cliff face, red pine forest, and mountain laurel understory. The red pine trees are old-growth and relatively undisturbed. The rare Canadian burnet can be found here.

Button Point Natural Area
Addison County
Located about 10 miles west of Vergennes along Lake Champlain, this 25-acre property encloses fossiliferous rock outcrops covered by a remnant of old-growth oak-hickory, white and red pine, and rare white cedar forest. Some of the larger trees are more than 100 feet tall, nearly 4 feet in diameter at breast height, and centuries old.

Cambridge Natural Area
Lamoille County
This 22-acre tract includes a stand of old-growth white pine and hemlock trees. The trees are up to 3 feet in diameter and more than 100 years old.

Camel's Hump Natural Area
Washington and Chittenden Counties
This 7,404-acre tract protects a vast area of rugged, forested terrain. Much of the subalpine boreal forest is untouched and contains large specimens of balsam fir, red spruce, and birch. The property also includes 10 acres of high-altitude alpine tundra, a delicate ecosystem that contains a number of rare and fragile plant species, such as mountain cranberry, alpine bilberry, and mountain sandwort.

Canfield-Fisher Memorial Pines
Bennington County
This 22-acre property about 3 miles west of Arlington protects the largest white pines in Vermont. The tallest specimens are about 150 feet tall and nearly 4 feet in diameter at breast height. An understory of birch-beech-maple-hemlock is gradually replacing the white pine stand.

Charles E. Smith Natural Area
Grand Isle County
Sand dunes and important forested wetlands are protected on this property within Alburg Dunes State Park near Alburg. Rare plants have been discovered within the dune community. A large black spruce swamp is relatively undisturbed and is an important breeding habitat for a variety of birds and animals.

Chickering Bog
Washington County
An excellent example of wetland fen is located on this Nature Conservancy property about 8 miles northeast of Montpelier. Among the diverse plant life are several rare species of orchids, northern white cedar, sedges, Joe-pye weed, and Jack-in-the-pulpit. The fen is also an important nesting and feeding habitat for a variety of birds.

Daniel's Notch Natural Area
Lamoille County
This 100-acre tract near Cambridge is one of the state's largest old-growth forest remnants. Large birch and maple trees are nearly 100 feet tall and more than 200 years old.

Emerald Lake Natural Area
Bennington County
Located about 5 miles northeast of Dorset, this 4-acre tract contains a stand of old-growth hemlock and maple trees. Several trees are hundreds of years old.

Eshqua Bog
Windsor County
Located just outside Woodstock, this Nature Conservancy preserve encloses a relatively undisturbed bog community. Several rare orchids grow in the bog. Early summer is a good time to see hundreds of pink lady's slippers blooming among the lowlands beside the bog.

Fisher-Scott Memorial Pines
Bennington County
A National Natural Landmark and state natural area, this 13-acre tract 2 miles west of Arlington, along Batten Kill, contains the largest white pines in Vermont. Nearly 120 feet tall and almost 4 feet in diameter at breast height, these impressive specimens are hundreds of years old. The remnant old-growth forest reflects presettlement conditions.

Gifford Woods
Rutland County
Located in Gifford Woods State Park just northwest of Sherburne Center, this 7-acre tract contains a remnant of old-growth northern hardwood forest. Spectacular specimens of sugar maple, beech, basswood, and hemlock are hundreds of years old. It is perhaps best known for dramatic displays of wildflowers among the understory.

Granville Gulf
Addison County
Located in Granville, this 20-acre stand contains an old-growth forest remnant. Spruce and hemlock trees are more than 100 years old but are unusually short due to the rock ledges and thin soil. The forest is also an important bird habitat.

Hazen's Notch
Orleans County
Part of Hazen's Notch State Park, this 273-acre parcel in the town of Westfield encloses some steep, forested, serpentinite cliffs. The cliff community supports some rare alpine and serpentinite-adapted plants, including ferns, lichens, and mosses. Peregrine falcons may also nest among the rock cliffs.

Highgate Cliffs
Franklin County
This 37-acre tract near Highgate Center is located along the shore of Lake Champlain. The forested, rocky hills contain several rare plant species. The site is also a nesting habitat for turkey vultures and peregrine falcons, two rare bird species. The lakeshore cobble community is undisturbed and represents one of the state's best examples.

Kingsland Bay Natural Area
Addison County
This 50-acre tract about 5 miles northwest of Ferrisburgh encloses two peninsulas that extend into Lake Champlain. The various forest and understory communities are relatively undisturbed and very diverse. Rocky cliffs shelter several rare plant species, including ferns and lichens. The mature hardwood forest contains large sugar maple, oak, and hickory trees.

La Platte River Marsh Natural Area
Chittenden County
This preserve is located near Hinesburg and encloses extensive swamp, marshland, and bottomland forest along La Platte River. It is an important stopover for migrating waterbirds and songbirds during spring and summer. Several species of waterfowl nest in the area.

Lake Carmi Bog
Franklin County
Part of Lake Carmi State Park about 5 miles southwest of East Franklin, this 140-acre property protects a relatively undisturbed black spruce–tamarack bog. Typical plants include mountain holly, sedges, sphagnum moss, and carnivorous plants. The bog is the largest and one of the most diverse undisturbed spruce bogs in the state.

Lake Willoughby State Park
Orleans and Caledonia Counties
Located about 10 miles southeast of Barton, Lake Willoughby is a long, deep, narrow, clean lake with a healthy fish population, including lake trout and landlocked salmon. Peregrine falcons and other raptors nest in the steep rock cliffs along the shore. The cliffs support rare arctic plants, such as yellow mountain saxifrage, green alder, and rare mosses and lichens. Rainbow trout spawn in nearby Willoughby River. The park is part of Willoughby State Forest.

Lords Hill
Caledonia County
Located in Groton State Forest southeast of Marshfield, this 25-acre plot contains tall old-growth trees. Large specimens of white ash, maple, birch, beech, and spruce are more than 100 years old and dominate the canopy. It is also an important nesting site for interior forest birds. Some state-record trees occur here, including white ash, yellow birch, sugar maple, and basswood.

Maidstone Lake State Park
Essex County
Located about 10 miles southeast of Bloomfield, this state park consists of spruce-fir forests that surround Maidstone Lake, which supports populations of landlocked salmon and lake trout. Several waterbirds nest here, including the state-endangered loon. It is also critical habitat for big mammals, such as moose, deer, bears, and bobcats. Uncommon boreal birds, such as spruce grouse, gray jays, and black-backed woodpeckers, inhabit the red spruce–balsam fir upland forest and black spruce wetlands.

Mallary Bugbee Natural Area
Caledonia County
Upland forest, lowland forest, and marsh-swamp communities are protected on this 12-acre parcel in the townships of Peacham and Danville. Remnants of a northern white cedar swamp and its related fen form the core of the property. Several rare or threatened plant species have been identified in the swamp and

fen. It is also well known for its waterflowers. The wetlands provide important bird habitat.

Marl Pond
Orleans and Caledonia Counties
Located in Willoughby State Forest, this 30-acre preserve in Sutton encloses a pond fed by calcium-rich groundwater. It is bordered by a northern white cedar swamp. The unique conditions of the calcium-rich pond and wetlands support several rare or endangered plant species.

Maynard Miller Black Gum Swamps
Windham County
Near Vernon, this 450-acre wetland complex protects relict species from an ancient warmer climate. A high number of rare southern plants grow here, such as tupelo, black gum, mountain laurel, Virginia chain fern, and the rare Massachusetts fern. Towering, 400-year-old tupelo trees grow in the deepest parts of the swamp. The surrounding communities are old-growth chestnut forest and red oak–red maple–white birch–beech forest, both important bird habitats.

Mendon Peak
Rutland County
This high-elevation, remote, 90-acre tract is located on Mendon Mountain in Coolidge State Forest. The spruce, fir, and birch trees are old and impressive. The property also includes a small alpine ecosystem with dwarf shrubs and other specialized plants. It is an important bird habitat.

Morristown Bog
Lamoille County
This 30-acre tract in Morrisville encloses an excellent remnant of undisturbed black spruce–tamarack bog. It has several well-developed peatland areas. The bog has a high plant diversity, including several rare bog plant species, such as pitcher plants and sundews.

Mount Mansfield Natural Area
Washington County
Located in the townships of Bolton, Cambridge, Underhill, Stowe, and Waterbury, this 3,850-acre tract northwest of Colbyville encloses some rare and diverse high-altitude ecosystems. Natural communities include alpine tundra, subalpine heath, subalpine spruce-fir forest, and steep cliff vegetation. Rare ferns, lichens, and other cliff plants are found clinging to the cold rock faces. It is also an important nesting habitat for peregrine falcons.

Peacham Bog
Caledonia County
This 748-acre property about 10 miles northwest of Groton is part of Groton State Forest. Its dominant natural feature is a raised peat bog (only two occur in the state). Several varieties of rare or unusual bog plant species are found in the wetlands, which are an important habitat for moose, beavers, and interior forest birds. Interesting or rare plant species include black spruce, tamarack, leatherleaf, sedges, and rhodora. More than 100 bird species have been seen, including the uncommon boreal chickadee and Wilson's warbler, which breed in the spruce-fir forest around the bog.

Shrewsbury Peak
Rutland County
Part of Coolidge State Forest, this 100-acre tract southeast of Rutland near Shrewsbury protects high-altitude undisturbed old-growth spruce-fir boreal forest. It is an important habitat for a variety of plant and animal species, including interior forest birds.

Smuggler's Notch
Lamoille County
Steep, forested cliffs are present on this property about 5 miles southeast of Pleasant Valley. Peregrine falcons nest on the rock faces, which also shelter alpine plants, including rare purple mountain saxifrage, carnivorous butterwort, ferns, mosses, and lichens.

Spectacle Pond Natural Area
Essex County
This 15-acre tract in Brighton State Park about 2 miles southeast of Island Pond is known for its old-growth stand of red pines, an unusual forest type in Vermont. Some of the trees are very tall. It is also a known raven nesting habitat.

Victory Bog
Essex County
Located about 25 miles northeast of Saint Johnsbury, this 18,600-acre wildlife management area borders the headwaters of the Moose River: The vast Victory Bog, a wetland complex of marsh, swamp, bog, and mixed hardwood-softwood bottomland forest. Rare communities that occur within the

wetlands are talus slope plants, calcareous wetlands complexes, and vernal woodland pools. This large site is critical habitat for moose, bears, coyotes, deer, otters, fishers, birds, and beavers; it is also the state's largest winter range for the white-tailed deer. Interesting or unusual plants include tamarack and cedar, rhodora azalea, Labrador tea, leatherleaf, and pitcher plant.

Weybridge Cave Natural Area
Addison County
This natural area protects what is considered to be the largest cave system in Vermont. Cave ecosystems are very delicate and support a variety of unique and specially adapted organisms, such as blind cave crickets, salamanders, and fish. It is also a bat habitat.

Willoughby Cliffs Natural Area
Orleans County
Located southeast of Westmore in Willoughby State Forest, this rugged, forested 950-acre tract encloses forest and cliff natural communities. The cool, moist cliff faces harbor rare plants, including ferns, mosses, and lichens. It is also an important nesting habitat for raptors such as the peregrine falcon.

VIRGINIA

Virginia can be divided into three major physiographic provinces, which trend roughly from northeast to southwest: the Coastal Plain, the Piedmont, and the western mountains of the Valley and Ridge and Blue Ridge.

The Coastal Plain occupies the eastern third of the state and includes subtidal, beach, dune-and-swale, and other coastal communities. Forests in the Coastal Plain tend to be mesic mixed hardwood (beech-tulip-poplar-oak-hickory) with a dense, junglelike understory of shrubs and vines. Dry to mesic oak-hickory forest is also abundant. Coastal communities include maritime grassland, maritime upland forest (beech–loblolly pine–hickory–American holly), dry longleaf pine forest, scrub oak forest, and maritime wetland forest. Rare communities include seasonal pond, longleaf pine–turkey oak woodland, pocosin, fen, and maritime live oak woodland.

The Piedmont occupies the central third of Virginia and consists of a series of rolling hills and valleys. Its forest cover is typically mesic mixed hardwood or basic mesic forest, including beech–tulip poplar–red oak. Dry white pine or Virginia pine, pitch pine, and chestnut oak heath forest can be found on rocky ridgelines, where soil is generally dry. Dry to mesic oak-hickory forests are very common on upland slopes and ridge tops. Wetlands in the Piedmont tend to be restricted to bottomland floodplains along rivers and streams (sycamore–river birch–aspen). Rare communities in the Piedmont include flatrock, Virginia pine serpentinite barren, and savanna.

The western mountains, in the western third of the state, are dominated by the Blue Ridge and Appalachian Mountains. Red spruce–Fraser fir and Fraser fir forests cloak the rugged slopes at elevations of 6,000 feet and higher. At the higher elevations one finds grassy and heath balds that form open, windswept meadows of mountain oatgrass, sedges, and low shrubs like rhododendron and mountain laurel. Northern hardwood forests (beech-birch-buckeye-maple) can be found at intermediate elevations, often on north-facing slopes. Forests of white oak, chestnut oak, and montane oak–hickory occur on lower slopes at montane elevations. Specialized communities, such as rock face plant assemblage, bog, wet prairie, glade, cave, and fen, can be found at various elevations.

Special Habitat Sites

Antioch Pines Natural Area
Isle of Wight
Located on 400 acres near Zuni, this state property protects mature to old-growth loblolly pine–turkey oak sandhill forest along Blackwater River. The rare blue jack, turkey, and sand post oak trees can all be found in the forest and represent a unique oak community for Virginia.

Berger Memorial Sanctuary
Spotsylvania and Caroline Counties
Located on 865 acres, this Nature Conservancy preserve encloses mature second-growth forest along the Rappahannock River southeast of Fredericksburg; it also contains Civil War earthworks of archeological significance. The upland forest contains large specimens of box

elder, oak, walnut, and cottonwood that are probably 200 years old or older. A relict rhododendron forest can be found along Snow Creek, an unusual occurrence in the Coastal Plain. The preserve is an excellent habitat for a variety of birds and animals, including herons and interior forest birds, such as pileated woodpeckers, cuckoos, and songbirds.

Bethel Beach Natural Area Preserve
Mathews County
Located 5 miles northeast of Onemo, this 50-acre site is located on the western shore of Chesapeake Bay. Dominant natural communities include sand beach, dune, and salt marsh. Sand dunes are anchored by switchgrass, sea rocket, beach grass, and the rare seabeach knotweed. Common loons, horned grebes, scoters, terns, peregrine falcons, eagles, and ospreys are some of the important birds that nest here. The sands are also a critical habitat for the rare northeastern beach tiger beetle.

Big Spring Bog
Grayson County
This rare cranberry glade 8 miles southeast of Galax grows in magnesium-rich soil. The cooler climate here supports sphagnum moss, bog cranberry, round-leaved sundew, largeleaf, grass-of-Parnassus, Gray's lily, and shortleaf sneezeweed. A rare butterfly (its larvae feed on cranberry bushes) has been sighted here.

Blackwater Ecological Preserve
Isle of Wight
A rare mature longleaf pine–turkey oak barren ecosystem is protected on this 318-acre property. Both longleaf pine–turkey oak and longleaf pine savanna communities cover the dry-mesic sand ridges. Nearly two dozen rare or special-concern plant and animal species can be found on the property.

Blackwater River Preserve
Southampton County
Located about 5 miles east of Wakefield, this Nature Conservancy property protects wetlands along the Blackwater River. The property is noted for large old-growth bald cypress and water tupelo trees.

Bottom Creek Gorge
Montgomery County
Located on 1,657 acres at the headwaters of Roanoke River, this Nature Conservancy project near Blacksburg protects a varied and spectacular section of wilderness. Natural communities include flowing stream, waterfall, old-growth forest, and very rare shale barren. A high number of rare, uncommon, or threatened species can be found here. Four rare fish endemic to Virginia can be found in Bottom Creek: orangefin madtom, bigeye jumprock, Roanoke darter, and riverweed darter. It also contains native brook trout. Due to the rugged terrain, logging did not reach the gorge, where huge Carolina hemlock trees hundreds of years old remain undisturbed. The globally rare chestnut lipfern grows on the shale barren.

Buffalo Mountain Natural Area
Floyd County
This 1,000-acre tract covers the summit of Buffalo Mountain. At nearly 4,000 feet, the summit is at subalpine elevation, yet windswept and treeless. Patchy meadows are dominated by big bluestem, blazing star, and many other wildflowers. The rocky summits harbor the rare three-toothed cinquefoil and the Rocky Mountain woodsia. Rare bog bluegrass, large-leaved grass-of-Parnassus, and an extremely rare mealybug species can be found in grassy seeps at the base of the mountain.

Bush Mill Stream Natural Area Preserve
Northumberland County
Located on 110 acres near Heathsville, this natural area protects the confluence of Bush Mill Stream and the Great Wicomico River. Natural communities include freshwater marsh, brackish tidal marsh, tidal mudflat, and bottomland riparian forest. The dense vegetation provides important nesting, feeding, and migratory habitats for a number of waterbirds, including great blue herons, wood ducks, and teals. Bald eagles can be found perched in the tall trees along the shoreline. Although most of the forest is second-growth, there are some old-growth trees, especially mountain laurel groves, in the steeper ravines that survived lumbering.

Cape Charles Coastal Habitat
Northampton County
This 29-acre property along the Atlantic Ocean preserves high-quality intertidal flats, coastal beach, sand dunes, and inland maritime forest. The rare coast bedstraw can be found anchoring some of the dunes. In addition to being a habitat for the federally endangered northeast

beach tiger beetle, the dunes and forest are important resting and feeding sites for migrating songbirds.

The Cedars
Lee County
A rare limestone glade is protected on this 50-acre property near Powell River in southwestern Virginia. Communities include xeric mixed pine-hardwood forest, xeric limestone glade, and various ponds and wetlands. Several globally rare plants have been identified here.

Chub Sandhill
Sussex County
This 387-acre state-owned tract covers some sandhills along Nottoway River. The dominant natural community is a remnant mature to old-growth sand post oak woodland with underlying scrub plants. Rare species that grow here are Queen's delight, golden puccoon, and hoary scurf-pea.

Cowbane Prairie
Augusta County
A rare wet prairie ecosystem is preserved on this 63-acre property on a western slope of the Blue Ridge Mountains. This virgin remnant is highly diverse and includes many rare plants and wildflowers, including queen-of-the-prairie, marsh-speedwell, and blueflag. The wildflowers are quite colorful during spring and summer. Many small grassland birds, mammals, and insects depend on this rare prairie habitat.

Cumberland Marsh
New Kent County
Located on 1,095 acres, this Nature Conservancy property is located along Pamunkey River about 30 miles east of Richmond. It protects a very high-quality, extensive freshwater marsh. Located along the Atlantic Flyway, it is an important staging area for tens of thousands of birds during spring and fall. Bald eagles, ospreys, herons, egrets, and terns nest here every year. The sensitive joint vetch can also be found in the marsh.

Falls Ridge Preserve
Montgomery County
Located about 7 miles southwest of Ironto, this 655-acre Nature Conservancy property encloses part of a steep ridge that flanks the Roanoke River. Natural communities include deciduous upland oak-hickory-pine forest, bottomland sycamore-cottonwood forest, and flowing stream and waterfall. Hemlock trees can be found in the narrow ravines. Rare or uncommon species include various raptors. pileated woodpeckers, the wild cherrystone snail, Allegheny plum, goldenseal, and Addison's leatherflower.

Fernbrook Natural Area
Albemarle County
This 63-acre Nature Conservancy tract encloses woodland in varying stages of succession along Rivanna River. Natural communities include mature upland oak-poplar-gum-hickory forest, lowland sycamore-ash-maple-walnut hardwood forest, and successional oak-pine forest, as well as riparian habitats along small streams that flow into the river. Some of the larger trees are probably 200 years old. The forest is an important nesting habitat for interior forest birds, including tanagers and woodpeckers. Rare or unusual plants on the property include southern adder's tongue. Colorful displays of wildflowers can be seen throughout the spring and summer.

Fraser Preserve
Fairfax County
Located on 220 acres, this Nature Conservancy project along the Potomac River about 4 miles northwest of Great Falls protects a representative example of Piedmont. Natural communities include meadow, wetland, flowing stream, bottomland forest, upland forest, and rocky cliff. The diversity of flora is high as a result and includes more than 300 species of wildflowers, some of them rare or unusual. The woodlands are an important nesting habitat for interior forest birds, such as pileated woodpeckers, gnatcatchers, tanagers, and other songbirds. Bald eagles also nest along the river.

Gasburg Granite Flatrocks Preserve
Brunswick County
Located near Gasburg, this property includes flat surfaces of granite bedrock that support rare plants, such as granite umbrella sedge, Small's purslane, and Small's stonecrop. Flatrock communities depend on rainwater that pools in shallow depressions in the bedrock. Other interesting species are red cedar, prickly pear cactus, and post oak.

Grafton Ponds
York County
Located on 375 acres, this property encloses Virginia's best Coastal Plain sinkhole pond system. The ponds are surrounded by a marshy wetland fringe and mixed pine-hardwood forest. Several rare plants, including Harper's fimbristylis, pond spice, and Cuthbert turtlehead, grow in the wetlands. The uncommon barking tree frog and Mabee's salamander have also been sighted on the preserve.

Helena's Island Preserve
Nelson County
This 57-acre Nature Conservancy tract is an island in the Saint James River about 2 miles east of Norwood. It is an excellent example of mature bottomland forest that floods periodically, creating high plant diversity. The forest contains mature black walnut, box elder, and basswood. The island is an important nesting site for a variety of birds, including waterfowl and songbirds. Evening primrose, tiger lily, and loosestrife are some of the colorful wildflowers that adorn the island during spring and summer.

Hughlett Point Natural Area Preserve
Northumberland County
This 204-acre natural area preserve is on the western shore of Chesapeake Bay, about 5 miles northeast of Kilmarnock. Natural communities protected here include a beach-dune complex, upland loblolly pine–mixed hardwood forest, salt pond, and tidal and nontidal wetlands. The sandy beach is critical habitat for the rare northeastern beach tiger beetle. Other interesting species include ospreys, bald eagles, and other raptors. The marshes are an important nursery for fish, shellfish, and other marine organisms that serve as a food source for waterbirds.

Johnson Creek Shale Barren
Alleghany County
A rare shale barren community exists on this 105-acre site near Callaghan. Within its borders, on a steep shale bluff overlooking Johnson Creek, is the well-developed shale barren community. Shale barren—slopes of eroding shale or rocky, shale-derived soil—generally form on south-facing slopes, are very dry, and are subjected to summer temperatures as high as 150 degrees. Some of the rare plants found here are white-haired leatherflower. At least a dozen other rare plants and animals inhabit this arid microclimate.

Kennedy Mountain Meadow
Augusta County
Located in the Blue Ridge Mountains, this 10-acre Nature Conservancy property protects part of a series of flooded sinkholes and specialized flora. The low areas are surrounded by oak–hickory–black gum forests. Some of the rare plants that grow in or near the sinkholes are black-fruited spike rush, Virginia sneezeweed, and asterlike boltonia.

MBC Cave
Tazewell County
This 5-acre property protects a cave in western Virginia that provides breeding habitat for the federally endangered big-eared bat. It is also one of only two locations in the world where a rare, cave-adapted millipede has been discovered. Access to the cave is restricted, as human visitation disrupts the delicate environmental conditions inside the cave. Cave exploration often results in the deaths of millions of infant bats that are dropped or abandoned by their parents.

Mountain Lake Wilderness
Giles County
Located about 15 miles northwest of Blacksburg in Jefferson National Forest, this 10,000-acre wilderness area includes high ridges, red spruce bogs, riparian stream corridors, and cliff ledges and outcrops at elevations ranging from 2,200 to 4,100 feet. Ridges are forested with red spruce, hemlock, and yellow birch; the cooler ravines carry stands of virgin white oak and hemlock that are 300–500 years old. Several rare plant assemblages grow on the rocky ridge tops and cliff faces.

The Narrows
Giles County
This special property on Peters Mountain overlooking the New River is the only place in the world where Peters Mountain mallow still grows. Its reproduction has been hampered by encroaching trees as well as browsing by animals.

New Point Comfort
Mathews County
Located on a peninsula that juts into Chesapeake Bay, this 105-acre property sits along the Atlantic Flyway and thus is an important stopover for tens of thousands of migrating songbirds and waterfowl. Its beach-dune-

maritime forest communities provide critical habitat for the nesting least tern, a rare waterbird. The federally threatened northeastern beach tiger beetle lives among the sand dunes.

North Landing River Preserve
City of Virginia Beach
Located about 12 miles from Virginia Beach, this 1,886-acre is one of the most pristine freshwater marshes on the East Coast. It includes Pocaty Creek and its wetlands, which flow into North Landing River. River, marsh, swamp, peatland, large old-growth cypress forest, and unique pocosin natural communities are among those found here. At least 30 rare plants and animals have been identified, including the Cyrano darner dragonfly. The property is managed by The Nature Conservancy.

Northwest River
City of Chesapeake
Located on 2,244 acres, this state natural area includes swamp, marsh, and mesic upland forest along the Lower Northwest River. Rare species include winged seedbox, sawgrass, the Dismal Swamp southeastern shrew, various sedges, the canebrake rattlesnake, and silky camelia.

Pedlar Hills
Montgomery County
This 81-acre property overlooking the South Fork of the Roanoke River is centered on a rare dolomite barren and calcareous woodland (stunted hardwoods and red cedar). The grassy barren supports the federally endangered smooth coneflower, tall gay-feather, Addison's leatherflower (a Virginia endemic), and several other rare species.

Pendleton Island
Scott County
This 35-acre island in Clinch River is actually a series of three smaller islands connected by shallow riffles and thick aquatic vegetation. The island is covered by a dense bottomland forest. The preserve is known for its high population of Cumberlandian mollusk species, eight of which are federally endangered; Pendleton Island is the only location for one of these species worldwide.

Pinnacle Natural Area Reserve
Russell County
Located on 68 acres about 7 miles northeast of Lebanon, this property includes cliff, waterfall, river, riparian corridor, rich cove woodland, and rocky glade–meadow communities. Several unique plants and wildflowers that are specialized to limestone bedrock and lime-rich soils can be seen here, including glade spurge, Carey's saxifrage, and smooth cliffbrake. Glade flowers include American harebell, tufted hairgrass, and the rare Canby's mountain clover. Rare mussels, including the globally rare birdwing pearly mussel, can be found in the clean waters of Clinch River and Big Cedar Creek. Several rare turtles and snails also inhabit the riparian corridor. Raptors are attracted to the cliff faces.

Poor Mountain Natural Area
Roanoke County
Located on 1,000 acres near Salem on the slopes of Poor Mountain, this state natural area protects high-quality examples of xeric Table Mountain pine–oak forest, mixed hardwood–pine forest on the lower slopes, and hemlock stands in the ravines. The globally rare shrub piratebush (which may contain cancer-fighting agents) can be found in the understories of all the forest types.

Potts Creek Shale Barren
Alleghany County
Shale barren ecosystems are rare in the United States. This 9-acre parcel in the rugged terrain of western Virginia protects a section of shale barren above Potts Creek. Some rare plant species here specially adapted to the shale substrate: chestnut lipfern, Kate's mountain clover, Alleghany rockcress, Allegheny stonecrop, mosspink, leatherflower, and Virginia nailwort. The surrounding trees are stunted cedar and pine. Owned and managed by The Nature Conservancy, access is restricted because of the delicate habitat and steep terrain.

Ramsey's Draft Wilderness Area
Augusta County
Part of George Washington and Jefferson National Forests about 12 miles west of Mount Solon, this 6,000-acre tract of virgin hemlock–white pine–oak–yellow poplar forest escaped the lumberman's saw due to the steep terrain. These majestic trees are centuries old and nearly 100 feet tall and provide important habitat for birds and mammals that depend on old-growth forest, such as woodpeckers, owls, and bears.

Savage Neck
Northampton County
Located on 148 acres on Chesapeake Bay, this state natural area includes high-quality tidal mudflat, beach, dune, and inland maritime forest communities. Several rare plant species, and the federally endangered northeastern beach tiger beetle, can be found here. Located on the Atlantic Flyway, the property is an important staging area for migrating songbirds.

Seashore State Park and Natural Area
City of Virginia Beach
Located in southeastern Virginia about 12 miles east of Norfolk, this coastal park includes high-quality cordgrass salt marsh along Long Creek. Other communities include loblolly pine and live oak stands on the sand ridges and old dunes, as well as freshwater interdunal bald cypress swamps. Several rare or unusual species live here, such as the chicken turtle and southern cricket frog. The bald cypress trees are draped in thick Spanish moss.

Unthanks Cave
Lee County
This large cave system in Lee County shelters a variety of extremely rare cave species. The Nature Conservancy owns the 23-acre tract, which contains the only known entrance to the cave. These troglodyte species include the Powell Valley planarium, Carabid beetle, and Hydrobiid snails, some of which are only endemic to a few caves in Virginia. Because of the sensitive nature of the cave ecosystems, visitation by the public is not permitted.

Virginia Coast Reserve
Northampton and Accomack Counties
Located on a sprawling, 35,000-acre tract, this Nature Conservancy site protects a series of 13 barrier islands along the Atlantic coast. It covers more than 40 miles of undisturbed coastline. Always shifting in form, the islands are probably 5,000 years old or older. Natural communities are delicate and specialized to the sandy coastline environments. The islands are surrounded by low saltwater cordgrass marshes, critical habitats for fish and shellfish species. Other communities include mudflat, salt flat, foredune, upland pine forest, and interior grassland. The reserve is a critical breeding habitat and migratory stopover for more than 200 species of birds, including the rare piping plover and peregrine falcon.

Voorhees Nature Preserve
Westmoreland County
This 729-acre Nature Conservancy preserve near Leedstown encloses mature Coastal Plain hardwood (oak-beech-pine-poplar) forest and freshwater tidal marsh habitats. Located along the Rappahannock River, the preserve is an important nesting site for bald eagles. Rare or endangered species, such as the puritan tiger beetle, cliff swallow, and sensitive joint vetch, are likely to occur on the property. Inventory studies are continuing.

Walnut Hill
Montgomery County
This tiny site in the Piedmont contains one of the world's largest documented populations of the rare smooth purple coneflower. The flower is threatened by growing land development and the destruction of grassy meadows in woodlands. The property is owned and maintained by The Nature Conservancy.

Wildcat Mountain
Fauquier County
Located on 655 acres, this Nature Conservancy preserve about 5 miles south of Marshall protects relatively undisturbed forest in the Blue Ridge foothills. The property encloses some old-growth stands of oak and hickory that are probably about 200 years old. Newer successional forests are marked by beech, pine, dogwood, and hackberry. The woodlands are an important habitat for mammals and interior forest birds such as woodpeckers. Raptors are also present.

William B. Trower Bayshore
Northampton County
A well-developed coastal beach and its associated dune system are protected on this 35-acre property on the eastern shore of Chesapeake Bay. Maritime scrub and forest communities occur inward from the dunes. This site is considered one of the best locations for the federally endangered northeastern beach tiger beetle and the rare coastal bedstraw. The dense scrub and forest lands are a staging area for migrating neotropical birds during spring and fall.

Wreck Island Natural Area Preserve
Northampton County
Located on 1,380 acres east of Cape Charles in the Atlantic Ocean, this barrier island preserves high-quality remnants of beach, dune, mar-

itime grassland and shrubland, salt flat, and salt marsh natural communities. It is a critical habitat to at least 12 bird species, including the rare least tern, black skimmer, and gull-billed tern. Egrets and herons nest in the salt marsh. The rare Wilson's plover and federally threatened piping plover occasionally nest on the beach. It is also an important resting and feeding stop for tens of thousands of birds during spring and fall migrations.

WASHINGTON

Washington can be divided into five community types: coastal rain forest, hemlock forest, subalpine fir forest, shrub steppe, and steppe.

Towering Sitka spruce trees dominate the strip of coastal rain forest along the Pacific coast. Only 2–5 miles wide, this zone is a result of mild temperatures and high precipitation. Other trees include western hemlock, western redcedar, and Douglas fir. Some of North America's most impressive old-growth timber grows here. The lush understory is graced by a variety of tall ferns and delicate cryptogams.

Interior to the coastal rain forest is western hemlock–mountain hemlock forest that covers much of the Cascade Range. Occupying roughly the western third of the state, this zone is more temperate and is subject to more variable precipitation. Other trees include Douglas fir and western hemlock. Grassy bald and talus slope communities are also part of this mountain landscape.

Subalpine forests cover a north-south strip in central Washington, as well as the northeastern corner. These forests occur at montane-subalpine elevations, within wetter and cooler conditions. Dominant trees are grand fir, Douglas fir, Pacific silver fir, subalpine fire, Shasta red fir, and, occasionally, mountain hemlock. Lush herbs and shrubs are found in the understory.

The shrub steppe and steppe zones occupy the southeastern third of the state. Shrub steppe grows in the rain shadows east of the Cascade Mountains and covers the Columbia Basin. Vegetation includes various bunchgrass-sagebrush associations, with some perennial grasses. The shrub steppe zone is ringed by steppe and dominated by wheatgrass-bluegrass and wheatgrass-fescue associations, with some sagebrush and bitterbush. The arid, flat-lying shrub steppe and steppe terrain is commonly referred to as desert and desert scrub.

Special Habitat Sites

Baird Basin Research Natural Area
Stevens County
Located on 65 acres about 20 miles east of Colville, this 160-acre property encloses high-quality ponderosa pine forest and Douglas fir forest on steep slopes. Elevations range roughly from 3,100 feet to 3,500 feet. Natural communities include mature to old-growth ponderosa pine–western larch–Douglas fir forest, Douglas fir–western larch forest, and lodgepole pine forest. This RNA is critical habitat for a variety of mammals and nesting bird species, some of them rare or unusual; migrating waterfowl also visit the site.

Black River Preserve
Thurston County
This Nature Conservancy property about 12 miles south of Olympia protects a stretch of Black River, unique in being fed by a lowland lake. It is lined by extensive wetlands and stands of red alder–Oregon ash bottomland. The very dense undergrowth of willow, dogwood, and Pacific ninebark creates a haven for important species, such as mink, ospreys, black bears, migratory songbirds, bitterns, herons, and other raptors.

Blackwater Island Research Natural Area
Cowlitz County
This 129-acre tract along the Lower Columbia River is a few miles northwest of Ridgefield. It protects floodplain communities along the river, shallow lakes, willow lowlands, and stands of Oregon white oak. The area is seasonally flooded. It contains excellent examples of mudflat, grassland, oak-ash-willow woodland, and mature stands of Oregon ash and Oregon white oak. Biodiversity is high with several rare or unusual plant and animal species, including golden and bald eagles and a variety of wildflowers.

Boston Glacier Research Natural Area
Skagit County
Located on 3,090 acres in North Cascades National Park, this RNA contains a geologic glacial cirque at the head of Skagit Queen Creek. Although almost two-thirds of the area is covered with glacial ice, natural communities include subalpine pond, subalpine meadow, rock face, shrub habitat, and forests of Pacific silver fir, subalpine fir, mountain hemlock, and subalpine larch. The pristine forest is an important habitat for a variety of birds and mammals, some of them rare or unusual, such as raptors, Vaux's swifts, blue grouse, bald eagles, and several vireos. Larger mammals, such as mountain lions and bighorn sheep, are also present.

Boulder River Wilderness
Snohomish County
Located 6 miles west of Darrington, this wilderness area encloses the last low-elevation old-growth forest in the Cascade Mountains. Scattered stands of trees are more than 500 years old and important habitats to old-growth–dependent birds and mammals.

Cedar River Watershed
King County
Located a few miles southwest of North Bend, this remnant old-growth forest is a core habitat for cougars, elk, bears, and nesting raptors and other birds. Although much of the area is covered in second-growth forest, some old-growth stands of Douglas fir are nearly 200 feet tall and 700 years old.

Chuckanut Island Cyrus Gates Memorial Preserve
Whatcom County
Located north of Bellingham and Governor's Point Island, this 5-acre Nature Conservancy tract is covered with mature to old-growth Douglas fir, grand fir, madrone, and western redcedar forests. Other communities include coastal habitat and an extensive salt marsh, which is an important nursery and spawning ground for fish, shellfish, and crustacean species and thus is an important food source for waterfowl. Bald eagles nest on the island.

Deception Pass State Park
Island County
Located in Puget Sound on the north end of Whidbey Island, this popular park is roughly 400 acres in size. Natural communities include beach, rocky headland, tidepool, sand dune, and old-growth coastal rain forest. Douglas fir trees with swordfern understories are 200 feet tall and 500 years old or older. The forest recesses are important habitats for bears, bald eagles, various birds, and the rare black-tailed deer.

Dishman Hills
Spokane County
Owned and managed by The Nature Conservancy, this preserve encloses a section of the lower slopes of Dishman Hills, just southeast of Spokane. Natural communities include Douglas fir and ponderosa pine forests, wetland, and woodland-grassland. The wildflowers are very colorful during spring and summer. Several rare plant and animal species have been found here, including Compton's tortoisehell butterfly; water howellia can be found on the grassy slopes.

Dungeness National Wildlife Refuge
Clallam County
Located along the Pacific Ocean on a spit nearly 5 miles long, this refuge is north of Sequim. Natural communities include submerged eelgrass meadow, sand beach, and grassland. The meadows and intertidal zone support large populations of fish, shellfish, and crustacean species and thus are an important food source for thousands of waterfowl that stop during fall and spring migrations; some of the bird species, such as the black brant, are rare or unusual. Other birds nest on the property.

Foulweather Bluff
Kitsap County
Located about 3 miles from Hansville, this 93-acre property encloses beach, freshwater and saltwater marshes, brackish marsh, lowland forest, upland forest, and sand levee natural communities. Tall coniferous forests contain swordfern and salal in the understory. Many insects, fishes, birds, and mammals depend on this habitat; these include herons, eagles, buffleheads, owls, and pileated woodpeckers. The freshwater and brackish marshes are among the best and least disturbed in Puget Sound. The property is owned and managed by The Nature Conservancy.

Gray's Bay Wetland Preserve
Wahkiakum County
This Nature Conservancy property just outside Rosburg encloses a very rare old-growth rem-

nant of tidally influenced freshwater spruce swamp along the Columbia River drainage. This pristine area comprises Sitka spruce–willow–red–osier dogwood forest. The dense woodland is an important habitat for nesting and migratory birds, including Swainson's thrush and the yellow warbler.

Hades Creek Research Natural Area
Jefferson County
A lush forest of Pacific silver fir and western hemlock is protected on this RNA in Olympic National Park. It is located on 560 acres south of Port Angeles. Natural communities include mature to old-growth mixed conifer forest, stream, riparian corridor, and rock face. The largest trees are almost 200 feet tall, with diameters reaching 7 feet. It is a critical habitat for species, especially certain birds, that depend on old-growth forest. Large mammals, like Roosevelt elk and mountain lions, also frequent this wilderness.

Higley Creek Research Natural Area
Grays Harbor County
Old-growth western hemlock forest is protected on this 480-acre RNA on the southwestern Olympic Peninsula near Lake Quinault. Natural communities include western hemlock forest with western redcedar, Douglas fir, Pacific silver fir, and Sitka spruce interspersed. The largest trees have diameters of nearly 10 feet, reach heights of more than 200 feet, and are at least 500 years old. The RNA is an important habitat for species that prefer old-growth forest, including the Roosevelt elk and black-tailed deer.

Hoh Rainforest Visitor Center
Jefferson County
Part of Olympic National Park, this coastal rain forest in Quinault Valley about 20 miles northeast of Humptulips is cloaked in majestic old-growth Sitka spruce, western redcedar, western hemlock, and Douglas fir. The understory is richly diverse in mosses, ferns, lichens, and wildflowers, including some rare plant species. These ancient trees are more than 200 feet tall and 6 feet in diameter at breast height. Elk, deer, and bears inhabit the site.

Jackson Creek Research Natural Area
Jefferson County
Located on the western Olympic Peninsula, this 160-acre RNA protects an old-growth Douglas fir forest along Hoh River about 50 miles southeast of Forks. Natural communities are dominated by Douglas fir–Sitka spruce–western hemlock–western redcedar forest. The tallest trees are nearly 250 feet tall and 9 feet in diameter. The forest is an important habitat for species that depend on old-growth stands, including certain birds and the Roosevelt elk.

Lake 22
Snohomish County
Subalpine lakes and an old-growth western redcedar–western hemlock forest are protected on this 790-acre RNA in Mount Baker National Forest. Located northeast of Silverton, the property encloses the 44-acre Lake 22, old-growth mixed coniferous forest, brushfield, rock face, and talus communities. Most of the communities have subalpine components at a much lower elevation (2,500 feet). Thus, several plant species in this cooler microclimate are unique for such a low elevation. The old-growth forest is an important habitat for birds and mammals that depend on old-growth forests, including owls, woodpeckers, and the Roosevelt elk.

Long Island
Pacific County
Part of Willapa National Wildlife Refuge, this 5,000-acre unique island northeast of Seaview is the largest estuarine island along the U.S. Pacific coast. Natural communities include beach, intertidal eelgrass meadow, saltgrass tidal marsh, tidal flat, and coastal forest. Forest trees include large western hemlock, Sitka spruce, and western redcedar, which shelter Roosevelt elk, black bears, otters, blue grouse, pileated woodpeckers, spotted owls, and various raptors. The wetlands on and around the island are critical habitats for spawning fish and shellfish species and thus are important sources of food for waterfowl. Cedar Grove, a 274-acre stand of towering old-growth cedar trees, is thousands of years old.

Meeks Table Research Natural Area
Yakima County
Located on 68 acres 40 miles northwest of Yakima in Snoqualmie National Forest, this RNA on the Eastern Slope of the Cascade Range encloses an ecotone between ponderosa pine–pinegrass forest and sagebrush steppe–grassland. Elevations range from 4,200 to 4,525 feet. Other species include Sandberg bluegrass, bottlebrush squirreltail, cheatgrass,

stiff sage, and bitterbrush. The area is very pristine and has never been logged or grazed. Biodiversity is quite high, and there are several plant and animal species of special concern found here.

Moran State Park
San Juan County
Part of the San Juan Islands, this park southwest of Bellingham contains old-growth conifer forest (Douglas fir, Sitka spruce, grand fir, and western hemlock). Other communities include dry and wet meadows, rock face, pond, and lake. Eagles and ospreys are abundant. The rare Vaux's swift nests in the old-growth trees. Other important species that depend on these communities are bats, elk, vultures, peregrine falcons, blue grouse, pileated woodpeckers, warblers, and finches.

Mount Pickett
San Juan County
Located on Orcas Island, this 1,255-acre state natural area protects the best remaining stand of old-growth forest in the Puget Trough. The impressive Douglas fir and western hemlock trees are 300–500 years old and nearly 200 feet tall. The understory comprises Oregon grape and swordfern, with a small area of Idaho fescue grassland. Several rare plants can be found here, including the arctic aster, few-flowered sedge, and pine broomrape. The forest is a critical habitat for species, especially birds, that depend on old-growth forest.

Norse Peak Wilderness Area
Pierce and Yakima Counties
Part of Wenatchee National Forest, the Norse Peak Wilderness Area contains forest of grand fir, subalpine fir, Pacific silver fir, and mountain hemlock. Other communities include rock face, talus, water, riparian, meadow, and shrubland assemblages. The conifer forests are typically mature to old-growth (150–500 years old). Several rare or unusual plants have been identified on the property. The forest is critical habitat for large mammals, such as bears, elk, and mountain lions; certain birds depend on the old-growth forest.

North Fork Nooksack Research Natural Area
Whatcom County
This RNA is located on 1,495 acres in the northern Cascades within Mount Baker National Forest about 20 miles east of Kendall. Natural communities include old-growth Douglas fir–western hemlock forest, western redcedar forest, and red alder forest. The largest trees are 6 feet in diameter and more than 600 years old; a subordinate forest consists of hemlock–Pacific silver fir–western redcedar. In addition to hosting the old-growth trees, the property contains a number of rare or unusual species, including various bats, moles, the Gapper red-backed vole, mountain lions, wolverine, elk, and mountain goats.

Okanogan National Forest
Okanogan County
The dense wilderness in this section of Okanogan National Forest 10 miles northwest of Republic shelters the largest surviving lynx population in the lower 48 states. A major food source for the lynx is the snowshoe hare, which feeds on lodgepole pine, a community that is being rapidly diminished by logging, old age, and fires. Thus, the extensive lodgepole pine forest here is a critical lynx habitat.

Pataha Bunchgrass Research Natural Area
Garfield County
Part of the Blue Mountains in southeastern Washington about 14 miles from Pomeroy, this 51-acre tract encloses high-quality bluebunch wheatgrass grassland and Douglas fir forest communities. Important mammals that use the tract are elk and black bears.

Pierce Island Preserve
Skamania County
A pristine island in the Columbia River gorge, this Nature Conservancy property protects one of the best examples of Columbia riparian habitat. The island supports a dense old-growth willow–black cottonwood forest interspersed with lush meadows of tufted hairgrass. The woodland is an important habitat for a variety of birds that nest in the canyon. The island's shoreline is graced with the extremely rare persistensepal yellowcress, which is known to occur in only ten locations in North America.

Puget Island Preserve
Wahkiakum County
Located a few miles southwest of Cathlamet on Puget Island, this preserve encloses a pristine section of bottomland forest and swamp in the

Columbia River floodplain. Natural communities include Sitka spruce forest, tidal marsh, wetland shrub thicket, and riparian edge. Biodiversity is high, and there are several unusual plant and animal species found on the property, including the endangered Columbia white-tailed deer.

Pyramid Lake Research Natural Area
Whatcom County
Located in montane forest in the northern Cascade Range near Newhalem, this 119-acre RNA covers part of the forested slopes of a cirque. Natural communities include cliff face, talus, marsh, lake, and mature to old-growth Pacific silver fir–hemlock forest and Douglas fir–Pacific silver fir forest. Lesser stands of Alaska-cedar and lodgepole pine also occur. The oldest trees have diameters of 5–7 feet and reach heights of nearly 200 feet. Several varieties of sensitive animal species occur here, including the northern rough-skinned newt, mountain lion, and pine martin. Several birds, including Vaux's swift, that live only within the rapidly vanishing old-growth forests nest here.

Quinault Research Natural Area
Grays Harbor County
Located on Olympic Peninsula, this 1,400-acre RNA in Olympic National Forest is about 40 miles north of Hoquiam. It was established to protect and study old-growth western cedar, Sitka spruce, Douglas fir, and western redcedar forests. The tallest Douglas fir trees are about 300 feet high, 400 years old, and 6 feet in diameter at breast height. The understory in this rain forest consists of lush ferns, mosses, and liverworts. The forest is an important habitat for the Roosevelt elk, black bear, mountain lion, and rare spotted owl.

Rainbow Creek Research Natural Area
Columbia County
This 420-acre RNA is located in the northern Blue Mountains of southeastern Washington about 25 miles southeast of Dayton. The RNA was established to protect virgin stands of grand fir–western white pine, Douglas fir–ponderosa pine, and western larch. The largest trees approach 175 feet in height and are more than 300 years old. Groundcover consists of the Pacific yew, thinleaf alder, and various forbs and grasses. It is critical summer habitat for the Rocky Mountain elk.

Rattlesnake Hills Research Natural Area
Benton County
Located about 17 miles northwest of Richland, this RNA encloses high-quality shrub steppe in arid southeastern Washington. Natural communities include big sagebrush–bluebunch wheatgrass, sagebrush–Sandberg bluegrass, saltbush-greasewood, Idaho fescue–perennial bunchgrass, and wheatgrass-bluegrass. The lower hills are covered by winterfat. Wildlife is diverse and includes a number of small mammals, as well as breeding birds such as sage thrashers and loggerhead shrikes.

Rose Creek Preserve
Whitman County
Located about 2 miles north of Albion, this 22-acre property along Rose Creek protects a very rare black hawthorn–cow parsnip plant community. Within this small property are nearly 400 species of plants, birds, and mammals. It is also one of the few places in North America where quaking aspen and black hawthorn coexist in a single location. This is an important breeding ground for several rare or unusual birds that are at home within the thickets. Rose Creek Preserve is owned and managed by The Nature Conservancy.

Ross Lake National Recreation Area
Whatcom County
The nation's largest stand of old-growth western redcedar grows on the western side of Ross Lake in Big Beaver Valley, about 5 miles north of Diablo. Many bird species nest here; the forest is critical habitat for the endangered gray wolf.

Salmo Research Natural Area
Pend Oreille County
Part of Colville National Forest, this RNA is located northeast of Metaline Falls on the steep slopes of Salmo Mountain. Natural communities include old-growth montane and subalpine coniferous forests of fir, larch, Engelmann spruce, western white pine, Douglas fir, and western redcedar. The largest trees are more than 150 feet tall and nearly 5 feet in diameter. Several shrubs and herb species comprise the understory. Arboreal lichens are abundant in the old-growth forests and are an important food source for some large mammals. The RNA is important habitat for the endangered mountain caribou, grizzly bear, gray wolf, lynx, and wolverine. Rare or unusual birds are the golden eagle, blue

grouse, rufous hummingbird, pileated woodpecker, and several warblers.

Salmo-Priest Wilderness Area
Pend Oreille County
Located along the U.S.-Canadian border, this rugged mountain wilderness includes old-growth western white pine forest, western larch forest, subalpine fir–Cascade azalea, western hemlock, and western redcedar–devil's club natural communities. Due to the high rainfall, a number of rare plants can be found here. The unbroken wilderness is critical habitat for lynx, cougars, pine martens, wolverine, moose, grizzly bears, and rare woodland caribous.

Silver Lake Research Natural Area
Whatcom County
Part of North Cascades National Park, this 1,687-acre RNA encloses Silver Lake and the northern slope and summit of Mount Spickard. Undisturbed natural communities include open lake, rock cliff, talus, and patchy alpine plant assemblages. Snowfields are always present. Plants that have anchored to the rocky surface include graminoids, mosses, lichens, and occasional meadows and shrublands. A variety of rare and fragile plants grows here.

Skagit River Bald Eagle Natural Area
Skagit County
Located about 2 miles east of Rockport along Skagit River, this Nature Conservancy property protects 6,000 acres of riverbank. Spawning salmon attract large populations of nesting and wintering bald eagles. This area has one of the largest eagle congregations in western North America. The old-growth trees along the river are favored as nesting sites by the eagles. Other species seen here are bears, elk, and mountain lions. The property is owned and managed by The Nature Conservancy.

Steamboat Mountain Research Natural Area
Skamania County
Located about 20 miles northwest of Trout Lake, this 1,377-acre RNA within Gifford Pinchot National Forest was established to protect pristine subalpine forests in the Cascade Range. Natural communities include open water, cliff face, talus, wet meadow, and undisturbed old-growth forests of mountain hemlock, Engelmann spruce, and subalpine fir with various shrubby understories. The forest is an important habitat for a variety of mammals and birds, including the uncommon or rare rough-skinned newt, Trowbridge shrew, several bats, Pacific jumping mouse, black bears, elk, mink, and mountain lions.

Thompson Clover Research Natural Area
Douglas County
Located on 200 acres in Wenatchee National Forest about 10 miles north of Wenatchee, this RNA protects a population of Thompson clover that is thought to be endemic only to this part of the state. Natural communities include ponderosa pine forest and grassland. Important animal species include cougars, bobcats, black bears, California big horn sheep, bald eagles, Vaux's swifts, spadefoot toads, and possibly elk.

Trapper Creek Wilderness Area
Skamania County
This pristine wilderness area about 30 miles northwest of Carson protects subalpine mountain hemlock–fir forest that is largely mature to old-growth. The old-growth trees that survive in steep gullies and ravines are nearly 150 feet tall and up to 500 years old. Other communities include rock face, wetland, meadow, and narrow riparian corridor. The unbroken wilderness is an important habitat for large mammals, such as elk, bears, and mountain lions. Raptors and the endangered spotted owl nest in the old-growth trees.

Turnbull Pine Research Natural Area
Spokane County
This 200-acre tract is part of Turnbull National Wildlife Refuge. Located about 5 miles southwest of Cheney, it encloses high-quality pristine ponderosa pine–bunchgrass savanna and wetland communities. The grasslands are dominated by bluebunch wheatgrass, Idaho fescue, and Sandberg bluegrass. The site is important habitat for a variety of birds, mammals, reptiles, amphibians, and insects, some of which are rare, threatened, or rapidly diminishing.

Twin Creek Research Natural Area
Jefferson County
Located on 1,000 acres in Olympic National Park, this RNA encloses old-growth Sitka spruce–western hemlock forest along Hoh River. The tallest trees are nearly 200 feet tall and 400–700 years old. This rain forest contains

a lush understory of vine maple, bigleaf maple, and various herbs. Cryptogams of moss-liverwort-lichen cover the ground. This is important habitat for bird and animal species that depend on old-growth forest, including owls and the Roosevelt elk.

Waldron Island Preserve
San Juan County
The relatively undisturbed Waldron Island contains a complete pristine island ecosystem. Natural communities include tidal flat, beach, dune, fresh and saltwater marshes, and forest. The tidal flats are an important food source for shorebirds, including plovers and dunlines. The highly diverse, clean freshwater marsh attracts thousands of birds, especially during spring and fall migrations. It is owned by the Nature Conservancy.

Wind River Research Natural Area
Skamania County
Located on 10,815 acres in Gifford Pinchot National Forest, this RNA is about 15 miles from Stevenson. Old-growth Douglas fir–western hemlock forest is the dominant natural community. Other communities include Pacific silver fir, various shrubby understories, and several ponds and wetlands. The old-growth forest is about 400 years old with specimens up to 200 feet tall. It is an important habitat for animal and bird species, such as elk and spotted owls, that depend on old-growth forest.

Yakima Canyon Preserve
Kitittas County
This important Nature Conservancy property protects a stretch of riparian corridor along the Yakima River canyon. The Yakima is one of the best trout habitats in the state and is sheltered by cliffs of basalt that are nearly 2,000 feet high. A variety of bird life is attracted to the large cottonwood and willow trees that grow along the river. The soaring cliffs are nesting sites for the rare osprey, bald eagle, golden eagle, peregrine falcon, and numerous other raptors. Specialized plant communities, including the very rare basalt daisy (endemic only to this canyon) cling to the rock faces.

Yellow Island Preserve
San Juan County
This 10-acre island off the Washington coast is part of the San Juan Archipelago and contains the highest biodiversity of any of the islands. Natural communities include fescue grassland, intertidal marine, oak woodland, and freshwater marsh. It contains the most complete representation of specialized flora and fauna among nearly 200 islands; this includes the rare hairy Indian paintbrush, brittle cactus, and broadleafed shooting star. Rare seabirds, rufous hummingbirds, and bald eagles nest here, and harbor seals use the shoreline as a birthing ground. The remarkably undisturbed grassy headland contains mostly intact native grasses.

WEST VIRGINIA

West Virginia is divided into three northeast-trending vegetation zones. From east to west: the Eastern Ridge and Valley; the Allegheny Mountain and Upland; and the Western Hills.

Eastern Ridge and Valley occupies the nine easternmost counties in the state; its western boundary with the Allegheny Mountain and Upland runs through Mineral, Grant, and Pendleton Counties. This series of steep ridges and valleys is forested dominantly with oak-pine and oak-hickory. Scrub oak, pitch pine, and scrub pine can be found on rocky south-facing slopes. Mixed mesophytic forest, including beech, basswood, maple, and hemlock, grows on the moister north-facing slopes. A variety of wildflowers covers the ground during spring.

Allegheny Mountain and Upland occupies the east-central section of the state. High mountains are covered by northern evergreen (red spruce, red pine, larch, and balsam fir) and northern hardwood forests. At midslope to lower elevations (2,500 to 3,000 feet), northern hardwoods, such as beech, basswood, red oak, and yellow birch, dominate the mountainsides. Old-growth stands survive in some of the narrower and harder-to-reach ravines that escaped logging. The rocky peaks often support heath barrens (mountain laurel–huckleberry–blueberry), sphagnum bogs, cliff face moss-lichen-fern assemblages, and grassy balds.

The Western Hills comprise the western two-thirds the state. This less rugged area is a mature plateau, with numerous streams that have dissected it into a series of hills. This terrain is covered by xeric oak-pine and oak-chestnut forest, with mesic mixed mesophytic and bottomland floodplain forests. Hardwoods include maple, beech, hickory, basswood, birch, and walnut.

Special Habitat Sites

Altona Piedmont Preserve
Jefferson County
This Nature Conservancy property a few miles west of Charles Town protects 90 acres of wetland communities along Evitts Run. Because it overlies limestone rocks, it is one of the few rare marl marshes in the United States. The calcium-rich water and soil support more than 24 rare plants that are specially adapted to this environment. This site provides important habitat for a variety of birds and mammals, some of them rare or elusive.

Bear Heaven Area
Randolph County
Located in Monongahela National Forest, this 147-acre area 9 miles northeast of Elkins comprises a number of natural communities, including red spruce–hemock forest, northern hardwood forest (cherry-birch-beech-maple), and wetland. It is a good example of the less common Appalachian red spruce forest. The rare Cheat Mountain salamander and northern flying squirrel inhabit the area.

Bear Rocks Bog
Tucker County
Part of Monongahela National Forest, this important bog-wetland complex contains a number of rare or endangered species. Comprising 66 acres northwest of Cabins, it includes open bog, heath shrub, and spruce forest natural communities. The bog contains a sphagnum moss mat as well as cottongrass, small cranberry, haircap moss, bog goldenrod, and the rare oceanorus. Shrubs are mostly mountain laurel. Rare insects, such as the pink-edged sulfur butterfly and bog copper butterfly, are found in the bog.

Big Run Bog
Tucker County
This 632-acre site about 4 miles west of Coketon is in the Monongahela National Forest. It encloses one of the most florally diverse bogs in the state, as well as spruce bog forest. There are also mature stands of maple–beech–yellow birch. The bog contains a sphagnum moss mat and seven rare or endangered plant species, including three orchids, bog bean, and northern sedge. The bog area is an important habitat for songbirds and the pink-edged sulfur butterfly.

Black Cherry
Randolph County
This 120-acre tract is an excellent example of the uncommon upland black cherry–maple forest community. Located about 6 miles southwest of Job, it is part of Monongahela National Forest. Mountain laurel and hay-scented fern are common in the understory. High-quality trout streams run through the area.

Blackwater Falls State Park
Tucker County
Located in Monongahela National Forest a few miles south of Davis, this area was heavily logged during the late 1800s. Natural communities include Blackwater River and its riparian corridor, hemlock forest, mixed hardwood forest, and lesser marsh and meadow. Unusual tree species include quaking aspens and hawthorn, relict species from the Ice Age. Rare or unusual animal species that depend on this habitat are the Virginia northern flying squirrel, Virginia big-eared bat, southern bog lemming, and northern water shrew. The forest is also important habitat for songbirds, raptors, and migratory bird species during spring and fall.

Brush Creek
Mercer County
This 124-acre Nature Conservancy property near Spanishburg is located in mountainous terrain along Bluestone River and includes Brush Creek Falls. The rocky gorge is lined with limestone and sandstone cliffs. Cooler microclimates along north-facing slopes harbor the rare Canby's Mountain lover and Canada yew relicts. Impressive wildflowers can be seen during spring and summer.

Cathedral State Park
Preston County
Located on 133 acres about 8 miles south of Terra Alta, this is a rare remnant virgin hemlock forest. The tall trees are more than 400 years old and nearly 100 feet tall. The Centennial Hemlock, the oldest tree in the park, is 123 feet tall, has a diameter at breast height of nearly 8

feet, and is about 500 years old. Other old-growth trees include large specimens of beech, oak, and maple. The understory contains a variety of colorful wildflowers, some of which are rare or unusual, during spring and summer. The park is an important habitat for woodland mammals, including the black bear.

Coopers Rock State Forest
Preston County
Located about 15 miles east of Morgantown along the Cheat River gorge, this 12,700-acre property is located in rugged terrain. It encloses some stands of mature to old-growth forest. A stand of towering hemlock trees is more than 300 years old.

Cranberry Glades
Pocahontas County
Nominated as an RNA by The Nature Conservancy in 1975, this 799-acre encloses a northern wetland system considered rare for this latitude. Located about 8 miles west of Buckeye, its important communities include bog, fen, shrub swamp, stream, and mixed spruce-deciduous forest. The wetland complex includes numerous rare plant and animal species, including bog rosemary, rose pogonia, buckbean, sundew, twayblade, chain fern, Jacob's ladder, serviceberry, Virginia northern flying squirrel, and various thrushes and warblers.

Cranesville Swamp Preserve
Preston County
This 400-acre Nature Conservancy property near Corinth contains excellent examples of northern boreal swamp, fen, and wooded upland communities. It is important habitat for rare northern relict boreal plants, such as American larch, red spruce, Canada yew, sundew, cranberry, and numerous ferns. Bog communities include some rare relict species from colder ancient climates, including thick mats of sphagnum moss. More than 200 different plant species grow here. Surrounding the swamp are larger older-growth trees, such as red spruce, hemlock, tamarack, and yellow birch—all northern species.

Dolly Sods Wilderness and Scenic Area
Tucker County
Located just east of Canaan Valley at the edge of Allegheny Plateau, this 10,000-acre property protects a mix of natural communities: upland bog, grass bald, scrub thicket, and bold rock outcrop covered with ferns, lichens, and mosses. A few remnants of presettlement hemlock-spruce forests remain. Wind-shaped rhododendrons and wiry red spruce cling to cracks in the steep rock faces; the flatter rock surfaces support mountain oat grass. Relict forest species include the yellow birch. More than 100 species of birds have been identified here, some of them rare or unusual. Raptors can be frequently seen gliding on the updrafts above the plateau.

Fisher Spring Run
Tucker County
Located on 553 acres in Monongahela National Forest, this property encloses a rare sphagnum-polytrichum bog. It is an example of a northern bog ecosystem. Other communities include red spruce, sugar maple–beech–yellow birch, and red pine forests. Rare plant species include cottongrass, cranberry, bracken fern, sedges, goldthread, and mannagrass. Other rare species are the Cheat Mountain salamander and the pink-edged sulfur, a boreal butterfly.

Gaudineer Scenic Area
Pocahontas County
Located about 6 miles northeast of Durbin, this 130-acre area is a Natural National Landmark. Old-growth red spruce and yellow birch, beech, and maple trees are hundreds of years old, nearly 100 feet tall, and display breast-height diameters of about 3 feet. The forest tract is an important habitat for birds and animals that prefer old-growth forests, such as owls and woodpeckers.

Greenland Gap
Grant County
Owned and managed by The Nature Conservancy, this 250-acre tract near Greenland protects a 400-foot-tall cliff face along the Allegheny Front. Several rare or unusual plant communities grow among the cool, wet rocks.

Hawthorn Area
Pocahontas County
This 13-acre tract in Monongahela National Forest northwest of Thornwood is rare because it is almost exclusively a hawthorn plant community. It was nominated by the U.S. Forest Service as an RNA during the 1970s.

Hungry Beech Preserve
Roane County
This Nature Conservancy property encompasses 124 acres of rare old-growth beech-oak-hickory forest. The trees are hundreds of years old, reach 4 feet or more in diameter at breast height, and are 100 feet tall or taller. Rapidly diminishing birds, such as woodpeckers and neotropical migrants, depend on these old-growth stands for food and shelter.

Ice Mountain
Hampshire County
Sparsely vegetated sandstone and shale talus slopes are protected on this rugged Nature Conservancy property. Pockets in the talus shelter ice and snow, which gradually melt during summer to keep the microclimate very cool. As a result, colder-climate plants grow here, several of them state-rare: bristly rose, twinflower, bunchberry, and evening primrose, which can be found only in shale barren communities.

Kanawha State Forest
Kanawha County
This 9,474-acre tract about 10 miles southwest of Charleston is one of the most biologically diverse areas in West Virginia. Natural communities include bottomland and upland conifer forests, mixed hardwood–pine forest, and mixed hardwood forest. Several species of ferns and wildflowers grow along narrow riparian corridors. More than 1,000 different species of trees and plants grow here—an incredibly high biodiversity—including some rare or unusual plants, some of them relicts from the Ice Age 10,000 years ago.

Meadow Creek Shale Barren
Greenbrier County
Located in very rugged terrain around Neola, this 14-acre property in the Monongahela National Forest protects a section of a very rare shale barren community. Shale barrens are unique because the vegetation has adapted specifically to the underlying shale. The forests are typically Virginia pine or chestnut oak. At least ten rare plants occur on these rocky slopes, including shale barren onion, shale barren rockcress, Kate's-mountain clover, and the Alleghany plum.

Morgan Woods
Morgan County
A rare shale barren community along Cacapon River is protected on this 60-acre tract. It supports a number of rare shale barren plants, the globally rare plant Harperella, and several rare butterflies and moths, such as the Olympia marble, northern metalmark, and grizzled skipper.

Mount Porte Crayon Area
Randolph and Pendleton Counties
Part of Monongahela National Forest, this tract 8 miles north of Onega was nominated by The Nature Conservancy as an RNA during the 1970s. Located in some of the most rugged mountain terrain in West Virginia, it protects good examples of red spruce forest and spruce–yellow birch–maple forest communities. It is excellent habitat for the rare Cheat Mountain salamander.

North Bend State Park
Richie County
Located just outside Cairo, this scenic park sits along North Fork River. It is known for majestic stands of sycamore-hemlock-oak-pine-poplar that are hundreds of years old; it represents one of the few remaining virgin forest remnants in the state. The wildflowers are abundant and colorful during spring. Several varieties of woodland birds and mammals, some of them rare or unusual, including the rapidly diminishing black bear, live in the park.

Panther Knob
Pendleton County
Owned by The Nature Conservancy, this rugged, forested, 500-acre plateau includes virgin red spruce forest, open subalpine cliff top, pine barren, and the world's largest population of variable sedge. Other rare species are Michaux's saxifrage, Canada mountain rice, and pink-edged sulfur. Several pristine trout streams also flow through the tract.

Red Spruce Area
Randolph and Pocahontas Counties
This 60-acre tract near Job protects a nearly pure stand of mature red spruce forest, spruce-hardwood forest, and northern hardwood forest. Some trees are large. It is an important site to study the forest dynamics of mature red spruce stands. It is a likely habitat for the rare northern flying squirrel and Cheat Mountain salamander.

Rohrbaugh Plains
Tucker County
Red spruce, red spruce–maple–beech, and maple–beech–yellow birch forests and mead-

ow are the natural communities preserved on this 149-acre tract within Monongahela National Forest. Located just southeast of Lanesville, it was recommended as an RNA by The Nature Conservancy during the 1970s. It is critical habitat for the rare Cheat Mountain salamander. Woolgrass is an unusual grass species that occurs in the meadow.

Slaty Mountain
Monroe County
Owned and managed by The Nature Conservancy, this 153-acre property protects high-quality dry hardwood forest, pine forest, and a very rare shale barren community. Located about 3 miles west of Sweet Springs, the property is known for an abundance of plant species that grow only on shale barrens, such as Kate's mountain clover, Steele's aster, and yellow buckwheat.

Smoke Hole
Grant County
Part of Monongahela National Forest, Smoke Hole is about 5 miles west of Cabins. Its primary feature is a series of cave openings that are critical habitat for the endangered Virginia big-eared bat. The endangered peregrine falcon and bald eagle also nest in the area. The rugged limestone bedrock is covered by dense mixed hardwood–conifer forest. Openings reveal cedar glades as well as the very rare limestone barren and shale barren natural communities, which support rare plant species specialized to those substrates. It also contains prairies of little bluestem and side oats grama. Endemic species include the Smoke Hole burgamot.

Stuart Knob Area
Randolph County
This 400-acre tract about 6 miles east of Elkins in Monongahela National Forest protects an excellent remnant of old-growth red spruce, an uncommon forest type in the state. The larger trees are probably more than 100 years old. It is also important habitat for the rare Cheat Mountain salamander, northern flying squirrel, and southern rock vole.

Whites Draft Shale Barren
Greenbrier County
Located in Monongahela National Forest southwest of White Sulphur Springs, this 74-acre parcel protects some very rare shale barren ecosystems. Because the ecosystem is so specialized, almost all the dominant species are extremely rare. These include shale barren onion, shale barren rockcress, Kates Mountain clover, shale barren bindweed, white-haired leatherflower, shale barren ragwort, heart-leaved skullcap, and Alleghany plum. The forest types are stunted chestnut oak and Virginia pine. Uncommon butterflies, such as the Olympic marble and checkered skipper, have been found here.

WISCONSIN

Wisconsin comprises six natural divisions. The Lake Superior Boreal Division is found in extreme northern Wisconsin along the Lake Superior shoreline. The gently rolling topography is covered by red clays and pink soils. It is characterized by coniferous forest (balsam fir and white spruce), with a few maples and birches; extensive peatlands can be found among some of the wetlands. The Northern Highland Division is mixed deciduous and coniferous forest and occupies most of northern Wisconsin. Forest types include pine-oak barren, maple–hemlock–birch–white pine, maple-basswood-birch, maple–oak–white pine, and maple-hemlock. Peat bogs with black spruce and tamarack are also common. The Lake Michigan Shoreland is dominated by southern and northern deciduous and coniferous mesic forest. A typical forest includes beech, maple, and hemlock trees to the north, maple-basswood-beech to the south. The Central Plains in central Wisconsin is composed of oak-pine barrens, oak forest, oak savanna, and wetlands. The Sandhills are glacial outwash deposits. The land is fairly level and covered by shallow peats and mucky soil. The Southeastern Moraine is covered by maple-basswood-oak-elm-hickory deciduous forests, oak savanna, and a variety of prairie types. The Southwestern Upland is covered by maple-basswood-oak forest, oak forest, savanna, and various prairie grasses. River-bottom forests along the Mississippi and Wisconsin Rivers are also prominent.

Special Habitat Sites

Abraham Woods
Green County
A cooperative project between The Nature Conservancy and the University of Wisconsin, this 64-acre property 35 miles south of Madison is a remnant of the old-growth "Big Woods" that once covered southern Wisconsin during presettlement times. Natural communities include upland oak-maple-elm forest and dry prairie. Some of the trees are nearly 200 years old. The groundcover offers brilliant displays of wildflowers, including the rare Goldier's fern, during spring and summer.

Archibald Lake Hemlock Forest
Oconto County
This 110-acre site is a cooperative project between The Nature Conservancy and the U.S. Forest Service. Located in Nicolet National Forest, this property contains huge hemlock trees that are hundreds of years old. Other old-growth trees are maple, beech, and white pine. It is an important woodland habitat for numerous species, including bald eagles, woodpeckers, tanagers, and pine martens.

Audubon Goose Pond
Columbia County
Located just south of Arlington, this 81-acre tract contains the open water Goose Pond and its surrounding wetlands. More than 200 bird species have been documented, including a variety of nesting ducks. Other visitors include tundra swans, herons, egrets, raptors, and peregrine falcons. The rare Blanding's turtle can also be found here.

Aurora Lake
Vilas County
Located on 260 acres about 3 miles northwest of Sayner, this lake is a nesting habitat for a state-threatened bird species. The lake contains a variety of aquatic vegetation, including an open bog with a floating sphagnum-sedge mat.

Avoca Prairie-Savanna
Iowa County
This 1,885-acre property contains the largest native tallgrass prairie east of the Mississippi River. Located about 2 miles east of Avoca, it also contains southwestern mesic forest and oak savanna communities. More than 200 vascular plant species grow here, including a state-threatened species.

Avon Bottoms
Rock County
This 168-acre tract about 2 miles southeast of Avon contains old-growth lowland hardwoods along Sugar River. Silver maple and white oak trees reach breast-height diameters of more than 3 feet. The rare cerulean warbler has been sighted in this area.

Bark Bay Slough
Bayfield County
This 225-acre site about 3 miles northeast of Herbster contains open bog, northern wet forest, and baymouth bar–beach communities. The baymouth bar sand and mud contain blueberry, bearberry, sweet gale, and beach grasses. A floating sedge bog occurs at the edge of the slough. It is an important nesting site and migratory stop for a variety of waterfowl.

Bass Lake Fen
Waushara County
This 77-acre site protects a 20-acre fen along the shore of Bass Lake. The fen contains more than 125 plant species and grades into a sedge meadow. The area is frequented by sandhill cranes.

Battle Bluff Prairie
Vernon County
The site of a historic battle, this 31-acre property contains a remnant dry prairie among rocky terrain. More than 80 prairie plants have been identified. The rare Canada yew and cliff-brake grow among the sandstone and limestone ledges and boulders.

Baxter's Hollow
Sauk County
About 5 miles south of Baraboo, this 1,995-acre site contains a representative portion of the "Big Woods," the presettlement southern dry-mesic forest. There are patches of unlogged old-growth timber. Tree species are typically oak, basswood, maple, and ironwood. It is critical habitat for birds that require large forests, such as the pileated woodpecker, ovenbird, and several warblers. More than 40 species breed in the forest; some are rare species, such as the worm-eating warbler and hooded warbler. Otter Creek, an undeveloped watershed, sustains very diverse aquatic insect fauna,

including more than 70 species of caddisflies. The rare pickerel frog can also be found here.

Bean Lake
Jefferson County
This 195-acre tract about 3 miles south of Lake Mills is a mix of dry and wet forest, shrub swamp, sedge meadow, and open water. Bean Lake is an alkaline seepage lake. Peatland is also located on the site. The wetland is an important nesting site for a variety of waterfowl species, including wood ducks, blue-winged teals, and rare sandhill cranes.

Belmont Mound Woods
Lafayette County
This 60-acre site 2 miles west of Belmont contains remnants of dry-mesic and mesic forests, which were rare in southwestern Wisconsin. The canopy is typically sugar maple–red oak–basswood–elm–black walnut. The site is relatively undisturbed, subjected to only minor logging during the early 1900s.

Beuhla Bog
Walworth County
This 78-acre tract in north-central Walworth County consists of mesic forest, oak savanna, and a bog that is host to threatened plant species. It also contains carnivorous plant species, such as the pitcher plant and sundew.

Black Earth Prairie
Dane County
This 16-acre Nature Conservancy site near Black Earth contains a relict dry-mesic prairie with an abundance of native vegetation, including big and little bluestem and Indian grass. Rare plant species are the aster hybrid, pomme-de-prairie, round-stemmed false foxglove, and white camas. The site has never been plowed.

Black Tern Bog
Vilas County
Located about 5 miles north of Woodruff, this 26-acre bog has two small seepage lakes and a quaking sphagnum bog. It includes a variety of sedges as well as the carnivorous sundew plant. Several rare or threatened wetland plants exist here. The bog is a nesting area for black terns, bitterns, and blackbirds.

Blue River Sand Barrens
Grant County
Located about 4 miles west of Muscoda, this 130-acre site contains the best example of sand barrens in Wisconsin. The loamy, fine glacial sand is stabilized by oak forest, oak savanna, and upland prairie communities. Unusual plants on the dunes include false heather, bearberry, spike moss, fame flower, and prickly pear cactus. There is an unusual abundance of ant lions. The state-endangered six-lined racerunner snake inhabits the site.

Bose Lake Hemlock Hardwoods
Forest County
Located in Nicolet National Forest east of Eagle River, this tract of virgin forest covers about 25 acres. It is remnant of the presettlement "Big Woods" and contains old-growth hemlock, sugar maple, yellow birch, and basswood. Hemlock diameters exceed 3 feet. It is bordered by a black spruce swamp. The property is an important habitat for a variety of birds and mammals, including the pileated woodpecker.

Brady's Bluff Prairie
Trempealeau County
Remnants of virgin dry-bluff prairie occur on this 10-acre site a mile west of Trempealeau. They are located on bluffs that face the Mississippi River to the southwest. Plant diversity is high, with more than 100 species recorded, including a few species of special concern. Great Plains species at the edge of their ranges include prairie larkspur and satin grass. Timber rattlesnakes may also be found.

Brant Brook Pines and Hardwoods
Burnett County
Old-growth pine forest occurs on this 190-acre tract about 4 miles northwest of Grantsburg. The tall white pines are more than 100 years old. A trout stream dissects the property.

Brule River Tract
Douglas County
Located about 20 miles east of Superior, this 4,958-acre tract is a cooperative effort between The Nature Conservancy and local landowners. The Nature Conservancy has secured conservation easements from property owners that will prevent development in these lands along the scenic Brule River. Natural communities include the pristine river, northern upland pine forest, lowland forest, and cedar swamp. This

is the only location in the state where old-growth red and white pines (over 300 years old) are found growing together. Rare, threatened, or endangered species include the bald eagle, gray wolf, black-backed woodpecker, Cape May warbler, Connecticut warbler, flycatchers, and white bog orchids. The Brule River is one the best trout habitats in the United States.

Brunsweiler River Gorge
Ashland County
This 40-acre tract in Chequamegon National Forest southwest of Mellen has been nominated for RNA status. It encloses a stretch of Brunsweiler River and its adjacent high cliffs. Natural communities include the river, associated riparian corridor, cliff face, and northern mesic forest. Two rare plants have been identified on the site. Old-growth trees that escaped logging can be found within the gorge.

Buena Vista Quarry Prairie
Portage County
About 10 miles southwest of Plover, this 40-acre tract contains remnant dry and dry-mesic prairie and patches of oak. The prairie plant species include little bluestem, Indian grass, and wormwood. It is also good habitat for grassland birds, including the state-threatened prairie chicken.

Cedarburg Bog
Ozaukee County
Located 5 miles southwest of Saukville, this 1,600-acre tract protects a well-developed, patterned, diverse bog system that is one of the largest in the state. There are six open water communities within the bog, as well as conifer swamp, sedge meadow, and bottomland hardwood forest. The fauna and flora are diverse and include tamarack, white cedar, pitcher plant, and some other rare bog plant species. The preserve is a cooperative project between The Nature Conservancy and the University of Wisconsin–Milwaukee.

Cherney Maribel Caves
Manitowoc County
Located about 2 miles east of Maribel, this 8-acre area is known for its shaded limestone cliff community, including the rare polypody, bulbet, and walking ferns. Impressively large white cedars occur at the base of the cliff.

Cherry Lake Sedge Meadow
Racine County
This 80-acre tract a mile south of Rochester contains a pristine sedge meadow and fen. The deep sedge mat has only a few openings. The meadow contains plants that grow in both alkaline and acidic soil conditions. Interesting and rare plant species include round-leaved sundew, shrubby cinquefoil, yellow twayblade, marsh Saint John's wort, sphagnum, and various ferns.

Chequamegon Hardwoods
Ashland County
Located in south-central Ashland County, this 80-acre parcel in Chequamegon National Forest encloses a remnant of the original "Big Woods" that once covered Wisconsin during presettlement times. Although selective logging has occurred in the past, many old-growth hemlock, sugar maple, and basswood trees remain. Large rock outcrops support unique rock face communities of lichens, mosses, ferns, and other small plants. The area is important habitat for interior forest birds as well as large mammals, such as the black bear.

Chiwaukee Prairie
Kenosha County
The state's most biologically diverse prairie occurs on this 354-acre site south of Kenosha. Owned by The Nature Conservancy, it is home to 11 state-threatened or state-endangered plant and animal species. Located near the Lake Michigan shoreline, it is characterized by a series of sandy beach ridges separated by wet swales. The ridge-and-swale topography hosts more than 400 plant species and provides critical habitat for several rare reptiles and birds (upland sandpiper, marsh wren). Rare or unusual plants include pink milkwort, smooth phlox, prairie Indian grass, and sand coreopsis. The property is known for splendid displays of wildflowers during spring and summer.

Clifford F. Messinger Bass Lake Preserve
Iron County
This 840-acre tract northwest of Lac du Flambeau is owned by The Nature Conservancy. It includes a pristine 15-acre lake surrounded by coniferous forest and wetland. Natural communities include open water, northern mesic forest (maple-birch), dry mesic forest (aspen-maple), swamp conifer forest

(tamarack–black spruce–white pine), and bog communities. The pristine bogs contain leatherleaf, various herbs, and the rare pitcher plant. Some of the white pine trees are quite old. The site is home to great blue herons, ospreys, bald eagles, and black bears. It is a valuable scientific resource, as it contains a complete lake watershed.

Comstock Bog
Marquette County
About 6 miles north of Montello, this 537-acre natural area consists of northern wet and dry forest and a sedge meadow containing a variety of rare or unusual plants, including the pitcher plant, bladderwort, and sundew. The marsh is also an important sandhill crane nesting area and a stopover for migrating birds.

Day Lake
Vilas County
This exceptionally scenic lake is located on a 117-acre tract about 5 miles north of Woodruff. The lake is undeveloped, containing very sterile but clear water. A sand bottom supports an excellent submerged sterile rosette flora of quillwort and the much rarer plantain shoreweed. Emergent vegetation types are sweet gale, alder, and three-way sedge.

Decorah Mounds
Trempealeau County
Located about 2 miles east of Galesville, this 40-acre Nature Conservancy site encloses a classic Mississippi River bluff. Natural communities on the property include mixed hardwood forest and a high-quality remnant of dry prairie. The prairie is covered by bluestem, lead plant, butterfly weed, and pasque flower. Oak, dogwood, and basswood dominate the forests on the slopes. Dutchman's breeches can be found among the understory. The limestone cliffs along the bluff are restricted habitats for columbine and polypody ferns. The white camas flower can also be found.

Dewey Heights Prairie
Grant County
This 27-acre tract contains the largest dry lime prairie in the state. It is located a mile west of Cassville on a bluff overlooking the Mississippi River. Typical prairie vegetation includes big and little bluestem, Indian grass, hairy grama, and porcupine grass. Compass plant can also be found. A state-threatened plant also inhabits the site.

Dewey Marsh
Portage County
A vast, undisturbed northern sedge meadow is located on this 926-acre tract about 8 miles northeast of Stevens Point. Characteristics of the meadow include sphagnum hummock, sedge mat, and minor peat development. It is an important nesting area for rare species, including the sharp-tailed grouse, sandhill crane, prairie chicken, and Henslow's sparrow.

Doering Tract
Price County
Located in Chequamegon National Forest in northeastern Price County, this 284-acre parcel encloses a stand of old-growth northern mesic forest along Flambeau River. The giant hemlock, sugar maple, white pine, and yellow birch trees are nearly 100 feet tall and centuries old. A forest subtype is black ash–American elm–red maple. This old-growth tract is an important habitat for old-growth–sensitive species, such as owls and woodpeckers.

Dunbar Barrens
Marinette County
This 240-acre tract about 10 miles east of Goodman is one of the few remaining excellent examples of presettlement pine barren, a rare community in northeastern Wisconsin. The acreage is covered by aspen and jack pine. The area is known for a variety of unusual lichen species.

Dunn Lake
Vilas County
This 562-acre tract is an excellent piece of virgin northern mesic forest at the south end of Dunn Lake. The supercanopy consists of very large white pine, hemlock, yellow birch, sugar maple, and basswood. The site is also home to nesting bald eagles.

Ennis Lake/Muir Park
Marquette County
Located about 11 miles north of Portage, this 150-acre state natural area contains a mix of oak-pine barren, sedge meadow, wet-mesic prairie, clearwater kettle lake (30 acres), open bog, and wet tamarack forest communities. Notable bog species include the carnivorous pitcher plant, grass-of-Parnassus, and golden Alexanders. The area was the boyhood home of naturalist John Muir, who founded the Sierra Club.

Flambeau River
Ashland and Iron Counties
Located northwest of Park Falls, this 1,112-acre cooperative project between The Nature Conservancy and the Wisconsin Department of Natural Resources protects a 12-mile stretch of Flambeau River. The banks are lined with old-growth upland boreal forest; some stands of white pine and red pine are more than 200 years old, with breast-height diameters reaching nearly 3 feet. This stretch of river is uncrossed by dam or road. Rare or special species that thrive here are bald eagles, red-shouldered hawks, wolves, lake sturgeons, river redhorses, and the small round-leaved orchis.

Fountain Creek
Green Lake County
This 50-acre wet prairie represents a very rare community in Wisconsin. Located west of Kingston, it is located beside the Grand River Marsh Flowage. The prairie is dominated by bluestem, northern dropseed, and cordgrass. The site is used by nesting and migratory waterfowl species, including geese, cranes, herons, harriers, and bitterns.

Frog Lake and Pines
Iron County
Although some selective logging has occurred here, tall old-growth red and white pines remain on this 193-acre tract about a mile south of Manitowish. The white pines are more than 100 feet tall and approach 3 feet in diameter. Most of Frog Lake's shoreline is wetland, which hosts bladderwort and Robbin's spike rush.

Genesee Oak Opening and Fen
Waukesha County
This 52-acre tract 2 miles northeast of North Prairie contains a fen and the best oak opening in the state. It is a classic bur oak opening in a rolling prairie-savanna complex. The fen hosts a number of uncommon plant species. The rare pickerel frog is found in a stream on the property.

Giant White Pine Grove
Forest County
This 30-acre tract in Nicolet National Forest is known for the very large white pine trees that overstory a mature northern mesic forest. It is an excellent example of hemlock hardwood with white pine supercanopy. The white pines are up to 3 feet in diameter. It is an important nesting habitat for pileated woodpeckers, warblers, vireos, and wrens.

Goose Pond Sanctuary
Columbia County
Owned the Audubon Society, this pond and surrounding wetland complex is about 3 miles northwest of Leeds. Natural communities include a prairie pothole lake, river bulrush wetland, and restored tallgrass prairie. More than 250 birds have been sighted here, including water and grassland species. Raptors, such as the northern harrier and rare peregrine falcon, hunt among the grasslands. The pond is also home to the eastern tiger salamander and the state-threatened Blanding's turtle.

Hemlock Draw
Sauk County
Located in the rustic Baraboo Hills, this 543-acre Nature Conservancy preserve is about 15 miles west of Sauk City. Natural communities on the property include cliff, glade, aquatic (stream), and five varieties of forests. Wet bottomland forests consist of hemlock, sugar maple, ironwood, aspen, and oak; larger hemlock trees are found in cooler, sheltered draws. Upland forests are dominated by oak. Rare plant and animal species include drooping sedge, kidney-leaved sullivantia, certain club mosses, the worm-eating warbler, and other are birds. It is an important nesting habitat for interior forest birds, such as woodpeckers, warblers, barred owls, and thrushes.

High Cliff Escarpment
Calumet County
Located on the edge of Lake Winnebago a mile southwest of Sherwood, this 125-acre tract contains cliff, wet-mesic, and littoral communities. The cliff faces support delicate ferns, leaf cup, cliff stickseed, and long-beaked sedge. Old-growth sugar maple, basswood, ash, and butternut trees of large size can be found on the talus slopes. The rare false rue anemone and Canada violet occur at the lake's edge.

Holmboe Conifer Forest
Oneida County
Located just southwest of Rhinelander, this 32-acre parcel contains some very old white pine trees that escaped logging during the early 1900s. Natural communities include hemlock forest, red and white pine forest, mixed conifer swamp, and open marsh. The hilly topography

results from ridges of sand and gravel deposited by melting glaciers about 10,000 years ago. Tamarack and white cedar form thick stands in the swamp forest. The rare yew tree can also be found. It is an important nesting habitat for a more than 20 birds, including herons, the rare spotted sandpiper, flycatchers, muartins, vireos, thrushes, and other interior forest birds. Pink lady's slippers cover the ground during spring. The property is owned by The Nature Conservancy.

Honey Creek
Sauk County
Located on 778 acres about 15 miles west of Sauk City, this Nature Conservancy property is part of the scenic Baraboo Hills. It protects a variety of natural communities, including rocky cliff, hemlock forest, mixed hardwood forest, wetland, and riparian habitat along Honey Creek. The upland maple-oak forests are important habitats for interior forest birds, including two rare warblers (cerulean and Kentucky). Turkey vultures nest along the cliff faces; black-and-white warblers nest in hemlock trees that grow among the cliff faces. The lowland ash–tamarack–alder–white pine forest shelters the rare bog bluegrass plant. Honey Creek and its tributaries contain trout, rare darters, and the rare pickerel frog. Waterfowl species breed in the wetlands, including herons and the Virginia rail. It is also an important resting and feeding site for thousands of migrating birds during spring and fall.

Ipswich Prairie
Lafayette and Grant Counties
This site is the only remnant of the extensive deep-soil mesic prairie that once covered southwestern Wisconsin. Located on 20 acres 3 miles southeast of Platteville, it contains more than 125 species of native prairie plants, including several uncommon or state-threatened species. The site is also known for its abundance of prairie mound–building ants.

Jackson Harbor Ridges
Door County
Located on Washington Island in Lake Michigan, this 70-acre tract protects sand beach, dune, and swale communities. The interdunal swale and undisturbed Great Lakes beach settings are globally threatened habitats. The swales shelter some rare and state-endangered or state-threatened species, including dwarf lake iris, Kalm's lobelia, false asphodel, Ohio goldenrod, cinquefoil, arctic primrose, low calamint, bladderworts, and various sedges. Sand coreopsis can also be found. The sheltered harbor is an important feeding and resting site for migrating birds. Jackson Harbor Ridges is a cooperative project between The Nature Conservancy and the Town of Washington Island, Door County.

Jung-Hemlock Forest
Shawano County
This 73-acre tract about 8 miles west of Shawano is the state's best example of northeastern mesic forest. An old-growth remnant, it contains hemlock, beech, and white pine trees that are about 200 years old. An open bog is also present. Thrushes, vireos, pewees, and grosbeaks nest in the area.

Kangaroo Lake
Door County
Located on the scenic Door Peninsula, this 117-acre Nature Conservancy property protects part of Kangaroo Lake about 2 miles southwest of Baileys Harbor. It shelters pristine shoreline along the lake, a flowing stream, emergent marsh, lowland forest, and upland forest. The large maple, beech, and oak trees of the upland forest cover a plateau with steep, rocky faces. The mixed conifer lowland forest shelters the Canada yew, a rapidly diminishing species. The area is well known for its display of wildflowers during spring. Piel Creek is one of only 15 sites in the world where the federally endangered Hine's emerald dragonfly breeds. Other important or rare species include the corcas copper butterfly, bald eagle, osprey, Caspian tern, and various warblers. The woodlands are an important nesting habitat for numerous interior forest birds, including woodpeckers.

Karcher Springs
Racine County
A spring, calcareous fen, and stream complex surrounded by forest and oak savanna occurs on this 32-acre tract just southwest of Burlington. The stream's upper reaches are pristine and undisturbed. Rare fen species include grass-of-Parnassus, Ohio horse mint, gerardia, Ohio goldenrod, and marsh blazing star.

Keller Whitcomb Creek Woods
Waupaca County
This 97-acre tract about 3 miles southwest of Big Falls is unique in that northern dry-mesic,

mesic, and wet-mesic forest communities coexist among rolling hills. Pine, oak, hemlock, birch, and maple compose the mesic and northern dry-mesic forests; white cedar dominates the wet-mesic forests. A world-class trout stream also runs through the property. The woods are habitat for the winter wren and veery.

Kenosha Sand Dunes
Kenosha County
This joint project between The Nature Conservancy and the Wisconsin Department of Natural Resources protects a 56-acre portion of one of the longest prairie wetland complexes in the eastern United States. Rapidly vanishing natural communities protected here include foredune and backdune, prairie, and wet swale. The complexity of the communities gives rise to a high diversity of plants; nearly 250 species have been identified, some of them rare or endangered. The dunes are an important hunting ground for raptors such as the bald eagle. Monarch butterflies congregate on the property during the fall migration to Mexico. Important archeological remains have also been found here.

Kettle Moraine Fen and Low Prairie
Waukesha County
Located 3 miles north of Eagle, this 250-acre remnant of the once-vast Scuppernong Prairie contains four state-endangered or special-concern plant and animal species. It contains fen, wet-mesic prairie, dry-mesic prairie, and lesser savanna communities. Rare plant species include cinquefoil, valerian, grass-of-Parnassus, and Ohio goldenrod.

Kissick Alkaline Bog Lake
Sawyer County
This undeveloped alkaline bog lake is enclosed by a 135-acre preserve about 3 miles west of Hayward. It contains open water, bog mat, and sedge meadow. More than 14 different kinds of orchid inhabit the site. The pileated woodpecker nests on the property.

Kohler Park Dunes
Sheboygan County
This 135-acre area about 4 miles south of Sheboygan protects excellent examples of lake dune, interdunal wetland, and beach communities along Lake Michigan. Some of the dunes are stabilized with small dry-mesic white pine forest. Four state-threatened plant species grow on the site. It is an important visiting site for migrating raptors.

Koshawago Springs
Sauk County
Near the southwest corner of Devil's Lake south of Baraboo, this 40-acre tract contains a diverse southern mesic forest, spring, and stream communities. The creek is bordered by sandstone cliffs. Oak and basswood trees dominate the forest canopy; ground plants include rare species of moss, algae, fern, and sedge. It is also an important habitat for summer birds, such as raptors, thrushes, and warblers.

Kurz Woods
Ozaukee County
Located just south of Saukville, this 31-acre parcel is owned and managed by The Nature Conservancy. It protects one of the few remnants of the "Big Woods," the immense forest that covered Wisconsin before European settlers arrived. The southern mesic forest, partially logged during the 1930s, is dominated by large maple and beech trees. Its old-growth status makes it an important forestry and ecology research site. Hepactica, bloodroot, spring beauty, and other wildflowers bloom impressively during spring.

Lulu Lake Preserve
Walworth and Waukesha Counties
Owned and managed by The Nature Conservancy, this 1,048-acre property near Mukwonago is located in the Kettle Moraine. Natural communities include a pristine portion of lake shoreline, open stands of oak trees in a prairie-savanna setting, wetlands, and the riparian habitats along Mukwonago River. Lulu Lake and its watershed contains almost half of all fish species endemic to Wisconsin. The diverse wetlands comprise kettle bogs, sedge meadows, and calcareous fens, which generate a high diversity of flora, including the uncommon lesser fringed gentian, Ohio goldenrod, and grass-of-Parnassus. One of the rarest plants is the northern kittentail, an unusual wildflower. Relict fish from the glacial age still live in the lake, including the blackfin shiner and Iowa darter. Other rare or endangered animal species are Blanchard's cricket frog, nesting sandhill cranes, and raptors. Rare insects have also been identified.

Mazomanie Bottoms
Dane County
Located northwest of Mazomanie along the Wisconsin River, this 160-acre tract consists of southern wet-mesic floodplain forest. The dominant trees are silver maple, elm, basswood, and swamp white oak. It is an important nesting and foraging ground for uncommon bird species, such as cerulean and Kentucky warblers, brown creepers, and pileated woodpeckers. Bald eagles nest here during winter.

McCarthy Lake and Cedars
Ashland County
This 188-acre tract in Chequamegon National Forest totally encloses a shallow, soft drainage lake. Natural communities around the lake include northern dry mesic forest, northern wet black spruce–tamarack forest, shrub carr (fen), and sedge meadow. The wetlands include cedar swamp, conifer swamp, and thick alder stands. A stand of old-growth white pine (more than 100 feet tall and hundreds of years old) is also found on the property. The lake is an important habitat for many birds, including nesting loons.

McCaslin Mountain
Forest County
Located 7 miles east of Lakewood in Nicolet National Forest, McCaslin Mountain is partially covered by old-growth northern hardwood forest (red oak–beech). Other trees are yellow birch and sugar maple. Rock outcrops are covered by pink corydalis, rock spikemoss, and bearberry. The forest serves as a critical habitat for the rare cerulean warbler, wood thrush, and yellow-throated vireo. Turkey vultures may nest on the rocky outcrops of the mountain.

Memorial Grove Hemlocks
Price County
Part of Chequamegon National Forest, this 68-acre parcel includes stands of virgin old-growth hemlock and yellow birch, a northern mesic forest type. Other trees include basswood and sugar maple. The large size of these trees indicate they are 200–400 years old; some are the largest specimens in the state.

Messinger Prairie and Preserve
Jefferson, Walworth, and Waukesha Counties
These scattered 246 acres in Kettle Moraine State Forest near Eagle and Palmyra are some of the best remnants of dry prairie and oak savanna remaining in the state. The prairie is dominated by big and little bluestem, Indian grass, and prairie dropseed; the savanna portions are dominated by bur oak. The prairie shelters a number of bird and reptile species, some of which are rare.

Mink River Estuary
Door County
This 1,465-acre preserve, located about 5 miles southeast of Ellison Bay on the Door Peninsula, protects one of the few remaining high-quality natural estuaries in the United States. Natural communities include the open water of the estuary, its spring-fed tributaries, wetlands, upland hardwood forest, and lowland mixed conifer-hardwood forest. The estuary is an important spawning ground for fish in the Great Lakes. It also attracts thousands of birds during spring and fall migrations, including teals, mallards, herons, bitterns, loons, and cormorants. Bald eagles and sandhill cranes have also been seen. Rare plant species include the dune thistle and dwarf lake iris. The area also supported prehistoric and Native American communities. The property is owned and managed by The Nature Conservancy.

Miscauno Cedar Swamp
Marinette County
This 555-acre wetland and wet-mesic conifer forest is a rare natural community in Wisconsin. Tree species include white cedar, balsam fir, black spruce, and tamarack. Several varieties of orchids, lichens, ferns, and mosses occur in the groundlayer. Tanagers, cuckoos, warblers, and thrushes use this area as a nesting site.

Mount Pisgah Hemlock Forest
Vernon County
A relict hemlock–yellow birch–white pine northern forest and cliff community occurs on this 65-acre tract about 2 miles east of Ontario. The cliff faces contain a number of delicate, rare plant species, including ferns and mosses.

Mud Lake–Radley Creek Savanna
Waupaca County
Mud Lake is a hardwater lake with abundant wild rice that is surrounded by undisturbed tamarack forest. It is located on a 100-acre tract about 3 miles south of Rural. The combination of natural communities is rare for the state; the absence of black spruce is also unusual for this environment. The carnivorous sundew and

pitcher plant are abundant here. The site is home to a state-threatened bird species.

Muskego Park Hardwoods
Waukesha County
Located a mile west of Muskego, this 60-acre site protects an old-growth southern dry-mesic forest. The large trees include sugar maple, hickory, walnut, cherry, and the rarer Kentucky coffee tree and blue ash. Several rare or endangered plant species occur on the site. It is home to the rare cerulean warbler.

Oakfield Ledge
Fond du Lac County
Located along the east side of Horicon Marsh southwest of Oakfield, this 40-acre tract contains cliffs of dolomite that protect delicate rock communities. The basswood, maple, elm, and oak trees form tall stands. Fragile species found in the cool, shaded cliff environments are Canada yew, walking fern, bulbet fern, fragile fern, cliffbrake, and rock cress.

Olson Oak Woods
Dane County
This large dry oak forest occupies 90 acres about 3 miles southwest of Verona. Scattered clusters of large, tall white oak trees that date to the mid-1700s indicate former savanna conditions. Other trees include black cherry and hickory. A fire burned the area in 1921.

Owen and Anne Gromme Preserve
Winnebago and Fond du Lac Counties
Owned and managed by The Nature Conservancy, this 608-acre preserve is located about 5 miles north of Ripon. The property borders Rush Lake, a shallow, marshy glacial lake. Natural communities include remnants of oak savanna, prairie, and sedge meadow. Rush Lake is critical breeding habitat and a migratory stopover for many bird species, including the threatened red-necked grebe and rare Forster's tern. The rapidly disappearing upland sandpiper and Henslow's sparrow can also be spotted here.

Page Creek Marsh
Marquette County
This 566-acre Nature Conservancy preserve about 15 miles north of Portage contains mixed woodlands, wetlands, remnant native prairie and oak savanna, sedge meadows, prairie fens, pine plantations, and old agricultural fields. The wetlands enclose Page Creek and three small, shallow lakes. The gently rolling terrain is underlain by ridges of sandy glacial soil. Tamarack and cedar dominated the swamps and bogs. Wild rice and numerous sedges line the edges of Page Creek. Rare wetland plant species include downy willow-herb and bog reed grass. Rare meadow birds inhabit the tall grasses of the prairie and savanna communities. The forests and wetlands are important nesting habitats and migratory stopovers for numerous bird species, including the rare sandhill crane.

Parfrey's Glen
Sauk County
This 480-acre tract about 3 miles southeast of Baraboo protects a deep sandstone gorge that protects a rare variety of cliff plants, including ferns and mosses. The cool, shady conditions harbor white pine and yellow birch. The uplands are surrounded by mature oak-maple-basswood forests.

Patterson Hemlocks
Oneida and Vilas Counties
A large block of old-growth northern mesic forest is located on this 310-acre tract about 14 miles west of Woodruff. There is one very old stand of hemlock–yellow birch–white pine that has yet to break up. Some of the white pines are more than 120 feet tall and exceed 3 feet in diameter. It is important habitat for veeries, pileated woodpeckers, warblers, and a state-threatened bird species.

Peat Lake
Kenosha County
Peat Lake is a state natural area, its shoreline uniquely undeveloped. The lake is less than 5 feet deep and supports only carp. It is an important nesting site for herons, rails, woodcocks, and bitterns; it is also a stopover for migrating birds during spring and fall.

Pecatonia River Woods
Iowa County
Located on 110 acres west of Mineral Point, this state natural area consists of excellent examples of floodplain and southern dry, dry-mesic, and mesic forest communities. It is the only known Wisconsin location for a state-endangered plant; a number of special concern plants are also found, including Short's

aster, great waterleaf, toothed cress, and grass sedge.

Pickerel Lake Fen

Walworth County

Owned and managed by The Nature Conservancy, this 47-acre preserve in southeastern Wisconsin is an excellent example of calcareous fen, one of the rarest wetland types in North America. In addition to the calcareous fen, natural communities include, marsh, oak savanna, and the shoreline of Pickerel Lake. The diverse wetland types support many rare or unusual species, including kittentails, pitcher plants, ciliated brome grass, Canada bluejoint grass, and the rare Blanding's turtle and starhead topminnow. The fen is well preserved due to the lack of nearby development, which would disrupt groundwater flow and starve the fen.

Pine Cliff Natural Area

Iowa County

Located 4 miles north of Dodgeville, this 20-acre site is a southern Wisconsin relic pine community that is unique because all three native pine species—white, red, and jack—occur here. Sandstone cliffs protect delicate fern-moss-lichen communities.

Plagge Woods Natural Area

Chippewa County

Located about 5 miles northeast of Holcombe, this 80-acre tract is dominated by virgin and old-growth maple-basswood-oak northern mesic forest. Some white pine trees exceed 3 feet in diameter.

Plum Lake Hemlock Forest

Vilas County

This 228-acre site about a mile south of Starlake is known for its large virgin old-growth stand of hemlock and associated hardwood trees. The last fire was in 1810, so the larger trees are nearly 200 years old. Yellow birch and basswood also occur. The forest is an important habitat for vireos, warblers, wrens, thrushes, veeries, and the northern parula.

Point Beach Ridges

Manitowoc County

About a mile north of Two Rivers, this 175-acre natural area protects eastern mesic forest and Lake Michigan shoreline communities, including sand dunes. The sand ridges are stabilized by trees and shrubs, including junipers and bearberry. The ridge nearest the beach is vegetated with rare Atlantic coastal flora, which remain undisturbed.

Pope Lake

Waupaca County

Located on 80 acres, this deep hardwater lake is protected simply because it is an undeveloped water body. The water is clear and supports diverse aquatic and emergent flora. It is surrounded by alder wetlands and tamaracks. Herons and flycatchers have been sighted here.

Quincy Bluff Wetlands

Adams County

Owned and managed by The Nature Conservancy, this 1,665-acre property about 15 miles north of Wisconsin Dells protects a steep bluff of sandstone. Natural communities include cliff face, dry oak–pine mesic forest, pine-oak barren, conifer swamp, prairie, and sedge meadow. Several varieties of rare or unusual species can be found. The tiger beetle resides in the pine-oak barrens. The federally endangered Karner blue butterfly feeds on the wild lupine on the property. Bristly sarsaparilla and harebell grow in the cliff faces. The wetlands host the rare meadow beauty, cross milkwort, and netted nut-rush. The rocky bluffs are a roosting site for turkey vultures.

Red Cedar Lake

Jefferson County

Located on 450 acres, this state natural area encloses a shallow lake and surrounding marsh areas. It contains a variety of submerged and emergent aquatic plant species, including sedges. It is an important nesting site for nine species of waterfowl, including grebes, bitterns, coots, black terns, sandhill cranes, herons, and marsh wrens.

Rice Lake–Thunder Lake Marsh

Oneida County

About 3 miles northeast of Three Lakes, this lake-marsh complex is well known for its wild rice, sedge meadows, and variety of bird life. Bird species include the great blue heron, black tern, and sandhill crane. The wetland is an important migratory stopover for waterfowl species during spring and fall. Rare bald eagles and ospreys nest and feed in the marsh.

Ridges Sanctuary

Door County
Located on 860 acres just north of Bailey's Harbor, this sanctuary has one of the greatest concentrations of endangered and threatened plants in the Midwest. Natural communities on the Lake Michigan beach ridges include spruce-pine forest, wet boggy swale, and boreal forest. It is also an important bird habitat.

Rush River Delta
Pierce County
This 325-acre site contains old-growth southern wet-mesic forest that has undergone only selective logging in the past. The larger trees are silver maple, elm, and cottonwood. Located along the Rush River, the sandy shore is used by turtles for a nesting site. Several varieties of birds, including some state-threatened species, such as the cerulean warbler, inhabit the woods.

Schluckebier Sand Prairie
Salk County
One of the few remnants of dry sand prairie remaining in Wisconsin, this 23-acre Nature Conservancy tract is located about 2 miles west of Prairie du Sac. Although cultivated more than 50 years ago and since abandoned, it has recovered many of its original prairie plants. Nearly 160 species have been identified, including the extremely rare prairie bush clover. Sections of the prairie are dominated by head-high big bluestem and Indian grass. Spectacular displays of wildflowers can be seen during spring and summer. The prairie is also an important habitat for rapidly disappearing prairie birds, including the bobolink.

Scuppernong Prairie
Waukesha County
Wet-mesic prairie with lesser oak savanna covers this 185 acres about 2 miles northwest of Eagle. Wet-mesic prairie is very rare in Wisconsin. The prairie, as well as a sedge meadow, are located at the edge of Scuppernong Marsh. Much of the prairie has been undisturbed. Prairie plant species are diverse and include a few rare or threatened species. It is also an important habitat for the rare upland sandpiper, Franklin's ground squirrel, and eastern hognose snake.

Shivering Sands
Door County
Part of an important 3,400-acre wetland complex, this 192-acre Nature Conservancy preserve in Door County is located along Lake Michigan. Several varieties of wetland habitats are protected here, including lake, stream, spring, tamarack-cedar swamp, fen, sedge meadow, and forested sand dune communities. Three lakes contain completely undeveloped shorelines. Rare or unusual species abound in this diverse flora, including various sedges, twig rush, tussock bulrush, dwarf lake iris, and orchids. It is an important breeding habitat and migratory stopover for more than 100 bird species, including sandhill cranes, black terns, and interior forest birds. The rare blue-spotted salamander is found here.

Snapper Memorial Prairie
Jefferson County
Owned and managed by The Nature Conservancy, this 28-acre property about 6 miles north of Lake Mills is all that remains of what was once a vast wet prairie during presettlement times. This virgin tract has retained its high diversity of prairie flora, including big and little bluestem, prairie dock, and prairie dropseed. The rare prairie Indian plantain and prairie milkweed also grow here. Prairie animals, such as the rapidly diminishing upland sandpiper and bobolink, live on the prairie. The brilliant displays of wildflowers are dominated by blazing star, coneflower, and compass plant.

Sohlberg Silver Lake
Adams County
This 174-acre tract just southwest of Adams contains a small seepage lake, open bogs, and tamarack swamps. Its sandy shores host rare Atlantic plant species, such as autumn sedge, golden pert, cross milkwort, beak rushes, yellow-eyed grass, and meadow beauty. Several varieties of songbirds nest in the surrounding woods.

Spring Green Reserve
Sauk County
Located about a mile northwest of Spring Green, this 590-acre site contains a mix of dry mesic oak forest, oak savanna, red cedar stands, rare black oak barrens, dry bluff prairie, and dry sand prairie. The rare prickly pear cactus grows on the dry sand prairie. Threatened

grassland birds, such as the dickcissel, lark sparrow, and western meadowlark, breed on the property. An old farm site, it shelters a variety of state-endangered or special concern species, including the six-lined racerunner, blue racer, black widows, wolf spiders, tiger beetles, certain cicadas and leafhoppers, and predatory wasps. The preserve is owned and managed by The Nature Conservancy.

Swenson Wet Prairie
Rock County
About 5 miles southeast of Brodhead, this 40-acre site contains an excellent example of wet-mesic prairie; savanna and sedge meadow communities also occur. The prairie is dominated by little and big bluestem. Two state-threatened species grow here, and the area is home to the rare Bells' vireo and cerulean warbler.

Thompson Memorial Prairie
Dane and Iowa Counties
Located about 2 miles south of Blue Mounds, this 175-acre tract preserves high-quality remnants of dry prairie. These sections of prairie escaped cultivation because the underlying soil is too rocky. It has retained a high diversity of flora, which include bluestem and porcupine grasses, bird-foot violet, and the more rare pomme de prairie and green milkweed. Dickcissels, upland sandpipers, and bobolinks are some of the grassland birds found here. Rare skipper and sulfur butterflies thrive on the wildflowers. Efforts are being made to restore the surrounding areas of degraded prairie. The property is owned and managed by The Nature Conservancy.

Toft Point
Door County
Excellent examples of northern mesic forest, boreal forest, and Lake Michigan shoreline are protected on this 703-acre site a mile east of Baileys Harbor. Tree species include balsam fir, white spruce, yellow birch, hemlock, and white pine. The site is home to 17 species of nesting warblers, area-sensitive birds.

Totagatic Highlands Hemlocks
Washburn County
This impressive old-growth mesic and wet-mesic forest occupies 160 acres about 15 miles east of Minong. The wet-mesic forest is virginal and dominated by a tall stand of white cedar. The uplands are covered by tall hemlock stands that were only lightly timbered in the past.

Tower Bottoms
Iowa County
This 25-acre tract about 2 miles southeast of Spring Green is an excellent example of undisturbed wet and wet-mesic floodplain forest. The dominant trees are large maple, ash, cottonwood, white oak, and American elm. It is an important habitat for the pileated woodpecker and prothonotary warbler.

Trenton Bluff Prairie
Pierce County
A remnant of dry prairie occurs on this 110-acre tract about 2 miles west of Hager City. Dominant prairie species include big and little bluestem, sideoats grama, muhly grass, and porcupine grass. Several state-endangered species can also be found. It is habitat for the blue racer and timber rattlesnake.

Trout Lake Conifer Swamp
Vilas County
This 25-acre tract about 6 miles north of Woodruff is a small but mature cedar-spruce-tamarack swamp with a hardwater spring run. Groundcover includes sphagnum moss, orchids, starflower, Canada mayflower, and other bog species. It is important habitat for warblers, phantom crane flies, and black bears.

Twin Lakes Bog
Taylor County
Located in north-central Taylor County in Chequamegon National Forest, this 51-acre tract encloses undisturbed open bog and northern wet forest communities. Biodiversity is high, and several unusual or rare plants can be found here, such as the pink moccasin flower. Large specimens of black spruce, balsam fir, and tamarack edge the bogs. The bogs are critical habitat for a variety of amphibians, reptiles, and birds.

Waubesa Wetlands
Dane County
This 194-acre parcel a few miles south of Madison is owned by The Nature Conservancy and protects about one-quarter of the larger Waubesa Wetlands, one of the most highly valued wetlands in the state. Natural communities include sedge meadow, floating sphagnum mat, spring, stream, swamp, and fen. The

springs originate from underground caves lined with filamentous algae and purple sulpher bacteria. The calcium-rich groundwater maintains the specialized fen vegetation, which includes sedges, brome grass, aster, goldenrod, and sage willow. Rarer species are the threatened less fringed gentian, small water parsnip, and yellow monkey flower. The wetlands are important habitats for a variety of birds, animals, and reptiles. Herons, sandhill cranes, and bitterns nest in the area. The threatened Blanding's turtle lives in the streams, which are important spawning corridors for the northern pike.

Waupun Maple Forest
Fond du Lac County
About 2 miles north of Waupun, this 40-acre old-growth southern mesic forest is an outstanding remnant of the forest type that once covered southern Wisconsin. The oldest trees are more than 100 feet tall and exceed 2 feet in diameter at breast height. The upper canopy is dominated by sugar maple, red oak, basswood, and walnut. It is an important habitat for flycatchers, pewees, wrens, thrushes, and vireos.

Whitefish Dunes
Door County
This 230-acre parcel protects a stretch of eastern mesic forest and Lake Michigan shoreline that contains the largest sand dunes in Wisconsin. The dunes are both stabilized and active, displaying a complete succession from open beach to northern mesic forest. The uncommon Canada yew is abundant, and several rare or threatened plant species occur on the dunes. It is a nesting sight for vireos and warblers.

Wyalusing Walnut Forest
Grant County
This 140-acre tract is an excellent example of southern dry, dry-mesic, and wet-mesic forest communities. Located north of Wyalusing, it contains a mature black walnut forest with lesser oaks; less common species on steeper slopes are yew, sullivantia, and amethyst shooting star. Other rare flora include walking fern, narrow-leaved spleenwort, and Goldier's fern. The site is critical habitat for a state-threatened bird species.

Zinn (Walter and Rose) Preserve
Washington County
This 227-acre tract near Hartford is roughly 35 miles from Milwaukee. It is owned and managed by The Nature Conservancy. The preserve shelters an undeveloped lake and tamarack swamp. Natural communities include wet mesic cedar-spruce forest, southern mesic (maple-oak) forest, bottomland forest, tamarack swamp, and sedge bog. Several varieties of rare or unusual plant species can be found on the property, including false Solonom's seal. It is an important nesting habitat and migratory stopover for a variety of bird species, including the blue-winged teal and various interior forest birds.

WYOMING

Wyoming can be subdivided into four major natural divisions: central short-grass prairie, northern Great Plains steppe, the Colorado Rocky Mountains and Utah-Wyoming Rocky Mountains, and the Wyoming Basin.

A small section of central short-grass prairie occurs in the extreme southeastern corner of Wyoming. Its dominant grasses are blue grama with lesser buffalo grass; a few zones of western wheatgrass and needle-and-thread (mixed-grass prairie) are scattered across the area.

The northeastern quarter of the state is covered with northern Great Plains steppe. This region is dominated by blue grama, western wheatgrass, needle-and-thread, threadleaf sedge, and prairie Junegrass. Interspersed with these grasslands is ponderosa pine–Rocky Mountain juniper woodland and lesser Wyoming big sagebrush shrub steppe. Stream corridors are lined with plains cottonwood and willows.

The Colorado Rocky Mountains occupy a portion of the southeastern corner of the state, and the Utah-Wyoming Rocky Mountains comprise most of the northwestern quarter. Major vegetation transitions are essentially the same in both ranges. Proceeding from lowest elevation to highest, these are: big sagebrush–Idaho fescue shrubland and western wheatgrass prairie, montane (5,000–7,500 feet) ponderosa pine, Douglas fir, or limber pine woodland, subalpine Engelmann spruce–subalpine fir, lodgepole pine, or whitebark pine forest, and alpine tundra consisting of various sedges and alpine avens.

The central third of the state is largely Wyoming Basin. The vegetation is very similar to that in the northern Great Plains steppe. Major communities include Wyoming big sagebrush shrub steppe, western wheatgrass–needle-and-thread–blue grama prairie, lesser saltbush desert shrub, and scattered patches of Utah juniper woodland.

Special Habitat Sites

Afton Front
Lincoln County
The upper reaches of Anderson and Blaney Canyons are protected on this 765-acre RNA in the Salt River Range about 3 miles northeast of Afton. Douglas fir forest, subalpine fir forest, xeric grassland, and bigtooth maple forest are the major community types. The steep canyons protect unique plant species. Rare or special-concern flora and fauna include the boreal draba, leavy dwarf knotweed, steer's head, Payson's bladderpod, three-toad woodpecker, and North American lynx.

Bull Elk Park Research Natural Area
Sheridan County
Located in Bighorn National Forest about 10 miles northwest of Burgess Junction, this 727-acre RNA includes lodgepole pine forest, grassy bald, and pristine remnants of native grassland communities. Grasses include bluebunch wheatgrass, smooth brome, mountain brome, Nelson needlegrass, Idaho fescue, and spike fescue. Other plants include snowberry, cinquefoil, sagebrush, and juniper. The area provides important habitats for black bears, mountain lions, and elk.

Carter Mountain
Park County
This 8,508-acre ACEC is located near Thomas Reservoir. Situated atop of Carter Mountain, it encloses high-quality alpine tundra. Unusual plant communities in this fragile environment include sedge–marsh marigold, silver lupine–bluegrass, alpine avens–bistort, and alpine avens–clover. Rare plant species include Evert's wafer parsnip, northwestern fleabane, twice-hairy groundsel, and William's rockcress.

Deep Creek Research Natural Area
Carbon County
This 4,596-acre tract straddles the Continental Divide and protects the headwaters of several streams. Natural communities include lodgepole pine forest, aspen woodland, sagebrush shrubland, and grassland. The pristine streams support healthy populations of the endemic Colorado River cutthroat trout. Engelmann spruce and thinleaf alder border the streams. Other rare or unusual species include the clustered lady's slipper, Rydberg twinpod, and northern goshawk.

Devils Tower National Monument
Crook County
This flat-topped pillar of volcanic rock about 10 miles east of Oshoto is nearly 1,000 feet tall. Grasslands and ponderosa pine forest encircle its base. Its high biodiversity results from the convergence of mountain, prairie, and river ecosystems. The tower hosts several rare flora and fauna, including nesting prairie falcons and rock doves. The top of the tower is covered with native grasses, forbs, and shrubs. Meadows, flowers, shrubs, and cottonwoods line the flatlands along Belle Fourche River, supporting nesting owls, red-headed woodpeckers, and cliff swallows.

Five-Springs Falls
Big Horn County
This 182-acre ACEC is located on the western slopes of the Big Horn Mountains about 20 miles east of Lovell. It protects part of Five Springs Creek Canyon where it cuts through limestone bedrock. Several rare plant communities exist on the rock faces, including some nearly endemic species.

Horse Creek
Teton County
Located on the western slopes of the Gros Ventre Range, this 6,565-acre property is a few miles from Camp Davis. Its central feature is the very narrow and steep Horse Creek Canyon, which preserves more than 30 different minimally disturbed alpine, forbland, forest, and riparian communities. Important animal species that inhabit the canyon are the wolverine, lynx, boreal toad, three-toed woodpecker, northern goshawk, and great gray owl. Rare plants are the Osterhout sagebrush, boreal draba, and Robbins milk vetch.

Jackson Canyon
Natrone County
Located on 13,745 acres along the North Platte River near Freeland, this ACEC protects a steep

canyon that is an important raptor habitat. Natural communities include rock face, juniper woodland, and limber pine forest. The rare bald eagle nests on the cliffs during winter, hunting in the nearby North Platte River. Several unusual plant species can be found on the rock faces. The property is also an important winter habitat for bighorn sheep and elk.

Kendall Warm Springs
Sublette County
Located near Klondike Hill on the east bank of the Green River, this 160-acre property protects warm and cold springs that support several unique plant species. Natural communities include montane shrubland, thermal wetland, wet meadow, calcareous fen, and riparian corridor. The springs are the only known site in the world where the endemic Kendall Warm Springs dace can be found. Other sensitive species that inhabit the area are the western boreal toda, boreal aster, alpine meadow-rue, hoary willow, and several rare sedges.

Osborn Mountain
Sublette County
Located on the western side of the Wind River Range about 5 miles west of the Continental Divide near Pinedale, this 2,830-acre property protects the headwaters of Mill Creek as well as alpine rock face, alpine talus-scree, alpine fellfield and felsemeere, alpine meadow, and riparian natural communities. Within these communities are rare plants, such as aromatic pussytoes, boreal wormwood, rockcress draba, large-flowered sandwort, and Rocky Mountain cinquefoil. This site also provides important habitat for large mammals, such as bighorn sheep and mountain lions.

Preacher Rock Bog
Sheridan County
This 124-acre parcel near Dome Lake is a rare spring-fed sphagnum bog in the Big Horn Mountains along Ranger Creek. It is surrounded by an edge of wetland vegetation and contains a slightly raised island of fir trees. A calcareous fen and evergreen swamp forest are the other natural communities. Rare species found here include various sedges, woodland horsetail, regal rush, and russet cottongrass. It is an important habitat for several bird and amphibian species.

Quartzite Peak
Carbon County
Located at elevations of 8,600–11,000 feet, this ACEC about 20 miles west of Riverside preserves high-quality natural communities, such as meadows and subalpine forests, near Bridger Peak on the Continental Divide. The rare Colorado cutthroat trout can be found in Haggerty Creek, its headwaters protected within the ACEC. Other rare or sensitive species found here include the northern goshawk, clustered lady's slipper, western boreal toad, pinnate fleabane, Colorado tansy-aster, and small-flowered trillium. The area is also an important habitat for large mammals, such as bighorn sheep and mountain lions.

Red Canyon Ranch
Fremont County
Located south of Lander, this 4,738-property is owned by The Nature Conservancy. Natural communities include sagebrush prairie, trout stream, and other riparian habitat. Antelope, black bears, mountain lions, and a variety of songbirds are common on the property. Rare plants, such as Barnbey's clover, Wyoming townsendia, and Rocky Mountain twinpod, are found on the rocky canyon walls. Cattle ranching is being explored experimentally to impose environmental controls, such as eliminating exotic species and preparing ground for future plantings.

Red Canyon/Table Mountain
Fremont County
This 15,000-acre ACEC includes Table Mountain and a portion of the southern Wind River Range. Located south of Lander, its natural communities include Douglas fir and aspen upland forest, sagebrush shrubland, willow shrubland, montane sedge meadow, and grassland. Riparian corridors are lined with chokecherry, willows, and sedges. Plant members include the globally rare Barnbey's clover, rubber boa, sword townsendia, Fremont bladderpod, Rocky Mountain twinpod, and beaver rim phlox. It is critical winter habitat for elk and other large mammals.

Sheep Mountain Research Natural Area
Albany County
Located near Lake Owen and Rex Lake, this tilted, 10-mile ridge is covered with pristine bluebunch wheatgrass grassland, sagebrush shrubland, and lodgepole pine forest. Covering

nearly 15,000 acres, this RNA is an important habitat for large animals such as bighorn sheep. Unique, rare, or endangered species found here include the arrow-leaved sweet coltsfoot, slender-leaved buckwheat, hoary willow, Colorado tansy-aster, peregrine falcon, and black-crowned night heron. Mountain lions have also been sighted.

Shell Canyon Research Natural Area
Big Horn County
Part of Bighorn National Forest, this 737-acre tract is located about 20 miles northwest of Greybull. Natural communities include Rocky Mountain juniper woodland, Douglas fir forest, black sagebrush flatland, and cliff face. The mature Douglas fir–limber pine forests are well developed on north-facing slopes. The Rocky Mountain juniper–bluebunch wheatgrass community is one of the most pristine examples in the United States. The area provides important habitat for elk and bighorn sheep; several rare endemic plant species grow here.

Snowy Range Research Natural Area
Albany County
This 312-acre RNA is part of Medicine Bow National Forest. Located northeast of Garrett, it encloses excellent examples of habitats at elevations between 9,000 and 10,000 feet. Natural communities include Engelmann spruce–subalpine fir forest, lodgepole pine forest, grassland, shrubland, and several shallow lakes. Having never been logged, its old-growth pine trees are of impressive size. The area supports large animals, such as mule deer, elk, and mountain lions.

Sweetwater River Project
Fremont County
This Nature Conservancy project south of Lander preserves a 6-mile stretch of varied riparian habitat along Sweetwater River, a high-quality trout habitat. Riparian community species include broadleaf cottonwood, silver sagebrush, greasewood, native geyer, yellow shrub willow, Baltic rush, and Nebraska sedge. Dry grasslands with sagebrush border the riparian corridor. The river attracts moose, pronghorn antelope, mule deer, eagles, prairie falcons, and sandhill cranes. Rare plants include pussytoes. The property is an important migratory stopover for raptors and is a critical protective corridor for elk, moose, and bears that travel from the Wind River Range.

Swift Creek
Lincoln County
Located on the western slopes of the Salt River Range, this 4,170-acre tract protects the upper portion of Swift Creek, including its headwaters. Natural communities include alpine talus, coniferous forest, forbland, upland shrub, and riparian corridor. Several varietis of rare or uncommon plant or animal species reside in the Swift Creek drainage basin, including the boreal draba, lynx, northern goshawk, boreal owl, three-toad woodpecker, mountain spiny-aster, Shultz's milk vetch, northern holly-fern, and Payson's bladderpod.

Sybille Wildlife Research and Conservation Area
Albany County
Located south of Wheatland, this conservation area was established to save the black-footed ferret, one of rarest mammals in North America. Development eliminated its primary source of food: prairie dogs. Nearly rendered extinct during the 1970s, 200 ferrets now live on this property.

Tensleep Preserve
Big Horn County
This 8,500-acre tract in the southwestern foothills of the Bighorn Mountains is northeast of Worland. It protects rocky canyons and upland forest along a 12-mile stretch of Canyon Creek. It is owned by The Nature Conservancy. Eight natural communities can be found on the preserve, including spring-fed stream, canyon, and ponderosa pine forest. Several rare plant species, 120 bird species, mountain lions, black bears, elk, and the rare spotted bat live on the property. Native American pictographs can also be seen on rock walls. Rare plants have been recorded here.

Thunder Basin National Grassland
Campbell and Weston Counties
Located southeast of Wright, dry short-grass prairie is preserved on these rolling plains and plateaus. Dominant species are blue grama, needle-and-thread, and western wheatgrass with lesser sagebrush and prickly pear cactus. Cottonwoods line the streams, and ponderosa pine and juniper woodlands are scattered across the higher slopes. This refuge contains one of the largest populations of pronghorn antelope in the country and also supports deer, mule deer, elk, sage grouse, prairie falcons, and mountain lions.

Twin Lakes

Park County

Located on 722 acres, this preserve protects part of Twin Lakes Basin, a unique glacial feature. The basin is surrounded by 700-foot-tall, ice-carved walls of bedrock. Because of its restrictive shape, it harbors rare alpine tundra and wetland communities, pristine lakes, and the headwaters of Chain Creek. More than 30 rare or threatened alpine and alpine-arctic plant species can be found here, including pink agoseris, fleabane, groundsel, white arctic whitlow-grass, arctic harebell, dwarf clover, Dane's gentian, koenigia, and various cinquefoils.

Whiskey Mountain Macrosite

Fremont County

This 34,324-acre ACEC near Dubois is an important bighorn sheep winter range habitat. Elevations vary from foothills (7,000 feet) to mountain peaks (11,000 feet); communities include subalpine and alpine environments, such as spruce and lodgepole pine forests. Other communities are wet sedge meadow and aspen stand. A herd of about 1,000 bighorn sheep winter in the area. In addition, the site harbors several rare or threatened species, including Weber's saw-wort, Jones's columbine, false uncinia sedge, and William's rockcress. The rare and elusive North American lynx also inhabits the area.

GLOSSARY OF SELECTED TERMS

ACEC (Area of Critical Environmental Concern). A designation under the Bureau of Land Management for areas with unique or special natural resource values that require special management.

Alpine. A biogeographic zone of high elevation, above timberline, that is typically cold and windy.

Alvar. Barrens and grassland vegetation that have developed on thin soils that overlie a flat bedrock surface.

Barrens. A plant community that is dominated by poorly developed or stunted individuals, usually as a result of extreme climatic conditions and toxic rock or soil chemistry.

Barrier island. A narrow island of sand and gravel that forms along ocean coastlines, usually from longshore currents.

Biodiversity. The variety of life and its ecological processes, usually in the context of a particular area. The term can be applied to large-scale ecosystems, different species, and the genetic variation within a species.

Bog. A restricted wetland that commonly forms in cool, boreal climates. Typically fed by rainwater, bogs are usually acidic and generate large volumes of peat moss.

Boreal. A term that describes the circumpolar forest region in the northern hemisphere that is dominated by conifer species. Boreal forests typically extend to the tundra to the north and to mixed conifer-deciduous forest or grasslands to the south.

Brackish. A term that describes a mix of salt water and freshwater, usually in an estuary near the coastline.

Climax community. A stable community that is the result of a successional series of communities in the same region over time.

DBH (diameter breast height). The diameter of a tree at a height of 4.5 feet above the ground.

Ecosystem. A specific biological community and its interaction with its surrounding physical environment.

Ecotone. The transition zone between two different communities or ecosystems, often defined by changes in climate or elevation.

Endangered species. A species that is in imminent danger of extinction.

Endemic. Originating in, and being restricted to, a particular region or environment.

Exotic. A species that has been introduced by humans into an area that is not native to the area.

Estuary. An aquatic transition zone between salt water and freshwater, usually at the mouth of a stream draining into the ocean.

Fen. A wetland subtype that is an open peatland, sometimes with scattered trees, and sustained by mineral-rich groundwater. Fens often support rare or unusual species of sedges, grasses, mosses, and shrubs.

Forest. A community dominated by trees greater than 15 feet in height, with at least 60 percent canopy-crown closure that may be interlocking.

Habitat. The place within an environment or ecosystem in which an organism lives.

Hydric. Wet, poorly drained soils.

Karst. An irregular ground surface that develops over limestone bedrock that is dotted with sinkholes, caves, fissures, and depressions.

Marsh. A wetland dominated by nonwoody plant species, typically cattails and rushes.

Mesic. Soils that are moist and well drained; intermediate between xeric (dry) and hydric (wet).

Mesophytic. A term for plants that grow on mesic soils, such as a mesophytic forest.

Monadnock. A steep-sided mass of bedrock (usually igneous) that rises abruptly over the surrounding flat countryside.

Montane. A biogeographic zone characterized by moist, cool slopes below timberline that are commonly forested by evergreens.

Native. Indigenous to, or originating in, a particular area and continuing to grow in an unaltered natural condition.

Pocosin. A raised area of peat that is covered by a dense mass of swampy evergreen vegetation. Common in the southeastern United States.

Riparian corridor. The strip of vegetation that forms along a stream or water body that is distinctive from surrounding ecosystems.

RNA (Research Natural Area). A region identified by the U.S. Forest Service as having outstanding natural resource values that need to be protected or studied.

Savanna. A vegetation type that is dominated by grassland with interspersed clumps of trees.

Scree. *See* Talus.

Subalpine. That biogeographic zone occuring above the upper limit of contiguous closed forest and below timberline; typically a mix of tree patches and open meadows.

Succession. The process by which species occupy an area and gradually change the environment so that other species can move in and eventually replace the original species.

Swamp. A type of wetland with trees.

Talus. Rock debris that accumulates at the bottom of a cliff or steep slope.

Tundra. A vegetation type dominated by short plants, such as mosses and lichens, that is typically found at high-elevation alpine zones; short growing season.

Vernal ponds. Small, shallow bodies of water that accumulate in the spring and have distinctive plant assemblages.

Wetlands. Permanent or seasonal ecosystems characterized by rooted vegetation surrounded by standing water. Types include swamp, marsh, bog, and fen.

Woodland. A vegetation type characterized by a sparse tree canopy cover of 10 percent to 60 percent, often with well-developed shrub layers.

Xeric. An area that is characterized by dry, well-drained soil (often a south-facing slope).

APPENDIX 1: NATIONAL ORGANIZATIONS AND OFFICES

Defenders of Wildlife
1101 14th Street Northwest
Suite 1400
Washington, DC 20005
202-202-9400

National Audubon Society
700 Broadway
New York, NY 10003
212-979-3000

National Fish and Wildlife Foundation
1120 Connecticut Avenue
Washington, DC 20036
202-857-0166

National Park Foundation
1101 17th Street Northwest
Suite 1102
Washington, DC 20036
202-785-4500

National Park Service
Main Interior building
P.O. Box 37127
Washington, DC 20013-7127
202-208-7520

Natural Areas Association
P.O. Box 1504
Bend, OR 97709
541-317-0199

The Nature Conservancy
1815 North Lynn Street
Arlington, VA 22209
703-841-5300

Research Natural Area Coordinator
USDA Forest Service
P.O. Box 96090
Washington, DC 20090
202-205-1149

Sierra Club
730 Polk Street
San Francisco, CA 94109
415-776-2211

U.S. Bureau of Land Management
Office of Public Affairs
1849 C Street
Washington, DC 20240
202-452-5125

U.S. Fish and Wildlife Service
Rt. 1, Box 166
Shepherdstown, WV 25443
800-344-9453

U.S. Department of Agriculture
Washington, DC 20250
800-245-6340

U.S.D.A. Forest Service
201 14th Street Northwest
P.O. Box 96090
Washington, DC 20090-6090
202-205-1760

The Wilderness Society
900 17th Street Northwest
Washington, DC 20006
202-833-2300

APPENDIX 2: SELECTED ORGANIZATIONS BY STATE

ALABAMA

Alabama Department of Conservation and Natural Resources
64 North Union Street
Montgomery, AL 36130
334-242-3334

Alabama Natural Heritage Program
Huntingdon College
1500 East Fairview Avenue
Montgomery, AL 36106-2148
334-834-4519

The Nature Conservancy
2821-C 2nd Avenue South
Birmingham, AL 35233
205-251-1155

ALASKA

Alaska Department of Fish and Game
P.O. Box 3-2000
Juneau, AK 99802
907-465-4107

Alaska Department of Natural Resources
400 Willoughby, 5th Floor
Juneau, AK 99801
907-465-2400

Alaska Natural Heritage Program
University of Alaska
Anchorage, AK 99501
907-257-2702

The Nature Conservancy
421 West First Avenue
Suite 200
Anchorage, AK 99501
907-276-3133

Research Natural Area Coordinator
Alaska Region
USDA Forest Service
P.O. Box 21628
Juneau, AK 99802-1628
907-586-8751

ARIZONA

Arizona Fish and Game Department
2222 West Greenway Road
Phoenix, AZ 85023
602-942-3000

Arizona Heritage Data Management System
Arizona Game and Fish Department
WM-H
2221 West Greenway Road
Phoenix, AZ 85023
602-789-3612

Arizona State Parks
1300 West Washington Street
Phoenix, AZ 85007
602-542-4174

The Nature Conservancy
300 East University Boulevard
Suite 230
Tucson, AZ 85705
520-622-3861

ARKANSAS

Arkansas Department of Parks and Tourism
1 Capitol Mall, 4th Floor
Little Rock, AR 72201
501-682-7777

Natural Heritage Program
Arkansas Natural Heritage Commission
Suite 1500, Tower Building
323 Center Street
Little Rock, AR 72201
501-324-9150

The Nature Conservancy
601 North University Avenue
Little Rock, AR 72205
501-663-6699

CALIFORNIA

Bureau of Land Management
2135 Butano Drive
Sacramento, CA 95825
916-979-2800

California Department of Fish and Game
1416 9th Street
Sacramento, CA 95814
916-322-2493

California Department of Parks and Recreation
1416 9th Street
Sacramento, CA 95814
916-653-4272

California Native Plant Society
1722 J Street, Suite 17
Sacramento, CA 95814
916-447-2677

Natural Heritage Division
Department of Fish and Game
1220 S Street
Sacramento, CA 95814
916-322-2493

The Nature Conservancy
201 Mission Street, 4th Floor
San Francisco, CA 94105
415-777-0487

Research Natural Area Coordinator
Pacific Southwest Region
USDA Forest Service
630 Sansome Street
San Francisco, CA 94111
415-705-1891

Research Natural Area Coordinator
Stanislaus National Forest
19777 Greenley Road
Sonora, CA 95370
209-532-3671

University of California Natural Reserve System
University of California
300 Lakeside Drive, 6th Floor
Oakland, CA 94612
510-987-0150

COLORADO

Bureau of Land Management
2850 Youngfield Street
Lakewood, CO 80215
303-239-3700

Colorado Department of Natural Resources
1313 Sherman Street, Room 718
Denver, CO 80203
303-866-3311

Colorado Natural Areas Program
1313 Sherman Street, Room 618
Denver, CO 80203
303-866-3203

Colorado State Parks
1313 Sherman Street, Room 618
Denver, CO 80203
303-866-3437

Natural Heritage Program
Colorado State University
254 General Services Building
Fort Collins, CO 80523
970-491-1309

The Nature Conservancy
1244 Pine Street
Boulder, CO 80302
303-444-2950

Research Natural Area Coordinator
Rocky Mountain Region
USDA Forest Service
P.O. Box 25127
Lakewood, CO 80225
303-275-5005

CONNECTICUT

Connecticut State Parks Division
79 Elm Street
Hartford, CT 06106
860-424-3200

Connecticut Wildlife Division
79 Elm Street
Hartford, CT 06106
203-424-3011

Natural Diversity Database
Natural Resources Center
Department of Environmental Protection
79 Elm Street, Store Level
Hartford, CT 06106-5127
860-424-3540

The Nature Conservancy
55 High Street
Middletown, CT 06457
860-344-0716

DELAWARE

Delaware Division of Fish and Wildlife
89 Kings Highway
Dover, DE 19903
302-739-5297

Natural Heritage Program
Division of Fish and Wildlife
Department of Natural Resources and Environmental Control
4876 Hay Point Landing Road
Smyrna, DE 19977
302-653-2880

The Nature Conservancy
260 Chapman Road
Suite 201D
Newark, DE 19702
302-369-4144

FLORIDA

Florida Division of Forestry
3125 Conner Boulevard
Tallahassee, FL 32399
904-488-4274

Florida Divison of Recreation and Parks
3900 Commonwealth Boulevard
Mail Station 500
Tallahassee, FL 32399
904-488-6131

Florida Division of Wildlife
620 South Meridian Street
Tallahassee, FL 32399
904-488-3831

Natural Areas Inventory
1018 Thomasville Road
Suite 200-C
Tallahassee, FL 32303
904-224-8207

The Nature Conservancy
222 South Westmonte Drive
Suite 300
Altamonte Springs, FL 32714
407-682-3664

GEORGIA

Georgia Parks and Recreation
205 Butler Street Southeast
1352 East Tower
Atlanta, GA 30334
404-656-2770

Georgia Wildlife Resources
2070 Route 278
Social Circle, GA 30279
770-918-6401

Natural Heritage Program
Wildlife Resources Division
Georgia Department of Natural Resources
2117 U.S. Highway 278 Southeast
Social Circle, GA 30279
706-557-3032

The Nature Conservancy
1401 Peachtree Street Northeast
Suite 236
Atlanta, GA 30309
404-873-6946

Research Natural Area Coordinator
USDA Forest Service
Southern Region
1720 Peachtree Street Northwest
Atlanta, GA 30367
404-347-4083

HAWAII

Hawaii Department of Land and Natural Resources
P.O. 621
Honolulu, HI 96809
808-587-0400

Hawaii Division of Forestry and Wildlife
1151 Punchbowl Street
Honolulu, HI 96813
808-587-0166

Hawaii Division of State Parks
P.O. 621
Honolulu, HI 96809
808-587-0300

The Nature Conservancy and the Natural Heritage Program
1116 Smith Street
Suite 201
Honolulu, HI 96817
808-537-4508

IDAHO

Idaho Conservation Data Center
Department of Fish and Game
600 South Walnut Street,
P.O. Box 25
Boise, ID 83707-0025
208-334-3402

Idaho Department of Parks and Recreation
7800 Fairview Avenue
Boise, ID 83704
208-327-7407

Idaho Fish and Game Department

600 South Walnut Street
P.O. Box 25
Boise, ID 83707
208-334-3700

The Nature Conservancy
P.O. Box 165
Sun Valley, ID 83353
208-726-3007

Research Natural Area Coordinator
USDA Forest Service
Idaho Conservation Data Center
600 South Walnut,
P.O. Box 25
Boise, ID 83707
208-334-3402

ILLINOIS

Natural Heritage Division
Department of Natural Resources
524 South Second Street
Springfield, IL 62701-1787
217-785-8774

The Nature Conservancy
8 South Michigan Avenue
Suite 900
Chicago, IL 60603
312-346-8166

INDIANA

Natural Heritage Data Center
Division of Natural Preserves
Department of Natural Resources
402 West Washington Street, Room W267
Indianapolis, IN 46204

The Nature Conservancy
1330 West 38th Street
Indianapolis, IN 46208
317-923-7547

IOWA

Natural Areas Inventory
Department of Natural Resources
Wallace State Office Building
East 9th/Grand Avenue, 4th Floor
Des Moines, IA 50319-0034
515-281-8524

The Nature Conservancy
431 East Locust Street
Suite 200
Des Moines, IA 50309
515-244-5044

KANSAS

Natural Heritage Inventory
Kansas Biological Survey
2041 Constant Avenue
Lawrence, KS 66047-2906
913-864-3453

The Nature Conservancy
820 Southeast Quincy Street
Suite 301
Topeka, KS 66612-1158
913-233-4400

KENTUCKY

Kentucky State Nature Preserves Commission
801 Schenkel Lane
Frankfort, KY 40601
502-573-2886

Natural Heritage Program
Kentucky State Nature Preserves Commission
801 Schenkel Lane
Frankfort, KY 40601
502-573-2886

The Nature Conservancy
642 West Main Street
Lexington, KY 40508
606-259-9655

LOUISIANA

Louisiana Department of Agriculture and Forestry
P.O. Box 631
Baton Rouge, LA 70821

Louisiana Department of Wildlife and Fisheries
P.O. Box 98000
Baton Rouge, LA 70898
504-765-2360

Louisiana Office of State Parks
P.O. Box 44426
Baton Rouge, LA 70804
504-379-1100

Louisiana Office of Tourism
P.O. Box 94291
Baton Rouge, LA 70804
504-342-8119

Natural Heritage Program
Department of Wildlife and Fisheries
P.O. Box 98000
Baton Rouge, LA 70898-9000
504-765-2821

The Nature Conservancy
P.O. Box 4125
Baton Rouge, LA 70821
504-338-1040

Research Natural Area Coordinator
USDA Forest Service
Southern Station
701 Loyola Avenue
Room T-10210
New Orleans, LA 70113
504-589-4551

MAINE

Maine Audubon Society
Gilsland Farm
P.O. Box 6009
Falmouth, ME 04105
207-781-2330

Maine Natural Areas Program
Department of Conservation
93 State House Station
Augusta, ME 04333-0093
207-287-8044

The Nature Conservancy
14 Maine Street
Suite 401
Brunswick, ME 04011
207-729-5181

MARYLAND

Maryland Heritage and Biodiversity Conservation Program
Department of Natural Resources
Tawes State Office Building, E-1
Annapolis, MD 21401
410-974-3195

Maryland State Forest and Park Service
580 Taylor Avenue, E-3
Annapolis, MD 21401
410-974-3771

Maryland Wildlife Division
580 Taylor Avenue, E-1
Annapolis, MD 21401
410-974-3195

The Nature Conservancy
Chevy Chase Metropolitan Building
2 Wisconsin Circle, Suite 300
Chevy Chase, MD 20815
301-656-8673

MASSACHUSETTS

Massachusetts Division of Forest and Parks
100 Cambridge Street, 19th Floor
Boston, MA 02202
617-727-3180

Natural Heritage and Endangered Species Program
Division of Fisheries and Wildlife
Route 135
Westborough, MA 01581
508-792-7270

The Nature Conservancy
79 Milk Street
Suite 300
Boston, MA 02109
617-423-2545

MICHIGAN

Michigan Natural Areas Council
5066 Elmhurst
Royal Oak, MI 48073
313-435-2070

Michigan Nature Association
7981 Beard Road
P.O. Box 102
Avoca, MI 48006
810-324-2626

Natural Features Inventory
Mason Building, 5th Floor
P.O. Box 30444
Lansing, MI 48909-7944
517-373-1552

The Nature Conservancy
2840 East Grand River Boulevard
Suite 5
East Lansing, MI 48823
517-332-1741

MINNESOTA

Minnesota Division of Forestry
500 Lafayette Road
St. Paul, MN 55155
612-296-4491

Minnesota Division of Parks and Recreation
500 Lafayette Road
St. Paul, MN 55155
612-296-9223

Natural Heritage and Nongame Research
Department of Natural Resources
500 Lafayette Road
St. Paul, MN 55155
612-297-4964

The Nature Conservancy
1313 Fifth Street Southeast
Suite 320
Minneapolis, MN 55414
612-331-0750

Research Natural Area Coordinator
USDA Forest Service
North Central Station
1992 Folwell Avenue
St. Paul, MN 55108
612-649-5249

MISSISSIPPI

Mississippi Department of Wildlife, Fisheries, and Parks
Natural Heritage Program
111 North Jefferson Street
Jackson, MS 39202
601-354-7303

Mississippi Department of Wildlife, Fisheries, and Parks
P.O. Box 451
Jackson, MS 39205
601-362-9212

Mississippi Division of Tourism
P.O. Box 1705
Ocean Springs, MS 39566
800-927-6378

The Nature Conservancy
P.O. Box 1028
Jackson, MS 39215
601-355-5357

MISSOURI

Missouri Natural Heritage Database
Missouri Department of Conservation
P.O. Box 180
Jefferson City, MO 65102-0180
573-751-4115

The Nature Conservancy
2800 South Brentwood Boulevard
St. Louis, MO 63144
314-968-1105

MONTANA

Montana Department of Fish, Wildlife, and Parks
1420 East 6th Avenue
Helena, MT 59620
406-444-1276

Natural Heritage Program
State Library Building
1515 East 6th Avenue
Helena, MT 59620
406-444-3009

The Nature Conservancy
32 South Ewing
Helena, MT 59601
406-443-0303

USDA Forest Service, Northern Region
Federal Building
P.O. Box 7669
Missoula, MT 59807
406-329-3511

NEBRASKA

Natural Heritage Program
Game and Parks Commission
2200 North 33rd Street
P.O. Box 30370
Lincoln, NE 68503
402-471-0641

The Nature Conservancy
1722 St. Mary's Avenue
Suite 403
Omaha, NE 68102
402-342-0282

NEVADA

Natural Heritage Program
Department of Conservation and Natural Resources
1550 East College Parkway, Suite 145
Carson City, NV 89710
702-687-4245

The Nature Conservancy
1771 East Flamingo Avenue
Suite 111B
Las Vegas, NV 89119
702-737-8744

Nevada Division of State Parks
123 West Nye Lane
Carson City, NV 89710
702-687-4384

NEW HAMPSHIRE

Audubon Society of New Hampshire
3 Silk Farm Road
P.O. Box 528B
Concord, NH 03301
603-224-9909

Natural Heritage Inventory
Department of Resources and Economic Development
172 Pembroke Street
P.O. Box 1856
Concord, NH 03302
603-271-3623

The Nature Conservancy
2 1/2 Beacon Street, Suite 6
Concord, NH 03301
603-224-5853

Research Natural Areas Coordinator
Northeastern Station
USDA Forestry Service
Forestry Science Lab
P.O. Box 640
Durham, NH 03824
603-868-7665

NEW JERSEY

Natural Heritage Program
Office of Natural Lands Management
22 South Clinton Avenue, CN404
Trenton, NJ 08625-0404
609-984-1339

The Nature Conservancy
200 Pottersville Road
Chester, NJ 07930
908-879-7262

New Jersey Department of Environmental Protection
Division of Parks and Forestry
501 East State Street, CN 404
Trenton, NJ 08625
609-984-0370

NEW MEXICO

Natural Heritage Program
University of New Mexico
2500 Yale Boulevard Southeast
Suite 100
Albuquerque, NM 87131-1091
505-277-1991

The Nature Conservancy
212 East Marcy Road
Suite 200
Santa Fe, NM 87501
505-988-3867

New Mexico State Parks
P.O. Box 1147
Santa Fe, NM 87504
1-888-667-2757

Research Natural Area Coordinator
Southwestern Region
USDA Forest Service
517 Gold Avenue
Albuquerque, NM 87102
505-842-3225

USDA Forest Service
517 Gold Avenue Southwest
Albuquerque, NM 87102
505-842-3898

NEW YORK

Natural Heritage Program
Department of Environmental Conservation
700 Troy-Schenectedy Road
Latham, NY 12110-2400
518-783-3932

The Nature Conservancy
91 Broadway
Albany, NY 12204
518-463-6133

New York State Department of Environmental Conservation
50 Wolf Road
Albany, NY 12233
518-457-3446

New York State Office of Parks, Recreation, and Historic Preservation
Empire State Plaza, Agency Building 1
Albany, NY 12238
518-474-0456

NORTH CAROLINA

The Nature Conservancy
4011 University Drive
Suite 201
Durham, NC 27707
919-403-8558

North Carolina Division of Forest Resources
P.O. Box 29581
Raleigh, NC 27626
919-733-2162

North Carolina Heritage Program
North Carolina Department of Environment, Health, and Natural Resources
Division of Parks and Recreation
P.O. Box 27687
Raleigh, NC 27611-7687
919-733-7701

Research Natural Area Coordinator
Southeastern Station
USDA Forestry Service
200 Weaver Boulevard
P.O. Box 2680
Asheville, NC 28802
704-257-4306

NORTH DAKOTA

Natural Heritage Inventory
North Dakota Parks and Recreation Department
1835 Bismarck Expressway
Bismarck, ND 58504
701-328-5357

The Nature Conservancy
Dakotas Field Office
2000 Schafer Street, Suite B
Bismarck, ND 58501
701-222-8484

North Dakota Parks and Recreation
1835 Bismarck Expressway
Bismarck, ND 58504
701-328-5370

North Dakota State Forest Service
First and Brander
Bottineau, ND 58318
701-228-5422

OHIO

Natural Heritage Data Base
Division of Natural Areas and Preserves
Department of Natural Resources
1889 Fountain Square, Building F-1
Columbus, OH 43224
614-265-6453

The Nature Conservancy
6375 Riverside Drive
Suite 50
Dublin, OH 43017
614-717-2770

OKLAHOMA

Natural Heritage Inventory
Oklahoma Biological Survey
111 East Chesapeake Street
University of Oklahoma
Norman, OK 73019-0575
405-325-1985

The Nature Conservancy
23 West Fourth Street
Suite 200
Tulsa, OK 74103
918-585-1117

Oklahoma Biological Survey
111 East Chesapeake Street
Norman, OK 73019
405-325-1985

Oklahoma Department of Tourism and Recreation
2401 North Lincoln, Room 505
Oklahoma City, OK 73105
800-652-6552

Oklahoma Department of Wildlife Conservation
P.O. Box 53465
Oklahoma City, OK 73152
405-521-3851

Oklahoma Division of State Parks
P.O. Box 52002
Oklahoma City, OK 73105
405-521-3411

OREGON

Department of Fish and Wildlife
P.O. Box 59
Portland, OR 97207
503-872-5268

Department of Forestry
2600 Salem Street
Salem, OR 97310
503-945-7200

Natural Heritage Program
821 Southeast 14th Avenue
Portland, OR 97214
503-731-3070

The Nature Conservancy
821 Southeast 14th Avenue
Portland, OR 97214
503-230-1221

Oregon Bureau of Land Management
1515 Southwest 5th Avenue
Portland, OR 97201
503-952-6001

Oregon Department of Fish and Wildlife
P.O. Box 59
Portland, OR 97207
503-872-5268

Oregon Parks and Recreation Department
1115 Commercial Street, Northeast
Salem, OR 97310
503-378-6305

Research Natural Area Coordinator
Pacific Northwest Region
USDA Forest Service
P.O. Box 3623
Portland, OR 87208
541-750-7360

PENNSYLVANIA

Lancaster County Conservancy
P.O. Box 716
Lancaster, PA 17608
717-392-7891

Natural Diversity Inventory
34 Airport Drive
Middletown, PA 17057
717-948-3962

The Nature Conservancy
1211 Chestnut Street, 12th Floor
Philadelphia, PA 19107
610-834-1323

Pennsylvania Department of Conservation and Natural Resources
P.O. Box 8767
Harrisburg, PA 17105
717-787-2869

Pennsylvania Game Commission
2001 Elmerton Avenue
Harrisburg, PA 17110
717-787-4250

Western Pennsylvania Conservancy
Natural Areas Program
316 Fourth Avenue
Pittsburgh, PA 15222
412-288-2777

RHODE ISLAND

Natural Heritage Program
Rhode Island Department of Environmental Management
Division of Planning and Development
235 Promenade Street
Providence, RI 02908
401-277-2776

The Nature Conservancy
45 South Angell Street
Providence, RI 02906
401-331-7110

Rhode Island Division of Fish and Wildlife
4808 Tower Hill Road
Wakefield, RI 02879
401-789-3094

Rhode Island Division of Parks
2321 Hartford Avenue
Johnston, RI 02919
401-277-2632

SOUTH CAROLINA

The Nature Conservancy
P.O. Box 5475
Columbia, SC 29250
803-254-9049

South Carolina Heritage Trust
South Carolina Department of Natural Resources
P.O. Box 167
Columbia, SC 29202
803-734-3893

South Carolina State Park System
1205 Pendleton Street
Columbia, SC 29201
803-734-0122

SOUTH DAKOTA

Natural Heritage Data Base
South Dakota Department of Game, Fish, and Parks
523 East Capitol Avenue
Pierre, SD 57501-3182
605-773-4227

The Nature Conservancy
Dakotas Field Office
2000 Schafer Street
Suite B
Bismarck, ND 58501
701-222-8484

South Dakota Division of Parks and Recreation
523 East Capitol Avenue
Pierre, SD 57501
605-773-3391

South Dakota Game, Fish, and Parks Department
523 East Capitol Avenue
Pierre, SD 57501
605-773-3387

TENNESSEE

The Nature Conservancy
50 Vantage Way
Suite 250
Nashville, TN 37228
615-255-0303

Tennessee Division of Natural Heritage
Department of Environment and Conservation
401 Church Street
8th Floor, L & C Tower
Nashville, TN 37243-0447
615-532-0431

TEXAS

The Nature Conservancy
P.O. Box 1440
San Antonio, TX 78295
210-224-8774

Texas Biological and Conservation Data System
3000 South IH-35, Suite 100
Austin, TX 78704
512-912-7011

Texas Parks and Wildlife Departmnet
4200 Smith School Road
Austin, TX 78744
512-389-4800

Texas Travel and Information Center
P.O. Box 5064
Austin, TX 78763
800-452-9292

U.S. Fish and Wildlife Service
Ecological Services
10711 Burnet Road, Suite 200
Austin, TX 78758
512-490-0057

UTAH

Natural Heritage Program
Utah Division of Wildlife Resources
1594 West North Temple, Suite 211
Salt Lake City, UT 84114-6301
801-538-4700

The Nature Conservancy
559 East South Temple
Salt Lake City, UT 84102
801-531-0999

Research Natural Area Coordinator
USDA Forest Service
Intermountain Region
324 25th Street
Ogden, UT 84401
801-625-5417

Utah Department of Natural Resources
1636 West North Temple, Suite 316
Salt Lake City, UT 84116
801-538-7200

Utah Office of the Bureau of Land Management
P.O. Box 45155
Salt Lake City, UT 84145
801-539-4010

VERMONT

The Nature Conservancy
27 State Street
Montpelier, VT 05602
802-229-4425

Vermont Agency of Natural Resources
103 S. Main Street
Waterbury, VT 05671
802-241-3636

Vermont Nongame and Natural Heritage Program
Vermont Fish and Wildlife Department
103 South Main Street
Waterbury, VT 05671-0501
802-241-3700

VIRGINIA

The Nature Conservancy
1223-A Cedars Court
Charlottesville, VA 22903
804-295-6106

Virginia Department of Conservation and Recreation
203 Governor Street, Suite 302
Richmond, VA 23219
804-786-7951

Virginia Division of Natural Heritage
Virginia Department of Conservation and Recreation
Main Street Station
1500 East Main Street, Suite 312
Richmond, VA 23219
804-786-7951

Virginia Division of State Parks
203 Governor Street, Suite 306
Richmond, VA 23219
804-786-1712

Virginia Wildlife Division
4010 West Broad Street
Richmond, VA 23230
804-376-9588

WASHINGTON

Natural Heritage Program
Department of Natural Resources
P.O. Box 47016
Olympia, WA 98504-7016
360-902-1340

The Nature Conservancy
217 Pine Street
Suite 1100
Seattle, WA 98101
206-343-4344

Washington State Parks and Recreation Commission
7150 Clearwater Lane, KY-11
Olympia, WA 98504
206-902-8509

WEST VIRGINIA

Natural Heritage Program
Division of Natural Resources
Ward Road, P.O. Box 67
Elkins, WV 26241
304-637-0245

The Nature Conservancy
723 Kanawha Boulevard East
Suite 500
Charleston, WV 25301
304-345-4350

WISCONSIN

Natural Heritage Program
Wisconsin Department of Natural Resources
101 South Webster Street
Madison, WI 53707
608-266-8916

The Nature Conservancy
633 West Main Street
Madison, WI, 53703
608-251-8140

Nicolet National Forest
Federal Building
68 South Stevens Street
Rhinelander, WI 54501
715-362-1300

Research Natural Area Coordinator
Eastern Region
USDA Forestry Service
310 Wisconsin Avenue, Room 500
Milwaukee, WI 53203
414-297-4125

WYOMING

The Nature Conservancy
258 Main Street
Suite 200
Lander, WY 82520
307-332-2971

Wyoming Game and Fish Department
5400 Bishop Boulevard
Cheyenne, WY 82006
307-777-4600

Wyoming Natural Diversity Database
1604 Grand Avenue, Suite 2
Laramie, WY 82070
307-745-5026

Wyoming State Parks and Historic Sites
2301 Central Avenue
Barrett Building
Cheyenne, WY 82002
307-777-6323

APPENDIX 3: FEDERALLY ENDANGERED AND FEDERALLY THREATENED FLORA AND FAUNA IN THE UNITED STATES

U.S. Fish and Wildlife Service Division of Endangered Species: Listed Species under Fish and Wildlife Service Jurisdiction By State/Territory as of December 31, 1998. Status Code: E - listed as Endangered; T - listed as Threatened)

Alabama—96 species
Animals—78 species
E — Acornshell, southern (*Epioblasma othcaloogensis*)
T — Bankclimber (mussel), purple (*Elliptoideus sloatianus*)
E — Bat, gray (*Myotis grisescens*)
E — Bat, Indiana (*Myotis sodalis*)
E — Blossom, turgid (pearlymussel) (*Epioblasma turgidula*)
E — Blossom, yellow (pearlymussel)(*Epioblasma florentina florentina*)
E — Catspaw (= purple cat's paw pearlymussel) (*Epioblasma obliquata obliquata*)
E — Cavefish, Alabama (*Speoplatyrhinus poulsoni*)
T — Chub, spotfin (= turquoise shiner) (*Cyprinella (= Hybopsis) monacha*)
E — Clubshell, black (= Curtus' mussel) (*Pleurobema curtum*)
E — Clubshell, ovate (*Pleurobema perovatum*)
E — Clubshell, southern (*Pleurobema decisum*)
E — Combshell, Cumberlandian (*Epioblasma brevidens*)
E — Combshell, southern (= penitent mussel) (*Epioblasma penita*)
E — Combshell, upland (*Epioblasma metastriata*)
E — Darter, boulder (= Elk River) (*Etheostoma wapiti*)
T — Darter, goldline (*Percina aurolineata*)
T — Darter, slackwater (*Etheostoma boschungi*)
T — Darter, snail (*Percina tanasi*)
E — Darter, watercress (*Etheostoma nuchale*)
T — Eagle, bald (*Haliaeetus leucocephalus*)
T — Elimia (snail), lacy (*Elimia crenatella*)
E — Falcon, American peregrine (*Falco peregrinus anatum*)
E — Fanshell (*Cyprogenia stegaria*)
T — Heelsplitter, Alabama (= inflated) (*Potamilus inflatus*)
E — Kidneyshell, triangular (*Ptychobranchus greeni*)
E — Lampmussel, Alabama (*Lampsilis virescens*)
E — Lilliput, pale (pearlymussel) (*Toxolasma cylindrellus*)
E — Lioplax (snail), cylindrical (*Lioplax cyclostomaformis*)
E — Manatee, West Indian (*Trichechus manatus*)
T — Moccasinshell, Alabama (*Medionidus acutissimus*)
E — Moccasinshell, Coosa (*Medionidus parvulus*)
E — Moccasinshell, Gulf (*Medionidus penicillatus*)
E — Monkeyface, Cumberland (pearlymussel)(*Quadrula intermedia*)
E — Mouse, Alabama beach (*Peromyscus polionotus ammobates*)

E — Mouse, Perdido Key beach (*Peromyscus polionotus trissyllepsis*)
T — Mucket, orangenacre (*Lampsilis perovalis*)
E — Mucket, pink (pearlymussel) (*Lampsilis abrupta*)
E — Mussel, oyster (*Epioblasma capsaeformis*)
E — Pearlymussel, cracking (*Hemistena lata*)
E — Pearlymussel, dromedary (*Dromus dromas*)
E — Pearlymussel, littlewing (*Pegias fabula*)
E — Wartyback, white (pearlymussel) (*Plethobasus cicatricosus*)
E — Pebblesnail, flat (*Lepyrium showalteri*)
E — Pigtoe, oval (*Pleurobema pyriforme*)
E — Pigtoe, dark (*Pleurobema furvum*)
E — Pigtoe, finerayed (*Fusconaia cuneolus*)
E — Pigtoe, flat (= Marshall's mussel) (*Pleurobema marshalli*)
E — Pigtoe, heavy (= Judge Tait's mussel) (*Pleurobema taitianum*)
E — Pigtoe, rough (*Pleurobema plenum*)
E — Pigtoe, shiny (*Fusconaia cor (= edgariana)*)
E — Pigtoe, southern (*Pleurobema georgianum*)
E — Pimpleback, orangefoot (pearlymussel) (*Plethobasus cooperianus*)
E — Pink, ring (mussel) (*Obovaria retusa*)
T — Plover, piping (*Charadrius melodus*)
E — Pocketbook, shinyrayed (*Lampsilis subangulata*)
T — Pocketbook, finelined (*Lampsilis altilis*)
E — Riversnail, Anthony's (*Athearnia anthonyi*)
T — Rocksnail, painted (*Leptoxis taeniata*)
E — Rocksnail, plicate (*Leptoxis plicata*)
T — Rocksnail, round (*Leptoxis ampla*)
T — Salamander, Red Hills (*Phaeognathus hubrichti*)
T — Sculpin, pygmy (*Cottus pygmaeus*)
T — Sea turtle, loggerhead (*Caretta caretta*)
T — Shiner, blue (*Cyprinella (= Notropis) caerulea*)
E — Shiner, Cahaba (*Notropis cahabae*)
E — Shiner, Palezone (*Notropis albizonatus*)
E — Shrimp, Alabama cave (*Palaemonias alabamae*)
T — Slabshell, Chipola (*Elliptio chipolaensis*)
E — Snail, tulotoma (= Alabama live-bearing) (*Tulotoma magnifica*)
T — Snake, eastern indigo (*Drymarchon corais couperi*)
E — Stirrupshell (*Quadrula stapes*)
E — Stork, wood (*Mycteria americana*)
T — Sturgeon, Gulf (*Acipenser oxyrinchus desotoi*)
T — Tortoise, gopher (*Gopherus polyphemus*)
E — Turtle, Alabama redbelly (*Pseudemys alabamensis*)
T — Turtle, flattened musk (*Sternotherus depressus*)
E — Woodpecker, red-cockaded (*Picoides borealis*)

Plants—18 species

T — Little amphianthus (*Amphianthus pusillus*)
T — Price's potato-bean (*Apios priceana*)
T — American hart's-tongue fern (*Asplenium scolopendrium* var. *americanum*)
E — Morefield's leather-flower (*Clematis morefieldii*)
E — Alabama leather-flower (*Clematis socialis*)
E — Leafy prairie-clover (*Dalea (= Petalostemum) foliosa*)
T — Eggert's sunflower (*Helianthus eggertii*)
T — Lyrate bladderpod (*Lesquerella lyrata*)
E — Pondberry (*Lindera melissifolia*)
T — Mohr's Barbara's buttons (*Marshallia mohrii*)
E — Harperella (*Ptilimnium nodosum (= fluviatile)*)

T — Kral's water-plantain (*Sagittaria secundifolia*)
E — Green pitcher-plant (*Sarracenia oreophila*)
E — Alabama canebrake pitcher-plant (*Sarracenia rubra* ssp. *alabamensis*)
E — Gentian pinkroot (*Spigelia gentianoides*)
T — Alabama streak-sorus fern (*Thelypteris pilosa* var. *alabamensis*)
E — Relict trillium (*Trillium reliquum*)
E — Tennessee yellow-eyed grass (*Xyris tennesseensis*)

Alaska—6 species
Animals—5 species
E — Curlew, Eskimo (*Numenius borealis*)
T — Eider, spectacled (*Somateria fischeri*)
T — Eider, Steller's (*Polysticta stelleri*)
E — Falcon, American peregrine (*Falco peregrinus anatum*)
T — Goose, Aleutian Canada (*Branta canadensis leucopareia*)
Plants—1 species
E — Aleutian shield-fern (= Aleutian holly-fern) (*Polystichum aleuticum*)

Arizona—51 species
Animals—34 species
E — Ambersnail, Kanab (*Oxyloma haydeni kanabensis*)
E — Bat, lesser (= Sanborn's) long-nosed (*Leptonycteris curasoae yerbabuenae*)
E — Bobwhite, masked (quail) (*Colinus virginianus ridgwayi*)
T — Catfish, Yaqui (*Ictalurus pricei*)
E — Chub, bonytail (*Gila elegans*)
E — Chub, humpback (*Gila cypha*)
T — Chub, Sonora (*Gila ditaenia*)
E — Chub, Virgin River (*Gila robusta semidnuda*)
E — Chub, Yaqui (*Gila purpurea*)
T — Eagle, bald (*Haliaeetus leucocephalus*)
E — Falcon, American peregrine (*Falco peregrinus anatum*)
E — Flycatcher, Southwestern willow (*Empidonax traillii extimus*)
E — Jaguar (*Panthera onca*)
E — Jaguarundi, Sinaloan (*Herpailurus (= Felis) yagouaroundi tolteca*)
T — Minnow, loach (*Rhinichthys (= Tiaroga) cobitis*)
E — Ocelot (*Leopardus (= Felis) pardalis*)
T — Owl, Mexican spotted (*Strix occidentalis lucida*)
E — Pronghorn, Sonoran (*Antilocapra americana sonoriensis*)
E — Pupfish, desert (*Cyprinodon macularius*)
E — Pygmy-owl, cactus ferruginous (*Glaucidium brasilianum cactorum*)
E — Rail, Yuma clapper (*Rallus longirostris yumanensis*)
E — Salamander, Sonoran tiger (*Ambystoma tigrinum stebbinsi*)
T — Shiner, beautiful (*Cyprinella (= Notropis) formosa*)
T — Spikedace (*Meda fulgida*).
T — Spinedace, Little Colorado (*Lepidomeda vittata*)
E — Squawfish, Colorado (*Ptychocheilus lucius*)
E — Squirrel, Mount Graham red (*Tamiasciurus hudsonicus grahamensis*)
E — Sucker, razorback (*Xyrauchen texanus*)
E — Topminnow, Gila (incl. Yaqui) (*Poeciliopsis occidentalis*)
T — Tortoise, desert (*Gopherus agassizii*)
T — Trout, Apache (= Arizona) (*Oncorhynchus (= Salmo) apache*)
E — Trout, Gila (*Oncorhynchus (= Salmo) gilae*)
E — Vole, Hualapai Mexican (*Microtus mexicanus hualpaiensis*)
E — Woundfin (*Plagopterus argentissimus*)
Plants—17 species

E — Arizona agave (*Agave arizonica*)
E — Kearney's blue-star (*Amsonia kearneyana*)
T — Welsh's milkweed (*Asclepias welshii*)
E — Sentry milk-vetch (*Astragalus cremnophylax* var. *cremnophylax*)
T — Navajo sedge (*Carex specuicola*)
T — Cochise pincushion cactus (*Coryphantha (= Escobaria) robbinsorum*)
E — Pima pineapple cactus (*Coryphantha scheeri* var. *robustispina*)
T — Jones cycladenia (*Cycladenia humilis* var. *jonesii*)
E — Nichol's Turk's head cactus (*Echinocactus horizonthalonius* var. *nicholii*)
E — Arizona hedgehog cactus (*Echinocereus triglochidiatus* var. *arizonicus*)
E — Huachuca water-umbel (*Lilaeopsis schaffneriana* ssp. *recurva*)
E — Brady pincushion cactus (*Pediocactus bradyi*)
E — Peebles Navajo cactus (*Pediocactus peeblesianus* var. *peeblesianus*)
T — Siler pincushion cactus (*Pediocactus sileri*)
E — Arizona cliffrose (*Purshia subintegra*)
T — San Francisco Peaks groundsel (*Senecio franciscanus*)
E — Canelo Hills ladies'-tresses (*Spiranthes delitescens*)

Arkansas—27 species
Animals—21 species
E — Bat, gray (*Myotis grisescens*)
E — Bat, Indiana (*Myotis sodalis*)
E — Bat, Ozark big-eared (*Corynorhinus (= Plecotus) townsendii ingens*)
E — Beetle, American burying (= giant carrion) (*Nicrophorus americanus*)
T — Cavefish, Ozark (*Amblyopsis rosae*)
E — Crayfish, cave [no common name] (*Cambarus aculabrum*)
E — Crayfish, cave [no common name] (*Cambarus zophonastes*)
T — Darter, leopard (*Percina pantherina*)
T — Eagle, bald (*Haliaeetus leucocephalus*)
E — Falcon, American peregrine (*Falco peregrinus anatum*)
T — Fatmucket, Arkansas (*Lampsilis powelli*)
E — Mucket, pink (pearlymussel) (*Lampsilis abrupta*)
E — Pearlymussel, Curtis (*Epioblasma (= Dysnomia) florentina curtisii*)
E — Pocketbook, fat (*Potamilus (= Proptera) capax*)
E — Pocketbook, Ouachita rock (= Wheeler's pearly mussel) (*Arkansia wheeleri*)
E — Pocketbook, speckled (*Lampsilis streckeri*)
T — Shagreen, Magazine Mountain (*Mesodon magazinensis*)
T — Shiner, Arkansas River (native pop. only) (*Notropis girardi*)
E — Sturgeon, pallid (*Scaphirhynchus albus*)
E — Tern, least (*Sterna antillarum*)
E — Woodpecker, red-cockaded (*Picoides borealis*)
Plants—6 species
T — *Geocarpon minimum* (Plant, no common name)
E — Missouri bladderpod (*Lesquerella filiformis*)
E — Pondberry (*Lindera melissifolia*)
T — Eastern prairie fringed orchid (*Platanthera leucophaea*)
E — Harperella (*Ptilimnium nodosum (= fluviatile)*)
E — Running buffalo clover (*Trifolium stoloniferum*)

California—251 species
Animals—87 species
T — Beetle, delta green ground (*Elaphrus viridis*)
E — Beetle, Mount Hermon June (*Polyphylla barbata*)
T — Beetle, valley elderberry longhorn (*Desmocerus californicus dimorphus*)
T — Butterfly, bay checkerspot (*Euphydryas editha bayensis*)

E — Butterfly, Behren's silverspot (*Speyeria zerene behrensii*)
E — Butterfly, callippe silverspot (*Speyeria callippe callippe*)
E — Butterfly, El Segundo blue (*Euphilotes battoides allyni*)
E — Butterfly, Lange's metalmark (*Apodemia mormo langei*)
E — Butterfly, lotis blue (*Lycaeides argyrognomon lotis*)
E — Butterfly, mission blue (*Icaricia icarioides missionensis*)
E — Butterfly, Myrtle's silverspot (*Speyeria zerene myrtleae*)
T — Butterfly, Oregon silverspot (*Speyeria zerene hippolyta*)
E — Butterfly, Palos Verdes blue (*Glaucopsyche lygdamus palosverdesensis*)
E — Butterfly, Quino checkerspot (*Euphydryas editha quino (= E. e. wrighti)*)
E — Butterfly, San Bruno elfin (*Callophrys mossii bayensis*)
E — Butterfly, Smith's blue (*Euphilotes enoptes smithi*)
E — Chub, bonytail (*Gila elegans*)
E — Chub, Mohave tui (*Gila bicolor mohavensis*)
E — Chub, Owens tui (*Gila bicolor snyderi*)
E — Condor, California (*Gymnogyps californianus*)
E — Crayfish, Shasta (= placid) (*Pacifastacus fortis*)
T — Eagle, bald (*Haliaeetus leucocephalus*)
E — Fairy shrimp, Conservancy (*Branchinecta conservatio*)
E — Fairy shrimp, longhorn (*Branchinecta longiantenna*)
E — Fairy shrimp, Riverside (*Streptocephalus woottoni*)
T — Fairy shrimp, vernal pool (*Branchinecta lynchi*)
E — Fairy shrimp, San Diego (*Branchinecta sandiegonensis*)
E — Falcon, American peregrine (*Falco peregrinus anatum*)
E — Fly, Delhi Sands flower-loving (*Rhaphiomidas terminatus abdominalis*)
E — Flycatcher, Southwestern willow (*Empidonax traillii extimus*)
E — Fox, San Joaquin kit (*Vulpes macrotis mutica*)
T — Frog , California red-legged (*Rana aurora draytonii*)
T — Gnatcatcher, coastal California (*Polioptila californica californica*)
E — Goby, tidewater (*Eucyclogobius newberryi*)
T — Goose, Aleutian Canada (*Branta canadensis leucopareia*)
E — Grasshopper, Zayante band-winged (*Trimerotropis infantilis*)
E — Kangaroo rat, Fresno (*Dipodomys nitratoides exilis*)
E — Kangaroo rat, giant (*Dipodomys ingens*)
E — Kangaroo rat, Morro Bay (*Dipodomys heermanni morroensis*)
E — Kangaroo rat, Stephens' (*Dipodomys stephensi* (incl. *D. cascus*))
E — Kangaroo rat, Tipton (*Dipodomys nitratoides nitratoides*)
E — Kangaroo rat, San Bernardino Merriam's (*Dipodomys merriami parvus*)
E — Lizard, blunt-nosed leopard (*Gambelia silus*)
T — Lizard, Coachella Valley fringe-toed (*Uma inornata*)
T — Lizard, Island night (*Xantusia riversiana*)
T — Moth, Kern primrose sphinx (*Euproserpinus euterpe*)
E — Mountain beaver, Point Arena (*Aplodontia rufa nigra*)
E — Mouse, Pacific pocket (*Perognathus longimembris pacificus*)
E — Mouse, salt marsh harvest (*Reithrodontomys raviventris*)
T — Murrelet, marbled (*Brachyramphus marmoratus marmoratus*)
T — Otter, southern sea (*Enhydra lutris nereis*)
T — Owl, northern spotted (*Strix occidentalis caurina*)
E — Pelican, brown (*Pelecanus occidentalis*)
T — Plover, western snowy (*Charadrius alexandrinus nivosus*)
E — Pupfish, desert (*Cyprinodon macularius*)
E — Pupfish, Owens (*Cyprinodon radiosus*)
E — Rail, California clapper (*Rallus longirostris obsoletus*)
E — Rail, light-footed clapper (*Rallus longirostris levipes*)
E — Rail, Yuma clapper (*Rallus longirostris yumanensis*)

E — Salamander, desert slender (*Batrachoseps aridus*)
E — Salamander, Santa Cruz long-toed (*Ambystoma macrodactylum croceum*)
E — Sheep, bighorn (Peninsular Ranges pop. in CA) (*Ovis canadensis*)
E — Shrike, San Clemente loggerhead (*Lanius ludovicianus mearnsi*)
E — Shrimp, California freshwater (*Syncaris pacifica*)
E — Skipper, Laguna Mountains (*Pyrgus ruralis lagunae*)
T — Smelt, delta (*Hypomesus transpacificus*)
E — Snail, Morro shoulderband (= banded dune) (*Helminthoglypta walkeriana*)
T — Snake, giant garter (*Thamnophis gigas*)
E — Snake, San Francisco garter (*Thamnophis sirtalis tetrataenia*)
T — Sparrow, San Clemente sage (*Amphispiza belli clementeae*)
E — Squawfish, Colorado (*Ptychocheilus lucius*)
E — Stickleback, unarmored threespine (*Gasterosteus aculeatus williamsoni*)
E — Sucker, Lost River (*Deltistes luxatus*)
E — Sucker, Modoc (*Catostomus microps*)
E — Sucker, razorback (*Xyrauchen texanus*)
E — Sucker, shortnose (*Chasmistes brevirostris*)
E — Tadpole shrimp, vernal pool (*Lepidurus packardi*)
E — Tern, California least (*Sterna antillarum browni*)
E — Toad, arroyo (*Bufo microscaphus californicus*)
T — Tortoise, desert (*Gopherus agassizii*)
T — Towhee, Inyo California (= brown) (*Pipilo crissalis eremophilus*)
T — Trout, Lahontan cutthroat (*Oncorhynchus (= Salmo) clarki henshawi*)
T — Trout, Little Kern golden (*Oncorhynchus (= Salmo) aguabonita whitei*)
T — Trout, Paiute cutthroat (*Oncorhynchus (= Salmo) clarki seleniris*)
E — Vireo, least Bell's (*Vireo bellii pusillus*)
E — Vole, Amargosa (*Microtus californicus scirpensis*)
T — Whipsnake (= striped racer), Alameda (*Masticophis lateralis euryxanthus*)

Plants—164 species

E — San Mateo thornmint (*Acanthomintha obovata* ssp. *duttonii*)
T — San Diego thornmint (*Acanthomintha ilicifolia*)
E — Munz's onion (*Allium munzii*)
E — Sonoma alopecurus (*Alopecurus aequalis* var. *sonomensis*)
E — Large-flowered fiddleneck (*Amsinckia grandiflora*)
E — Hoffmann's rock-cress (*Arabis hoffmannii*)
E — McDonald's rock-cress (*Arabis mcdonaldiana*)
T — Morro manzanita (*Arctostaphylos morroensis*)
E — Presidio (= Raven's) manzanita (*Arctostaphylos hookeri* var. *ravenii*)
E — Del Mar manzanita (*Arctostaphylos glandulosa* ssp. *crassifolia*)
E — Santa Rosa Island manzanita (*Arctostaphylos confertiflora*)
T — Pallid manzanita (*Arctostaphylos pallida*)
E — Marsh sandwort (*Arenaria paludicola*)
T — Bear Valley sandwort (*Arenaria ursina*)
E — Cushenbury milk-vetch (*Astragalus albens*)
E — Braunton's milk-vetch (*Astragalus brauntonii*)
E — Clara Hunt's milk-vetch (*Astragalus clarianus*)
E — Lane Mountain (= Coolgardie) milk-vetch (*Astragalus jaegerianus*)
E — Coachella Valley milk-vetch (*Astragalus lentiginosus* var. *coachellae*)
T — Fish Slough milk-vetch (*Astragalus lentiginosus* var. *piscinensis*)
T — Peirson's milk-vetch (*Astragalus magdalenae* var. *peirsonii*)
E — Coastal dunes milk-vetch (*Astragalus tener* var. *titi*)
E — Triple-ribbed milk-vetch (*Astragalus tricarinatus*)
E — San Jacinto Valley crownscale (= saltbush) (*Atriplex coronata* var. *notatior*)
T — Encinitis baccharis (= Coyote bush), (*Baccharis vanessae*)
E — Nevin's barberry (*Berberis nevinii*)

E — Island barberry (*Berberis pinnata ssp. insularis*)
E — Truckee barberry (*Berberis sonnei*)
E — Sonoma sunshine (= Baker's stickyseed) (*Blennosperma bakeri*)
T — Thread-leaved brodiaea (*Brodiaea filifolia*)
T — Chinese Camp brodiaea (*Brodiaea pallida*)
T — Tiburon mariposa lily (*Calochortus tiburonensis*)
T — Mariposa pussypaws (*Calyptridium pulchellum*)
E — Stebbins' morning-glory (*Calystegia stebbinsii*)
T — San Benito evening-primrose (*Camissonia benitensis*)
E — White sedge (*Carex albida*)
E — Tiburon paintbrush (*Castilleja affinis* ssp. *neglecta*)
T — Fleshy owl's-clover (*Castilleja campestris* ssp. *succulenta*)
T — Ash-gray paintbrush (*Castilleja cinerea*)
E — San Clemente Island Indian paintbrush (*Castilleja grisea*)
E — Soft-leaved paintbrush (*Castilleja mollis*)
E — California jewelflower (*Caulanthus californicus*)
E — Coyote ceanothus (= Coyote Valley California-lilac) (*Ceanothus ferrisae*)
T — Vail Lake ceanothus (*Ceanothus ophiochilus*)
E — Pine Hill ceanothus (*Ceanothus roderickii*)
T — Spring-loving centaury (*Centaurium namophilum*)
E — Catalina Island mountain-mahogany (*Cercocarpus traskiae*)
T — Hoover's spurge (*Chamaesyce hooveri*)
E — Howell's spineflower (*Chorizanthe howellii*)
E — Orcutt's spineflower (*Chorizanthe orcuttiana*)
E — Ben Lomond spineflower (*Chorizanthe pungens* var. *hartwegiana*)
T — Monterey spineflower (*Chorizanthe pungens* var. *pungens*)
E — Robust spineflower (includes Scotts Valley spineflower) (*Chorizanthe robusta*)
E — Sonoma spineflower (*Chorizanthe valida*)
E — Chorro Creek bog thistle (*Cirsium fontinale obispoense*)
E — Fountain thistle (*Cirsium fontinale var. fontinale*)
E — Suisun thistle (*Cirsium hydrophilum* var. *hydrophilum*)
E — Presidio clarkia (*Clarkia franciscana*)
E — Vine Hill clarkia (*Clarkia imbricata*)
E — Pismo clarkia (*Clarkia speciosa* ssp. *immaculata*)
T — Springville clarkia (*Clarkia springvillensis*)
E — Salt marsh bird's-beak (*Cordylanthus maritimus* ssp. *maritimus*)
E — Palmate-bracted bird's-beak (*Cordylanthus palmatus*)
E — Pennell's bird's-beak (*Cordylanthus tenuis* ssp. *capillaris*)
E — Soft bird's-beak (*Cordylanthus mollis* ssp. *mollis*)
E — Santa Cruz cypress (*Cupressus abramsiana*)
T — Gowen cypress (*Cupressus goveniana* ssp. *goveniana*)
E — San Clemente Island larkspur (*Delphinium variegatum* ssp. *kinkiense*)
E — Slender-horned spineflower (*Dodecahema leptoceras*)
T — Conejo dudleya (*Dudleya abramsii* ssp. *parva*)
T — Marcescent dudleya (*Dudleya cymosa* ssp. *marcescens*)
T — Santa Monica Mountains dudleya (*Dudleya cymosa* ssp. *ovatifolia*)
T — Santa Cruz Island dudleya (*Dudleya nesiotica*)
E — Santa Clara Valley dudleya (*Dudleya setchellii*)
T — Laguna Beach liveforever (*Dudleya stolonifera*)
E — Santa Barbara Island liveforever (*Dudleya traskiae*)
T — Verity's dudleya (*Dudleya verityi*)
E — Kern mallow (*Eremalche kernensis*)
E — Santa Ana River woolly-star (*Eriastrum densifolium* ssp. *sanctorum*)
T — Hoover's woolly-star (*Eriastrum hooveri*)
T — Parish's daisy (*Erigeron parishii*)

E — Indian Knob mountain balm (*Eriodictyon altissimum*)
T — Southern mountain wild-buckwheat (*Eriogonum kennedyi* var. *austromontanum*)
E — Cushenbury buckwheat (*Eriogonum ovalifolium* var. *vineum*)
E — San Mateo woolly sunflower (*Eriophyllum latilobum*)
E — San Diego button-celery (*Eryngium aristulatum* var. *parishii*)
E — Loch Lomond coyote-thistle (*Eryngium constancei*)
E — Contra Costa wallflower (*Erysimum capitatum* var. *angustatum*)
E — Menzies' wallflower (*Erysimum menziesii*)
E — Ben Lomond wallflower (*Erysimum teretifolium*)
E — Pine Hill flannelbush (*Fremontodendron californicum* ssp. *decumbens*)
E — Mexican flannelbush (*Fremontodendron mexicanum*)
E — Island bedstraw (*Galium buxifolium*)
E — El Dorado bedstraw (*Galium californicum* ssp. *sierrae*)
E — Monterey gilia (*Gilia tenuiflora* ssp. *arenaria*)
E — Hoffmann's slender-flowered gilia (*Gilia tenuiflora ssp. hoffmannii*)
T — Ash Meadows gumplant (*Grindelia fraxino-pratensis*)
T — Island rush-rose (*Helianthemum greenei*)
T — Otay tarplant (*Hemizonia conjugens*)
T — Marin dwarf-flax (*Hesperolinon congestum*)
T — Water howellia (*Howellia aquatilis*)
E — Burke's goldfields (*Lasthenia burkei*)
E — Contra Costa goldfields (*Lasthenia conjugens*)
E — Beach layia (*Layia carnosa*)
E — San Joaquin wooly-threads (*Lembertia congdonii*)
E — San Bernardino Mountains bladderpod (*Lesquerella kingii* ssp. *bernardina*)
E — San Francisco lessingia (*Lessingia germanorum* (= *L. g.* var. *germanorum*)
E — Western lily (*Lilium occidental*)
E — Pitkin Marsh lily (*Lilium pardalinum* ssp. *pitkinense*)
E — Butte County meadowfoam (*Limnanthes floccosa* ssp. *californica*)
E — Sebastopol meadowfoam (*Limnanthes vinculans*)
E — San Clemente Island woodland-star (*Lithophragma maximum*)
E — San Clemente Island broom (*Lotus dendroideus* ssp. *traskiae*)
E — Clover lupine (*Lupinus tidestromii*)
E — San Clemente Island bush-mallow (*Malacothamnus clementinus*)
E — Santa Cruz Island bush-mallow (*Malacothamnus fasciculatus var. nesioticus*)
E — Santa Cruz Island malocothrix (*Malacothrix indecora*)
E — Island malacothrix (*Malacothrix squalida*)
E — Willowy monardella (*Monardella linoides* ssp. *viminea*)
T — Navarretia, spreading (= prostrate, = no-named) (*Navarretia fossalis*)
E — Navarretia, few-flowered (*Navarretia leucocephala* ssp. *pauciflora* (= N. pauciflora))
E — Navarretia, many-flowered (*Navarretia leucocephala* ssp. *plieantha*)
T — Colusa grass (*Neostapfia colusana*)
E — Amargosa niterwort (*Nitrophila mohavensis*)
E — Eureka Valley evening-primrose (*Oenothera avita* ssp. *eurekensis*)
E — Antioch Dunes evening-primrose (*Oenothera deltoides* ssp. *howellii*)
E — Bakersfield cactus (*Opuntia treleasei*)
E — California Orcutt grass (*Orcuttia californica*)
T — San Joaquin Valley Orcutt grass (*Orcuttia inaequalis*)
E — Hairy (= pilose) Orcutt grass (*Orcuttia pilosa*)
T — Slender Orcutt grass (*Orcuttia tenuis*)
E — Sacramento Orcutt grass (*Orcuttia viscida*)
E — Cushenbury oxytheca (*Oxytheca parishii* var. *goodmaniana*)
E — Lake County stonecrop (*Parvisedum leiocarpum*)
E — White-rayed pentachaeta (*Pentachaeta bellidiflora*)
E — Lyon's pentachaeta (*Pentachaeta lyonii*)

E — Island phacelia (*Phacelia insularis ssp. insularis*)
E — Yadon's piperia (*Piperia yadonii*)
E — Calistoga allocarya (*Plagiobothrys strictus*)
E — San Bernardino bluegrass (*Poa atropurpurea*)
E — Napa bluegrass (*Poa napensis*)
E — San Diego mesa mint (*Pogogyne abramsii*)
E — Otay mesa mint (*Pogogyne nudiuscula*)
E — Hickman's potentilla (*Potentilla hickmanii*)
E — Hartweg's golden sunburst (*Pseudobahia bahiifolia*)
T — San Joaquin adobe sunburst (*Pseudobahia peirsonii*)
E — Gambel's watercress (*Rorippa gambellii*)
T — Layne's butterweed (*Senecio layneae*)
E — Santa Cruz Island rockcress (*Sibara filifolia*)
E — Kenwood Marsh checker-mallow (*Sidalcea oregana* ssp. *valida*)
E — Pedate checker-mallow (*Sidalcea pedata*)
E — Metcalf Canyon jewelflower (*Streptanthus albidus* ssp. *albidus*)
E — Tiburon jewelflower (*Streptanthus niger*)
E — California seablite (*Suaeda californica*)
E — Eureka Dune grass (*Swallenia alexandrae*)
E — California taraxacum (*Taraxacum californicum*)
E — Slender-petaled mustard (*Thelypodium stenopetalum*)
T — Hidden Lake bluecurls (*Trichostema austromontanum* ssp. *compactum*)
E — Showy Indian clover (*Trifolium amoenum*)
E — Monterey (= Del Monte) clover (*Trifolium trichocalyx*)
E — Greene's orcutt grass (*Tuctoria greenei*)
E — Solano grass (*Tuctoria mucronata*)
T — Red Hills vervain (*Verbena californica*)
T — Big-leaved crownbeard (*Verbesina dissita*)

Colorado—30 species
Animals—18 species
T — Bear, grizzly (*Ursus arctos*)
E — Butterfly, Uncompahgre fritillary (*Boloria acrocnema*)
E — Chub, bonytail (*Gila elegans*)
E — Chub, humpback (*Gila cypha*)
E — Crane, whooping (*Grus americana*)
T — Eagle, bald (*Haliaeetus leucocephalus*)
E — Falcon, American peregrine (*Falco peregrinus anatum*)
E — Ferret, black-footed (*Mustela nigripes*)
E — Flycatcher, Southwestern willow (*Empidonax traillii extimus*)
T — Mouse, Preble's meadow jumping (*Zapus hudsonius preblei*)
T — Owl, Mexican spotted (*Strix occidentalis lucida*)
T — Plover, piping (*Charadrius melodus*)
T — Skipper, Pawnee montane (*Hesperia leonardus (= pawnee) montana*)
E — Squawfish, Colorado (*Ptychocheilus lucius*)
E — Sucker, razorback (*Xyrauchen texanus*)
E — Tern, least (*Sterna antillarum*)
T — Trout, greenback cutthroat (*Oncorhynchus (= Salmo) clarki stomias*)
E — Wolf, gray (*Canis lupus*)
Plants—12 species
E — Mancos milk-vetch (*Astragalus humillimus*)
E — Osterhout milk-vetch (*Astragalus osterhoutii*)
E — Clay-loving wild-buckwheat (*Eriogonum pelinophilum*)
T — Penland alpine fen mustard (*Eutrema penlandii*)
T — Dudley Bluffs bladderpod (*Lesquerella congesta*)

E — Knowlton cactus (*Pediocactus knowltonii*)
E — Penland beardtongue (*Penstemon penlandii*)
E — North Park phacelia (*Phacelia formosula*)
T — Dudley Bluffs twinpod (*Physaria obcordata*)
T — Uinta Basin hookless cactus (*Sclerocactus glaucus*)
T — Mesa Verde cactus (*Sclerocactus mesae-verdae*)
T — Ute ladies'-tresses (*Spiranthes diluvialis*)

Connecticut—10 species
Animals—8 species
T — Beetle, northeastern beach tiger (*Cicindela dorsalis dorsalis*)
T — Beetle, Puritan tiger (*Cicindela puritana*)
T — Eagle, bald (*Haliaeetus leucocephalus*)
E — Falcon, American peregrine (*Falco peregrinus anatum*)
T — Plover, piping (*Charadrius melodus*)
E — Tern, roseate (*Sterna dougallii dougallii*)
T — Turtle, bog (= Muhlenberg) (*Clemmys muhlenbergii*)
E — Wedgemussel, dwarf (*Alasmidonta heterodon*)
Plants—2 species
E — Sandplain gerardia (*Agalinis acuta*)
T — Small whorled pogonia (*Isotria medeoloides*)

Delaware—9 species
Animals—5 species
T — Eagle, bald (*Haliaeetus leucocephalus*)
E — Falcon, American peregrine (*Falco peregrinus anatum*)
T — Plover, piping (*Charadrius melodus*)
E — Squirrel, Delmarva Peninsula fox (*Sciurus niger cinereus*)
T — Turtle, bog (= Muhlenberg) (*Clemmys muhlenbergii*)
Plants—4 species
T — Swamp pink (*Helonias bullata*)
T — Small whorled pogonia (*Isotria medeoloides*)
E — Canby's dropwort (*Oxypolis canbyi*)
T — Knieskern's beaked-rush (*Rhynchospora knieskernii*)

Florida—100 species
Animals—46 species
T — Bankclimber (mussel), purple (*Elliptoideus sloatianus*)
E — Bat, gray (*Myotis grisescens*)
E — Butterfly, Schaus swallowtail (*Heraclides (= Papilio) aristodemus ponceanus*)
T — Caracara, Audubon's crested (*Polyborus plancus audubonii*)
E — Crocodile, American (*Crocodylus acutus*)
E — Darter, Okaloosa (*Etheostoma okaloosae*)
E — Deer, key (*Odocoileus virginianus clavium*)
T — Eagle, bald (*Haliaeetus leucocephalus*)
E — Falcon, American peregrine (*Falco peregrinus anatum*)
T — Scrub-jay, Florida (*Aphelocoma coerulescens*)
E — Kite, Everglade snail (*Rostrhamus sociabilis plumbeus*)
E — Manatee, West Indian (*Trichechus manatus*)
E — Moccasinshell, Gulf (*Medionidus penicillatus*)
E — Moccasinshell, Ochlockonee (*Medionidus simpsonianus*)
E — Mouse, Anastasia Island beach (*Peromyscus polionotus phasma*)
E — Mouse, Choctawhatchee beach (*Peromyscus polionotus allophrys*)
E — Mouse, Key Largo cotton (*Peromyscus gossypinus allapaticola*)
E — Mouse, Perdido Key beach (*Peromyscus polionotus trissyllepsis*)

T — Mouse, southeastern beach (*Peromyscus polionotus niveiventris*)
E — Mouse, St. Andrew beach (*Peromyscus polionotus peninsularis*)
E — Panther, Florida (*Puma (= Felis) concolor coryi*)
E — Pigtoe, oval (*Pleurobema pyriforme*)
T — Plover, piping (*Charadrius melodus*)
E — Pocketbook, shinyrayed (*Lampsilis subangulata*)
E — Rabbit, Lower Keys (*Sylvilagus palustris hefneri*)
E — Rice rat, silver (*Oryzomys palustris natator*)
E&T — Sea turtle, green (*Chelonia mydas*)
E — Sea turtle, hawksbill (*Eretmochelys imbricata*)
E — Sea turtle, leatherback (*Dermochelys coriacea*)
T — Sea turtle, loggerhead (*Caretta caretta*)
T — Shrimp, Squirrel Chimney Cave (= Florida cave) (*Palaemonetes cummingi*)
T — Skink, bluetail mole (*Eumeces egregius lividus*)
T — Skink, sand (*Neoseps reynoldsi*)
T — Slabshell, Chipola (*Elliptio chipolaensis*)
T — Snail, Stock Island tree (*Orthalicus reses* (not incl. *nesodryas*))
T — Snake, Atlantic salt marsh (*Nerodia clarkii taeniata*)
T — Snake, eastern indigo (*Drymarchon corais couperi*)
E — Sparrow, Cape Sable seaside (*Ammodramus maritimus mirabilis*)
E — Sparrow, Florida grasshopper (*Ammodramus savannarum floridanus*)
E — Stork, wood (*Mycteria americana*)
T — Sturgeon, Gulf (*Acipenser oxyrinchus desotoi*)
T — Tern, roseate (*Sterna dougallii dougallii*)
E — Three-ridge (mussel), fat (*Amblema neislerii*)
E — Vole, Florida salt marsh (*Microtus pennsylvanicus dukecampbelli*)
E — Woodpecker, red-cockaded (*Picoides borealis*)
E — Woodrat, Key Largo (*Neotoma floridana smalli*)

Plants—54 species

E — Crenulate lead-plant (*Amorpha crenulata*)
E — Four-petal pawpaw (*Asimina tetramera*)
T — Florida bonamia (*Bonamia grandiflora*)
E — Brooksville (= Robins') bellflower (*Campanula robinsiae*)
E — Fragrant prickly-apple (*Cereus eriophorus* var. *fragrans*)
E — Deltoid spurge (*Chamaesyce deltoidea* ssp. *deltoidea*)
T — Garber's spurge (*Chamaesyce garberi*)
E — Pygmy fringe-tree (*Chionanthus pygmaeus*)
E — Florida golden aster (*Chrysopsis floridana*)
E — Florida perforate cladonia (*Cladonia perforata*)
T — Pigeon wings (*Clitoria fragrans*)
E — Short-leaved rosemary (*Conradina brevifolia*)
E — Etonia rosemary (*Conradina etonia*)
E — Apalachicola rosemary (*Conradina glabra*)
E — Avon Park harebells (*Crotalaria avonensis*)
E — Okeechobee gourd (*Cucurbita okeechobeensis* ssp. *okeechobeensis*)
E — Beautiful pawpaw (*Deeringothamnus pulchellus*)
E — Rugel's pawpaw (*Deeringothamnus rugelii*)
E — Garrett's mint (*Dicerandra christmanii*)
E — Longspurred mint (*Dicerandra cornutissima*)
E — Scrub mint (*Dicerandra frutescens*)
E — Lakela's mint (*Dicerandra immaculata*)
T — Scrub buckwheat (*Eriogonum longifolium* var. *gnaphalifolium*)
E — Snakeroot (*Eryngium cuneifolium*)
T — Telephus spurge (*Euphorbia telephioides*)
E — Small's milkpea (*Galactia smallii*)

E — Harper's beauty (*Harperocallis flava*)
E — Highlands scrub hypericum (*Hypericum cumulicola*)
E — Beach jacquemontia (*Jacquemontia reclinata*)
E — Cooley's water-willow (*Justicia cooleyi*)
E — Scrub blazingstar (*Liatris ohlingerae*)
E — Pondberry (*Lindera melissifolia*)
E — Scrub lupine (*Lupinus aridorum*)
T — White birds-in-a-nest (*Macbridea alba*)
E — Britton's beargrass (*Nolina brittoniana*)
T — Papery whitlow-wort (*Paronychia chartacea*)
E — Key tree-cactus (*Pilosocereus robinii (= Cereus r.)*)
T — Godfrey's butterwort (*Pinguicula ionantha*)
E — Lewton's polygala (*Polygala lewtonii*)
E — Tiny polygala (*Polygala smallii*)
E — Wireweed (*Polygonella basiramia*)
E — Sandlace (*Polygonella myriophylla*)
E — Scrub plum (*Prunus geniculata*)
E — Chapman rhododendron (*Rhododendron chapmanii*)
T — Miccosukee gooseberry (*Ribes echinellum*)
E — American chaffseed (*Schwalbea americana*)
T — Florida skullcap (*Scutellaria floridana*)
E — Fringed campion (*Silene polypetala*)
E — Gentian pinkroot (*Spigelia gentianoides*)
E — Cooley's meadowrue (*Thalictrum cooleyi*)
E — Florida torreya (*Torreya taxifolia*)
E — Wide-leaf warea (*Warea amplexifolia*)
E — Carter's mustard (*Warea carteri*)
E — Florida ziziphus (*Ziziphus celata*)

Georgia—55 species
Animals—33 species
E — Acornshell, southern (*Epioblasma othcaloogensis*)
T — Bankclimber (mussel), purple (*Elliptoideus sloatianus*)
E — Bat, gray (*Myotis grisescens*)
E — Bat, Indiana (*Myotis sodalis*)
E — Clubshell, ovate (*Pleurobema perovatum*)
E — Clubshell, southern (*Pleurobema decisum*)
E — Combshell, upland (*Epioblasma metastriata*)
E — Darter, amber (*Percina antesella*)
T — Darter, Cherokee (*Etheostoma scotti*)
E — Darter, Etowah (*Etheostoma etowahae*)
T — Darter, goldline (*Percina aurolineata*)
T — Darter, snail (*Percina tanasi*)
T — Eagle, bald (*Haliaeetus leucocephalus*)
E — Falcon, American peregrine (*Falco peregrinus anatum*)
E — Kidneyshell, triangular (*Ptychobranchus greeni*)
E — Logperch, Conasauga (*Percina jenkinsi*)
E — Manatee, West Indian (*Trichechus manatus*)
T — Moccasinshell, Alabama (*Medionidus acutissimus*)
E — Moccasinshell, Coosa (*Medionidus parvulus*)
E — Moccasinshell, Gulf (*Medionidus penicillatus*)
E — Moccasinshell, Ochlockonee (*Medionidus simpsonianus*)
E — Pigtoe, oval (*Pleurobema pyriforme*)
E — Pigtoe, southern (*Pleurobema georgianum*)
T — Plover, piping (*Charadrius melodus*)

E — Pocketbook, shinyrayed (*Lampsilis subangulata*)
T — Pocketbook, finelined (*Lampsilis altilis*)
T — Sea turtle, loggerhead (*Caretta caretta*)
T — Shiner, blue (*Cyprinella (= Notropis) caerulea*)
T — Snake, eastern indigo (*Drymarchon corais couperi*)
E — Stork, wood (*Mycteria americana*)
T — Tern, roseate (*Sterna dougallii dougallii*)
E — Three-ridge (mussel), fat (*Amblema neislerii*)
E — Woodpecker, red-cockaded (*Picoides borealis*)

Plants—22 species

T — Little amphianthus (*Amphianthus pusillus*)
E — Hairy rattleweed (*Baptisia arachnifera*)
E — Smooth coneflower (*Echinacea laevigata*)
T — Swamp pink (*Helonias bullata*)
E — Black-spored quillwort (*Isoetes melanospora*)
E — Mat-forming quillwort (*Isoetes tegetiformans*)
T — Small whorled pogonia (*Isotria medeoloides*)
E — Pondberry (*Lindera melissifolia*)
T — Mohr's Barbara's buttons (*Marshallia mohrii*)
E — Canby's dropwort (*Oxypolis canbyi*)
E — Harperella (*Ptilimnium nodosum (= fluviatile)*)
E — Michaux's sumac (*Rhus michauxii*)
T — Kral's water-plantain (*Sagittaria secundifolia*)
E — Green pitcher-plant (*Sarracenia oreophila*)
E — American chaffseed (*Schwalbea americana*)
E — Large-flowered skullcap (*Scutellaria montana*)
E — Fringed campion (*Silene polypetala*)
T — Virginia spiraea (*Spiraea virginiana*)
E — Florida torreya (*Torreya taxifolia*)
E — Persistent trillium (*Trillium persistens*)
E — Relict trillium (*Trillium reliquum*)
E — Tennessee yellow-eyed grass (*Xyris tennesseensis*)

Hawaii—297 species

Animals—34 species

E — \`Akepa, Hawaii (honeycreeper) (*Loxops coccineus coccineus*)
E — \`Akepa, Maui (honeycreeper) (*Loxops coccineus ochraceus*)
E — \`Akialoa, Kauai (honeycreeper) (*Hemignathus procerus*)
E — \`Akiapola\`au (honeycreeper) (*Hemignathus munroi*)
E — Bat, Hawaiian hoary (*Lasiurus cinereus semotus*)
E — Coot, Hawaiian (= \`alae-ke\`oke\`o) (*Fulica americana alai*)
E — Creeper, Hawaii (*Oreomystis mana*)
E — Creeper, Molokai (= kakawahie) (*Paroreomyza flammea*)
E — Creeper, Oahu (= alauwahio) (*Paroreomyza maculata*)
E — Crow, Hawaiian (= \`alala) (*Corvus hawaiiensis*)
E — Duck, Hawaiian (= koloa) (*Anas wyvilliana*)
E — Duck, Laysan (*Anas laysanensis*)
E — Finch, Laysan (honeycreeper) (*Telespyza cantans*)
E — Finch, Nihoa (honeycreeper) (*Telespyza ultima*)
E — Goose, Hawaiian (= nene) (*Branta (= Nesochen) sandvicensis*)
E — Hawk, Hawaiian (= io) (*Buteo solitarius*)
E — Honeycreeper, crested (= \`akohekohe) (*Palmeria dolei*)
E — Millerbird, Nihoa (old world warbler) (*Acrocephalus familiaris kingi*)
E — Moorhen (= gallinule), Hawaiian common (*Gallinula chloropus sandvicensis*)
E — Nukupu\`u (honeycreeper) (*Hemignathus lucidus*)

E — \`O\`o, Kauai (= \`o\`o \`a\`a) (honeyeater) (*Moho braccatus*)
E — \`O\`u (honeycreeper) (*Psittirostra psittacea*)
E — Palila (honeycreeper) (*Loxioides bailleui*)
E — Parrotbill, Maui (honeycreeper) (*Pseudonestor xanthophrys*)
E — Petrel, Hawaiian dark-rumped (*Pterodroma phaeopygia sandwichensis*)
E — Po\`ouli (honeycreeper) (*Melamprosops phaeosoma*)
T — Sea turtle, green (*Chelonia mydas*)
E — Sea turtle, hawksbill (*Eretmochelys imbricata*)
T — Shearwater, Newell's Townsend's (formerly Manx) (= \`a\`o) (*Puffinus auricularis newelli*)
E — Snails, Oahu tree (*Achatinella* spp.)
E — Stilt, Hawaiian (= ae\`o) (*Himantopus mexicanus knudseni*)
E — Thrush, large Kauai (*Myadestes myadestinus*)
E — Thrush, Molokai (= oloma\`o) (*Myadestes lanaiensis rutha*)
E — Thrush, small Kauai (= puaiohi) (*Myadestes palmeri*)

Plants—263 species

E — *Abutilon eremitopetalum* (Plant, no common name)
E — Ko\`oloa\`ula (*Abutilon menziesii*)
E — *Abutilon sandwicense* (Plant, no common name)
E — Liliwai (*Acaena exigua*)
E — *Achyranthes mutica* (Plant, no common name)
E — Round-leaved chaff-flower (*Achyranthes splendens* var. *rotundata*)
E — Pendant kihi fern (*Adenophorus periens*)
E — Mahoe (*Alectryon macrococcus*)
E — *Alsinidendron obovatum* (Plant, no common name)
E — *Alsinidendron trinerve* (Plant, no common name)
E — Kuawawaenohu (*Alsinidendron lychnoides*)
E — *Alsinodendron viscosum* (Plant, no common name)
E — *Amaranthus brownii* (Plant, no common name)
E — Mauna Loa (= Ka\`u) silversword (*Argyroxiphium kauense*)
T — \`Ahinahina (= Haleakala silversword) (*Argyroxiphium sandwicense* ssp. *macrocephalum*)
E — \`Ahinahina (= Mauna Kea silversword) (*Argyroxiphium sandwicense* ssp. *sandwicense*)
E — *Asplenium fragile* var. *insulare* (Fern, no common name)
E — Ko\`oko\`olau (*Bidens micrantha* ssp. *kalealaha*)
E — Ko\`oko\`olau (*Bidens wiebkei*)
E — *Bonamia menziesii* (Plant, no common name)
E — \`Olulu (*Brighamia insignis*)
E — Pua \`ala (*Brighamia rockii*)
E — Uhiuhi (*Caesalpinia kavaiense*)
E — \`Awikiwiki (*Canavalia molokaiensis*)
E — Kamanomano (*Cenchrus agrimonioides*)
E — \`Awiwi (*Centaurium sebaeoides*)
E — \`Akoko (*Chamaesyce celastroides* var. *kaenana*)
E — \`Akoko (*Chamaesyce deppeana (= Euphorbia d.)*)
E — *Chamaesyce halemanui* (Plant, no common name)
E — \`Akoko (*Chamaesyce herbstii*)
E — \`Akoko (*Chamaesyce kuwaleana*)
E — \`Akoko (*Chamaesyce rockii*)
E — \`Ewa Plains \`akoko (*Chamaesyce skottsbergii* var. *kalaeloana*)
E — \`Oha wai (*Clermontia drepanomorpha*)
E — \`Oha wai (*Clermontia lindseyana*)
E — \`Oha wai (*Clermontia oblongifolia* ssp. *brevipes*)
E — \`Oha wai (*Clermontia oblongifolia* ssp. *mauiensis*)
E — \`Oha wai (*Clermontia peleana*)
E — \`Oha wai (*Clermontia pyrularia*)
E — Kauila (*Colubrina oppositifolia*)

E — Pauoa (*Ctenitis squamigera*)
E — Haha (*Cyanea acuminata*)
E — Haha (*Cyanea asarifolia*)
E — Haha (*Cyanea copelandii* ssp. *copelandii*)
E — *Cyanea (= Rollandia) crispa* (Plant, no common name)
E — Haha (*Cyanea dunbarii*)
E — Haha (*Cyanea grimesiana* ssp. *obatae*)
E — Haha (*Cyanea grimesiana ssp. grimesiana*)
E — Haha (*Cyanea hamatiflora* ssp. *carlsonii*)
E — Haha (*Cyanea (= Rollandia) humboldtiana*)
E — Haha (*Cyanea koolauensis*)
E — Haha (*Cyanea lobata*)
E — Haha (*Cyanea longiflora*)
E — *Cyanea macrostegia* ssp. *gibsonii* (Plant, no common name)
E — Haha (*Cyanea mannii*)
E — Haha (*Cyanea mceldowneyi*)
E — Haha (*Cyanea pinnatifida*)
E — Haha (*Cyanea platyphylla*)
E — Haha (*Cyanea procera*)
T — Haha (*Cyanea recta*)
E — Haha (*Cyanea remyi*)
E — Haha (*Cyanea shipmannii*)
E — Haha (*Cyanea (= Rollandia) st-johnii*)
E — Haha (*Cyanea stictophylla*)
E — *Cyanea superba* (Plant, no common name)
E — Haha (*Cyanea truncata*)
E — *Cyanea undulata* (Plant, no common name)
E — Pu`uka`a (*Cyperus trachysanthos*)
E — Ha`iwale (*Cyrtandra crenata*)
E — Mapele (*Cyrtandra cyaneoides*)
E — Ha`iwale (*Cyrtandra dentata*)
E — Ha`iwale (*Cyrtandra giffardii*)
T — Ha`iwale (*Cyrtandra limahuliensis*)
E — Ha`iwale (*Cyrtandra munroi*)
E — Ha`iwale (*Cyrtandra polyantha*)
E — Ha`iwale (*Cyrtandra subumbellata*)
E — Ha`iwale (*Cyrtandra tintinnabula*)
E — Ha`iwale (*Cyrtandra viridiflora*)
E — *Delissea rhytidosperma* (Plant, no common name)
E — `Oha (*Delissea rivularis*)
E — `Oha (*Delissea subcordata*)
E — *Delissea undulata* (Plant, no common name)
E — Asplenium-leaved diellia (*Diellia erecta*)
E — *Diellia falcata* (Fern, no common name)
E — *Diellia pallida* (Fern, no common name)
E — *Diellia unisora* (Fern, no common name)
E — *Diplazium molokaiense* (Fern, no common name)
E — Na`ena`e (*Dubautia herbstobatae*)
E — Na`ena`e (*Dubautia latifolia*)
E — Na`ena`e (*Dubautia pauciflorula*)
E — Fosberg's love grass (*Eragrostis fosbergii*)
E — Nioi (*Eugenia koolauensis*)
E — `Akoko (*Euphorbia haeleeleana*)
E — Heau (*Exocarpos luteolus*)
E — Mehamehame (*Flueggea neowawraea*)

E — *Gahnia lanaiensis* (Plant, no common name)
E — Na`u or Hawaiian gardenia (*Gardenia brighamii*)
E — Nanu (*Gardenia mannii*)
E — Hawaiian red-flowered geranium (*Geranium arboreum*)
E — Nohoanu (*Geranium multiflorum*)
E — *Gouania hillebrandii* (Plant, no common name)
E — *Gouania meyenii* (Plant, no common name)
E — *Gouania vitifolia* (Plant, no common name)
E — *Haplostachys haplostachya* (Plant, no common name)
E — `Awiwi (*Hedyotis cookiana*)
E — Kio`ele (*Hedyotis coriacea*)
E — *Hedyotis degeneri* (Plant, no common name)
E — Pilo (*Hedyotis mannii*)
E — *Hedyotis parvula* (Plant, no common name)
E — Na Pali beach hedyotis (*Hedyotis st.-johnii*)
E — *Hesperomannia arborescens* (Plant, no common name)
E — *Hesperomannia arbuscula* (Plant, no common name)
E — *Hesperomannia lydgatei* (Plant, no common name)
E — Kauai hau kuahiwi (*Hibiscadelphus distans*)
E — Hau kuahiwi (*Hibiscadelphus giffardianus*)
E — Hau kuahiwi (*Hibiscadelphus hualalaiensis*)
E — Hau kuahiwi (*Hibiscadelphus woodii*)
E — Koki`o ke`oke`o (*Hibiscus arnottianus* ssp. *immaculatus*)
E — Ma`o hau hele (*Hibiscus brackenridgei*)
E — Clay's hibiscus (*Hibiscus clayi*)
E — Koki`o ke`oke`o (*Hibiscus waimeae ssp. hannerae*)
E — Hilo ischaemum (*Ischaemum byrone*)
E — Aupaka (*Isodendrion hosakae*)
E — Aupaka (*Isodendrion laurifolium*)
T — Aupaka (*Isodendrion longifolium*)
E — Wahine noho kula (*Isodendrion pyrifolium*)
E — Cooke's koki`o (*Kokia cookei*)
E — Koki`o ( = hau-hele`ula or Hawaii tree cotton) (*Kokia drynarioides*)
E — Kaua`i Koki`o (*Kokia kauaiensis*)
E — Kamakahala (*Labordia cyrtandrae*)
E — Kamakahala (*Labordia lydgatei*)
E — Kamakahala (*Labordia tinifolia* var. *wahiawaensis*)
E — `Anaunau (*Lepidium arbuscula*)
E — Nehe (*Lipochaeta fauriei*)
E — Nehe (*Lipochaeta kamolensis*)
E — Nehe (*Lipochaeta lobata* var. *leptophylla*)
E — Nehe (*Lipochaeta micrantha*)
E — Nehe (*Lipochaeta tenuifolia*)
E — *Lipochaeta venosa* (Plant, no common name)
E — Nehe (*Lipochaeta waimeaensis*)
E — *Lobelia gaudichaudii* ssp. *koolauensis*, (Plant, no common name)
E — *Lobelia monostachya* (Plant, no common name)
E — *Lobelia niihauensis* (Plant, no common name)
E — *Lobelia oahuensis* (Plant, no common name)
E — *Lysimachia filifolia* (Plant, no common name)
E — *Lysimachia lydgatei* (Plant, no common name)
E — *Lysimachia maxima (= ternifolia)* (Plant, no common name)
E — *Mariscus fauriei* (Plant, no common name)
E — *Mariscus pennatiformis* (Plant, no common name)
E — `Ihi`ihi (*Marsilea villosa*)

E — Alani (*Melicope adscendens*)
E — Alani (*Melicope balloui*)
E — Alani (*Melicope haupuensis*)
E — Alani (*Melicope knudsenii*)
E — Alani (*Melicope lydgatei*)
E — Alani (*Melicope mucronulata*)
E — Alani (*Melicope ovalis*)
E — Alani (*Melicope pallida*)
E — Alani (*Melicope quadrangularis*)
E — Alani (*Melicope reflexa*)
E — Alani (*Melicope saint-johnii*)
E — Alani (*Melicope zahlbruckneri*)
E — *Munroidendron racemosum* (Plant, no common name)
E — Kolea (*Myrsine juddii*)
T — Kolea (*Myrsine linearifolia*)
E — *Neraudia angulata* (Plant, no common name)
E — *Neraudia ovata* (Plant, no common name)
E — *Neraudia sericea* (Plant, no common name)
E — `Aiea (*Nothocestrum breviflorum*)
E — `Aiea (*Nothocestrum peltatum*)
E — Kulu`i (*Nototrichium humile*)
E — Holei (*Ochrosia kilaueaensis*)
E — Carter's panicgrass (*Panicum fauriei* var. *carteri*)
E — Lau `ehu (*Panicum niihauense*)
T — Makou (*Peucedanum sandwicense*)
E — Wawae`iole (*Phlegmariurus (= Lycopodium, = Huperzia) mannii*))
E — Wawae`iole (*Phlegmariurus (= Lycopodium) nutans*)
E — *Phyllostegia glabra* var. *lanaiensis* (Plant, no common name)
E — *Phyllostegia mannii* (Plant, no common name)
E — *Phyllostegia mollis* (Plant, no common name)
E — *Phyllostegia waimeae* (Plant, no common name)
E — Kiponapona (*Phyllostegia racemosa*)
E — *Phyllostegia velutina* (Plant, no common name)
E — *Phyllostegia warshaueri* (Plant, no common name)
E — *Phyllostegia hirsuta* (Plant, no common name)
E — *Phyllostegia knudsenii*, (Plant, no common name)
E — *Phyllostegia parviflora* (Plant, no common name)
E — *Phyllostegia wawrana*, (Plant, no common name)
E — *Phyllostegia kaalaensis* (Plant, no common name)
E — Laukahi kuahiwi (*Plantago hawaiensis*)
E — Laukahi kuahiwi (*Plantago princeps*)
E — *Platanthera holochila* (Plant, no common name)
E — Hala pepe (*Pleomele hawaiiensis*)
E — Mann's bluegrass (*Poa mannii*)
E — Hawaiian bluegrass (*Poa sandvicensis*)
E — *Poa siphonoglossa* (Plant, no common name)
E — Po`e (*Portulaca sclerocarpa*)
E — Loulu (*Pritchardia affinis*)
E — Wahane (= Hawane or lo`ulu) (*Pritchardia aylmer-robinsonii*)
E — Loulu (*Pritchardia kaalae*)
E — Loulu (*Pritchardia munroi*)
E — Loulu (*Pritchardia napaliensis*)
E — Lo`ulu (*Pritchardia remota*)
E — Loulu (*Pritchardia schattaueri*)
E — Loulu (*Pritchardia viscosa*)

E — Kaulu (*Pteralyxia kauaiensis*)
E — *Pteris lidgatei* (Fern, no common name)
E — *Remya kauaiensis* (Plant, no common name)
E — Maui remya (*Remya mauiensis*)
E — *Remya montgomeryi* (Plant, no common name)
E — *Sanicula mariversa* (Plant, no common name)
E — *Sanicula purpurea* (Plant, no common name)
E — Lanai sandalwood or `iliahi (*Santalum freycinetianum* var. *lanaiense*)
E — Dwarf naupaka (*Scaevola coriacea*)
E — Diamond Head schiedea (*Schiedea adamantis*)
E — Ma`oli`oli (*Schiedea apokremnos*)
E — *Schiedea haleakalensis* (Plant, no common name)
E — *Schiedea helleri*, (Plant, no common name)
E — *Schiedea hookeri* (Plant, no common name)
E — *Schiedea kaalae* (Plant, no common name)
E — *Schiedea kauaiensis* (Plant, no common name)
E — Ma`oli`oli (*Schiedea kealiae*)
E — *Schiedea lydgatei* (Plant, no common name)
E — *Schiedea membranacea*, (Plant, no common name)
E — *Schiedea nuttallii* (Plant, no common name)
E — *Schiedea sarmentosa* (Plant, no common name)
E — *Schiedea spergulina* var. *leiopoda* (Plant, no common name)
T — *Schiedea spergulina* var. *spergulina* (Plant, no common name)
E — Laulihilihi (*Schiedea stellarioides*)
E — *Schiedea verticillata* (Plant, no common name)
E — `Ohai (*Sesbania tomentosa*)
E — `Anunu (*Sicyos alba*)
E — *Silene alexandri* (Plant, no common name)
T — *Silene hawaiiensis* (Plant, no common name)
E — *Silene lanceolata* (Plant, no common name)
E — *Silene perlmanii* (Plant, no common name)
E — Popolo ku mai (*Solanum incompletum*)
E — `Aiakeakua, popolo (*Solanum sandwicense*)
E — *Spermolepis hawaiiensis* (Plant, no common name)
E — *Stenogyne angustifolia var. angustifolia* (Plant, no common name)
E — *Stenogyne bifida* (Plant, no common name)
E — *Stenogyne campanulata* (Plant, no common name)
E — *Stenogyne kanehoana* (Plant, no common name)
E — *Tetramolopium arenarium* (Plant, no common name)
E — Pamakani (*Tetramolopium capillare*)
E — *Tetramolopium filiforme* (Plant, no common name)
E — *Tetramolopium lepidotum* ssp. *lepidotum* (Plant, no common name)
E — *Tetramolopium remyi* (Plant, no common name)
T — *Tetramolopium rockii* (Plant, no common name)
E — `Ohe`ohe (*Tetraplasandra gymnocarpa*)
E — *Trematolobelia singularis* (Plant, no common name)
E — Opuhe (*Urera kaalae*)
E — Hawaiian vetch (*Vicia menziesii*)
E — *Vigna o-wahuensis* (Plant, no common name)
E — Pamakani (*Viola chamissoniana* ssp. *chamissoniana*)
E — *Viola helenae* (Plant, no common name)
E — Nani wai`ale`ale, (*Viola kauaiensis var. wahiawaensis*)
E — *Viola lanaiensis* (Plant, no common name)
E — *Viola oahuensis* (Plant, no common name)
E — Dwarf iliau (*Wilkesia hobdyi*)

E — *Xylosma crenatum* (Plant, no common name)
E — A`e (*Zanthoxylum dipetalum var. tomentosum*)
E — A`e (*Zanthoxylum hawaiiense*)

Idaho—17 species
Animals—14 species
T — Bear, grizzly (*Ursus arctos*)
E — Caribou, woodland (*Rangifer tarandus caribou*)
E — Crane, whooping (*Grus americana*)
T — Eagle, bald (*Haliaeetus leucocephalus*)
E — Falcon, American peregrine (*Falco peregrinus anatum*)
E — Limpet, Banbury Springs (*Lanx* sp.)
T — Snail, Bliss Rapids (*Taylorconcha serpenticola*)
E — Snail, Snake River physa (*Physa natricina*)
E — Snail, Utah valvata (*Valvata utahensis*)
E — Springsnail, Bruneau Hot (*Pyrgulopsis bruneauensis*)
E — Springsnail, Idaho (*Fontelicella idahoensis*)
E — Sturgeon, white (Kootenai River pop.) (*Acipenser transmontanus*)
T&E — Trout, bull (*Salvelinus confluentus*)
E — Wolf, gray (*Canis lupus*)
Plants—3 species
T — Water howellia (*Howellia aquatilis*)
T — MacFarlane's four-o'clock (*Mirabilis macfarlanei*)
T — Ute ladies'-tresses (*Spiranthes diluvialis*) Illinois—25 species
Animals—16 species
E — Amphipod, Illinois cave (*Gammarus acherondytes*)
E — Bat, gray (*Myotis grisescens*)
E — Bat, Indiana (*Myotis sodalis*)
E — Butterfly, Karner blue (*Lycaeides melissa samuelis*)
E — Dragonfly, Hine's emerald (*Somatochlora hineana*)
T — Eagle, bald (*Haliaeetus leucocephalus*)
E — Falcon, American peregrine (*Falco peregrinus anatum*)
E — Fanshell (*Cyprogenia stegaria*)
E — Higgins eye (pearlymussel) (*Lampsilis higginsii*)
E — Mucket, pink (pearlymussel) (*Lampsilis abrupta*)
E — Pimpleback, orangefoot (pearlymussel) (*Plethobasus cooperianus*)
E&T — Plover, piping (*Charadrius melodus*)
E — Pocketbook, fat (*Potamilus (= Proptera) capax*)
E — Snail, Iowa Pleistocene (*Discus macclintocki*)
E — Sturgeon, pallid (*Scaphirhynchus albus*)
E — Tern, least (*Sterna antillarum*)
Plants—9 species
T — Price's potato-bean (*Apios priceana*)
T — Mead's milkweed (*Asclepias meadii*)
T — Decurrent false aster (*Boltonia decurrens*)
T — Pitcher's thistle (*Cirsium pitcheri*)
E — Leafy prairie-clover (*Dalea (= Petalostemum) foliosa*)
T — Lakeside daisy (*Hymenoxys herbacea*)
T — Small whorled pogonia (*Isotria medeoloides*)
T — Prairie bush-clover (*Lespedeza leptostachya*)
T — Eastern prairie fringed orchid (*Platanthera leucophaea*)

Indiana—24 species
Animals—21 species
E — Bat, gray (*Myotis grisescens*)

E — Bat, Indiana (*Myotis sodalis*)
E — Blossom, tubercled (pearlymussel) (*Epioblasma torulosa torulosa*)
E — Butterfly, Karner blue (*Lycaeides melissa samuelis*)
E — Butterfly, Mitchell's satyr (*Neonympha mitchellii mitchellii*)
E — Catspaw, white (pearlymussel) (*Epioblasma obliquata perobliqua (= sulcata delicata)*)
E — Clubshell (*Pleurobema clava*)
T — Eagle, bald (*Haliaeetus leucocephalus*)
E — Falcon, American peregrine (*Falco peregrinus anatum*)
E — Fanshell (*Cyprogenia stegaria*)
E — Mucket, pink (pearlymussel) (*Lampsilis abrupta*)
E — Pearlymussel, cracking (*Hemistena lata*)
E — Wartyback, white (pearlymussel) (*Plethobasus cicatricosus*)
E — Pigtoe, rough (*Pleurobema plenum*)
E — Pimpleback, orangefoot (pearlymussel) (*Plethobasus cooperianus*)
E — Pink, ring (mussel) (*Obovaria retusa*)
E&T — Plover, piping (*Charadrius melodus*)
E — Pocketbook, fat (*Potamilus (= Proptera) capax*)
E — Riffleshell, northern (*Epioblasma torulosa rangiana*)
T — Snake, copperbelly water (northern population) (*Nerodia erythrogaster neglecta*)
E — Tern, least (*Sterna antillarum*)

Plants—3 species

T — Mead's milkweed (*Asclepias meadii*)
T — Pitcher's thistle (*Cirsium pitcheri*)
E — Running buffalo clover (*Trifolium stoloniferum*)

Iowa—14 species

Animals—9 species

E — Bat, Indiana (*Myotis sodalis*)
T — Eagle, bald (*Haliaeetus leucocephalus*)
E — Falcon, American peregrine (*Falco peregrinus anatum*)
E — Higgins eye (pearlymussel) (*Lampsilis higginsii*)
T — Plover, piping (*Charadrius melodus*)
E — Shiner, Topeka (*Notropis topeka (= tristis)*)
E — Snail, Iowa Pleistocene (*Discus macclintocki*)
E — Sturgeon, pallid (*Scaphirhynchus albus*)
E — Tern, least (*Sterna antillarum*)

Plants—5 species

T — Northern wild monkshood (*Aconitum noveboracense*)
T — Mead's milkweed (*Asclepias meadii*)
T — Prairie bush-clover (*Lespedeza leptostachya*)
T — Eastern prairie fringed orchid (*Platanthera leucophaea*)
T — Western prairie fringed orchid (*Platanthera praeclara*)

Kansas—16 species

Animals—14 species

E — Bat, gray (*Myotis grisescens*)
E — Bat, Indiana (*Myotis sodalis*)
E — Crane, whooping (*Grus americana*)
E — Curlew, Eskimo (*Numenius borealis*)
T — Eagle, bald (*Haliaeetus leucocephalus*)
E — Falcon, American peregrine (*Falco peregrinus anatum*)
E — Ferret, black-footed (*Mustela nigripes*)
T — Madtom, Neosho (*Noturus placidus*)
T — Plover, piping (*Charadrius melodus*)

T — Shiner, Arkansas River (native pop. only) (*Notropis girardi*)
E — Shiner, Topeka (*Notropis topeka (= tristis)*)
E — Sturgeon, pallid (*Scaphirhynchus albus*)
E — Tern, least (*Sterna antillarum*)
E — Vireo, black-capped (*Vireo atricapillus*)
Plants—2 species
T — Mead's milkweed (*Asclepias meadii*)
T — Western prairie fringed orchid (*Platanthera praeclara*)

Kentucky—42 species
Animals—33 species
E — Bat, gray (*Myotis grisescens*)
E — Bat, Indiana (*Myotis sodalis*)
E — Bat, Virginia big-eared (*Corynorhinus (= Plecotus) townsendii virginianus*)
E — Bean, Cumberland (pearlymussel)(*Villosa trabalis*)
E — Blossom, tubercled (pearlymussel) (*Epioblasma torulosa torulosa*)
E — Catspaw (= purple cat's paw pearlymussel) (*Epioblasma obliquata obliquata*)
E — Clubshell (*Pleurobema clava*)
E — Combshell, Cumberlandian (*Epioblasma brevidens*)
T — Dace, blackside (*Phoxinus cumberlandensis*)
E — Darter, relict (*Etheostoma (Catonotus) chienense*)
T — Eagle, bald (*Haliaeetus leucocephalus*)
E — Elktoe, Cumberland (*Alasmidonta atropurpurea*)
E — Falcon, American peregrine (*Falco peregrinus anatum*)
E — Fanshell (*Cyprogenia stegaria*)
E — Mapleleaf, winged (mussel) (*Quadrula fragosa*)
E — Mucket, pink (pearlymussel) (*Lampsilis abrupta*)
E — Mussel, oyster (*Epioblasma capsaeformis*)
E — Pearlymussel, cracking (*Hemistena lata*)
E — Pearlymussel, dromedary (*Dromus dromas*)
E — Pearlymussel, littlewing (*Pegias fabula*)
E — Wartyback, white (pearlymussel) (*Plethobasus cicatricosus*)
E — Pigtoe, rough (*Pleurobema plenum*)
E — Pimpleback, orangefoot (pearlymussel) (*Plethobasus cooperianus*)
E — Pink, ring (mussel) (*Obovaria retusa*)
T — Plover, piping (*Charadrius melodus*)
E — Pocketbook, fat (*Potamilus (= Proptera) capax*)
E — Riffleshell, northern (*Epioblasma torulosa rangiana*)
E — Riffleshell, tan (*Epioblasma florentina walkeri*)
E — Shiner, Palezone (*Notropis albizonatus*)
E — Shrimp, Kentucky cave (*Palaemonias ganteri*)
E — Sturgeon, pallid (*Scaphirhynchus albus*)
E — Tern, least (*Sterna antillarum*)
E — Woodpecker, red-cockaded (*Picoides borealis*)
Plants—9 species
T — Price's potato-bean (*Apios priceana*)
E — Braun's rock cress (*Arabis perstellata*)
E — Cumberland sandwort (*Arenaria cumberlandensis*)
T — Cumberland rosemary (*Conradina verticillata*)
T — Eggert's sunflower (*Helianthus eggertii*)
T — White-haired goldenrod (*Solidago albopilosa*)
E — Short's goldenrod (*Solidago shortii*)
T — Virginia spiraea (*Spiraea virginiana*)
E — Running buffalo clover (*Trifolium stoloniferum*)

Louisiana—21 species
Animals—17 species
T — Bear, Louisiana black (*Ursus americanus luteolus*)
T — Eagle, bald (*Haliaeetus leucocephalus*)
E — Falcon, American peregrine (*Falco peregrinus anatum*)
T — Heelsplitter, Alabama (= inflated) (*Potamilus inflatus*)
E — Manatee, West Indian (*Trichechus manatus*)
E — Mucket, pink (pearlymussel) (*Lampsilis abrupta*)
T — Pearlshell, Louisiana (*Margaritifera hembeli*)
E — Pelican, brown (*Pelecanus occidentalis*)
T — Plover, piping (*Charadrius melodus*)
T — Sea turtle, loggerhead (*Caretta caretta*)
T — Sturgeon, Gulf (*Acipenser oxyrinchus desotoi*)
E — Sturgeon, pallid (*Scaphirhynchus albus*)
E — Tern, least (*Sterna antillarum*)
T — Tortoise, gopher (*Gopherus polyphemus*)
T — Turtle, ringed map (= sawback) (*Graptemys oculifera*)
E — Vireo, black-capped (*Vireo atricapillus*)
E — Woodpecker, red-cockaded (*Picoides borealis*)
Plants—4 species
T — *Geocarpon minimum* (Plant, no common name)
E — Louisiana quillwort (*Isoetes louisianensis*)
E — Pondberry (*Lindera melissifolia*)
E — American chaffseed (*Schwalbea americana*)

Maine—7 species
Animals—4 species
T — Eagle, bald (*Haliaeetus leucocephalus*)
E — Falcon, American peregrine (*Falco peregrinus anatum*)
T — Plover, piping (*Charadrius melodus*)
E — Tern, roseate (*Sterna dougallii dougallii*)

Plants—3 species
T — Small whorled pogonia (*Isotria medeoloides*)
E — Furbish lousewort (*Pedicularis furbishiae*)
T — Eastern prairie fringed orchid (*Platanthera leucophaea*)

Maryland—16 species
Animals—10 species
E — Bat, Indiana (*Myotis sodalis*)
T — Beetle, northeastern beach tiger (*Cicindela dorsalis dorsalis*)
T — Beetle, Puritan tiger (*Cicindela puritana*)
E — Darter, Maryland (*Etheostoma sellare*)
T — Eagle, bald (*Haliaeetus leucocephalus*)
E — Falcon, American peregrine (*Falco peregrinus anatum*)
T — Plover, piping (*Charadrius melodus*)
E — Squirrel, Delmarva Peninsula fox (*Sciurus niger cinereus*)
T — Turtle, bog (= Muhlenberg) (*Clemmys muhlenbergii*)
E — Wedgemussel, dwarf (*Alasmidonta heterodon*)
Plants—6 species
T — Sensitive joint-vetch (*Aeschynomene virginica*)
E — Sandplain gerardia (*Agalinis acuta*)
T — Swamp pink (*Helonias bullata*)
E — Canby's dropwort (*Oxypolis canbyi*)

E — Harperella (*Ptilimnium nodosum (= fluviatile)*)
E — Northeastern (= Barbed bristle) bulrush (*Scirpus ancistrochaetus*)

Massachusetts—13 species
Animals—10 species
E — Beetle, American burying (= giant carrion) (*Nicrophorus americanus*)
T — Beetle, northeastern beach tiger (*Cicindela dorsalis dorsalis*)
T — Beetle, Puritan tiger (*Cicindela puritana*)
T — Eagle, bald (*Haliaeetus leucocephalus*)
E — Falcon, American peregrine (*Falco peregrinus anatum*)
T — Plover, piping (*Charadrius melodus*)
E — Tern, roseate (*Sterna dougallii dougallii*)
T — Turtle, bog (= Muhlenberg) (*Clemmys muhlenbergii*)
E — Turtle, Plymouth redbelly (*Pseudemys rubriventris bangsi*)
E — Wedgemussel, dwarf (*Alasmidonta heterodon*)
Plants—3 species
E — Sandplain gerardia (*Agalinis acuta*)
T — Small whorled pogonia (*Isotria medeoloides*)
E — Northeastern (= Barbed bristle) bulrush (*Scirpus ancistrochaetus*)

Michigan—21 species
Animals—13 species
E — Bat, Indiana (*Myotis sodalis*)
E — Beetle, American burying (= giant carrion) (*Nicrophorus americanus*)
E — Beetle, Hungerford's crawling water (*Brychius hungerfordi*)
E — Butterfly, Karner blue (*Lycaeides melissa samuelis*)
E — Butterfly, Mitchell's satyr (*Neonympha mitchellii mitchellii*)
E — Clubshell (*Pleurobema clava*)
T — Eagle, bald (*Haliaeetus leucocephalus*)
E — Falcon, American peregrine (*Falco peregrinus anatum*)
E — Plover, piping (*Charadrius melodus*)
E — Riffleshell, northern (*Epioblasma torulosa rangiana*)
T — Snake, copperbelly water (northern population) (*Nerodia erythrogaster neglecta*)
E — Warbler, Kirtland's (*Dendroica kirtlandii*)
E — Wolf, gray (*Canis lupus*)
Plants—8 species
T — American hart's-tongue fern (*Asplenium scolopendrium* var. *americanum*)
T — Pitcher's thistle (*Cirsium pitcheri*)
T — Lakeside daisy (*Hymenoxys herbacea*)
T — Dwarf lake iris (*Iris lacustris*)
T — Small whorled pogonia (*Isotria medeoloides*)
E — Michigan monkey-flower (*Mimulus glabratus* var. *michiganensis*)
T — Eastern prairie fringed orchid (*Platanthera leucophaea*)
T — Houghton's goldenrod (*Solidago houghtonii*)

Minnesota—12 species
Animals—8 species
E — Butterfly, Karner blue (*Lycaeides melissa samuelis*)
T — Eagle, bald (*Haliaeetus leucocephalus*)
E — Falcon, American peregrine (*Falco peregrinus anatum*)
E — Higgins eye (pearlymussel) (*Lampsilis higginsii*)
E — Mapleleaf, winged (mussel) (*Quadrula fragosa*)
E&T — Plover, piping (*Charadrius melodus*)
E — Shiner, Topeka (*Notropis topeka (= tristis)*)
T — Wolf, gray (*Canis lupus*)

Plants—4 species
E — Minnesota dwarf trout lily (*Erythronium propullans*)
T — Prairie bush-clover (*Lespedeza leptostachya*)
T — Western prairie fringed orchid (*Platanthera praeclara*)
T — Leedy's roseroot (*Sedum integrifolium* ssp. *leedyi*)

Mississippi—33 species
Animals—29 species
E — Bat, Indiana (*Myotis sodalis*)
T — Bear, Louisiana black (*Ursus americanus luteolus*)
E — Clubshell, black (= Curtus' mussel) (*Pleurobema curtum*)
E — Clubshell, ovate (*Pleurobema perovatum*)
E — Clubshell, southern (*Pleurobema decisum*)
E — Combshell, southern (= penitent mussel) (*Epioblasma penita*)
E — Crane, Mississippi sandhill (*Grus canadensis pulla*)
T — Darter, bayou (*Etheostoma rubrum*)
T — Eagle, bald (*Haliaeetus leucocephalus*)
E — Falcon, American peregrine (*Falco peregrinus anatum*)
T — Heelsplitter, Alabama (= inflated) (*Potamilus inflatus*)
E — Manatee, West Indian (*Trichechus manatus*)
T — Moccasinshell, Alabama (*Medionidus acutissimus*)
T — Mucket, orangenacre (*Lampsilis perovalis*)
E — Pelican, brown (*Pelecanus occidentalis*)
E — Pigtoe, flat (= Marshall's mussel) (*Pleurobema marshalli*)
E — Pigtoe, heavy (= Judge Tait's mussel) (*Pleurobema taitianum*)
T — Plover, piping (*Charadrius melodus*)
E — Pocketbook, fat (*Potamilus (= Proptera) capax*)
T — Sea turtle, loggerhead (*Caretta caretta*)
T — Snake, eastern indigo (*Drymarchon corais couperi*)
E — Stirrupshell (*Quadrula stapes*)
T — Sturgeon, Gulf (*Acipenser oxyrinchus desotoi*)
E — Sturgeon, pallid (*Scaphirhynchus albus*)
E — Tern, least (*Sterna antillarum*)
T — Tortoise, gopher (*Gopherus polyphemus*)
T — Turtle, ringed map (= sawback) (*Graptemys oculifera*)
T — Turtle, yellow-blotched map (= sawback) (*Graptemys flavimaculata*)
E — Woodpecker, red-cockaded (*Picoides borealis*)
Plants—4 species
T — Price's potato-bean (*Apios priceana*)
E — Louisiana quillwort (*Isoetes louisianensis*)
E — Pondberry (*Lindera melissifolia*)
E — American chaffseed (*Schwalbea americana*)

Missouri—23 species
Animals—16 species
E — Bat, gray (*Myotis grisescens*)
E — Bat, Indiana (*Myotis sodalis*)
E — Bat, Ozark big-eared (*Corynorhinus (= Plecotus) townsendii ingens*)
T — Cavefish, Ozark (*Amblyopsis rosae*)
T — Darter, Niangua (*Etheostoma nianguae*)
T — Eagle, bald (*Haliaeetus leucocephalus*)
E — Falcon, American peregrine (*Falco peregrinus anatum*)
E — Higgins eye (pearlymussel) (*Lampsilis higginsii*)
T — Madtom, Neosho (*Noturus placidus*)
E — Mucket, pink (pearlymussel) (*Lampsilis abrupta*)

E — Pearlymussel, Curtis (*Epioblasma (= Dysnomia) florentina curtisii*)
T — Plover, piping (*Charadrius melodus*)
E — Pocketbook, fat (*Potamilus (= Proptera) capax*)
E — Shiner, Topeka (*Notropis topeka (= tristis)*)
E — Sturgeon, pallid (*Scaphirhynchus albus*)
E — Tern, least (*Sterna antillarum*)
Plants—7 species
T — Mead's milkweed (*Asclepias meadii*)
T — Decurrent false aster (*Boltonia decurrens*)
T — *Geocarpon minimum* (Plant, no common name)
E — Missouri bladderpod (*Lesquerella filiformis*)
E — Pondberry (*Lindera melissifolia*)
T — Western prairie fringed orchid (*Platanthera praeclara*)
E — Running buffalo clover (*Trifolium stoloniferum*)

Montana—14 species
Animals—12 species
T — Bear, grizzly (*Ursus arctos*)
E — Crane, whooping (*Grus americana*)
E — Curlew, Eskimo (*Numenius borealis*)
T — Eagle, bald (*Haliaeetus leucocephalus*)
E — Falcon, American peregrine (*Falco peregrinus anatum*)
E — Ferret, black-footed (*Mustela nigripes*)
T — Plover, piping (*Charadrius melodus*)
E — Sturgeon, pallid (*Scaphirhynchus albus*)
E — Sturgeon, white (Kootenai River pop.) (*Acipenser transmontanus*)
E — Tern, least (*Sterna antillarum*)
T — Trout, bull (*Salvelinus confluentus*)
E — Wolf, gray (*Canis lupus*)
Plants—2 species
T — Water howellia (*Howellia aquatilis*)
T — Ute ladies'-tresses (*Spiranthes diluvialis*)

Nebraska—13 species
Animals—10 species
E — Beetle, American burying (= giant carrion) (*Nicrophorus americanus*)
E — Crane, whooping (*Grus americana*)
E — Curlew, Eskimo (*Numenius borealis*)
T — Eagle, bald (*Haliaeetus leucocephalus*)
E — Falcon, American peregrine (*Falco peregrinus anatum*)
E — Ferret, black-footed (*Mustela nigripes*)
T — Plover, piping (*Charadrius melodus*)
E — Shiner, Topeka (*Notropis topeka (= tristis)*)
E — Sturgeon, pallid (*Scaphirhynchus albus*)
E — Tern, least (*Sterna antillarum*)
Plants—3 species
E — Blowout penstemon (*Penstemon haydenii*)
T — Western prairie fringed orchid (*Platanthera praeclara*)
T — Ute ladies'-tresses (*Spiranthes diluvialis*)

Nevada—34 species
Animals—26 species
E — Chub, bonytail (*Gila elegans*)
E — Chub, Pahranagat roundtail (= bonytail) (*Gila robusta jordani*)
E — Chub, Virgin River (*Gila robusta semidnuda*)

E — Cui-ui (*Chasmistes cujus*)
E — Dace, Ash Meadows speckled (*Rhinichthys osculus nevadensis*)
E — Dace, Clover Valley speckled (*Rhinichthys osculus oligoporus*)
T — Dace, desert (*Eremichthys acros*)
E — Dace, Independence Valley speckled (*Rhinichthys osculus lethoporus*)
E — Dace, Moapa (*Moapa coriacea*)
T — Eagle, bald (*Haliaeetus leucocephalus*)
E — Falcon, American peregrine (*Falco peregrinus anatum*)
T — Naucorid, Ash Meadows (*Ambrysus amargosus*)
E — Poolfish (= killifish), Pahrump (*Empetrichthys latos*)
E — Pupfish, Ash Meadows Amargosa (*Cyprinodon nevadensis mionectes*)
E — Pupfish, Devils Hole (*Cyprinodon diabolis*)
E — Pupfish, Warm Springs (*Cyprinodon nevadensis pectoralis*)
T — Spinedace, Big Spring (*Lepidomeda mollispinis pratensis*)
E — Spinedace, White River (*Lepidomeda albivallis*)
E — Springfish, Hiko White River (*Crenichthys baileyi grandis*)
T — Springfish, Railroad Valley (*Crenichthys nevadae*)
E — Springfish, White River (*Crenichthys baileyi baileyi*)
E — Sucker, razorback (*Xyrauchen texanus*)
T — Tortoise, desert (*Gopherus agassizii*)
E — Trout, bull (*Salvelinus confluentus*)
T — Trout, Lahontan cutthroat (*Oncorhynchus (= Salmo) clarki henshawi*)
E — Woundfin (*Plagopterus argentissimus*)

Plants—8 species

T — Ash Meadows milk-vetch (*Astragalus phoenix*)
T — Spring-loving centaury (*Centaurium namophilum*)
T — Ash Meadows sunray (*Enceliopsis nudicaulis* var. *corrugata*)
E — Steamboat buckwheat (*Eriogonum ovalifolium* var. *williamsiae*)
T — Ash Meadows gumplant (*Grindelia fraxino-pratensis*)
T — Ash Meadows ivesia (*Ivesia kingii* var. *eremica*)
T — Ash Meadows blazing-star (*Mentzelia leucophylla*)
E — Amargosa niterwort (*Nitrophila mohavensis*)

New Hampshire—10 species

Animals—6 species

T — Beetle, Puritan tiger (*Cicindela puritana*)
E — Butterfly, Karner blue (*Lycaeides melissa samuelis*)
T — Eagle, bald (*Haliaeetus leucocephalus*)
E — Falcon, American peregrine (*Falco peregrinus anatum*)
T — Plover, piping (*Charadrius melodus*)
E — Wedgemussel, dwarf (*Alasmidonta heterodon*)

Plants—4 species

E — Jesup's milk-vetch (*Astragalus robbinsii* var. *jesupi*)
T — Small whorled pogonia (*Isotria medeoloides*)
E — Robbins' cinquefoil (*Potentilla robbinsiana*)
E — Northeastern (= Barbed bristle) bulrush (*Scirpus ancistrochaetus*)

New Jersey—13 species

Animals—8 species

E — Bat, Indiana (*Myotis sodalis*)
T — Beetle, northeastern beach tiger (*Cicindela dorsalis dorsalis*)
T — Eagle, bald (*Haliaeetus leucocephalus*)
E — Falcon, American peregrine (*Falco peregrinus anatum*)
T — Plover, piping (*Charadrius melodus*)
E — Tern, roseate (*Sterna dougallii dougallii*)

T — Turtle, bog (= Muhlenberg) (*Clemmys muhlenbergii*)
E — Wedgemussel, dwarf (*Alasmidonta heterodon*)
Plants—5 species
T — Sensitive joint-vetch (*Aeschynomene virginica*)
T — Swamp pink (*Helonias bullata*)
T — Small whorled pogonia (*Isotria medeoloides*)
T — Knieskern's beaked-rush (*Rhynchospora knieskernii*)
E — American chaffseed (*Schwalbea americana*)

New Mexico—39 species
Animals—25 species
E — Bat, lesser (= Sanborn's) long-nosed (*Leptonycteris curasoae yerbabuenae*)
E — Bat, Mexican long-nosed (*Leptonycteris nivalis*)
T — Chub, Chihuahua (*Gila nigrescens*)
E — Crane, whooping (*Grus americana*)
T — Eagle, bald (*Haliaeetus leucocephalus*)
E — Falcon, American peregrine (*Falco peregrinus anatum*)
E — Flycatcher, Southwestern willow (*Empidonax traillii extimus*)
E — Gambusia, Pecos (*Gambusia nobilis*)
E — Isopod, Socorro (*Thermosphaeroma (= Exosphaeroma) thermophilus*)
E — Jaguar (*Panthera onca*)
T — Minnow, loach (*Rhinichthys (= Tiaroga) cobitis*)
E — Minnow, Rio Grande silvery (*Hybognathus amarus*)
T — Owl, Mexican spotted (*Strix occidentalis lucida*)
T — Rattlesnake, New Mexican ridge-nosed (*Crotalus willardi obscurus*)
T — Shiner, Arkansas River (native pop. only) (*Notropis girardi*)
T — Shiner, beautiful (*Cyprinella (= Notropis) formosa*)
T — Shiner, Pecos bluntnose (*Notropis simus pecosensis*)
T — Spikedace (*Meda fulgida*)
E — Springsnail, Alamosa (*Tryonia alamosae*)
E — Springsnail, Socorro (*Pyrgulopsis neomexicana*)
E — Sucker, razorback (*Xyrauchen texanus*)
E — Tern, least (*Sterna antillarum*)
E — Topminnow, Gila (incl. Yaqui) (*Poeciliopsis occidentalis*)
E — Trout, Gila (*Oncorhynchus (= Salmo) gilae*)
E — Woundfin (*Plagopterus argentissimus*)
Plants—14 species
E — Sacramento prickly-poppy (*Argemone pleiacantha* ssp. *pinnatisecta*)
E — Mancos milk-vetch (*Astragalus humillimus*)
T — Sacramento Mountains thistle (*Cirsium vinaceum*)
T — Lee pincushion cactus (*Coryphantha sneedii* var. *leei*)
E — Sneed pincushion cactus (*Coryphantha sneedii* var. *sneedii*)
E — Kuenzler hedgehog cactus (*Echinocereus fendleri* var. *kuenzleri*)
E — Lloyd's hedgehog cactus (*Echinocereus lloydii*)
T — Lloyd's Mariposa cactus (*Echinomastus (= Sclerocactus) mariposensis*)
T — Zuni (= rhizome) fleabane (*Erigeron rhizomatus*)
T — Gypsum wild-buckwheat (*Eriogonum gypsophilum*)
E — Todsen's pennyroyal (*Hedeoma todsenii*)
E — Holy Ghost ipomopsis (*Ipomopsis sancti-spiritus*)
E — Knowlton cactus (*Pediocactus knowltonii*)
T — Mesa Verde cactus (*Sclerocactus mesae-verdae*)

New York—15 species
Animals—9 species
E — Bat, Indiana (*Myotis sodalis*)

E — Butterfly, Karner blue (*Lycaeides melissa samuelis*)
T — Eagle, bald (*Haliaeetus leucocephalus*)
E — Falcon, American peregrine (*Falco peregrinus anatum*)
E&T — Plover, piping (*Charadrius melodus*)
T — Snail, Chittenango ovate amber (*Succinea chittenangoensis*)
E — Tern, roseate (*Sterna dougallii dougallii*)
T — Turtle, bog (= Muhlenberg) (*Clemmys muhlenbergii*)
E — Wedgemussel, dwarf (*Alasmidonta heterodon*)

Plants—6 species

T — Northern wild monkshood (*Aconitum noveboracense*)
E — Sandplain gerardia (*Agalinis acuta*)
T — Seabeach amaranth (*Amaranthus pumilus*)
T — American hart's-tongue fern (*Asplenium scolopendrium* var. *americanum*)
T — Leedy's roseroot (*Sedum integrifolium* ssp. *leedyi*)
T — Houghton's goldenrod (*Solidago houghtonii*)

North Carolina—49 species

Animals—23 species

E — Bat, Indiana (*Myotis sodalis*)
E — Bat, Virginia big-eared (*Corynorhinus (= Plecotus) townsendii virginianus*)
E — Butterfly, Saint Francis' satyr (*Neonympha mitchellii francisci*)
T — Chub, spotfin (= turquoise shiner) (*Cyprinella (= Hybopsis) monacha*)
T — Eagle, bald (*Haliaeetus leucocephalus*)
E — Elktoe, Appalachian (*Alasmidonta raveneliana*)
E — Falcon, American peregrine (*Falco peregrinus anatum*)
E — Heelsplitter, Carolina (*Lasmigona decorata*)
E — Manatee, West Indian (*Trichechus manatus*)
E — Pearlymussel, littlewing (*Pegias fabula*)
T — Plover, piping (*Charadrius melodus*)
T — Sea turtle, loggerhead (*Caretta caretta*)
E — Shiner, Cape Fear (*Notropis mekistocholas*)
T — Shrew, Dismal Swamp southeastern (*Sorex longirostris fisheri*)
T — Silverside, Waccamaw (*Menidia extensa*)
T — Snail, noonday (*Mesodon clarki nantahala*)
E — Spider, spruce-fir moss (*Microhexura montivaga*)
E — Spinymussel, Tar River (*Elliptio steinstansana*)
E — Squirrel, Carolina northern flying (*Glaucomys sabrinus coloratus*)
T — Tern, roseate (*Sterna dougallii dougallii*)
E — Wedgemussel, dwarf (*Alasmidonta heterodon*)
E — Wolf, red (*Canis rufus*)
E — Woodpecker, red-cockaded (*Picoides borealis*)

Plants—26 species

T — Sensitive joint-vetch (*Aeschynomene virginica*)
T — Seabeach amaranth (*Amaranthus pumilus*)
E — Small-anthered bittercress (*Cardamine micranthera*)
E — Smooth coneflower (*Echinacea laevigata*)
E — Spreading avens (*Geum radiatum*)
E — Rock gnome lichen (*Gymnoderma lineare*)
E — Roan Mountain bluet (*Hedyotis purpurea* var. *montana*)
E — Schweinitz's sunflower (*Helianthus schweinitzii*)
T — Swamp pink (*Helonias bullata*)
T — Dwarf-flowered heartleaf (*Hexastylis naniflora*)
T — Mountain golden heather (*Hudsonia montana*)
T — Small whorled pogonia (*Isotria medeoloides*)
T — Heller's blazingstar (*Liatris helleri*)

E — Pondberry (*Lindera melissifolia*)
E — Rough-leaved loosestrife (*Lysimachia asperulaefolia*)
E — Canby's dropwort (*Oxypolis canbyi*)
E — Harperella (*Ptilimnium nodosum (= fluviatile)*)
E — Michaux's sumac (*Rhus michauxii*)
E — Bunched arrowhead (*Sagittaria fasciculata*)
E — Green pitcher-plant (*Sarracenia oreophila*)
E — Mountain sweet pitcher-plant (*Sarracenia rubra* ssp. *jonesii*)
E — American chaffseed (*Schwalbea americana*)
E — White irisette (*Sisyrinchium dichotomum*)
T — Blue Ridge goldenrod (*Solidago spithamaea*)
T — Virginia spiraea (*Spiraea virginiana*)
E — Cooley's meadowrue (*Thalictrum cooleyi*)

North Dakota—10 species
Animals—9 species
E — Crane, whooping (*Grus americana*)
E — Curlew, Eskimo (*Numenius borealis*)
T — Eagle, bald (*Haliaeetus leucocephalus*)
E — Falcon, American peregrine (*Falco peregrinus anatum*)
E — Ferret, black-footed (*Mustela nigripes*)
T — Plover, piping (*Charadrius melodus*)
E — Sturgeon, pallid (*Scaphirhynchus albus*)
E — Tern, least (*Sterna antillarum*)
E — Wolf, gray (*Canis lupus*)
Plants—1 species
T — Western prairie fringed orchid (*Platanthera praeclara*)

Ohio—22 species
Animals—16 species
E — Bat, Indiana (*Myotis sodalis*)
E — Beetle, American burying (= giant carrion) (*Nicrophorus americanus*)
E — Butterfly, Karner blue (*Lycaeides melissa samuelis*)
E — Butterfly, Mitchell's satyr (*Neonympha mitchellii mitchellii*)
E — Catspaw (= purple cat's paw pearlymussel) (*Epioblasma obliquata obliquata*)
E — Catspaw, white (pearlymussel) (*Epioblasma obliquata perobliqua (= sulcata delicata)*)
E — Clubshell (*Pleurobema clava*)
E — Dragonfly, Hine's emerald (*Somatochlora hineana*)
T — Eagle, bald (*Haliaeetus leucocephalus*)
E — Falcon, American peregrine (*Falco peregrinus anatum*)
E — Fanshell (*Cyprogenia stegaria*)
E — Madtom, Scioto (*Noturus trautmani*)
E — Mucket, pink (pearlymussel) (*Lampsilis abrupta*)
E&T — Plover, piping (*Charadrius melodus*)
E — Riffleshell, northern (*Epioblasma torulosa rangiana*)
T — Snake, copperbelly water (northern population) (*Nerodia erythrogaster neglecta*)
Plants—6 species
T — Northern wild monkshood (*Aconitum noveboracense*)
T — Lakeside daisy (*Hymenoxys herbacea*)
T — Small whorled pogonia (*Isotria medeoloides*)
T — Eastern prairie fringed orchid (*Platanthera leucophaea*)
T — Virginia spiraea (*Spiraea virginiana*)
E — Running buffalo clover (*Trifolium stoloniferum*)

Oklahoma—18 species

Animals—17 species
E — Bat, gray (*Myotis grisescens*)
E — Bat, Indiana (*Myotis sodalis*)
E — Bat, Ozark big-eared (*Corynorhinus (= Plecotus) townsendii ingens*)
E — Beetle, American burying (= giant carrion) (*Nicrophorus americanus*)
T — Cavefish, Ozark (*Amblyopsis rosae*)
E — Crane, whooping (*Grus americana*)
E — Curlew, Eskimo (*Numenius borealis*)
T — Darter, leopard (*Percina pantherina*)
T — Eagle, bald (*Haliaeetus leucocephalus*)
E — Falcon, American peregrine (*Falco peregrinus anatum*)
T — Madtom, Neosho (*Noturus placidus*)
T — Plover, piping (*Charadrius melodus*)
E — Pocketbook, Ouachita rock (= Wheeler's pearly mussel) (*Arkansia wheeleri*)
T — Shiner, Arkansas River (native pop. only) (*Notropis girardi*)
E — Tern, least (*Sterna antillarum*)
E — Vireo, black-capped (*Vireo atricapillus*)
E — Woodpecker, red-cockaded (*Picoides borealis*)
Plants—1 species
T — Western prairie fringed orchid (*Platanthera praeclara*)

Oregon—27 species
Animals—18 species
T — Butterfly, Oregon silverspot (*Speyeria zerene hippolyta*)
E — Chub, Borax Lake (*Gila boraxobius*)
T — Chub, Hutton tui (*Gila bicolor* ssp.)
E — Chub, Oregon (*Oregonichthys (= Hybopsis) crameri*)
T — Dace, Foskett speckled (*Rhinichthys osculus* ssp.)
E — Deer, Columbian white-tailed (*Odocoileus virginianus leucurus*)
T — Eagle, bald (*Haliaeetus leucocephalus*)
E — Falcon, American peregrine (*Falco peregrinus anatum*)
T — Goose, Aleutian Canada (*Branta canadensis leucopareia*)
T — Murrelet, marbled (*Brachyramphus marmoratus marmoratus*)
T — Owl, northern spotted (*Strix occidentalis caurina*)
E — Pelican, brown (*Pelecanus occidentalis*)
T — Plover, western snowy (*Charadrius alexandrinus nivosus*)
E — Sucker, Lost River (*Deltistes luxatus*)
E — Sucker, shortnose (*Chasmistes brevirostris*)
T — Sucker, Warner (*Catostomus warnerensis*)
T — Trout, bull (*Salvelinus confluentus*)
T — Trout, Lahontan cutthroat (*Oncorhynchus (= Salmo) clarki henshawi*)
Plants—9 species
E — Marsh sandwort (*Arenaria paludicola*)
E — Applegate's milk-vetch (*Astragalus applegatei*)
T — Golden paintbrush (*Castilleja levisecta*)
T — Water howellia (*Howellia aquatilis*)
E — Western lily (*Lilium occidental*)
E — Bradshaw's desert-parsley (= lomatium) (*Lomatium bradshawii*)
T — MacFarlane's four-o'clock (*Mirabilis macfarlanei*)
T — Nelson's checker-mallow (*Sidalcea nelsoniana*)
E — Malheur wire-lettuce (*Stephanomeria malheurensis*)

Pennsylvania—16 species
Animals—13 species
E — Bat, Indiana (*Myotis sodalis*)

E — Clubshell (*Pleurobema clava*)
T — Eagle, bald (*Haliaeetus leucocephalus*)
E — Falcon, American peregrine (*Falco peregrinus anatum*)
E — Mucket, pink (pearlymussel) (*Lampsilis abrupta*)
E — Pearlymussel, cracking (*Hemistena lata*)
E — Pigtoe, rough (*Pleurobema plenum*)
E — Pimpleback, orangefoot (pearlymussel) (*Plethobasus cooperianus*)
E — Pink, ring (mussel) (*Obovaria retusa*)
E&T — Plover, piping (*Charadrius melodus*)
E — Riffleshell, northern (*Epioblasma torulosa rangiana*)
T — Turtle, bog (= Muhlenberg) (*Clemmys muhlenbergii*)
E — Wedgemussel, dwarf (*Alasmidonta heterodon*)
Plants—3 species
T — Small whorled pogonia (*Isotria medeoloides*)
E — Northeastern (= Barbed bristle) bulrush (*Scirpus ancistrochaetus*)
T — Virginia spiraea (*Spiraea virginiana*)

Rhode Island—8 species
Animals—6 species
E — Beetle, American burying (= giant carrion) (*Nicrophorus americanus*)
T — Beetle, northeastern beach tiger (*Cicindela dorsalis dorsalis*)
T — Eagle, bald (*Haliaeetus leucocephalus*)
E — Falcon, American peregrine (*Falco peregrinus anatum*)
T — Plover, piping (*Charadrius melodus*)
E — Tern, roseate (*Sterna dougallii dougallii*)
Plants—2 species
E — Sandplain gerardia (*Agalinis acuta*)
T — Small whorled pogonia (*Isotria medeoloides*)

South Carolina—30 species
Animals—11 species
E — Bat, Indiana (*Myotis sodalis*)
T — Eagle, bald (*Haliaeetus leucocephalus*)
E — Falcon, American peregrine (*Falco peregrinus anatum*)
E — Heelsplitter, Carolina (*Lasmigona decorata*)
E — Manatee, West Indian (*Trichechus manatus*)
T — Plover, piping (*Charadrius melodus*)
T — Sea turtle, loggerhead (*Caretta caretta*)
T — Snake, eastern indigo (*Drymarchon corais couperi*)
E — Stork, wood (*Mycteria americana*)
T — Tern, roseate (*Sterna dougallii dougallii*)
E — Woodpecker, red-cockaded (*Picoides borealis*)
Plants—19 species
T — Seabeach amaranth (*Amaranthus pumilus*)
T — Little amphianthus (*Amphianthus pusillus*)
E — Smooth coneflower (*Echinacea laevigata*)
E — Schweinitz's sunflower (*Helianthus schweinitzii*)
T — Swamp pink (*Helonias bullata*)
T — Dwarf-flowered heartleaf (*Hexastylis naniflora*)
E — Black-spored quillwort (*Isoetes melanospora*)
T — Small whorled pogonia (*Isotria medeoloides*)
E — Pondberry (*Lindera melissifolia*)
E — Rough-leaved loosestrife (*Lysimachia asperulaefolia*)
E — Canby's dropwort (*Oxypolis canbyi*)
E — Harperella (*Ptilimnium nodosum (= fluviatile)*)

E — Michaux's sumac (*Rhus michauxii*)
T — Miccosukee gooseberry (*Ribes echinellum*)
E — Bunched arrowhead (*Sagittaria fasciculata*)
E — Mountain sweet pitcher-plant (*Sarracenia rubra* ssp. *jonesii*)
E — American chaffseed (*Schwalbea americana*)
E — Persistent trillium (*Trillium persistens*)
E — Relict trillium (*Trillium reliquum*)

South Dakota—11 species
Animals—11 species
E — Beetle, American burying (= giant carrion) (*Nicrophorus americanus*)
E — Crane, whooping (*Grus americana*)
E — Curlew, Eskimo (*Numenius borealis*)
T — Eagle, bald (*Haliaeetus leucocephalus*)
E — Falcon, American peregrine (*Falco peregrinus anatum*)
E — Ferret, black-footed (*Mustela nigripes*)
T — Plover, piping (*Charadrius melodus*)
E — Shiner, Topeka (*Notropis topeka (= tristis)*)
E — Sturgeon, pallid (*Scaphirhynchus albus*)
E — Tern, least (*Sterna antillarum*)
E — Wolf, gray (*Canis lupus*)
Plants—0 species

Tennessee—88 species
Animals—68 species
E — Acornshell, southern (*Epioblasma othcaloogensis*)
E — Bat, gray (*Myotis grisescens*)
E — Bat, Indiana (*Myotis sodalis*)
E — Bean, Cumberland (pearlymussel)(*Villosa trabalis*)
E — Bean (mussel), purple (*Villosa perpurpurea*)
E — Blossom, green (pearlymussel)(*Epioblasma torulosa gubernaculum*)
E — Blossom, tubercled (pearlymussel) (*Epioblasma torulosa torulosa*)
E — Blossom, turgid (pearlymussel) (*Epioblasma turgidula*)
E — Blossom, yellow (pearlymussel)(*Epioblasma florentina florentina*)
E — Catspaw (= purple cat's paw pearlymussel) (*Epioblasma obliquata obliquata*)
T — Chub, slender (*Erimystax (= Hybopsis) cahni*)
T — Chub, spotfin (= turquoise shiner) (*Cyprinella (= Hybopsis) monacha*)
E — Clubshell, ovate (*Pleurobema perovatum*)
E — Clubshell, southern (*Pleurobema decisum*)
E — Combshell, Cumberlandian (*Epioblasma brevidens*)
E — Combshell, upland (*Epioblasma metastriata*)
E — Crayfish, Nashville (*Orconectes shoupi*)
T — Dace, blackside (*Phoxinus cumberlandensis*)
E — Darter, amber (*Percina antesella*)
E — Darter, bluemask (= jewel) (*Etheostoma (Doration)* sp.)
E — Darter, boulder (= Elk River) (*Etheostoma wapiti*)
E — Darter, duskytail (*Etheostoma percnurum*)
T — Darter, slackwater (*Etheostoma boschungi*)
T — Darter, snail (*Percina tanasi*)
T — Eagle, bald (*Haliaeetus leucocephalus*)
E — Elktoe, Appalachian (*Alasmidonta raveneliana*)
E — Elktoe, Cumberland (*Alasmidonta atropurpurea*)
E — Falcon, American peregrine (*Falco peregrinus anatum*)
E — Fanshell (*Cyprogenia stegaria*)
E — Kidneyshell, triangular (*Ptychobranchus greeni*)

E — Lampmussel, Alabama (*Lampsilis virescens*)
E — Lilliput, pale (pearlymussel) (*Toxolasma cylindrellus*)
E — Logperch, Conasauga (*Percina jenkinsi*)
E — Madtom, pygmy (*Noturus stanauli*)
E — Madtom, Smoky (*Noturus baileyi*)
T — Madtom, yellowfin (*Noturus flavipinnis*)
E — Mapleleaf, winged (mussel) (*Quadrula fragosa*)
E — Marstonia (snail), royal (= obese) (*Pyrgulopsis (= Marstonia) ogmorhaphe*)
T — Moccasinshell, Alabama (*Medionidus acutissimus*)
E — Moccasinshell, Coosa (*Medionidus parvulus*)
E — Monkeyface, Appalachian (pearlymussel)(*Quadrula sparsa*)
E — Monkeyface, Cumberland (pearlymussel)(*Quadrula intermedia*)
E — Mucket, pink (pearlymussel) (*Lampsilis abrupta*)
E — Mussel, oyster (*Epioblasma capsaeformis*)
E — Pearlymussel, birdwing (*Conradilla caelata*)
E — Pearlymussel, cracking (*Hemistena lata*)
E — Pearlymussel, dromedary (*Dromus dromas*)
E — Pearlymussel, littlewing (*Pegias fabula*)
E — Wartyback, white (pearlymussel) (*Plethobasus cicatricosus*)
E — Pigtoe, Cumberland (= Cumberland pigtoe mussel) (*Pleurobema gibberum*)
E — Pigtoe, finerayed (*Fusconaia cuneolus*)
E — Pigtoe, rough (*Pleurobema plenum*)
E — Pigtoe, shiny (*Fusconaia cor (= edgariana)*)
E — Pigtoe, southern (*Pleurobema georgianum*)
E — Pimpleback, orangefoot (pearlymussel) (*Plethobasus cooperianus*)
E — Pink, ring (mussel) (*Obovaria retusa*)
T — Pocketbook, finelined (*Lampsilis altilis*)
E — Rabbitsfoot, rough (*Quadrula cylindrica strigillata*)
E — Riffleshell, tan (*Epioblasma florentina walkeri*)
E — Riversnail, Anthony's (*Athearnia anthonyi*)
T — Shiner, blue (*Cyprinella (= Notropis) caerulea*)
T — Snail, painted snake coiled forest (*Anguispira picta*)
E — Spider, spruce-fir moss (*Microhexura montivaga*)
E — Squirrel, Carolina northern flying (*Glaucomys sabrinus coloratus*)
E — Sturgeon, pallid (*Scaphirhynchus albus*)
E — Tern, least (*Sterna antillarum*)
E — Wolf, red (*Canis rufus*)
E — Woodpecker, red-cockaded (*Picoides borealis*)

Plants—20 species

T — Price's potato-bean (*Apios priceana*)
E — Braun's rock cress (*Arabis perstellata*)
E — Cumberland sandwort (*Arenaria cumberlandensis*)
T — American hart's-tongue fern (*Asplenium scolopendrium* var. *americanum*)
E — Pyne's (= Guthrie's) ground-plum (*Astragalus bibullatus*)
T — Cumberland rosemary (*Conradina verticillata*)
E — Leafy prairie-clover (*Dalea (= Petalostemum) foliosa*)
E — Tennessee purple coneflower (*Echinacea tennesseensis*)
E — Spreading avens (*Geum radiatum*)
E — Rock gnome lichen (*Gymnoderma lineare*)
E — Roan Mountain bluet (*Hedyotis purpurea* var. *montana*)
T — Eggert's sunflower (*Helianthus eggertii*)
T — Small whorled pogonia (*Isotria medeoloides*)
E — Spring Creek bladderpod (*Lesquerella perforata*)
E — Ruth's golden aster (*Pityopsis (= Heterotheca = Chrysopsis) ruthii*)
E — Green pitcher-plant (*Sarracenia oreophila*)

E — Large-flowered skullcap (*Scutellaria montana*)
T — Blue Ridge goldenrod (*Solidago spithamaea*)
T — Virginia spiraea (*Spiraea virginiana*)
E — Tennessee yellow-eyed grass (*Xyris tennesseensis*)

Texas—73 species
Animals—46 species
E — Amphipod, Peck's cave (*Stygobromus (= Stygonectes) pecki*)
E — Bat, Mexican long-nosed (*Leptonycteris nivalis*)
T — Bear, Louisiana black (*Ursus americanus luteolus*)
E — Beetle, Coffin Cave mold (*Batrisodes texanus*)
E — Beetle, Comal Springs riffle (*Heterelmis comalensis*)
E — Beetle, Comal Springs dryopid (*Stygoparnus comalensis*)
E — Beetle, Kretschmarr Cave mold (*Texamaurops reddelli*)
E — Beetle, Tooth Cave ground (*Rhadine persephone*)
E — Crane, whooping (*Grus americana*)
E — Curlew, Eskimo (*Numenius borealis*)
E — Darter, fountain (*Etheostoma fonticola*)
T — Eagle, bald (*Haliaeetus leucocephalus*)
E — Falcon, American peregrine (*Falco peregrinus anatum*)
E — Falcon, northern aplomado (*Falco femoralis septentrionalis*)
E — Flycatcher, Southwestern willow (*Empidonax traillii extimus*)
E — Gambusia, Big Bend (*Gambusia gaigei*)
E — Gambusia, Clear Creek (*Gambusia heterochir*)
E — Gambusia, Pecos (*Gambusia nobilis*)
E — Gambusia, San Marcos (*Gambusia georgei*)
E — Harvestman, Bee Creek Cave (*Texella reddelli*)
E — Harvestman, Bone Cave (*Texella reyesi*)
E — Jaguar (*Panthera onca*)
E — Jaguarundi, Gulf Coast (*Herpailurus (= Felis) yagouaroundi cacomitli*)
E — Manatee, West Indian (*Trichechus manatus*)
E — Minnow, Rio Grande silvery (*Hybognathus amarus*)
E — Ocelot (*Leopardus (= Felis) pardalis*)
T — Owl, Mexican spotted (*Strix occidentalis lucida*)
E — Pelican, brown (*Pelecanus occidentalis*)
T — Plover, piping (*Charadrius melodus*)
E — Prairie-chicken, Attwater's greater (*Tympanuchus cupido attwateri*)
E — Pseudoscorpion, Tooth Cave (*Tartarocreagris (= Microcreagris) texana*)
E — Pupfish, Comanche Springs (*Cyprinodon elegans*)
E — Pupfish, Leon Springs (*Cyprinodon bovinus*)
E — Salamander, Barton Springs (*Eurycea sosorum*)
T — Salamander, San Marcos (*Eurycea nana*)
E — Salamander, Texas blind (*Typhlomolge rathbuni*)
E — Sea turtle, Kemp's (= Atlantic) ridley (*Lepidochelys kempii*)
T — Sea turtle, loggerhead (*Caretta caretta*)
T — Shiner, Arkansas River (native pop. only) (*Notropis girardi*)
T — Snake, Concho water (*Nerodia paucimaculata*)
E — Spider, Tooth Cave (*Neoleptoneta (= Leptoneta) myopica*)
E — Tern, least (*Sterna antillarum*)
E — Toad, Houston (*Bufo houstonensis*)
E — Vireo, black-capped (*Vireo atricapillus*)
E — Warbler, golden-cheeked (*Dendroica chrysoparia*)
E — Woodpecker, red-cockaded (*Picoides borealis*)
Plants—27 species
E — Large-fruited sand-verbena (*Abronia macrocarpa*)

E — South Texas ambrosia (*Ambrosia cheiranthifolia*)
E — Tobusch fishhook cactus (*Ancistrocactus tobuschii*)
E — Star cactus (*Astrophytum asterias*)
E — Texas ayenia (*Ayenia limitaris*)
E — Texas poppy-mallow (*Callirhoe scabriuscula*)
E — Nellie cory cactus (*Coryphantha (= Escobaria) minima*)
T — Bunched cory cactus (*Coryphantha ramillosa*)
E — Sneed pincushion cactus (*Coryphantha sneedii* var. *sneedii*)
E — Terlingua Creek cats-eye (*Cryptantha crassipes*)
T — Chisos Mountain hedgehog cactus (*Echinocereus chisoensis* var. *chisoensis*)
E — Lloyd's hedgehog cactus (*Echinocereus lloydii*)
E — Black lace cactus (*Echinocereus reichenbachii (= melanocentrus)* var. *albertii*)
E — Davis' green pitaya (*Echinocereus viridiflorus* var. *davisii*)
T — Lloyd's Mariposa cactus (*Echinomastus (= Sclerocactus) mariposensis*)
E — Johnston's frankenia (*Frankenia johnstonii*)
E — Slender rush-pea (*Hoffmannseggia tenella*)
E — Texas prairie dawn-flower (= Texas bitterweed) (*Hymenoxys texana*)
E — White bladderpod (*Lesquerella pallida*)
E — Walker's manioc (*Manihot walkerae*)
E — Texas trailing phlox (*Phlox nivalis* ssp. *texensis*)
E — Little Aguja pondweed (*Potamogeton clystocarpus*)
T — Hinckley's oak (*Quercus hinckleyi*)
E — Navasota ladies'-tresses (*Spiranthes parksii*)
E — Texas snowbells (*Styrax texanus*)
E — Ashy dogweed (*Thymophylla tephroleuca*)
E — Texas wild-rice (*Zizania texana*)

Utah—39 species
Animals—18 species
E — Ambersnail, Kanab (*Oxyloma haydeni kanabensis*)
E — Chub, bonytail (*Gila elegans*)
E — Chub, humpback (*Gila cypha*)
E — Chub, Virgin River (*Gila robusta semidnuda*)
E — Crane, whooping (*Grus americana*)
T — Eagle, bald (*Haliaeetus leucocephalus*)
E — Falcon, American peregrine (*Falco peregrinus anatum*)
E — Ferret, black-footed (*Mustela nigripes*)
E — Flycatcher, Southwestern willow (*Empidonax traillii extimus*)
T — Owl, Mexican spotted (*Strix occidentalis lucida*)
T — Prairie dog, Utah (*Cynomys parvidens*)
E — Snail, Utah valvata (*Valvata utahensis*)
E — Squawfish, Colorado (*Ptychocheilus lucius*)
E — Sucker, June (*Chasmistes liorus*)
E — Sucker, razorback (*Xyrauchen texanus*)
T — Tortoise, desert (*Gopherus agassizii*)
T — Trout, Lahontan cutthroat (*Oncorhynchus (= Salmo) clarki henshawi*)
E — Woundfin (*Plagopterus argentissimus*)
Plants—21 species
E — Dwarf bear-poppy (*Arctomecon humilis*)
T — Welsh's milkweed (*Asclepias welshii*)
T — Heliotrope milk-vetch (*Astragalus montii*)
T — Navajo sedge (*Carex specuicola*)
T — Jones cycladenia (*Cycladenia humilis* var. *jonesii*)
T — Maguire daisy (*Erigeron maguirei*)
E — Barneby ridge-cress (= peppercress) (*Lepidium barnebyanum*)

E — Kodachrome bladderpod (*Lesquerella tumulosa*)
E — San Rafael cactus (*Pediocactus despainii*)
T — Siler pincushion cactus (*Pediocactus sileri*)
T — Winkler cactus (*Pediocactus winkleri*)
E — Clay phacelia (*Phacelia argillacea*)
T — Maguire primrose (*Primula maguirei*)
E — Autumn buttercup (*Ranunculus acriformis* var. *aestivalis*)
T — Clay reed-mustard (*Schoenocrambe argillacea*)
E — Barneby reed-mustard (*Schoenocrambe barnebyi*)
E — Shrubby reed-mustard (= toad-flax cress) (*Schoenocrambe suffrutescens*)
T — Uinta Basin hookless cactus (*Sclerocactus glaucus*)
E — Wright fishhook cactus (*Sclerocactus wrightiae*)
T — Ute ladies'-tresses (*Spiranthes diluvialis*)
T — Last Chance townsendia (*Townsendia aprica*)

Vermont—7 species
Animals—5 species
E — Bat, Indiana (*Myotis sodalis*)
T — Beetle, Puritan tiger (*Cicindela puritana*)
T — Eagle, bald (*Haliaeetus leucocephalus*)
E — Falcon, American peregrine (*Falco peregrinus anatum*)
E — Wedgemussel, dwarf (*Alasmidonta heterodon*)
Plants—2 species
E — Jesup's milk-vetch (*Astragalus robbinsii* var. *jesupi*)
E — Northeastern (= Barbed bristle) bulrush (*Scirpus ancistrochaetus*)

Virginia—53 species
Animals—40 species
E — Bat, gray (*Myotis grisescens*)
E — Bat, Indiana (*Myotis sodalis*)
E — Bat, Virginia big-eared (*Corynorhinus (= Plecotus) townsendii virginianus*)
E — Bean (mussel), purple (*Villosa perpurpurea*)
T — Beetle, northeastern beach tiger (*Cicindela dorsalis dorsalis*)
E — Blossom, green (pearlymussel)(*Epioblasma torulosa gubernaculum*)
T — Chub, slender (*Erimystax (= Hybopsis) cahni*)
T — Chub, spotfin (= turquoise shiner) (*Cyprinella (= Hybopsis) monacha*)
E — Combshell, Cumberlandian (*Epioblasma brevidens*)
E — Darter, duskytail (*Etheostoma percnurum*)
T — Eagle, bald (*Haliaeetus leucocephalus*)
E — Falcon, American peregrine (*Falco peregrinus anatum*)
E — Fanshell (*Cyprogenia stegaria*)
E — Isopod, Lee County cave (*Lirceus usdagalun*)
T — Isopod, Madison Cave (*Antrolana lira*)
E — Logperch, Roanoke (*Percina rex*)
T — Madtom, yellowfin (*Noturus flavipinnis*)
E — Monkeyface, Appalachian (pearlymussel)(*Quadrula sparsa*)
E — Monkeyface, Cumberland (pearlymussel)(*Quadrula intermedia*)
E — Mucket, pink (pearlymussel) (*Lampsilis abrupta*)
E — Mussel, oyster (*Epioblasma capsaeformis*)
E — Pearlymussel, birdwing (*Conradilla caelata*)
E — Pearlymussel, cracking (*Hemistena lata*)
E — Pearlymussel, dromedary (*Dromus dromas*)
E — Pearlymussel, littlewing (*Pegias fabula*)
E — Pigtoe, finerayed (*Fusconaia cuneolus*)
E — Pigtoe, rough (*Pleurobema plenum*)

E — Pigtoe, shiny (*Fusconaia cor (= edgariana)*)
T — Plover, piping (*Charadrius melodus*)
E — Rabbitsfoot, rough (*Quadrula cylindrica strigillata*)
E — Riffleshell, tan (*Epioblasma florentina walkeri*)
E — Salamander, Shenandoah (*Plethodon shenandoah*)
T — Shrew, Dismal Swamp southeastern (*Sorex longirostris fisheri*)
E — Snail, Virginia fringed mountain (*Polygyriscus virginianus*)
E — Spinymussel, James (= Virginia) (*Pleurobema collina*)
E — Squirrel, Delmarva Peninsula fox (*Sciurus niger cinereus*)
E — Squirrel, Virginia northern flying (*Glaucomys sabrinus fuscus*)
E — Tern, roseate (*Sterna dougallii dougallii*)
E — Wedgemussel, dwarf (*Alasmidonta heterodon*)
E — Woodpecker, red-cockaded (*Picoides borealis*)
Plants—13 species
T — Sensitive joint-vetch (*Aeschynomene virginica*)
E — Shale barren rock-cress (*Arabis serotina*)
T — Virginia round-leaf birch (*Betula uber*)
E — Small-anthered bittercress (*Cardamine micranthera*)
E — Smooth coneflower (*Echinacea laevigata*)
T — Virginia sneezeweed (*Helenium virginicum*)
T — Swamp pink (*Helonias bullata*)
E — Peter's Mountain mallow (*Iliamna corei*)
T — Small whorled pogonia (*Isotria medeoloides*)
T — Eastern prairie fringed orchid (*Platanthera leucophaea*)
E — Michaux's sumac (*Rhus michauxii*)
E — Northeastern (= Barbed bristle) bulrush (*Scirpus ancistrochaetus*)
T — Virginia spiraea (*Spiraea virginiana*)

Washington—19 species
Animals—13 species
T — Bear, grizzly (*Ursus arctos*)
T — Butterfly, Oregon silverspot (*Speyeria zerene hippolyta*)
E — Caribou, woodland (*Rangifer tarandus caribou*)
E — Deer, Columbian white-tailed (*Odocoileus virginianus leucurus*)
T — Eagle, bald (*Haliaeetus leucocephalus*)
E — Falcon, American peregrine (*Falco peregrinus anatum*)
T — Goose, Aleutian Canada (*Branta canadensis leucopareia*)
T — Murrelet, marbled (*Brachyramphus marmoratus marmoratus*)
T — Owl, northern spotted (*Strix occidentalis caurina*)
E — Pelican, brown (*Pelecanus occidentalis*)
T — Plover, western snowy (*Charadrius alexandrinus nivosus*)
T — Trout, bull
E — Wolf, gray (*Canis lupus*)
Plants—6 species
E — Marsh sandwort (*Arenaria paludicola*)
T — Golden paintbrush (*Castilleja levisecta*)
T — Water howellia (*Howellia aquatilis*)
E — Bradshaw's desert-parsley (= lomatium) (*Lomatium bradshawii*)
T — Nelson's checker-mallow (*Sidalcea nelsoniana*)
T — Ute ladies'-tresses (*Spiranthes diluvialis*)

West Virginia—20 species
Animals—14 species
E — Bat, gray (*Myotis grisescens*)
E — Bat, Indiana (*Myotis sodalis*)

E — Bat, Virginia big-eared (*Corynorhinus (= Plecotus) townsendii virginianus*)
E — Blossom, tubercled (pearlymussel) (*Epioblasma torulosa torulosa*)
E — Clubshell (*Pleurobema clava*)
T — Eagle, bald (*Haliaeetus leucocephalus*)
E — Falcon, American peregrine (*Falco peregrinus anatum*)
E — Fanshell (*Cyprogenia stegaria*)
E — Mucket, pink (pearlymussel) (*Lampsilis abrupta*)
E — Riffleshell, northern (*Epioblasma torulosa rangiana*)
T — Salamander, Cheat Mountain (*Plethodon nettingi*)
T — Snail, flat-spired three-toothed (*Triodopsis platysayoides*)
E — Spinymussel, James (= Virginia) (*Pleurobema collina*)
E — Squirrel, Virginia northern flying (*Glaucomys sabrinus fuscus*)

Plants—6 species

E — Shale barren rock-cress (*Arabis serotina*)
T — Small whorled pogonia (*Isotria medeoloides*)
E — Harperella (*Ptilimnium nodosum (= fluviatile)*)
E — Northeastern (= Barbed bristle) bulrush (*Scirpus ancistrochaetus*)
T — Virginia spiraea (*Spiraea virginiana*)
E — Running buffalo clover (*Trifolium stoloniferum*)

Wisconsin—15 species

Animals—9 species

E — Butterfly, Karner blue (*Lycaeides melissa samuelis*)
E — Dragonfly, Hine's emerald (*Somatochlora hineana*)
T — Eagle, bald (*Haliaeetus leucocephalus*)
E — Falcon, American peregrine (*Falco peregrinus anatum*)
E — Higgins eye (pearlymussel) (*Lampsilis higginsii*)
E — Mapleleaf, winged (mussel) (*Quadrula fragosa*)
E&T — Plover, piping (*Charadrius melodus*)
E — Warbler, Kirtland's (*Dendroica kirtlandii*)
E — Wolf, gray (*Canis lupus*)

Plants—6 species

T — Northern wild monkshood (*Aconitum noveboracense*)
T — Pitcher's thistle (*Cirsium pitcheri*)
T — Dwarf lake iris (*Iris lacustris*)
T — Prairie bush-clover (*Lespedeza leptostachya*)
T — Fassett's locoweed (*Oxytropis campestris* var. *chartacea*)
T — Eastern prairie fringed orchid (*Platanthera leucophaea*)

Wyoming—12 species

Animals—11 species

T — Bear, grizzly (*Ursus arctos*)
E — Crane, whooping (*Grus americana*)
E — Dace, Kendall Warm Springs (*Rhinichthys osculus thermalis*)
T — Eagle, bald (*Haliaeetus leucocephalus*)
E — Falcon, American peregrine (*Falco peregrinus anatum*)
E — Ferret, black-footed (*Mustela nigripes*)
T — Mouse, Preble's meadow jumping (*Zapus hudsonius preblei*)
E — Squawfish, Colorado (*Ptychocheilus lucius*)
E — Sucker, razorback (*Xyrauchen texanus*)
E — Toad, Wyoming (*Bufo hemiophrys baxteri*)
E — Wolf, gray (*Canis lupus*)

Plants—1 species

T — Ute ladies'-tresses (*Spiranthes diluvialis*)

American Samoa—2 species
Animals—2 species
T — Sea turtle, green (*Chelonia mydas*)
E — Sea turtle, hawksbill (*Eretmochelys imbricata*)
Plants—0 species

Commonwealth of the Northern Mariana Islands—9 species
Animals—8 species
E — Crow, Mariana (*Corvus kubaryi*)
E — Mallard, Mariana (*Anas oustaleti*)
E — Megapode, Micronesian (= La Perouse's) (*Megapodius laperouse*)
T — Monarch, Tinian (*Monarcha takatsukasae*)
E — Moorhen (= gallinule), Mariana common (*Gallinula chloropus guami*)
T — Sea turtle, green (*Chelonia mydas*)
E — Swiftlet, Mariana gray (= vanikoro) (*Aerodramus vanikorensis bartschi*)
E — Warbler, nightingale reed (*Acrocephalus luscinia*)
Plants—1 species
E — Hayun lagu (Guam), Tronkon guafi (Rota) (*Serianthes nelsonii*)

District of Columbia—3 species
Animals—3 species
E — Amphipod, Hay's Spring (*Stygobromus hayi*)
T — Eagle, bald (*Haliaeetus leucocephalus*)
E — Falcon, American peregrine (*Falco peregrinus anatum*)
Plants—0 species

Guam—13 species
Animals—12 species
E — Bat, little Mariana fruit (*Pteropus tokudae*)
E — Bat, Mariana fruit (*Pteropus mariannus mariannus*)
E — Broadbill, Guam (*Myiagra freycineti*)
E — Crow, Mariana (*Corvus kubaryi*)
E — Kingfisher, Guam Micronesian (*Halcyon cinnamomina cinnamomina*)
E — Mallard, Mariana (*Anas oustaleti*)
E — Moorhen (= gallinule), Mariana common (*Gallinula chloropus guami*)
E — Rail, Guam (*Rallus owstoni*)
T — Sea turtle, green (*Chelonia mydas*)
E — Sea turtle, hawksbill (*Eretmochelys imbricata*)
E — Swiftlet, Mariana gray (= vanikoro) (*Aerodramus vanikorensis bartschi*)
E — White-eye, bridled (*Zosterops conspicillatus conspicillatus*)
Plants—1 species
E — Hayun lagu (Guam), Tronkon guafi (Rota) (*Serianthes nelsonii*)

Puerto Rico—71 species
Animals—23 species
E — Anole, Culebra Island giant (*Anolis roosevelti*)
E — Blackbird, yellow-shouldered (*Agelaius xanthomus*)
T — Boa, Mona (*Epicrates monensis monensis*)
E — Boa, Puerto Rican (*Epicrates inornatus*)
E — Boa, Virgin Islands tree (*Epicrates monensis granti*)
T — Coqui, golden (*Eleutherodactylus jasperi*)
E — Falcon, American peregrine (*Falco peregrinus anatum*)
E — Gecko, Monito (*Sphaerodactylus micropithecus*)
T — Guajón
E — Hawk, Puerto Rican broad-winged (*Buteo platypterus brunnescens*)

E — Hawk, Puerto Rican sharp-shinned (*Accipiter striatus venator*)
T — Iguana, Mona ground (*Cyclura stejnegeri*)
E — Manatee, West Indian (*Trichechus manatus*)
E — Nightjar, Puerto Rican (= whip-poor-will) (*Caprimulgus noctitherus*)
E — Parrot, Puerto Rican (*Amazona vittata*)
E — Pelican, brown (*Pelecanus occidentalis*)
E — Pigeon, Puerto Rican plain (*Columba inornata wetmorei*)
T — Plover, piping (*Charadrius melodus*)
T — Sea turtle, green (*Chelonia mydas*)
E — Sea turtle, hawksbill (*Eretmochelys imbricata*)
E — Sea turtle, leatherback (*Dermochelys coriacea*)
T — Tern, roseate (*Sterna dougallii dougallii*)
T — Toad, Puerto Rican crested (*Peltophryne lemur*)

Plants—48 species

E — *Adiantum vivesii* (Fern, no common name)
E — *Aristida chaseae* (Plant, no common name)
E — Pelos del diablo (*Aristida portoricensis*)
E — *Auerodendron pauciflorum* (Plant, no common name)
E — Palo de Ramón (*Banara vanderbiltii*)
E — Vahl's boxwood (*Buxus vahlii*)
E — Capá rosa (= péndula cimarrona) (*Callicarpa ampla*)
E — *Calyptranthes thomasiana* (Plant, no common name)
T — Palma de manaca or manac palm (*Calyptronoma rivalis*)
E — *Chamaecrista glandulosa* var. *mirabilis (= Cassia mirabilis)* (Plant, no common name)
E — *Cordia bellonis* (Plant, no common name)
E — Palo de nigua (= capá jugüerilla) (*Cornutia obovata*)
E — *Cranichis ricartii* (Plant, no common name)
E — Higuero de Sierra (*Crescentia portoricensis*)
E — Elfin tree fern (*Cyathea dryopteroides*)
E — *Daphnopsis hellerana* (Plant, no common name)
E — *Elaphoglossum serpens* (Fern, no common name)
E — Uvillo (*Eugenia haematocarpa*)
E — *Eugenia woodburyana* (Plant, no common name)
T — *Gesneria pauciflora* (Plant, no common name)
E — Beautiful goetzea or matabuey (*Goetzea elegans*)
T — Higo chumbo (*Harrisia portoricensis*)
E — Cook's holly (*Ilex cookii*)
E — *Ilex sintenisii* (Plant, no common name)
E — Nogal or West Indian walnut (*Juglans jamaicensis*)
E — *Lepanthes eltoroensis* (Plant, no common name)
E — *Leptocereus grantianus* (Plant, no common name)
E — *Lyonia truncata* var. *proctorii* (Plant, no common name)
E — *Mitracarpus maxwelliae* (Plant, no common name)
E — *Mitracarpus polycladus* (Plant, no common name)
E — *Myrcia paganii* (Plant, no common name)
E — Palo de rosa (*Ottoschulzia rhodoxylon*)
E — Wheeler's peperomia (*Peperomia wheeleri*)
E — Chupacallos (= Chupagallo) (*Pleodendron macranthum*)
E — *Polystichum calderonense* (Fern, no common name)
T — *Schoepfia arenaria* (Plant, no common name)
E — Erubia (*Solanum drymophilum*)
T — Cóbana negra (*Stahlia monosperma*)
E — Palo de jazmín (*Styrax portoricensis*)
E — *Tectaria estremerana* (Fern, no common name)
E — Palo colorado (*Ternstroemia luquillensis*)

E — *Ternstroemia subsessilis* (Plant, no common name)
E — *Thelypteris inabonensis* (Fern, no common name)
E — *Thelypteris verecunda* (Fern, no common name)
E — *Thelypteris yaucoensis* (Fern, no common name)
E — Bariaco (= guayabacón) (*Trichilia triacantha*)
E — *Vernonia proctorii* (Plant, no common name)
E — St. Thomas prickly-ash (*Zanthoxylum thomasianum*)

U.S. Virgin Islands—11 species
Animals—8 species
E — Falcon, American peregrine (*Falco peregrinus anatum*)
E — Lizard, St. Croix ground (*Ameiva polops*)
E — Manatee, West Indian (*Trichechus manatus*)
E — Pelican, brown (*Pelecanus occidentalis*)
T — Sea turtle, green (*Chelonia mydas*)
E — Sea turtle, hawksbill (*Eretmochelys imbricata*)
E — Sea turtle, leatherback (*Dermochelys coriacea*)
T — Tern, roseate (*Sterna dougallii dougallii*)
Plants—3 species
E — Vahl's boxwood (*Buxus vahlii*)
E — *Calyptranthes thomasiana* (Plant, no common name)
E — St. Thomas prickly-ash (*Zanthoxylum thomasianum*)U.S. Fish and Wildlife Service Division of Endangered Species: Listed Species under Fish and Wildlife Service Jurisdiction By State/Territory as of December 31, 1998. Status Code: E - listed as Endangered; T - listed as Threatened)

REFERENCES

SELECTED GENERAL REFERENCES ON ECOLOGY AND PRESERVATION

Agee, J. K., and D. R. Johnston. *Ecosystem Management for Parks and Wilderness.* Seattle. University of Washington Press, 1988.

Arno, Stephen F., and Ramona P. Hammerly. *Northwest Trees.* Seattle: The Mountaineers, 1977.

Ausubel, K. *Restoring the Earth.* Tiburon, CA: H. J. Kramer, 1998.

Baer-Brown, L., and B. Rhein. *Earth Keepers.* San Francisco: Mercury House, 1995.

Bailey, R. G., P. Avers, T. King, and W. H. McNab. *Ecoregions and Subregions of the United States* (map and text). Washington, DC: USDA Forest Service. 1995.

Balee, William, ed. *Advances in Historical Ecology.* New York: Columbia University Press, 1998.

Barnes, B. V. 1993. "The Landscape Ecosystem Approach and Conservation of Endangered Spaces." In *Endangered Species Update* 10: 13–19.

Bast, J. L., P. J. Hill, and R. C. Rue. *Eco-Sanity: A Common-Sense Guide to Environmentalism.* Lanham, MD: Madison Books, 1994.

Berger, J. J. *Understanding Forests.* San Francisco: Sierra Club, 1998.

Beynus, Janine M. *The Field Guide to Wildlife Habitats of the Western United States.* New York: Simon and Schuster, 1989.

Bormann, S. H., and G. E. Likens. *Pattern and Process in a Forested Ecosystem.* New York: Springer-Verlag, 1977.

Boyce, M. S., and A. Haney. *Ecosystem Management.* New Haven: Yale University Press, 1997.

Bradley, G. A., ed. *Land Use and Forest Resources in a Changing Environment.* Seattle: University of Washington Press, 1984.

Brandon, K., K. H. Rudford, and S. E. Sanderson. *Parks in Peril.* Washington, DC: Island Press, 1998.

Brown, D. E., S. E. Franson, and F. Reichenbacher. *Classification of North American Biotic Communities.* Salt Lake City: University of Utah Press, 1998.

Brown, J. H. *Macroecology.* Chicago: University of Chicago Press, 1998.

Brown, Lauren. *Grasslands.* New York: Knopf, 1985.

Caduto, M. J. *Pond and Brook.* Englewood Cliffs, NJ: Prentice-Hall, 1995.

Callicott, J. B., and M. P. Nelson, eds. *The Great New Wilderness Debate.* Athens: University of Georgia Press, 1998.

Camp, W. B., and T. B. Daugherty. *Managing Our National Resources.* 3d ed. Albany, NY: Delmar, 1998.

Castillon, D. A. *Conservation of Natural Resources: A Resource Management Approach.* Dubuque, IA: Brown and Benchmark, 1996.

Chertow, M. R., and D. C. Esty, eds. *Thinking Ecologically.* New Haven: Yale, 1997.

Clark, J. S. "The Development of a Tidal Marsh: Upland and Oceanic Influences." *Ecology Monography* 55(2) (1985): 189–217.

Collins, S. L., and L. L. Wallace, eds. *Fire in North American Tallgrass Prairies.* Norman: University of Oklahoma Press.

Costello, D. F. *The Prairie World.* Minneapolis: University of Minnesota Press, 1969.

Cowardin, L., V. Carter, F. Golet, and E. LaRoe. *Classification of Wetlands and Deepwater Habitats of the United States.* Washington, DC: U.S. Department of Interior, 1979.

Crum, H. A. *A Focus on Peatlands and Peat Mosses.* Ann Arbor: University of Michigan Press, 1992.

Cunningham, W. P., and B. Woodworth Saigo. *Environmental Science: A Global Concern.* Dubuque, IA: Brown, 1997.

Damman, A. W. H., and T. W. French. *The Ecology of Peat Bogs of the Glaciated Northeastern United States: A Community Profile.* U.S. Fish and Wildlife Service Biological Report 85(7.16), 1987.

Delcourt, P. A., and H. R. Delcourt. *Long-Term Forest Dynamics of the Temperate Zone.* New York: Springer-Verlag, 1988.

Dennis, John V. *The Great Cypress Swamps.* Baton Rouge: Louisiana State University Press, 1988.

Devevan, W. M. "The Pristine Myth: The Landscape of the Americas in 1492." *Annals of the Association of American Geographers* 82(3) (1992): 369–385.

Dobson, A. P. *Conservation and Biodiversity.* New York: W. H. Freeman, 1998.

Drengson, A., and D. Taylor, eds. *Ecoforestry.* Stony Creek, CT: New Society Press, 1997.

Dwelley, Marilyn J. *Trees and Shrubs of New England.* Camden, ME: Down East Books, 1980.

Eldredge, N. *Life in the Balance.* Princeton: University of Princeton Press, 1998.

Eyre, F. H. *Forest Cover Types of the United States and Canada.* Washington, DC: Society of American Foresters, 1980.

Falk, D. A., and K. E. Holsinger. *Genetics and Conservation of Rare Plants.* New York: Oxford University Press, 1991.

Fielder, P., and S. Jains, eds. *Conservation Biology: The Theory and Practice of Nature Conservation, Preservation, and Management.* New York: Chapman and Hall, 1992.

Finch, Robert. *Cape Cod.* Washington, DC: National Park Service, 1994.

Forman, R. T. T., ed. *Pine Barrens: Ecosystems and Landscape.* New York: Academic Press, 1979.

Franklin, J. F. 1988. "Pacific Northwest Forests." In M. G. Barbour and W. D. Billings, eds., *North American Terrestrial Vegetation.* Cambridge: Cambridge University Press, 1988.

Freese, C. H. *Wild Species as Commodities.* Washington, DC: Island Press, 1998.

Futuyma, D. *Evolutionary Biology.* Sunderland, MA: Sinauer Associates, 1979.

Gleason, H. A., and A. Cronquist. *Manual of Vascular Plants of Northeastern United States and Adjacent Canada.* 2d ed. Bronx, New York: New York Botanical Garden, 1991.

Gómez-Pompa, A., and A. Kaus. "Taming the Wilderness Myth." *BioScience* 42(4) (1992): 271–279.

Gould, F. W. *Grasses of the Southwestern United States.* Tucson: University of Arizona Press, 1951.

Grossman, D. H. *Rare Plant Communities of the Conterminous United States.* Arlington, VA: The Nature Conservancy, 1994.

Guruswamy, L. D., and J. A. McNeely, eds. *Protection of Global Biodiversity.* Durham: Duke University Press, 1998.

Harris, D. *The Last Stand.* San Francisco: Sierra Club Books, 1998.

Harris, L. D. *The Fragmented Forest: Island Biogeographic Theory and the Preservation of Biotic Diversity.* Chicago: University of Chicago Press, 1984.

Heywood, V. H., ed. *Global Biodiveristy Assessment.* New York: Cambridge University Press, 1995.

Hitchcock, C. L., and A. Cronquist. *Flora of the Pacific Northwest.* Seattle: University of Washington Press, 1976.

Hunt, Charles B. *Natural Regions of the United States and Canada.* New York: W. H. Freeman, 1974.

Hunt, Charles B. *Physiography of the United States.* New York: W. H. Freeman, 1967.

Jensen, M. E., and P. S. Bourgeron. *Ecosystem Management: Principles and Applications.* Missoula, MT: USDA Forest Service, Northern Region, 1993.

Johnson, C. E. *Bogs of the Northeast.* Hanover: University Press of New England, 1985.

Jorgensen, Neil. A *Sierra Club Naturalist's Guide: Southern New England.* San Francisco: Sierra Club Books, 1978.

Kadlec, R. H., and R. L. Knight. *Treatment of Wetlands.* Boca Raton, FL: CRC, 1996.

Kellogg, E., ed. *Coastal Temperate Rain Forests: Ecological Characteristics, Status, and Distribution Worldwide.* Portland, OR, and Washington, DC: Ecotrust and Conservation International, 1992.

Kemp, D. D. *The Environmental Dictionary.* New York: Rutledge Press, 1998.

Kohm, K., ed. *Balancing on the Brink of Extinction.* Washington, DC: Island Press, 1991.

Kuchler, A. W. 1964. *Potential Natural Vegetation of the Conterminous United States.* New York: American Geographic Society, 1964.

MacKintosh, Gay, ed. *Preserving Communities and Corridors.* Washington, DC: Defenders of Wildlife, 1989.

Marchand, Peter J. *North Woods.* Boston: Appalachian Mountain Club, 1987.

Marsh, G. P. *Man and Nature.* Cambridge: Harvard University Press, 1998.

Marzluff, J. M., and R. Sallabanks, eds. *Avian Conservation: Resources and Managment.* Washington, DC: Island Press, 1998.

McNab, W. H., and P. E. Avers. *Ecological Subregions of the United States.* Washington, DC: USDA Forest Service, 1994.

National Audubon Society. *Wildlife Sanctuaries.* Sharon, CT: National Audubon Society, 1991.
National Geographic Society. *Birds of North America.* Washington, DC: National Geographic Society, 1983.
National Research Council. *Beach Nourishment and Protection.* Washington, DC: National Academy of Science, 1995.
Niering, W. A., ed. *Wetlands.* New York: Knopf, 1998.
Noss, R. "Protecting Natural Areas in Fragmented Landscapes." *Natural Areas Journal* 7 (1987): 2–13.
Noss, R. F. 1987. "From Plant Communities to Landscapes in Conservation Inventories: A Look at The Nature Conservancy." In *Biological Conservation* 41: 11–37.
Noss, R. F. "A Regional Landscape Approach to Maintain Diversity." *BioScience* 3 (1983): 700–706.
Noss, R. F., and A. Y. Cooperrider. *Saving Nature's Legacy: Protecting and Restoring Biodiversity.* Washington, DC: Island Press, 1994.
Noss, R. F., E. T. LaRoe III, and J. M. Scott. *Endangered Ecosystems of the United States: A Preliminary Assessment of Loss and Degradation.* Biological Report 28. Washington, DC: National Biological Service, 1995.
Prance, G. T., and T. S. Elias. *Extinction Is Forever: Threatened and Endangered Species of Plants in the Americas and Their Significance in Ecosystems Today and in the Future.* Bronx, NY: New York Botanical Garden, 1977.
Preston Jr., R. J. *North American Trees.* Cambridge: Massachusetts Institute of Technology Press, 1975.
Primack, R. B. *A Primer of Conservation Biology.* Sunderland, MA: Sinauer Associates, 1995.
Reed, C. F. *Floras of the Serpentinite Formations of Eastern North America, with Descriptions of Geomorphology and Mineralogy of the Formations.* Baltimore: Reed Herbarium, 1986.
Robbins, C. S., B. Bruun, and H. S. Zim. *Birds of North America.* New York: Golden Press, 1983.
Robinson, S. K. "Regional Forest Fragmentation and the Nesting Success of Migratory Birds." *Science* 267 (March 1995): 1987–1990.
Rockwell, D. *The Nature of Nature.* New York: Berkeley Press, 1998.
Sampson, F. B., and F. L. Knopf, eds. *Ecosystems Managment.* New York: Springer, 1996.
Sauer, L. J. *The Once and Future Forest.* Washington, DC: Island Press, 1998.
Scott, J. M., et al. *Gap Analysis: A Geographic Approach to Protection of Biological Diversity.* Wildlife Monographs No. 123, 1993. Bethesda, MD: Wildlife Society, 1993.
Shafer, C. L. *Nature Preserves: Island Theory and Conservation Practice.* Washington, DC: Smithsonian Institution Press, 1990.
Shelford, V. E. *The Ecology of North America.* Urbana: University of Illinois Press, 1963.
Smithsonian Institution. *The Smithsonian Guides to Natural America.* Washington, DC: Smithsonian Institution, 1996.
Spurr, S. H., and B. V. Barnes. *Forest Ecology.* 3d ed. New York: John Wiley, 1980.
Synge, H., ed. *The Biological Aspects of Rare Plant Conservation.* New York: John Wiley, 1981.
Takacs, D. *The Idea of Biodiversity.* Baltimore: John Hopkins Press, 1996.
The Nature Conservancy. *The Magazine of the Nature Conservancy.* Arlington, VA: various issues.
Trimble, S. *The Sage Brush Ocean: A Natural History of the Great Basin.* Reno: University of Nevada Press, 1989.
USDA Forest Service. *Research Natural Areas: A Living Library.* Informational brochure. USDA Forest Service, Intermountain Research Station, Missoula, MT.
Weaver, J. E. *The North American Prairie.* Lincoln: University of Nebraska Press, 1954.
Weaver, J. E., and F. W. Albertson. *The Grasslands of the Great Plains.* Lincoln: University of Nebraska Press, 1956.
Wildlife Habitat Council. *Longleaf Legacy: An Ecosystem Formed By Fire.* Silver Spring, MD: Wildlife Habitat Council, 1988.
Zwinger, A. H., and B. E. Willard. *Land above the Trees: A Guide to American Alpine Tundra.* New York: Harper and Row, 1972.

SELECTED REFERENCES BY STATE

ALABAMA

Dean, B. F., A. Mason, and J. L. Thomas. *Wildflowers of Alabama and Adjoining States.* Tuscaloosa: University of Alabama Press, 1973.
Duncan, Julia C., and Ed Malles. *Alabama.* Portland, OR: Graphic Arts Center Publishing, 1983.
Freeman, J. D., A. S. Causey, and J. W. Short.

"Endangered, Threatened, and Special Concern Plants of Alabama." *Journal of the Alabama Academy of Science* 50 (1979): 1–26.
Harper, R. M. *Economic Botany of Alabama.* Tuscaloosa: University of Alabama, 1928.
Harper. R. M. *Forests of Alabama.* Tuscaloosa: University of Alabama, 1943.
Mohr, C. *Plant Life of Alabama.* Washington, DC: U.S. Department of Agriculture, 1901.
Mount, R. H. *The Reptiles and Amphibians of Alabama.* Auburn, AL: Auburn University, 1975.
Vaughan, R. *Birder's Guide to Alabama and Mississippi.* Houston: Gulf Publishing Company, 1994.

ALASKA

Alaska Department of Fish and Game. *A Guide to Wildlife Viewing in Alaska.* Juneau: Alaska Department of Fish and Game, 1983.
Alaska Department of Fish and Game. Series of pamphlets describing game refuges and areas of critical habitat. Juneau: Department of Fish and Game, 1990.
Armstrong, R. A. *A Field Guide to the Birds of Alaska.* Anchorage: Alaska Northwest Publishing Company, 1980.
Armstrong, Robert H. *The Nature of Southeast Alaska.* Portland: Alaska Northwest Books, 1992.
Brown, William E. *This Last Treasure.* Anchorage: Alaska Natural History Association, 1982.
Fleming, M. D. *Alaska Vegetation-Land Cover Classification.* Berkeley: University of California, 1997.
Gallant, A. L., E. F. Binnian, J. M. Omernik, and M. B. Shasby. *Ecoregions of Alaska.* Geological Survey Professional Paper 1567. Reston, VA: U.S. Geological Survey, 1995.
Heacox, Kim. *Alaska's National Parks.* Portland, OR: Graphic Arts Center, 1990.
Hodge, J. P. *Reptiles and Amphibians in Alaska, the Yukon, and Northwest Territories.* Anchorage: Alaska Northwest Publishing, 1976.
Hulten, E. *Flora of Alaska and Neighboring Territories.* Stanford: Stanford University Press, 1968.
Ketchum, R. G., and C. D. Ketchum. *The Tongass: Alaska's Vanishing Rain Forest.* New York: Aperture Foundation, 1994.
Matz, G. "The Tanana Valley State Forest." *Alaska Geographic Society* 22 (1995): 74–78.
Reardon, J. *Mammals of Alaska.* Anchorage: Alaska Geographic, 1981.
Sydeman, M., and A. Lund. *Alaska Viewing Guide.* Helena, MT: Falcon Press, 1996.
The Nature Conservancy. *Alaska Newsletter.* Anchorage: The Nature Conservancy, 1998.
The Nature Conservancy. Internal documents, fact sheets, and information packets. Anchorage: The Nature Conservancy, 1995–1997.
Yaeger, L., and A. Channdonnet. *The Alaska Almanac.* 17th ed. Anchorage: Alaska Northwest Books, 1993.

ARIZONA

Dames and Moore, (Consulting Firm). *An Inventory of Natural Areas in the State of Arizona.* Phoenix: Dames and Moore, 1988.
Arizona Rivers Coalition. *Arizona Rivers: Lifeblood of the Desert.* Phoenix: Arizona Rivers Coalition, 1991.
Arizona State Parks. *Arizona Rivers and Streams Guide.* Phoenix: Arizona State Parks, 1989.
Bahre, C. J. *A Legacy of Change: Historic Human Impact on Vegetation of the Arizona Borderlands.* Tucson: University of Arizona Press, 1991.
Brown, D. E. *Natural Vegetation Communities of Arizona.* Phoenix: Arizona Resources Information System, 1973.
Carr, J. N. *Arizona Wildlife Viewing Guide.* Helena, MT: Falcon Press, 1996.
Kearny, T. H., and R. H. Peebles. *Arizona Flora.* 2d ed. Berkeley: University of California Press, 1960.
Krause, Steve. *Streamside Trails: Day Hiking Central Arizona's Lakes, Rivers, and Creeks.* Tempe, AZ: Pinyon Publishing, 1994.
Lowe, C. H. *Arizona's Natural Environment.* Tucson: University of Arizona Press, 1964.
Smith, E. L. *Established Natural Areas in Arizona: A Guidebook for Scientists and Educators.* Phoenix: Arizona Office of Economic Planning and Development, 1974.

ARKANSAS

Arkansas Natural Heritage Commission. *Guide to State Natural Areas.* Little Rock: Arkansas Natural Heritage Commission, 1997.
Foti, T., and G. Hudson. *Arkansas and the Land.* Fayetteville, AR: University of Arkansas Press, 1992.
Hunter, C. G. *Wildflowers of Arkansas.* 2d ed. Little Rock: Ozark Society Foundation, 1988.
Robinson, H. W., and R. T. Allen. *Only in Arkansas: A Study of the Endemic Plants and Animals of the State.* Fayetteville: University of Arkansas Press, 1995.

Sealander, J., and G. Heidt. *Arkansas Mammals: Their Natural History, Classification, and Distribution.* Fayetteville: University of Arkansas Press, 1990.

Shepherd, B., ed. *Arkansas's Natural Heritage.* Little Rock: August House, 1984.

Smith, E. B. *Keys to the Flora of Arkansas.* Fayetteville: University of Arkansas Press, 1994.

The Nature Conservancy. *A Guide to The Nature Conservancy's Preserves and Projects in Arkansas.* Little Rock: The Nature Conservancy, 1997.

The Nature Conservancy. *Arkansas Field Office Newsletter.* Little Rock, Arkansas.

The Nature Conservancy. *The Big Woods of Arkansas.* Little Rock: Arkansas Nature Conservancy, 1998.

CALIFORNIA

Barbor, M., B. Pavlik, F. Drysdale, and S. Lindstrom. *California's Changing Landscapes: Diversity and Conservation of California Vegetation.* Sacramento: California Native Plant Society, 1993.

Barbour, M. G., and J. Major, eds. *Terrestrial Vegetation of California.* New York: Wiley-Interscience, 1977.

California Department of Fish and Game. *List of California Terrestrial Natural Communities Recognized by the Natural Diversity Data Base.* Unpublished report. Sacramento: California Department of Fish and Game, 1997.

Clark, Jeanne L. *California Wildlife Viewing Guide.* Helena, MT: Falcon Press, 1992.

Coughman, Madge, and Joanne S. Ginsburg. *California Coastal Resources Guide.* Berkeley: University of California Press, 1987.

Faber, Phyllis M. *California's Wild Gardens.* Sacramento: California Native Plant Society Press, 1997

Herring, M. L. "The Old-growth Forest: A Community of Equals." *Pacific Discovery* 44(4) (1991): 8–17.

Holing, Dwight. *California Wild Lands: A Guide to the Nature Conservancy Preserves.* San Francisco: Chronicle Books, 1988.

Jones and Stokes Associates. *Sliding Toward Extinction: The State of California's Natural Heritage.* San Francisco: California Nature Conservancy, 1987.

Keeler-Wolf, Todd. *Ecological Surveys of Forest Service Research Natural Areas in California.* General Technical Report PSW-125. USDA Forest Service, Pacific Southwest Research Station, Berkeley, California, 1990.

Kruckeberg, A. R. *California Serpentines: Flora, Vegetation, Geology, Soils, and Management Problems.* Berkeley: University of California Press, 1984.

Outdoor California. A monthly magazine published by the California Department of Fish and Game, Sacramento, 1940–1998.

Sawyer, John O., and Todd Keeler-Wolf. *A Manual of California Vegetation.* Sacramento, CA: California Native Plant Society, 1995.

Steinhart, Peter. *California's Wild Heritage.* San Francisco: Sierra Club Books, 1990.

The Nature Conservancy of California. *Guide to Preserves.* San Francisco: The Nature Conservancy, 1997.

Thelander, C. G., ed. *Life on the Edge. A Guide to California's Endangered Natural Resources.* Berkeley, CA: Heyday Press, 1994.

University of California. *Guides to the University of California Natural Reserve System.* Oakland: University of California, 1998.

COLORADO

Bailey, A. M., and R. J. Niedrach. *Birds of Colorado.* Two volumes. Denver: Denver Museum of Natural History, 1965.

Gray, M. T. *Colorado Wildlife Viewing Guide.* Helena, MT: Falcon Press, 1992.

Hammerson, G. A. *Amphibians and Reptiles in Colorado.* Denver: Colorado Division of Wildlife, 1982.

Livingston, R. B. "Relict True Prairie Communities in Central Colorado." *Ecology* 33 (1952): 72–86.

Mutel, C. F., and J. C. Emerick. *From Grassland to Glacier: The Natural History of Colorado.* Boulder: Johnson Publishing Company, 1984.

Nelson, R. A. *Handbook of Rocky Mountain Plants.* Estes Park, CO: Skyland Publishers, 1979.

Ryan, M., L. Joyce, T. Andrews, and K. Jones. *Research Natural Areas in Colorado, Nebraska, South Dakota, and Parts of Wyoming.* Fort Collins, CO: USDA Forest Service, 1994.

Young, R. G., and J. W. Young. *Colorado West: Land of Geology and Wildflowers.* Grand Junction, CO: Wheelwright Press, 1977.

CONNECTICUT

Connecticut Department of Environmental Protection. *Connecticut's Endangered, Threatened, and Special Concern Species.* Hartford: Connecticut Department of Environmental Protection, 1998.

Connecticut Division of Conservation and Preservation. *Connecticut Natural Area Preserves.* Hartford: Connecticut Division of

Conseration and Preservation, 1998.
The Nature Conservancy. *Preserve Sampler: A Guide to Nature Conservancy Lands in Connecticut.* Middletown, CT: The Nature Conservancy, 1997.

DELAWARE
Clancy, K. *A Preliminary Classification of the Natural Communities of Delaware.* Dover: Delaware Natural Heritage Inventory, 1993.
Delaware State Planning Office. *Delaware Natural Resource Inventory.* Dover: Delaware State Planning Office, 1970.
Fleming, Lorraine. *Delaware's Outstand Natural Areas and Their Preservation.* Hockessin, MD: Delaware Nature Education Society, 1978.
Scott, Jane. *Between Ocean and Bay: A Natural History of Delmarva.* Centerville, MD: Tidewater Publications, 1991.

FLORIDA
Bechtold, W. A., M. J. Brown, and R. M. Sheffield. *Florida's Forests, 1987.* Asheville, NC: USDA Forest Service, 1990.
Blanchard, Jon, and Sally Jue. *Florida Conservation Lands.* Tallahassee: Florida Natural Areas Inventory, 1997.
Cerulean, S., and A. Morrow. *Florida Wildlife Viewing Guide.* Helena, MT: Falcon Press, 1998.
Cox, J., R. Kautz, M. MacLaughlin, and T. Gilbert. *Closing the Gaps in Florida's Wildlife Habitat Conservation System.* Tallahassee: Florida Game and Fresh Water Fish Commission, 1994.
Douglas, Marjorie S. *The Everglades: River of Grass.* New York: Rinehart, 1947.
Florida Natural Areas Inventory. *A Guide to the Natural Communities of Florida.* Tallahassee: Florida Natural Areas Inventory and the Florida Department of Natural Resources, 1990.
Humphrey, S. R., ed. *Rare and Endangered Biota of Florida.* Gainesville: University Press of Florida, 1992.
Kautz, R. S. "Land, Climate, and Vegetation of Florida." Unpublished report. Tallahassee: Florida Game and Fresh Water Fish Commission, 1998.
Kautz, R. S. "Trends in Florida Wildlife Habitat, 1937–1987." *Florida Scientist* 56(1) (1993): 7–24.
Muller, J. W., E. Hardin, D. Jackson, S. Gatewood, and N. Caire. *Summary Report on the Vascular Plants, Animals, and Plant Communities Endemic to Florida.* Tallahassee: Florida Game and Fresh Water Fish Commission, 1989.
Myers, Ronald L., and John H. Ewel, eds. *Ecosystems of Florida.* Orlando: University of Central Florida Press, 1990.
The Nature Conservancy. *A Guide to Florida Preserves.* Altamonte Springs, FL: The Nature Conservancy, 1997.
The Nature Conservancy. "Managed Conservation Lands in Florida." Unpublished report. Altamonte Springs, FL: The Nature Conservancy, 1993.

GEORGIA
Brown, Fred. *The Georgia Conservancy's Guide to the North Georgia Mountains.* Atlanta: Longstreet Press, 1990.
Georgia Department of Natural Resources. *Natural Area Fact Sheets.* Social Circle: Georgia Department of Natural Resources, Wildlife Resources Division, Natural Heritage Program, 1997.
McKee, Gwen, ed. *The Georgia Conservancy's Guide to the Georgia Coast.* Savannah: The Georgia Conservancy, 1984.
The Nature Conservancy. *Georgia Wildlands: A Guide to Lands Protected by The Nature Conservancy of Georgia.* Atlanta: The Nature Conservancy, 1996.
Wharton, C. H. *The Natural Environments of Georgia.* Atlanta: Georgia State University, 1978.

HAWAII
Ball Jr., Stuart M. *The Hiker's Guide to Oahu.* Honolulu: University of Hawaii Press, 1993.
Berger, Andrew J. *Birdlife in Hawaii.* Aiea, HI: Island Heritage, 1983.
Department of Land and Natural Resources. *The Hawaii Natural Area Reserves System.* Honolulu: Division of Forestry and Wildlife, 1998.
Department of Land and Natural Resources. *More than Trees and Trails.* Honolulu: Division of Forestry and Wildlife, 1994.
Scott, Susan. *Plants and Animals of Hawaii.* Honolulu: Bess Press, 1991.
Sohmer, S. H., and Robert Gustafson. *Plants and Flowers of Hawaii.* Honolulu: University of Hawaii Press, 1987.
Stone, C. P., and D. B. Stone, eds. *Conservation Biology in Hawaii.* Honolulu: University of Hawaii Press, 1989.
Stone, C. P., and J. M. Scott, eds. *Hawaii's Terrestrial Ecosystems: Preservation and Management.* Honolulu: University of Hawaii Press, 1985.

The Nature Conservancy. *Preserve Guides and Fact Sheets.* Honolulu: The Nature Conservancy, 1997.

IDAHO

Carpenter, L. B. *Idaho Wildlife Viewing Guide.* Helena, MT: Falcon Press, 1990.

Cooper, S. V., K. E. Newman, and D. W. Roberts. *Forest Habitat Types of Northern Idaho.* Ogden, UT: USDA Forest Service, 1991.

Johnson, F. D. *Wild Trees of Idaho.* Moscow: University of Idaho Press, 1995.

Steele, R., R. D. Pfister, R. A. Ryker, and J. A. Kittams. *Forest Habitat Types of Central Idaho.* Ogden, UT: USDA Forest Service, 1981.

Steele, R., S. V. Cooper, D. M. Ondov, D. W. Roberts, and R. D. Pfister. *Forest Habitat Types of Eastern Idaho and Western Wyoming.* Ogden, UT: USDA Forest Service, 1983.

The Nature Conservancy of Idaho. *Idaho Field Notes* (newsletter). Sun Valley, ID: The Nature Conservancy, 1998.

The Nature Conservancy. *Preserve and Project Guide.* Sun Valley, ID: The Nature Conservancy, 1996.

ILLINOIS

Evers, R. A., and L. M. Page. *Some Unusual Natural Areas in Illinois.* Springfield: Illinois Department of Energy and Natural Resources, 1977.

Illinois Natural History Survey. *The Changing Illinois Environment: Critical Trends.* Springfield: Illinois Department of Energy and Natural Resources, 1994.

Jeffords, M. R., S. L. Post, and K. R. Robertson. *Illinois Wilds.* Urbana: Phoenix Press, 1995.

Oliver, J. S., and R. W. Graham. *Preliminary Inventory of Natural Resources in Select Caves in Illinois.* Springfield: Illinois State Museum, 1988.

Schwegman, J. E. 1973. T*he Natural Divisions of Illinois.* Springfield: Illinois Nature Preserves Commission.

Sullivan, J. *Chicago Wilderness: An Atlas of Biodiversity.* Chicago: Chicago Regional Biodiversity Council, 1997.

Suloway, L., and M. Hubbell. *Wetland Resources of Illinois: An Analysis and Atlas.* Springfield: Illinois Natural History Survey, 1994.

The Nature Conservancy. *Preserve Guide.* Chicago: The Nature Conservancy, 1991.

INDIANA

Barton, J. D., and D. V. Schmelz. 1987. "Thirty Years of Growth Records in Donaldson's Woods." *Indiana Academy of Science* 96: 209–214.

Deam, C. C. *Flora of Indiana.* Indianapolis: Indiana Department of Conservation, 1940.

Jackson, Marion T. *The Natural Heritage of Indiana.* Bloomington: Indiana University Press, 1997.

Lindsey, A. A., D. V. Schmelz, and S. A. Nichols. *Natural Areas in Indiana and Their Preservation.* Indianapolis: Indiana Academy of Science, 1969.

Lindsey, A. A., ed. *Natural Features of Indiana.* Indianapolis: Indiana Academy of Science, 1966.

Mumford, R. E., and J. O. Whitaker. *Mammals of Indiana.* Bloomington: Indiana University Press, 1982.

Mumford, R. E., and C. E. Keller. *The Birds of Indiana.* Bloomington: Indiana University Press, 1984.

Seng, P. T., and D. J. Chase. *Indiana Wildlife Viewing Guide.* Helena, MT: Falcon Press, 1992.

Shull, E. M. *The Butterflies of Indiana.* Indianapolis: Indiana Academy of Science, 1987.

The Nature Conservancy. *The Nature Conservancy Guide to Indiana Preserves and Projects.* Indianapolis: The Nature Conservancy, 1997.

Wampler, M., and F. Wampler. *Wildflowers of Indiana.* Bloomington: Indiana University Press, 1988.

Whitaker, J. O., and J. R. Gammon. *Endangered and Threatened Vertebrate Animals of Indiana: Their Distribution and Abundance.* Indianapolis: Indiana Academy of Science, 1988.

IOWA

Connard, H. S. "The Vegetation of Iowa." *University of Iowa Studies of Natural History* 19(4) (1952): 1–166.

Eilers, L. J., and D. M. Roosa. *The Vascular Plants of Iowa.* Iowa City: University of Iowa Press, 1994.

Gilly, C. L., and M. E. MacDonald. "Rare and Unusual Plants from Southeastern Iowa." *Proceedings of the Iowa Academy of Science* 43 (1936): 143–149.

Great Plains Association. *Flora of the Great Plains.* Lawrence: University of Kansas Press, 1986.

Pusateri, W. P., and D. M. Roosa. "Habitat and Distribution of Plants Special to Iowa's Driftless Area." *Journal of the Iowa Academy of Science* 100 (1993): 29–53.

Roosa, D. M. *Distribution of Iowa's Endangered and Threatened Vascular Plants.* Des Moines: Iowa Department of Natural Resources, 1989.

Smith, D. D. "Iowa Prairie: An Endangered Ecosystem." *Proceedings of the Iowa Academy of Science* 88 (1981): 7–10.

KANSAS

Bare, Janet E. *Wildflowers and Weeds of Kansas.* Lawrence: Regents Press of Kansas, 1979.

Buchanan, R. C., and J. A. McCauley. *Roadside Kansas: A Traveler's Guide to Its Geology and Landmarks.* Lawrence: University Press of Kansas, 1987.

Collins, J. T. *Kansas Wildlife.* Lawrence: University Press of Kansas, 1991.

Collins, J. T., ed. *Natural Kansas.* Lawrence: University Press of Kansas, 1985.

Collins, J. T., S. L. Collins, and B. Gress. *Kansas Wetlands.* Lawrence: University Press of Kansas, 1994.

Freeman, C. C., and E. K. Scholfield. *Roadside Wildflowers of the Southern Great Plains.* Lawrence: University Press of Kansas, 1991.

Gress, Bob, and George Potts. *Watching Kansas Wildlife.* Lawrence: University Press of Kansas, 1993.

Küchler, A. W. *A New Vegetation Map of Kansas.* Ecology 55 (1974): 586–604.

Reichman, O. J. *Living Landscapes of Kansas.* Lawrence: University Press of Kansas, 1995.

Zimmerman, John L. *Cheyenne Bottoms: Wetland in Jeopardy.* Lawrence: University Press of Kansas, 1990.

KENTUCKY

Kentucky State Nature Preserves Commission. *Directory of Kentucky State Nature Preserves.* Lexington, KY: Nature Preserves Commission, 1998.

Lander, A. B. *A Guide to Kentucky Outdoors.* Ann Arbor, MI: Thomas, 1978.

Lynn, C. H. *Kentucky Wildlife Viewing Guide.* Helena, MT: Falcon Press, 1994.

McGrain, P. *The Geologic Story of Kentucky.* Lexington: University of Kentucky, 1983.

Monroe, B. L. *The Birds of Kentucky.* Bloomington, IN: Indiana University Press, 1994.

Taylor, D. J., ed. *Kentucky Alive: A Report of the Biodiversity of Kentucky Task Force.* Frankfort: Commonwealth of Kentucky, 1995.

Wharton, M. E., and R. W. Barbour. *The Wildflowers and Ferns of Kentucky.* Lexington: University of Kentucky Press, 1979.

LOUISIANA

Brown, C. A. *Wildflowers of Louisiana.* Baton Rouge: Louisiana State University Press, 1984.

Lockwood, C. C. *Discovering Louisiana.* Baton Rouge: Louisiana State University Press, 1986.

Louisiana Department of Wildlife and Fisheries. *Wildlife Management Area Guide.* Baker, LA: Baker Printing Company, 1998.

The Nature Conservancy. *The Nature Conservancy of Louisiana Preserves Guidebook.* Baton Rouge: The Nature Conservancy, 1997.

MAINE

Doyle, K. M., T. K. Fahey, and R. D. Paratley. "Subalpine Eathlands of the Mahoosuc Range, Maine." *Bulletin of the Torrey Botanica Club* 114(4) (1987): 429–436.

King, Brian. *Appalachian Trail Guide to Maine.* Harpers Ferry, WV: Appalachian Trail Conference, 1993.

Maine Natural Heritage Program. *Natural Landscapes of Maine: A Classification of Ecosystems and Natural Communities.* Augusta, ME: Office of Comprehensive Planning, State House Station 130, 1991.

Pierson, J. E., P. D. Vickery, and E. C. Pierson. *A Birder's Guide to Maine.* Camden, ME: Down East Books, 1998.

MARYLAND

Berdine, A. *Maryland Vegetation Classification.* Annapolis: Maryland Department of Natural Resources, 1998.

Brown, M. L., and R. G. Brown. *Herbaceous Plants of Maryland.* Baltimore: Port City Press, 1984.

Brush, G. C. "The Natural Forests of Maryland." *Ecology Monograph* 50 (1980): 77–92.

The Nature Conservancy. *Mountains to Marshes: The Nature Conservancy Preserves in Maryland.* Chevy Chase, MD: The Nature Conservancy, 1997.

MASSACHUSETTS

Davis, W. J. *Massachusetts Wildlife Viewing Guide.* Helena, MT: Falcon Press, 1996.

Geist, Margaret A., ed. *Waquoit Bay National Estuarine Research Reserve.* Boston: Massachusetts Department of Environmental Management, 1996.

Motzkin, Glenn. 1990. "Atlantic White Cedar Wetlands of Massachusetts." Unpublished report. Massachusetts Natural Heritage and Endangered Species Program.

The Nature Conservancy. "The Nature Conservancy of Massachusetts" (brochure). Boston: The Nature Conservancy, 1997.

Whitney, G. G. 1991. "Relation of Plant Species to Substrate, Landscape Position, and Aspect in North Central Massachusetts." *Canadian Journal of Forest Resources* 21: 1245–1252.

MICHIGAN

Albert, D. A., S. R. Denton, and B. V. Barnes. 1986. *Regional Landscape Ecosystems of Michigan.* Ann Arbor: University of Michigan.

Crispin, S. R. "Nature Preserves in Michigan." *Michigan Botanist* 19(3) (1980): 99–242.

Holzman, R. W., B. A. Daubendick, L. Rizor, and F. Sibley. *Nature Sanctuary Guidebook.* Avoca: Michigan Nature Association, 1994.

Powers, T. *A Complete Guide to Michigan State and National Parks.* Davison, MI: Friede Publications, 1993.

Powers, T. *Natural Michigan: A Guide to 165 Michigan Natural Attractions.* Davison, MI: Friede Publications, 1987.

Seng, P. T. *Michigan Wildlife Viewing Guide.* East Lansing: Michigan State University Press, 1994.

The Natural Heritage Program. *The Spotting Scope* (quarterly newsletter). Lansing, MI: The Natural Heritage Program, Wildlife Division, Department of Natural Resources, 1997.

The Nature Conservancy. *Project Directory.* East Lansing: The Nature Conservancy, 1997.

MINNESOTA

Albert, D. A. *Regional Landscape Ecosystems of Michigan, Minnesota, and Wisconsin: A Working Map and Classification.* St. Paul: USDA Forest Service, North Central Forest Experiment Station, 1995.

Anderson, J. L., and D. F. Girgal. *Soils and Landscapes of Minnesota.* St. Paul: University of Minnesota, 1984.

Coffin, B., and L. Pfannmuller, eds. *Minnesota's Endangered Flora and Fauna.* Minneapolis: University of Minnesota Press, 1988.

Flaccus, E., and L. F. Ohmann. "Old Growth Northern Hardwood Forests in Northeastern Minnesota." *Ecology* 45 (1964): 448–459.

Grimm, E. C., "Fire and Other Factors Controlling the Big Woods Vegetation of Minnesota in the Mid-nineteenth Century." *Ecological Monographs* 54(3) (1984): 291–311.

Minnesota Department of Natural Resources. *A Guide to Minnesota's Scientific and Natural Areas.* St. Paul: Minnesota Department of Natural Resources, 1995.

Sansome, C. J. *Minnesota Underfoot.* Stillwater, MN: Voyageur Press, 1983.

Tester, John R. *Minnesota's Natural Heritage: An Ecological Perspective.* Minneapolis: University of Minnesota Press, 1995.

The Nature Conservancy. *Preserve Fact Sheets.* Minneapolis: The Nature Conservancy, 1997.

Wendt, Keith M. *A Guide to Minnesota Prairies.* St. Paul: Minnesota Natural Heritage Program, 1984.

Wright, H. E., Jr. *Physiography of Minnesota.* In P. Sims and G. B. Morey, eds. *Geology of Minnesota.* St. Paul: Minnesota Geological Survey, 1972.

MISSISSIPPI

Dennett, D. *Louisiana's Wildlife Worth Watching.* Baton Rouge: Louisiana Department of Wildlife and Fisheries, 1997.

Larke, Julio O., and Latimore Smith. *Rare Plants of Pine-Hardwood Forests in Louisiana.* Baton Rouge: Louisiana Department of Wildlife and Fisheries, 1994.

Lowery, George H. *Louisiana Birds.* Baton Rouge: Louisiana State University Press, 1974.

Lowery, George. *Mammals of Louisiana and Its Adjacent Waters.* Baton Rouge: Louisiana State University Press, 1974.

Rossman, Douglas A., and Harold A. Dundee. *The Amphibians and Reptiles of Louisiana.* Baton Rouge: Louisiana State University Press, 1989.

The Nature Conservancy. *Guidebook to Nature Conservancy Preserves in Mississippi.* Jackson: The Nature Conservancy, 1997.

MISSOURI

Jacbos, B., and J. D. Wilson. *Missouri Breeding Bird Atlas, 1986–1992.* Jefferson City: Missouri Department of Conservation, 1997.

Johnson, T. R. *The Amphibians and Reptiles of Missouri.* Jefferson City: Missouri Department of Conservation, 1987.

Missouri Department of Conservation. *Directory of Missouri Natural Areas.* Jefferson City: Missouri Department of Conservation, 1996.

Nelson, P. W. *The Terrestrial Natural Communities of Missouri.* Jefferson City: Missouri Department of Natural Resources, 1985.

Richard, J., and J. E. Heitzman. *Butterflies and Moths of Missouri.* Jefferson City: Missouri Department of Conservation, 1987.

Robbins, M. B., and D. A. Easterla. *Birds of Missouri: Their Distribution and Abundance.* Columbia: University of Missouri Press, 1987.

Schwartz, C. W., and E. R. Schwartz. *About Mammals and How They Live.* Jefferson City: Missouri Department of Conservation, 1993.

Schwartz, C. W., and E. R. Schwartz. *The Wild Mammals of Missouri.* Columbia: University of Missouri Press, 1981.

Steyermark, J. A. *Flora of Missouri.* Ames: Iowa State University Press, 1975.

The Nature Conservancy. *Discover Natural Missouri: A Guide to Exploring The Nature Conservancy Preserves.* St. Louis: The Nature Conservancy, 1997.

MONTANA

Chadde, S. W., S. F. Kimball, and A. G. Evenden. *Research Natural Areas of the Northern Region: Status and Needs Assessment.* Washington, DC: USDA Forest Service, 1996.

Dorn, R. D. *Vascular Plants of Montana.* Cheyenne, WY: Mountain West Publishing, 1984.

Fischer, H., and C. Fischer. *Montana Wildlife Viewing Guide.* Helena, MT: Falcon Press, 1995.

Pfister, R. D. and B. Kovalchik. *Forest Habitat Types of Montana.* Ogden, UT: USDA Forest Service, 1977.

The Nature Conservancy. *Preserve Fact Sheets.* Helena, MT: The Nature Conservancy.

The Nature Conservancy. *The Newsletter of the Nature Conservancy of Montana.* Helena, MT: various issues.

NEBRASKA

Bleed, A., and C. Flowerday. *An Atlas of the Sandhills.* Lincoln: University of Nebraska, Institute of Agriculture and Natural Resources, 3rd ed., 1998.

Ducey, J. E. *Nebraska Birds: Breeding Status and Distribution.* Omaha: Simmons-Boardman, 1988.

Farrar, John. *Field Guide to Wildflowers of Nebraska and the Great Plains.* Lincoln: Nebraskaland Magazine, 1990.

Johnsgard, P. A. *This Fragile Land: A Natural History of the Nebraska Sandhills.* Lincoln: University of Nebraska Press, 1995.

Jones J. K., Jr. *Distribution and Taxonomy of Mammals of Nebraska.* Lawrence: University of Kansas Publications, 1964.

Knue, J. *Nebraska Wildlife Viewing Guide.* Helena, MT: Falcon Press, 1997.

Nagel, H. G., and V. Plambeck, eds. *The Loess Hills Prairies of Central Nebraska.* Kearney: University of Nebraska, 1998.

Ryan, M. G., L. A. Joyce, T. Andrews, and K. Jones. *Research Natural Areas in Colorado, Nebraska, North Dakota, South Dakota, and Parts of Wyoming.* Fort Collins, CO: USDA Forest Service, 1994.

The Nature Conservancy. *Preserve Guide.* Omaha: The Nature Conservancy, 1997.

The Nature Conservancy. *The Plains Messenger* (newsletter). Omaha: The Nature Conservancy, 1997.

Weaver, J. E. *Native Vegetation of Nebraska.* Lincoln: University of Nebraska Press, 1965.

NEVADA

Bureau of Land Management. *Nevada Watchable Wildlife.* Carson City, NV: Bureau of Land Management, 1997.

Clark, Jeanne L. *Nevada Wildlife Viewing Guide.* Helena, MT: Falcon Press, 1993.

Lanier, Ronald M. *Trees of the Great Basin.* Reno: University of Nevada Press, 1984.

Taylor, Ronald J. *Sagebrush Country: A Wildflower Sanctuary.* Missoula, MT: Mountain Press, 1992.

White, Michael C. *Nevada Wilderness Areas and Great Basin National Park.* Berkeley, CA: Wilderness Press, 1997.

NEW HAMPSHIRE

King, Brian. *Appalachian Trail Guide to New Hampshire and Vermont.* Harpers Ferry, WV: Appalachian Trail Conference, 1992.

Lyons, C. J. *New Hampshire Natural Areas.* Concord: New Hampshire Office of State Planning, 1971.

Perry, John, and Jane G. Perry. *The Sierra Club Guide to the Natural Areas of New England.* San Francisco: Sierra Club Books, 1990.

Short, Frederick T. *The Ecology of the Great Bay Estuary, New Hampshire and Maine.* Durham: University of New Hampshire Press, 1992.

Silverberg, J. K. *New Hampshire Viewing Guide.* Helena, MT: Falcon Press, 1997.

Sperduto, D. D. *A Classification of the Natural Communities of New Hampshire.* Concord, NH: New Hampshire Natural Heritage Inventory, 1997.

The Nature Conservancy. *Trail Maps and Guides.* North Conway, NH: The Nature Conservancy, 1997.

NEW JERSEY

Boyd, Howard P. *A Field Guide to the Pine Barrens of New Jersey.* Medford, NJ: Plexus, 1971.

Boyle, William J. *A Guide to Bird Finding in New Jersey.* New Brunswick, NJ: Rutgers University Press, 1986.

Karlin, E. F., ed. *New Jersey's Rare and Endangered Plants and Animals.* Mahwah, NJ: Ramapo College, 1989.

McPhee, John. *The Pine Barrens.* New York: Farrar, Straus, and Giroux, 1968.

Mitchell, A. E. *The New Jersey Highlands: Treasures at Risk.* Morristown: New Jersey Conservation Foundation, 1992.

New Jersey Department of Environmental Protection. *The Guide to New Jersey's State Parks and Forests.* Trenton: New Jersey Department of Environmental Protection, 1998.

New Jersey Department of Environmental Protection. *New Jersey Natural Areas Program Newsletter.* Trenton: New Jersey Department of Environmental Protection, Division of Parks and Forestry, Office of Natural Lands Management, 1997.

New Jersey Department of Environmental Protection. *State Natural Area Management Plans.* Trenton: New Jersey Department of Environmental Protection and Energy, Division of Parks and Forestry, 1994–1997.

Pettigrew, L. *New Jersey Wildlife Viewing Guide.* Helena, MT: Falcon Press, 1998.

The Nature Conservancy. *New Jersey Preserve Guide.* Chester, NJ: The Nature Conservancy, 1997.

Widmer, Kemble. *The Geology and Geography of New Jersey.* Princeton: Van Nostrand, 1964.

Wolf, P. E. *The Geology and Landscapes of New Jersey.* New York: Crane, Russak, 1977.

NEW MEXICO

Castetter, E. F. "The Vegetation of New Mexico." *New Mexico Quarterly* 26 (Sept. 1956): 256–288.

Dick-Peddie, W. A. *New Mexico Vegetation: Past, Present, and Future.* Albuquerque: University of New Mexico Press, 1993.

Dick-Peddie, W. A. *Unique Wildlife Ecosystem Plan for the State of New Mexico.* Albuquerque: USDA Fish and Wildlife Service, 1978.

MacCarter, J. S. *New Mexico Wildlife Viewing Guide.* Helena, MT: Falcon Press, 1994.

Martin, W. C., and C. R. Hutchins. *A Flora of New Mexico.* Vaduz, NM: J. Cramer, 1980.

Parent, Lawrence. *The Hiker's Guide to New Mexico.* Helena, MT: Falcon Press, 1994.

Peterson, R. S., and E. Rasmussen. *Research Natural Areas in New Mexico.* Fort Collins, CO: USDA Forest Service, 1986.

Szaro, R. C. "Riparian Forest and Scrubland Community Types of Arizona and New Mexico." *Desert Plants* 9 (3–4) (1989): 69–138.

NEW YORK

Barnett, Lincoln. *The Ancient Adirondacks.* New York: Time-Life Books, 1974.

Dickinson, N. R. "A Division of Southern and Western New York State into Ecological Zones." Unpublished Report. Delmar: New York Department of Environmental Conservation, 1979.

Drennan, Susan. *Where to Find Birds in New York State.* Syracuse, NY: Syracuse University Press, 1981.

Johnson, C. W. *Bogs of the Northeast.* Hanover, NH: University Press of New England, 1985.

Leopold, D. J., C. Reschke, and D. Smith. "Old-growth Forests of Adirondack Park, New York." *Natural Areas Journal* 8(3) (1988): 166–189.

Reschke, C. *Ecological Communities of New York State.* Latham: New York Natural Heritage Program, 1990.

The Nature Conservancy. *Nature Preserves of The Nature Conservancy's Lower Hudson Chapter.* Mount Kisco, NY: The Nature Conservancy, 1997.

The Nature Conservancy. *Preserve Guide, Adirondack Chapter.* Keene Valley, NY: The Nature Conservancy, 1996.

The Nature Conservancy. *Preserve Guide, Eastern New York Chapter.* Troy, NY: The Nature Conservancy, 1995.

The Nature Conservancy. *Preserve Guide, South Fork/Shelter Island Chapter.* East Hampton, NY: The Nature Conservancy, 1997.

The Nature Conservancy. *Sanctuary* (newsletter of the Long Island chapter of The Nature Conservancy). Cold Spring Harbor, NY: The Nature Conservancy, 1997.

NORTH CAROLINA

LeGrand, H. E. *Natural Heritage Program List of the Rare Animal Species of North Carolina.* Raleigh: Division of Parks and Recreation, Department of Environment, Health, and Natural Resources, 1996.

North Carolina Cooperative Extension Service. *Natural Resources Inventory of North Carolina.* Raleigh: North Carolina Cooperative Extension Service, North Carolina State University, 1997.

North Carolina Division of Parks. *North Carolina Natural Heritage Program.* Raleigh: Division of Parks and Recreation, Department of

Environment, Health, and Natural Resources, 1993.

North Carolina Division of Parks. *North Carolina Natural Heritage Program Biennial Protection Plan.* Raleigh: Division of Parks and Recreation, Department of Environment, Health, and Natural Resources, 1993.

Radford, A. E., H. E. Ahles, and C. R. Bell. *Manual of the Vascular Flora of the Carolinas.* Chapel Hill: University of North Carolina Press, 1976.

Roe, C. E. *North Carolina Wildlife Viewing Guide.* Helena, MT: Falcon Press, 1992.

Schafale, M. P., and A. S. Weakley. *Classification of the Natural Communities of North Carolina.* Raleigh: Division of Parks and Recreation, Department of Environment, Health, and Natural Resources, 1996.

Weakley, Alan S. *Natural Heritage Program List of the Rare Plant Species of North Carolina.* Raleigh: Division of Parks and Recreation, Department of Environment, Health, and Natural Resources, 1996.

NORTH DAKOTA

Johnson, M. D., and J. Knue. *Feathers from the Prairie: A Short History of Upland Game Birds.* Bismarck: North Dakota Game and Fish Department, 1989.

Kannowski, P. B. *Wildflowers of North Dakota.* Grand Forks: University of North Dakota Press, 1989.

Knue, Joseph. *Big Game in North Dakota: A Short History.* Bismarck: North Dakota Game and Fish Department, 1991.

Knue, Joseph. *North Dakota Wildlife Viewing Guide.* Helena, MT: Falcon Press, 1992.

Knue, Joseph. *Of Time and the Prairie.* Bismarck: North Dakota Game and Fish Department, 1988.

Royer, R. A. *Butterflies of North Dakota: An Atlas and Guide.* Minot, ND: Minot State University, 1988.

Ryan, Michael G., L. A. Joyce, T. Andrews, and K. Jones. *Research Natural Areas in Colorado, Nebraska, North Dakota, South Dakota, and Parts of Wyoming.* Fort Collins, CO: USDA Forest Service, 1994.

Stevens, O. A. *Handbook of North Dakota Plants.* Fargo: North Dakota Institute for Regional Studies, 1963.

OHIO

Cooperrider, T. S., ed. *Endangered and Threatened Plants of Ohio.* Columbus: Ohio State University, 1982.

Downhower, J. F., ed. *Biography of the Island Region of Western Lake Erie.* Columbus: Ohio State University, 1988.

Gordon, R. B. "The Natural Vegetation of Ohio in Pioneer Days." *Ohio Biological Survey New Series Bulletin* 3(2) (1966): 1–109.

Gottschang, J. L. *A Guide to the Mammals of Ohio.* Columbus: Ohio Academy of Science, 1979.

Gross, W. H. *Ohio Wildlife Viewing Guide.* Helena, MT: Falcon Press, 1996.

Lafferty, M. B., ed. *Ohio's Natural Heritage.* Columbus: Ohio Academy of Science, 1979.

Ohio Department of Natural Resources. *Directory of Ohio's State Nature Preserves.* Columbus: Division of Natural Areas and Preserves, 1996.

Trautman, M. B. *The Ohio Country from 1750 to 1977: A Naturalist's View.* Columbus: Ohio State University, 1977.

OKLAHOMA

Bruner, W. E. "The Vegetation of Oklahoma." *Ecology Monograph* 1 (1931): 99–188.

Caire, W., J. D. Taylor, B. P. Glass, and M. A. Mares. *Mammals of Oklahoma.* Norman: University of Oklahoma Press, 1989.

Duck, L. G., and J. B. Fletcher. *A Game Type Map of Oklahoma.* Norman: Division of Wildlife Restoration, 1943.

Duck, L. G., and J. B. Fletcher. *A Survey of the Game and Furbearing Animals in Oklahoma.* Oklahoma City: Oklahoma Game and Fish Commission, 1943.

Miller, R. J., and H. W. Robison. *The Fishes of Oklahoma.* Norman: Oklahoma University Press, 1973.

Murray, N. L., ed. *Oklahoma's Biodiversity Plan: A Shared Vision for Conserving Our Natural Heritage.* Oklahoma City: Oklahoma Department of Wildlife Conservation, 1996.

Oklahoma Cooperative Extension Service. *Endangered and Threatened Species of Oklahoma.* Stillwater: Oklahoma Cooperative Extension Service, Oklahoma State University, 1993.

Oklahoma Department of Wildlife Conservation. *Oklahoma Watchable Wildlife Viewing Guide.* Oklahoma City: Oklahoma Department of Wildlife Conservation.

Wood, D. S., and G. D. Schnell. *Distribution of Oklahoma Birds.* Stillwater: Oklahoma State University Press, 1984.

OREGON

Bureau of Land Management. *Research Natural Areas in Washington and Oregon.* 2d ed. Portland, OR:

Bureau of Land Management, 1996.
Franklin, J. F., and C. T. Dyrness. *Natural Vegetation of Oregon and Washington.* Portland, OR: USDA Forest Service, 1973.
Mathews, Daniel. *Cascade-Olympic Natural History: A Trailside Reference.* Portland, OR: Raven Editions, 1988.
Yuskavitch, James A. *Oregon Wildlife Viewing Guide.* Helena, MT: Falcon Press, 1994.

PENNSYLVANIA

Erdman, K. S., and P. G. Wiegman. *Preliminary List of Natural Areas in Pennsylvania.* Harrisburg: Western Pennsylvania Conservancy, 1974.
Fike, Jean. *Terrestrial and Palustrine Plant Communities of Pennsylvania.* Harrisburg: Pennsylvania Natural Diversity Inventory, 1998.
Korber, Kathy, and Hall Korber. *A Viewer's Guide to Pennsylvania Wildlife.* Lemoune, PA: Northwoods Publications, 1994.
Lancaster County Conservancy. *Guide to the Nature Preserves of the Lancaster County Conservancy.* Lancaster, PA: Lancaster County Conservancy, 1998.
Majumdar, S. K., R. P. Brooks, and F. J. Brenner. *Wetland Ecology and Conservation: Emphasis in Pennsylvania.* Easton: Pennsylvania Academy of Science, 1989.
Smith, T. *Natural Communities of Pennsylvania.* Unpublished. Harrisburg: Pennsylvania Natural Diversity Inventory, 1985.
Smith, T. L. *Natural Ecological Communities of Pennsylvania.* Middletown: Pennsylvania Science Office of the Nature Conservancy, 1991.
The Nature Conservancy. *Penn's Woods: The Newsletter of the Pennsylvania Chapter of the Nature Conervancy.* Conshohocken, PA: The Nature Conservancy, 1998.
The Nature Conservancy. *Preserve Fact Sheets.* Conshohocken, PA: The Nature Conservancy, 1998.
Western Pennsylvania Conservancy and The Nature Conservancy. *A Study of Calcareous Fen Communities in Pennsylvania.* Harrisburg: Western Pennsylvania Conservancy, 1995.

SOUTH CAROLINA

Barry, J. M. *Natural Vegetation of South Carolina.* Columbia: University of South Carolina, 1980.
Holt, Perry C., ed. *The Distributional History of the Biota of the Southern Appalachians.* Blacksburg: Virginia Polytechnical Institute, 1970.
Johnson, Douglas. *The Origin of the Carolina Bays.* New York: Columbia University Press, 1942.
Leifermann, Henry. *South Carolina.* New York: Fodor, 1995.
Manning, Phillip. *Palmetto Journal: Walks in the Natural Areas of South Carolina.* Winston-Salem: Blair, 1995.
Penfound, W. T. "Southern Swamps and Marshes." *Botanical Review* 18 (1952): 414–446.
South Carolina Department of Natural Resources. *Preserve Guide.* Columbia: South Carolina Department of Natural Resources, Heritage Trust Program, 1997.
South Carolina Department of Natural Resources. *South Carolina Heritage Preserves.* Columbia: South Carolina Department of Natural Resources, Heritage Trust Program, 1997.
South Carolina Wildlife and Marine Resources Department. *Heritage Trust Program Preserve Guide.* Columbia: South Carolina Wildlife and Marine Resources Department, 1997.
The Nature Conservancy. *Preserve Fact Sheets.* Columbia: The Nature Conservancy, 1997.
Wells, B. W. "Plant communities of the Coastal Plain of North Carolina and Their Successional Relations." *Ecology* 9: 230–243.

SOUTH DAKOTA

Froiland, S. G., and R. R. Weedon. *Natural History of the Black Hills and Badlands.* Sioux Falls: Center for Western Studies, Augusta College, 1990.
Holden, David J. *Dakota Visions: A County Approach.* Sioux Falls: Center for Western Studies, Augusta College, 1982.
Over, W. H. *Flora of South Carolina.* Vermillion: University of South Dakota, 1932.
Ryan, M. G., L. A. Joyce, T. Andrews, and K. Jones. *Research Natural Areas in Colorado, Nebraska, North Dakota, South Dakota, and Parts of Wyoming.* Fort Collins, CO: USDA Forest Service, 1994.
Shepard, Lansing. *The Smithsonian Guides to Natural America: The Northern Plains.* Washington, DC: Smithsonian Press, 1996.

TENNESSEE

Etnier, D. A., and W. C. Starnes. *Fishes of Tennessee.* Knoxville: University of Tennessee Press, 1993.
Hamel, P. *Tennessee Wildlife Viewing Guide.* Helena, MT: Falcon Press, 1993.
Nicholson, C. P. *Atlas of Breeding Birds of Tennessee.* Knoxville: University of Tennessee Press, 1997.
Smith, R. *Wildflowers of the Southern Mountains.* Knoxville: University of Tennessee Press, 1998.

Southen Appalachian Man and Biosphere. *The Southern Appalachian Assessment.* Washington, DC: U.S. Department of Agriculture, 1996.

TEXAS

Fritz, Edward C. *Realms of Beauty: A Guide to the Wilderness Areas of East Texas.* Austin: University of Texas Press, 1993.

Frye, R. G., K. L. Brown, and C. A. McMahan. *The Vegetation Types of Texas.* Austin: Texas Parks and Wildlife Department, 1984.

Gould, F. W. *The Grasses of Texas.* College Station: Texas A&M University Press, 1975.

Graham, Gary L. *Texas Wildlife Viewing Guide.* Helena, MT: Falcon Press, 1992.

Hatch, S. L. *Checklist of the Vascular Plants of Texas.* College Station: Texas A&M University Press, 1990.

Holt, Harold R. *A Birder's Guide to the Texas Coast.* Colorado Springs, CO: American Birding Association, 1993.

Texas Parks and Wildlife Department. *Managed Area Data Sheets.* Austin: Texas Biological and Conservation Data System, Texas Parks and Wildlife Department, 1997.

Tharp, B. C. *The Vegetation of Texas.* Houston: Anson Jones Press, 1939.

UTAH

Aitchison, Stewart. *Utah Wildlands.* Salt Lake City: Utah Geographic, 1987.

Mauk, R. L., and J. A. Henderson. *Coniferous Forest Habitat Types of Northern Utah.* Ogden, UT: USDA Forest Service, 1984.

Mutel, Cornelia, and John Emerick. *From Grassland to Glacier: The Natural History of Colorado and the Surrounding Region.* Boulder, CO: Johnson Books, 1992.

United States Department of Agriculture. *Research Natural Areas Cooperative Fact Sheet.* Ogden, UT: USDA Forest Service, Intermountain Research Station, 1994.

Welsh, S. L., N. D. Atwood, S. Goodrich, and L. C. Higgins. *A Utah FLora.* Provo, UT: Brigham Young University, 1993.

Zwinger, Ann. *Wind in the Rock.* Tucson: University of Arizona Press, 1978.

VERMONT

Brown, C. K. *Vermont Wildlife Viewing Guide.* Helena, MT: Falcon Press, 1994.

Fitzgerald, Brian, and Robert Lindemann, eds. *Day Hiker's Guide to Vermont.* Waterbury, VT: Green Mountain Club, 1992.

Johnson, Charles W. *The Nature of Vermont: Introduction and Guide to a New England Environment.* Hanover, NH: University Press of New England, 1980.

Klein, R. *Technical Report: Vermont Natural Areas Project (Phase II).* Montpelier: Vermont Natural Resources Council, 1976.

Thompson, L. 1989. "Natural Communities of Vermont." Unpublished report. Waterbury: Vermont Natural Heritage and Nongame Program.

VIRGINIA

Duda, M. D. *Virginia Wildlife Viewing Guide.* Helena, MT: Falcon Press, 1994.

Terwilliger, K., and J. Tate, eds. *A Guide to Endangered and Threatened Species in Virginia.* Blacksburg, VA: McDonald and Woodward, 1995.

The Nature Conservancy. *Preserve Guide.* Charlottesville, VA: The Nature Conservancy, 1997.

Virginia Department of Conservation. *Natural Area Fact Sheets.* Richmond: Virginia Department of Conservation, 1998.

Winegar, G., and D. Winegar. *Natural Wonders of Virginia: A Guide to Parks, Preserves, and Wild Places.* Castine, ME: Country Roads Press, 1994.

WASHINGTON

Daubenmire, R. F. *Steppe Vegetation of Washington.* Olympia: Washington Agricultural Experiment Station, 1970.

Douglas, G. W., and L. C. Bliss. "Alpine and High Subalpine Plant Communities of the North Cascades Range, Washington, and British Columbia." *Ecological Monographs* 47: 113–150.

Franklin, J. F., F. C. Hall, C. T. Dyrness, and C. Maser. *Federal Research Natural Areas in Oregon and Washington: A Guidebook for Scientists and Educators.* Portland, OR: USDA Forest Service, 1972.

Kuntze, L. M. and L. C. Cornelius. *Baseline Inventory of Rare, Threatened, and Endangered Plant Species/Communities along Washington's Pacific Coast.* Olympia: Washington Natural Heritage Program, 1982.

La Tourrette, Joe. *Washington Wildlife Viewing Guide.* Helena, MT:

The Nature Conservancy. *Guide to Washington Nature Preserves.* Seattle: The Nature Conservancy, 1998.

Washington Department of Natural Resources. *Endangered, Threatened, and Sensitive*

Vascular Plants of Washington. Olympia: Washington State Department of Natural Resources, 1997.
Washington Department of Natural Resources. *Natural Area Preserves.* Olympia: Washington State Department of Natural Resources, Forest Resources Division, 1997.
Washington Department of Natural Resources. *Natural Heritage Plan.* Olympia: Washington Department of Natural Resources, 1995.
Western Forests. *The Audubon Society Nature Guides.* New York: Knopf, 1985.

WEST VIRGINIA
Core, Earl L. *Vegetation of West Virginia.* Parsons, WV: McClain, 1974.
Core, Earl L. *The Wonderous Year: West Virginia through the Seasons.* Grantsville, WV: Seneca, 1975.
Hall, G. A. *West Virginia Birds.* Pittsburgh, PA: Carnegie Museum of Natural History, 1983.
Stephenson, Steve, ed. *Upland Forests of West Virginia.* Parsons, WV: McClain, 1993.
Strausbaugh, P. D., and E. L. Core. *Flora of West Virginia.* Grantsville, WV: Seneca Books, 1993.
The Nature Conservancy. *West Virginia Chapter News.* Charleston, WV: The Nature Conservancy, 1996–1997.
Wood, John M., and Petra B. Wood. *Bibliography of the Flora and Fauna of West Virginia.* Elkins: West Virginia Division of Natural Resources, 1997.

WISCONSIN
Curtis, J. T. *The Vegetation of Wisconsin. An Ordination of Plant Communities.* Madison: University of Wisconsin Press, 1959.
Finley, R. W. *Original Vegetation Cover of Wisconsin* (map and text). St. Paul, MN: USDA Forest Service, 1976.
Hole, F. D., and C. E. Germain. *Natural Divisions of Wisconsin* (map and text). Madison: Wisconsin Department of Natural Resources, 1994.
Judd, M. K. *Wisconsin Wildlife Viewing Guide.* Helena, MT: Falcon Press, 1995.
The Nature Conservancy. *The Places We Save: A Guide to The Nature Conservancy's Preserves in Wisconsin.* Minocqua, WI: NorthWord Press, 1997.
United States Forest Service. *Candidate Research Natural Areas.* Park Falls, WI: Chequamagon National Forest Headquarters, no date.

WYOMING
Baxter, G. T., and M. D. Stone. *Amphibians and Reptiles of Wyoming.* Cheyenne: Wyoming Game and Fish Department, 1985.
Baxter, G. T., and M. D. Stone. *Fishes of Wyoming.* Cheyenne: Wyoming Game and Fish Department, 1995.
Clark, T. W., and M. R. Stromberg. *Mammals in Wyoming.* Lawrence: University of Kansas Press, 1987.
Dorn, J. L., and R. D. Dorn. *Wyoming Birds.* Cheyenne: Mountain West, 1990.
Dorn, R., ed. *Vascular Plants of Wyoming.* Cheyenne: Mountain West, 1992.
Fertig, W., C. Refsdal, and J. Whipple. *Wyoming Rare Plant Field Guide.* Cheyenne: Wyoming Rare Plant Committee, 1994.
Knight, Dennis H. *Mountains and Plains: The Ecology of Wyoming Landscapes.* New Haven, CT: Yale University Press, 1994.
Long, Charles A. *The Mammals of Wyoming.* Lawrence: University of Kansas Press, 1965.
Ryan, Michael G., L. A. Joyce, T. Andrews, and K. Jones. *Research Natural Areas in Colorado, Nebraska, North Dakota, South Dakota, and Parts of Wyoming.* Fort Collins, CO: USDA Forest Service, 1994.
Weaver, J. E., and F. W. Albertson. *Grasslands of the Great Plains: Their Nature and Use.* Lincoln, NE: Johnsen, 1956.
Wyoming Game and Fish Department. *Wyoming Wildlife Viewing Tour Guide.* Cheyenne: Wyoming Game and Fish Department, 1995.

INDEX